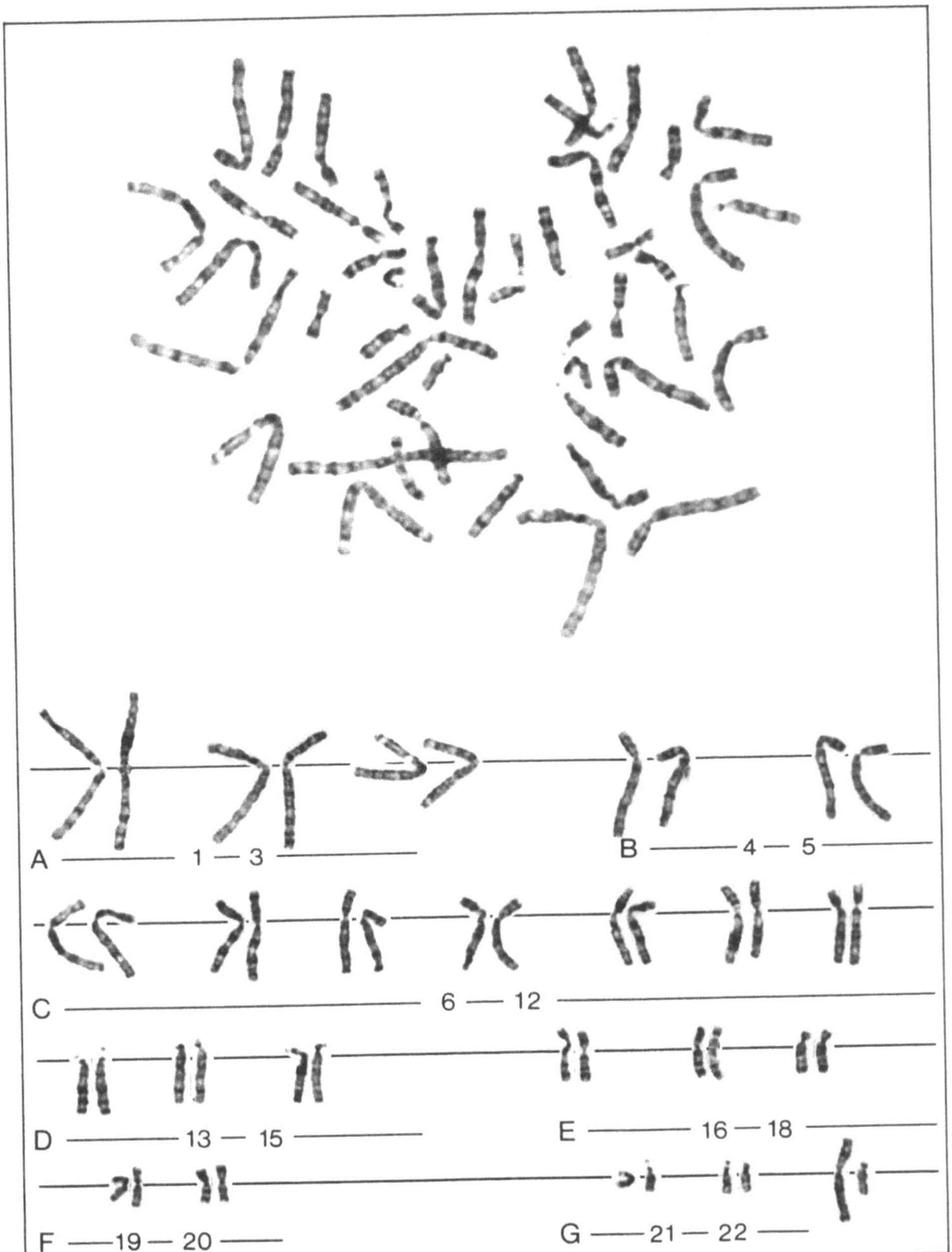

C-metaphase from a normal human male peripheral lymphocyte (2n = 46,XY) with Giemsa (G)-banding (approx. 600 bands). The chromosomes are cut out from the photograph of the C-metaphase and the homologous chromosomes are ordered according to length and banding pattern (karyotype). There are seven groups of chromosomes (A–G) and the two sex chromosomes (X and Y) to the right of chromosome pair 22. (Kindly provided by Dr. R.-D. Wegner, Institut für Humangenetik, Freie Universität Berlin, FRG)

Cytogenetics

Basic and Applied Aspects

Edited by
Günter Obe and Armin Basler

With 128 Figures

Springer-Verlag Berlin Heidelberg GmbH

Professor Dr. GÜNTER OBE
Institut für Genetik
Freie Universität Berlin
Arnimallee 5–7
1000 Berlin 33

Direktor und Professor Dr. ARMIN BASLER
Bundesgesundheitsamt
Postfach 330013
1000 Berlin 33

ISBN 978-3-642-72804-4 ISBN 978-3-642-72802-0 (eBook)
DOI 10.1007/978-3-642-72802-0

Library of Congress Cataloging in Publication Data. Cytogenetics: basic and applied aspects. Includes index. 1. Cytogenetics. I. Obe, G. II. Basler, Armin.
QH430.C96 1987 575.1 87-20690.

Softcover reprint of the hardcover 1st edition 1987

2131/3130-543210

Preface

Nearly 100 years ago, in 1888, Wilhelm von Waldeyer coined the term chromosome in his paper entitled *"Über Karyokinese und ihre Beziehungen zu den Befruchtungsvorgängen"* published in *"Archiv für mikroskopische Anatomie,* Vol. 32, pp. 1–122. The eukaryotic chromosome is a complex structure containing mainly DNA and proteins. An enormous amount of DNA is packed in the chromosomes which can be easily analyzed in a metaphase-like stage resulting from treatment of dividing cells with cholchicine (C-metaphases). The instrument of choice to analyze chromosomes is still the light microscope which may be adjusted to computers to automatize some aspects of the analyses. Special staining procedures lead to bands in the chromosomes which are indispensable tools in fields such as chromosome evolution and tumor cytogenetics. Structural and numerical chromosomal alterations are relatively frequent in the human population and high frequencies of such anomalies are found in mitotic chromosomes of human eggs and spermatozoa. It is of utmost importance to understand how such aberrations arise and how they can be prevented. A great variety of agents are known to induce chromosome alterations, such as DNA nucleases, radiation, and man-made as well as naturally occurring chemicals with widely differing structures. These agents induce a plethora of changes in the chromosomal DNA which eventually may lead to sister-chromatid exchanges, chromosome aberrations, and micronuclei. Analyses of chromosome alterations in peripheral blood lymphocytes of persons exposed to environmental agents may give an impression of whether such exposures lead to mutations in vivo in man. Cytogenetic tests in which an agent is analyzed with respect to its possible chromosome-damaging capacity in vitro or in vivo are indispensable in genetic toxicology. All these analyses may give the impression that chromosomes are merely containers of chromosomal DNA whose chemical changes can be seen in the light microscope: this is not true. Chromosomes are biological structures and are part of the complex network of interactions between cellular components.

In this book, different aspects of what we know about the beautiful and still magic chromosomes are described. The topics discussed in the 19 chapters are the following: Biochemical aspects of chromosome structure and staining (Chaps. 1 and 2), computerized systems for chromosome analyses (Chap. 3), chromosome evolution (Chaps. 4 and 5), chromosomes and cancer (Chap. 6), chromosome alterations in human populations (Chaps. 7 and 8) and in human eggs and spermatozoa (Chaps. 9 and 10), restriction endonucleases and their actions on chromosomes (Chaps. 11–13), effects of DNase I on chromosomes (Chap. 14), chemical modifications of DNA leading to aberrations and sister-chromatid exchanges (Chaps. 15 and 16), chromosome alterations in occupationally exposed persons (Chap. 17), chromosome alterations induced by chemicals produced by spiders (Chap. 18), and the importance of cytogenetic tests in genetic toxicology (Chap. 19).

We thank the authors of the book for their contributions and the staff of the Springer-Verlag, especially Dr. Dieter Czeschlik, Mrs. Antonella Cerri and Ms. C. Grössl for their help.

Berlin-Dahlem, August 1987

Günter Obe
Armin Basler

Contents

List of Contributors

1 Towards Understanding the Structure of Eukaryotic Chromosomes

P. JEPPESEN and D. J. BOWER[1]

1 Introduction

Since it was realized that a pattern of only four different bases within the nucleic acid genome of each biological species is sufficient to determine the life cycle of that species and, in being replicated, is capable of passing on that information to its progeny, the mechanisms by which nucleic acid sequences are read and converted into life processes have fascinated biologists in general and been at the forefront of research by molecular biologists in particular. Because of their relatively small size and the ease with which large numbers of genetic recombinants could be generated, allowing fine analysis of small genomic regions, the genomes of viruses (bacteriophage), prokaryotes and lower eukaryotes were the first to receive intensive study. With the advent of new technologies to supplement classical genetics, progress has been rapid in understanding the function of these "simple" genomes and, although a great deal remains to be learned, many basic principles which govern how individual genes are converted into protein molecules, and how total genomes replicate have been established for these lower life forms. Based on the available experimental evidence and on the principle that higher eukaryotes have evolved from lower forms, it seemed likely that these basic mechanisms would hold true for the expression of genetic content in plants and mammals. As we shall describe, the discoveries over the last decade of techniques for restriction mapping, cloning, amplifying and sequencing fragments of higher plant and mammalian genomes have led to a reassessment of this assumption. Although our understanding of gene function in higher eukaryotes still lags behind the situation in prokaryotes, especially at the level of control, it is already clear that while the basic tenets of molecular biology hold true, namely DNA replication, transcription into RNA and translation of RNA into protein via the genetic code, a number of fundamental differences exist between prokaryotes and higher eukaryotes in genome organization and expression.

1 Medical Research Council, Clinical and Population Cytogenetics Unit, Western General Hospital, Crewe Road, Edinburgh EH4 2XU, Scotland, Great Britain

Cytogenetics. Ed. by G. Obe and A. Basler

1.1 Eukaryotic Genome Size

One striking difference has been noted since it was realized that DNA is the genetic material: namely, the large quantity of DNA in eukaryotes in general, and higher plants and animals in particular, compared to prokaryotes (see, for example, Gall 1981). The step from prokaryote to the simplest fungae and algae involves an increase in DNA content of around tenfold. The amount of DNA in higher eukaryotes varies considerably, and may be as much as 50–100 pg in some monocot plants and amphibians. Among mammals the variation is less, and most species contain 3–4 pg DNA per haploid genome, i.e. approximately 3×10^6 kilobase pairs (kb). This exceeds the DNA content of a typical bacterium, e.g. *E. coli* with 4.5×10^3 kb, by around 1000-fold and even allowing for a considerable overestimate in the likely number of genes coding for structural proteins in mammals compared to bacteria, say 10×, this still leaves an apparent 100× "excess" DNA for which to account.

An obvious difference between prokaryotes and eukaryotes, especially higher plants and animals, is the capacity for differentiation of cell types during development of the latter, leading to complex multicellular organisms in which cells from different tissues and organs of the same individual may appear, by morphological and biochemical criteria, to be more dissimilar than, e.g., cells of two distantly related bacterial species. Genetically, however, the two cell types from the same individual are indistinguishable, and both, in theory, contain the information necessary to code for a complete organism. At the risk of oversimplifying the complex processes of development, it seems reasonable to assume that the information enabling higher eukaryotes to differentiate resides in at least a substantial fraction of this "excess" DNA, an assumption justified by the almost direct relationship between DNA content and position in the phylogenetic series (Britten and Davidson 1969).

The tendency to larger genome size in eukaryotes must have been paralleled by the evolution of novel ways to cope with ever increasing quantities of DNA. Evidence for an initial divergence in genome organization provides the distinguishing feature between prokaryotes and eukaryotes – the presence in the latter of a compartmentalized nucleus separated from the cytoplasm in non-dividing cells by a nuclear membrane, which disappears at cell division with the simultaneous condensation of the genetic material into discrete chromosomes. Apart from this unifying feature among eukaryotes, divergence in nuclear organization between species is evidenced at the macroscopic level by the seemingly limitless variation in chromosome morphologies and properties which has been described in eukaryotes. Such diversity of eukaryotic chromosome structure requires a volume to itself, and has been reviewed many times elsewhere (see, for example, Bostock and Sumner 1978). The aim of this chapter is to describe some recent advances in our understanding of the structural organization of eukaryotic chromosomes at a molecular level, with special reference to mammals, but drawing on evidence from other eukaryotes where relevant. Since much speculation exists about the possible involvement of specific DNA base sequences in chromosome structure, it seems appropriate to start by summarizing what is currently known about DNA sequence organization in eukaryotes.

2 Repeated DNA in Eukaryotes

The genomes of higher eukaryotes are distinguished from lower orders not only by their size, but also by the presence of a large proportion of repeated DNA sequences (Britten and Kohne 1968). Copy numbers range from above 10^6 for highly repetitious DNA, through 10^3-10^5 for moderately repeated sequences, down to a few copies for "unique" DNA.

2.1 Satellite DNA

Satellite components having different buoyant densities and hence separating from main band DNA during centrifugation on CsCl density gradients have long been recognized (Kit 1961). It has subsequently been shown that most mammalian species contain satellite DNAs, which in some cases require the addition of Ag^+ and/or Hg^{2+} to density gradients of Cs_2SO_4 in order to separate them from the remaining genomic DNA (reviewed by Walker 1971). Reassociation kinetics suggested that satellite DNAs comprised very high copy numbers of relatively short sequences (Waring and Britten 1966), and these sequences were shown by in situ hybridization often to correspond to blocks of nuclear constitutive heterochromatin and C-bands in metaphase chromosomes (Walker 1971). Although the average is about 10%, different species vary enormously in their content of highly repeated sequences. In the kangaroo rat (*Dipodomys ordii*), for instance, over 50% of the genome is highly repetitious (Hatch and Mazrimas 1970), whereas in Chinese hamsters (*Cricetulus griseus*) the figure is only 2–3% (Benz and Burki 1978). The first satellite DNA to be characterized by sequence analysis was the α-satellite component of guinea pig (Southern 1970) which was shown to be based on tandem repetition of a 6-bp unit CCCTAA with superimposed random mutations. Many other satellite DNA sequences have since been characterized from a variety of species: all are derived from tandem repetitions of simple sequences, but with different repeating units (reviewed by Brutlag 1980).

The frequent, though not invariable, association of satellite DNA and heterochromatin with metaphase chromosome centromeres, and to a lesser extent telomeres (see succeeding sections, and Brutlag 1980) prompts the suggestion that satellite DNAs might be involved in the structural organization of these subchromosomal organelles. The possibility that related sequences might also be distributed more homogeneously through chromosomes extended this hypothesis to include satellite involvement in chromosome structure in general (Razin et al. 1978). While this may be true for certain satellite-like sequences, the wide variation in satellite DNA content among species argues that the bulk of this class of DNA in genomes containing a high proportion of highly repetitious sequences is probably not involved in chromosome structure. This view is reinforced by the observed lack of inter-species sequence homologies between major satellite DNA families, except in closely related species, indicating that the events leading to simple sequence amplification and resulting in satellite DNAs are of evolutionarily recent occurrence, long post-dating the emergence of the present structure of higher eukaryotic chromosomes. Where homologous satellite sequences are present in closely related species, e.g. primates (Mitchell et al. 1981), their high rate of divergence suggests no selective advantage in the sequences per se. The significance of satellites may therefore be that they reflect a successful mechanism for increasing total DNA which has evolved in eukaryotes, and which has contributed to increasing genome size through the randomization by mutation of successive acquisitions of satellite sequences.

2.2 Intermediate Dispersed Repeats

Quantities of moderately repetitious DNA vary less between species, amounting to about 20% in humans. It is now known that a high proportion of DNA of this class belongs to only a few families of interspersed repeated sequences. For example, four families probably comprise the major portion of human non-satellite repetitious DNA: Alu, L1, THE1 and poly(CA) (reviewed by Weiner et al. 1986).

The best characterized of these is the Alu family, named from its generation by restriction endonuclease *Alu*I (Schmid and Jelinek 1982). The Alu family comprises around 500,000 single copies of an approximately 300-bp repeating unit interspersed throughout the genome, and constituting about 5% of total DNA. Sequence studies of Alu family members have established that the repeat unit is a dimeric derivative of 7 SL RNA, a cytoplasmic component of the protein translation apparatus (Walter and Blobel 1982). Alu repeats are "retropseudogenes", that is, they have arisen by insertion into the genome of cDNA copies of processed cellular RNA (Weiner et al. 1986). Although the exact mechanisms of "retroposition" are obscure, "retroposons" are distinguished as inserted elements by the presence of short (typically 5–20 bp) duplications at the sites of insertion. Rodents possess related families of short interspersed elements, or "SINES" (Singer 1982), the B1 superfamily, which are monomeric derivatives of 7 SL RNA (Weiner et al. 1986). Why 7 SL RNA should have given rise to such high copy numbers of retropseudogenes in man and rodents is a mystery, although the presence of a conserved RNA polymerase III promoter within the repeated unit may be significant. Families of SINES derived from different cellular RNAs have also been described in other species (Weiner et al. 1986).

The L1 or "LINE" 1 family (long interspersed elements, Singer 1982), also known as the Kpn family from its digestion pattern with restriction endonuclease *Kpn*I, is another example of a highly repeated and inserted retropseudogene (reviewed by Singer and Skowronski 1985; Weiner et al. 1986). The repeating unit has a maximum size of 6.5 kb, but is often truncated at the 5′ end, with a copy number of approximately $2–5 \times 10^5$. The RNA precursor appears to have been mRNA, as evidenced by the 3′ poly(A) tails and upstream polyadenylation signal found in many members, although the gene product is obscure. Related L1 families have also been detected in other mammals, e.g. mice, rats and monkeys.

The THE1 family numbers about 10^4 copies of a 2–3-kb insertion, which has many features in common with the integrated provirus of RNA retroviruses (Paulson et al. 1985). In particular, THE1 family members have 350-bp-long terminal repeats (LTRs), generate a 5-bp duplication at the site of insertion and encode a 2-kb polyadenylated RNA polymerase II transcript. THE1 sequences, however, bear no obvious homology to any known retrovirus. Analogous, but unrelated, families of transposable elements are known in yeast (Ty) and *Drosophila* (copia family).

Sequences hybridizing with poly(CA):poly(GT) probes are interspersed throughout the human genome in about 5×10^4 copies, and may therefore be classed as moderately repetitious interspersed repeats (Hamada et al. 1982). However, in their simple sequence structure they resemble satellite DNAs, and their mode of amplification and insertion are unclear (Rogers 1983). The discovery of poly(CA) hybridizing sequences in a wide range of species and the propensity of alternating pyrimidine-purine sequences to take up a left-handed helical conformation (Z-DNA) under certain conditions in vitro, have led to considerable speculation regarding the possible regulatory role of poly(CA) and/or Z-DNA in gene expression (Hamada et al. 1982; Nordheim et al. 1981).

The recognition of dispersed repetitous DNA sequences in eukaryotes led to the hypothesis that they might play a role in gene regulation (Britten and Davidson 1969). It has also been suggested that interspersed repeated sequences might be important in organizing chromosome structure (Jeppesen et al. 1978). However, the characterization of the major families of human interspersed repeats summarized above has provided no confirmatory evidence for these assertions. On the contrary, it appears that a substantial fraction of this class of DNA has arisen by adventitious amplification of a small number of sequences, and their insertion at random sites within the genome. This does not exclude the possibility that other minor families of interspersed repeats remain to be discovered which might have regulatory or structural functions.

2.3 "Unique" DNA

Most eukaryotic structural genes that have been studied occur in the low copy number or "unique" DNA fraction. However, even here, a major difference between the organization of higher eukaryotic genes compared to prokaryotes and lower eukaryotes has become apparent. With a few exceptions, the structural genes in higher eukaryotes that have been analyzed to date contain intervening sequences or "introns" which split the coding region into a number of separate "exons". While the size distribution of exons has been shown to be fairly constant in most genes studied, around a mean of 140 bp, intron sizes vary considerably and, in total, intron DNA considerably exceeds exon DNA (Naora and Deacon 1982), which helps to account for some of the "excess" DNA present in higher eukaryotes. Together with flanking sequences, introns are removed from transcribed RNA during processing in the nucleus before being transported to the cytoplasm and translated into protein (reviewed by Padgett et al. 1986).

The origin and possible function of introns has promoted considerable debate. The fact that they have not been found in prokaryotic and lower eukaryotic genes even though the coding regions are in many instances well conserved (see, for example, Lonberg and Gilbert 1985) suggests that they have become inserted since the evolutionary divergence of eukaryotes and prokaryotes, possibly as transposable elements. This view is supported by the recent finding in *Drosophila* of a whole gene, including its own introns, "nested" entirely within an intron of a different gene (Henikoff et al. 1986), and by other instances where introns appear to be relatively recent insertions. However, it has been noted in a number of cases that exons seem to correspond to different structural units or domains in the encoded protein (Lonberg and Gilbert 1985; Stone et al. 1985). These observations have led to speculation that introns may provide a means for isolating polypeptide sequences that have evolved to fulfil "useful" functions, which can then, by exon duplication and translocation, be incorporated into other genes, transferring their function with them. Such a model of exon "shuffling" predicts more rapid evolution than if each gene mutates independently. If this hypothesis is correct, sequence homologies between prokaryotic and eukaryotic genes imply that introns must have been present before the divergence of eukaryotes and prokaryotes, and have been lost from the latter during evolution.

Although referred to as "unique", a number of mammalian genes that have been studied occur as clusters of related genes, for example the globins (reviewed by Maniatis et al. 1980) and immunoglobulins (reviewed by Honjo 1983). In addition, many genes possess unexpressed "pseudogene" homologues, sometimes genetically linked, but often apparently randomly distributed about the genome. Many of the unlinked pseudogenes are "processed retropseudogenes", i.e. retroposons derived from processed mRNA by reverse transcription and integration (see above). The derivation of retropseudogenes has been inferred from their lack of introns and flanking sequences, and by the presence of a poly(A) 3′ tail (reviewed by Weiner et al. 1986). The functional significance, if any, of pseudogenes is unknown.

Certain other genes, such as those for ribosomal-RNA (reviewed by Long and Dawid 1980) and histones in some species (reviewed by Maxson et al. 1983), appear as amplified tandem repeats, often lacking introns. This may reflect a primitive form of gene organization which has been retained by these evolutionarily ancient genes, in which gene expression is regulated partly by a gene dosage effect. In birds and mammals, although histone genes are clustered, they are not exact tandem repeats.

2.4 "Selfish" DNA

The 10-fold step in genome size from prokaryote to simple eukaryote, and the further 100-fold jump in the evolution of mammals, is accompanied by a corresponding increase in the complexity of sequence organization. In the mammalian genome single or low copy number "unique" DNA is regularly interspersed with families of moderately repetitious sequences and, occasionally, by larger blocks of highly repetitive simple sequence satellite DNAs. Only a small fraction of the total DNA codes for protein, and only a fraction of this is ever expressed in any given cell. Until the advent of methods for DNA sequence analysis, it was intuitively assumed by many that the genomes of higher eukaryotes would be "efficient", i.e. all the non-coding DNA would perform functional roles in gene regulation, chromosome structure, etc. However, the large differences in DNA content observed between some closely related species – the "C-value paradox" (Gall 1981) – suggested that eukaryotic genomes might be capable of accommodating large amounts of "redundant" DNA. Following the clear demonstration from molecular studies of large quantities of non-coding DNA in intergenic and intervening sequences with no obvious functions, together with the high proportion of apparently randomly inserted repeated sequences, a novel concept has arisen, namely that of "selfish" DNA (Doolittle and Sapienza 1980; Orgel and Crick 1980). In essence, it is argued that a proportion of non-coding DNA, particularly repetitious sequences, may fulfil no useful purpose, but is present because certain DNA sequences have evolved the ability for self-propagation and insertion within the genome, and are tolerated because the "host" has concurrently evolved mechanisms to cope with them. There are as yet no convincing arguments against this hypothesis, since it is clear that higher eukaryotes have the ability to organize a large quantity of DNA, and mechanisms exist for excising unwanted sequences and resplicing both RNA and DNA. However, although insertion of mobile DNA elements at many sites may be "silent", at other sites, such as coding or control regions, insertion might be expected to be lethal or severely selected against, while at yet other sites, insertion of additional promoters, transposition of whole genes or gene segments, and other numerous subtle effects modulating gene expression, might confer a selective advantage. Thus the combined presence of selfish DNA elements and the ability of host genomes to tolerate many of their effects might have contributed to the rapidity of evolution in eukaryotes.

Whatever the truth of these contentions, the ability to organize a large quantity of DNA in a highly specific manner within chromosomes distinguishes higher eukaryotes from lower orders, and it is the structure of individual chromosomes with which the remainder of this chapter will be concerned. Not only must DNA be packaged in an inactive state within highly condensed and discrete chromosomes at mitotic and meiotic cell division for segregation among progeny cells, but it must also be able to decondense during interphase in order to fulfil its active roles in transcription and replication.

3 Structural Organization of Metaphase Chromosomes

The structural organization of eukaryotic chromatin is at its most evident during cell division, when it becomes condensed, both at mitosis and meiosis, into the particles recognizable in the light microscope as chromosomes. At mitosis, following DNA replication in S phase and a lag during G2 phase, nuclear chromatin progressively condenses until by prophase individual chromosomes can be distinguished and the nuclear envelope has disappeared. At metaphase, condensation and contraction reach their maximum, and chromosomes become aligned at the metaphase plate. Because they possess a clearly defined morphology, lack the nuclear envelope and are relatively stable, allowing the isolation of suspensions essentially free of other cellular components, metaphase chromosomes have provided the starting material for many studies on eukaryotic chromosome structure. It should be stressed, however, that metaphase represents a very special case of chromosome organization, occupying only a small fraction of a cell's life cycle, in which virtually inactive chromatin is tightly packaged for the purpose of segregation between divided cells. Attempts to explain the structural organization of eukaryotic chromosomes based on the study of metaphase must be compatible with the more general features of chromosome organization observed at other stages of the life cycle.

3.1 Chromatin Structure

DNA probably seldom, if ever, occurs in the free state in eukaryotic cells, but is complexed with histones and other non-histone proteins in the form of chromatin. The basic structural unit of chromatin in all eukaryotic cells studied is the nucleosome (reviewed by Felsenfeld 1978), which is repeated about once every 200 bp of DNA, the exact spacing depending on the source of the chromatin (e.g. active or inactive). The nucleosome consists of approximately two turns of double-stranded DNA wound as a left-handed helix on the surface of a roughly spherical octamer comprising two molecules each of the four core histones, H2a, H2b, H3 and H4. The core particle possesses dyad symmetry about the mid-point of the attached DNA, which has a mean length in limit digests with micrococcal nuclease of 145–146 bp. The variable region between nucleosomes is less well defined at present, but consists of the linker DNA associated with histone H1 (or H5 in avian erythrocytes) and possibly other proteins such as the high mobility group (HMG) non-histone proteins (Goodwin et al. 1978). The whole structure of nucleosomes plus spacer regions forms a unit fibre of diameter 10 nm. This unit fibre itself has the capacity to coil into a solenoid containing six to seven nucleosomes per turn, and having a diameter of 30 nm (Finch and Klug 1976). The 30-nm fibre has been detected by X-ray diffraction in interphase and metaphase chromosomes (Langmore and Paulson 1983), although whether all chromatin exists in this form, or just inactive chromatin, is not clear at present.

The next level of organization, the folding of the 30-nm fibre within the interphase nucleus, or the more condensed chromosomes visible at cell division, is not currently amenable to the exact analytical methods which have helped to define the structures of the nucleosome and 30-nm fibre. To condense approximately 1 m of DNA double-helix which comprises a haploid mammalian genome into around 100 μm total metaphase chromosome length requires a reduction of roughly $10^4:1$. The 30-nm chromatin solenoid provides a packing ratio of around 40:1, thus a further 250-fold compaction in structure is required at metaphase. Although, as will be shown, the organization of chromatin in metaphase chromosomes is highly ordered and non-random, no regular periodicity has yet been demonstrated which might enable a crystallographic structural approach to be attempted. Instead, the elucidation of higher levels of chromatin organization at present requires the interpretation of morphological and biochemical data.

3.2 The Unineme Chromatid

A typical mammalian metaphase chromosome appears in the light microscope as two separated sister chromatids joined at the primary constriction or centromere. The centromere may be approximately centrally located between the chromatid extremities or telomeres (metacentric chromosomes), or at or near one pair of telomeres (telocentric and acrocentric chomosomes respectively).

It is now generally accepted that a chromatid is a unineme structure, i.e. it corresponds to a single DNA double-helix linearly organized with respect to the chromatid axis (for a review of the evidence for and against uninemy see Prescott 1970). Because of the fragility of long, naked DNA, no chromosome-sized molecules have been measured in higher eukaryotes, although such measurements have been made for yeast and *Drosophila* chromosomes (Kavenoff et al. 1974; Petes et al. 1974). However, almost all of the available evidence is consistent with the unineme hypothesis, whereas many observations, including the exact correspondence between genetic linkage groups and chromosome number, the semi-conservative segregation of newly synthesized DNA between daughter chromosomes, sister-chromatid exchanges, chromosome translocations and genetic recombination at meiosis, are difficult to explain by models involving multiple DNA strands. The evidence suggesting polynemy or binemy is less compelling, and in most cases is not necessarily in conflict with the unineme hypothesis, or can be discarded as an artefact.

3.3 Telomeres

The unineme hypothesis implies that the linear DNA molecule comprising a chromatid should terminate at the telomeres, which raises two questions: firstly, how are the 5′ ends of the DNA replicated, since initiation of DNA synthesis requires a 5′ RNA priming sequence (Sugino et al. 1972); and secondly, why are telomeres stable when there is considerable evidence that artificial termini created, for example, by radiation damage (Resnick and Martin 1976), are inherently unstable and tend to fuse with each other? Bateman (1975) postulated that 3′ and 5′ DNA ends might be continuous at the telomeres, forming a hairpin structure which is nicked by a specific endonuclease after replication, followed by religation of individual chromatid ends. This resembles the structure which has since been found at the ends of some viral genomes (Baroudy et al. 1982). Such a model can resolve the DNA replication problem, and might also explain telomere stability. Recent evidence from the study of DNA termini in lower eukaryotes confirms the hairpin model in these species, and further demonstrates the importance of the base sequences at the ends for functional telomeres. These sequences consist of tandem repeats of short, simple, GC-rich units, varying slightly between species, and varying in number of repeats within a population (reviewed by Blackburn and Szostak 1984). *Tetrahymena* ends (5′–$[CCCCAA]_n$) can function as telomeres in yeast (Szostak and Blackburn 1982), but yeast-specific ends (5′–$[C_{1-3}A]_n$) are added to them in vivo (Blackburn and Szostak 1984). A possible mechanism for the addition of telomere ends is suggested by the demonstration of a terminal transferase activity in extracts of *Tetrahymena* which adds TTGGGG single-stranded oligomers to the 3′ ends of both *Tetrahymena* and yeast telomeres in vitro (Greider and Blackburn 1985). A study of yeast mutants with altered telomere structure suggests that there is a combination of untemplated addition and semi-conservative replication (Lustig and Petes 1986).

In mammals, telomeres are often associated at the macroscopic level with blocks of constitutive heterochromatin, and there is also evidence for an association of telomeres with the nuclear envelope during interphase (reviewed by Blackburn and Szostak 1984). Sequences isolated from these regions show some similarity in organization to the telomeres of lower eukaryotes (Blackburn and Szostak 1984) but it cannot be assumed that these blocks lie at the extreme termini. Using a probe which maps within 15–20 kb of the telomeres of human X and Y chromosomes, the sequence organization at the chromosome ends has been studied indirectly (Cooke et al. 1985; Cooke and Smith 1986). Southern blot analysis of partial restriction digests of genomic DNA, and timecourse digestions using restriction enzymes combined with *Bal*31 (one of whose activities is directed at DNA termini), led to the conclusion that telomeres are heterogeneous in length and consist of short, direct repeats, as in lower eukaryotes. Comparing telomere length in sperm and blood cells from the same individual, in six separate cases sperm telomeres were 5 kb longer on average than blood. B or T lymphocytes showed the same degree of heterogeneity as whole blood, but lymphoblastoid cell lines and blood from individuals with B-cell leukaemia showed very homogeneous telomere length. This suggested that terminal transferase activity on telomeres may be restricted to germ-line or early embryonic stages, periods to which "healing" of broken chromosomes in maize and coccids is confined (Hughes-Schrader and Ris 1941; McClintock 1950).

3.4 Centromeres

Centromeres are the sites of attachment of chromosomes to the mitotic or meiotic spindles. In the electron microscope microtubules can be seen attached to the kinetochores, which are trilaminar plate structures on either side of the centromeric constriction (Jokelainen 1967; Comings and Okada 1971; Roos 1973; Rieder 1982). In mouse kinetochores, a series of fibrillar loops, 25–30 nm in diameter, are arranged in a parallel array along the plane of the plate; 20–50 microtubules in parallel rows are seen perpendicular to the outer plate, spaced at 60-nm intervals (Rattner 1986). In some non-vertebrate species kinetochores are holocentric – elongated trilaminar plates with attached microtubules running the entire length of the mitotic chromosomes (Blackburn and Szostak 1984). The molecular components of mammalian kinetochores include DNA (Ris and Witt 1981; Rattner 1986) and protein (Moroi et al. 1980).

Making use of anti-kinetochore antibodies found in human autoimmune sera from patients suffering from the CREST syndrome (Moroi et al. 1980), discrete kinetochore structures have been shown to persist through interphase doubling very late in G2 ("presumptive" kinetochores, or pre-kinetochores, Brenner et al. 1981; Moroi et al. 1981). Indirect immunofluorescence studies suggest that pre-kinetochores are associated with nucleoli and/or the nuclear periphery (Moroi et al. 1981; unpublished observations of authors), although others have failed to observe this in the electron microscope (Brenner et al. 1981). Using anti-kinetochore autoantibodies, several groups have attempted to identify kinetochore-specific polypeptides after separation by SDS/polyacrylamide gel electrophoresis and transfer to nitrocellulose ("Western blots"), and by indirect immunofluorescence with selectively pre-absorbed antibodies. These studies have detected antibody binding to polypeptides of 14, 20, 23, 34 kD (Cox et al. 1983), 77, 110 kD (Earnshaw et al. 1984), 19.5 kD (Guldner et al. 1984; Spowart et al. 1985), 18, 80 kD (Valdivia and Brinkley 1985) and 17, 80, 140 kD (Earnshaw and Rothfield 1985). Not all of the discrepancies in these somewhat conflicting results can be attributed to differences in molecular weight estimation, and it may be that different autoimmune sera recognize different kinetochore antigens, although this explanation seems unlikely since the results in any one study with a number of different patient sera are consistent. Another possibility is that antibody binds to proteolytic degradation products derived from the intact antigen(s), and that alternative chromosomal protein isolation procedures lead to different patterns of degradation.

This explanation would imply that the higher molecular weight polypeptides detected might be more meaningful. The problems of identifying antigens using autoimmune sera and Western blotting are compounded by the observed reduction in antigenicity of a number of nuclear antigens as a result of protein denaturation during solubilization (Palmer and Margolis 1985; Bower and Jeppesen 1986), and the observation that some autoimmune sera possessing anti-kinetochore antibody also contain other activities against nuclear and cytoplasmic components (Bower and Jeppesen 1986; Jeppesen and Nicol 1986) and it cannot be ruled out that some of the polypeptides identified using human sera might not be associated with kinetochores.

DNA sequences at the kinetochores of vertebrates have not yet been determined. In the yeast *Saccharomyces cerevisiae*, a single microtubule is associated with the centromeric region of each of the very small chromosomes. Sequences with inferred centromeric activity, i.e. they stabilize copy number and cause orderly mitotic segregation of plasmids transfected into yeast, have been cloned from ten *S. cerevisiae* chromosomes and their molecular structure determined. They do not cross-hybridize, but they contain some similar sequences and common structural features (reviewed by Clarke and Carbon 1985). They contain an AT-rich region of 78–86 bp on one side of which is a partially conserved 25-bp sequence and on the other an 8-bp conserved sequence. Thus it appears that functional *S. cerevisiae* centromeres may be less than 200 bp long. A similar attempt to identify the centromeric sequences of the fission yeast *Schizosaccharomyces pombe* has been less successful (Nakaseko et al. 1986). The *S. pombe* chromosomes are about five times longer than those from *S. cerevisiae*, and show some condensation at metaphase. 210 kb of chromosome II and 60 kb of chromosome I centromeric DNA have been cloned, but none of the individual clones, which contain relatively short sequences, conferred mitotic stability. It may be that the greater size of *S. pombe* chromosomes requires larger kinetochore structures, formed over a greater continuous length of DNA than was present in any of the individual cloned sequences.

3.5 Nucleolus Organizing Regions (NORs)

In mammals, it has been demonstrated by in situ hybridization that the multiple copies of the genes coding for 18 S and 28 S ribosomal-RNA are clustered at a small number of chromosomal locations (Henderson et al. 1972; Evans et al. 1974). These sites are often apparent as secondary constrictions of the chromatids at metaphase, and are implicated in nucleolus organization (reviewed by Wilson 1982). Not all rRNA gene clusters are necessarily active in transcription, the number varying with the individual and cell type within a given species in a heritable way. Only actively transcribing NORs form nucleoli, which are evident during interphase as roughly spherical, electron dense structures within the nucleus. Multiple nucleoli tend to fuse during interphase leading to a close association between non-homologous chromosomes at their NORs. Although normally fully condensed at metaphase and anaphase, nucleoli have been seen attached to secondary constrictions in the large prophase chromosomes of the rat kangaroo, *Potorous tridactylis* (Hsu et al. 1967), and it may be that secondary constrictions represent incomplete chromosomal condensation at sites of late rRNA transcription.

Cytological staining techniques specific for NORs, in particular, silver staining (Goodpasture and Bloom 1975), have been developed for metaphase chromosomes. It has been shown that silver staining depends on interaction with specific proteins located at NORs which were actively involved in rRNA transcription and nucleolus formation in the preceding interphase (Miller et al. 1976), indicating that proteins specifically involved in nucleolus structure and function remain associated with the NOR through mitotic interphase. Matsui et al. (1986), using immunological methods, have found proteins of 41, 47, 51, 55, 57, 60, 70 and 75 kD preferentially associated with interphase nucleoli. The 41- and 55-kD polypeptides were also found associated with NORs in metaphase chromosomes. RNA polymerase I, which transcribes rDNA, has also been found specifically associated with nucleoli and metaphase NORs (Scheer and Rose 1984).

3.6 Other Secondary Constrictions

In addition to the NORs, other secondary constrictions have been observed in metaphase chromosomes. Some are consistently found at certain positions in the karyotype of a given species, and many of these occur at locations of C-banding constitutive heterochromatin. Other sites are distributed among the population as fixed and heritable chromosomal markers linked to genetic loci. A number of these are only apparent under conditions of artificial cell culture in deprived medium, where they appear to be particularly prone to mechanical shear in fixed preparations, and hence have been called "fragile" sites (reviewed by Sutherland and Hecht 1986). The structural significance of fragile sites and other secondary constrictions not associated with nucleoli is not clear, but by analogy with NORs, some may represent regions of incomplete chromosome condensation, possibly as a result of late RNA transcription.

3.7 Chromosome Bands

The appearance of the total metaphase chromosome set (karyotype) is highly characteristic for a given species, with relative positions of centromeres and chromatid arm length ratios constant for all tissues. In addition, various differential staining procedures have been developed which reveal that chromatin is organized in a highly ordered and non-random manner, resulting in banding patterns which are also constant within a species (reviewed by Bostock and Sumner 1978). Indeed, banding shows up similarities between species, where chromosome translocations and rearrangements may be deduced from the constancy of banding over the whole karyotype (Yunis and Sawyer 1980). A banded karyotype allows the positions of chromosome deletions, translocations, inversions, etc., to be mapped with much higher accuracy than was previously possible with unbanded chromosomes, the resolution being limited by the degree of contraction of the chromosomes. In this way it has been possible to compare genetic linkage maps with physical chromosome banding maps, for example, in the human genome. In all instances, studies of this type have shown an exact correspondence between genetic loci and chromosome location defined by banding. Since genetic loci are defined by DNA base sequence, these results imply that chromosome bands are also a direct consequence of primary DNA base organization, although the precise mechanisms of chromosome banding are still a matter for debate, and probably result from a combination of several factors which vary along the length of the chromosome, such as DNA concentration and base composition, association of non-histone proteins, etc. (see Chapter 2).

It is hard to envisage how chromatin organized as a regular 30-nm solenoid as described in Section 3.1 could fold in such an ordered yet non-periodic structure as the metaphase chromatid without the involvement of information encoded in the DNA along the chromatid. There is evidence that nucleosomes may occupy preferred sites along the DNA sequence (Thoma and Simpson 1985) and that bending of the DNA double-helix is probably also non-random (Drew and Travers 1985), and it is therefore conceivable that ordered folding or coiling of the 30-nm fibre occurs as a result of a non-regular primary chromatin structure. However, it seems more likely that specific base sequences are recognized by agents, probably proteins, which then mediate the higher level organization of the chromatid.

4 A Looped Model for Chromosome Structure

In addition to DNA and histones, isolated metaphase chromosomes also contain non-histone protein and RNA. The significance of RNA in metaphase chromosome structure is unclear, and most of it can be accounted for by rRNA (18 S and 28 S), nucleolar RNA (45 S rRNA precursor) and tRNA (4 S) (reviewed by Prescott 1970), which are presumably present as cytoplasmic contaminants. Whereas treatment of metaphase chromosomes with proteases destroys their structure, treatment with RNase has no detectable effect, which confirms the view that RNA plays no role in maintaining metaphase organization. On the other hand, as will become clear, non-histone protein appears to be an integral component of metaphase chromosome structure.

4.1 The Metaphase Chromosome Core or "Scaffold"

The idea of a protein core or "backbone" organizing the metaphase chromosome is not new (Dounce et al. 1973), and was originally based upon the observed instability of metaphase chromosomes to agents which specifically affect protein structure, although early models postulating protein linkers joining subchromosome lengths of DNA are now known to be incorrect. The protein core hypothesis received impetus with the publication by Paulson and Laemmli (1977) of electron micrographs of metaphase chromosomes depleted of histones by treatment with polyvalent cations or 2 M NaCl, and spread on a water surface by the Kleinschmidt procedure, in which a "halo" of DNA loops 50–150 μm in circumference was seen to emanate from a central residual protein/DNA complex. The bulk of the DNA could be liberated from histone-depleted chromosomes by nuclease digestion, leaving a stable, predominantly hon-histone, protein core or "scaffold" (Adolph et al. 1977). A similar result was obtained if 0.2 N HCl was used instead to remove histones from Chinese hamster metaphase chromosomes (Jeppesen et al. 1978).

These early results were criticized on the grounds that they might reflect an artefactual deposition of insoluble or poorly soluble proteins on aggregated or collapsed DNA fibres formed as a consequence of the high DNA concentration following dehistonization (Okada and Comings 1980; Hadlaczky et al. 1981). In interpreting the dense, axial, protein-rich structures evident in some light and electron micrographs, this view is probably partially correct, but the argument centres largely on whether the arrangement of non-histone protein observed by microscopy of histone-depleted chromosomes corresponds to an organization that exists in vivo. Subsequent studies have provided considerable support for a structure in which a core of non-histone protein, intimately associated with DNA, organizes chromatin as a series of loops in metaphase chromosomes. The exact form that this protein substructure takes in native, non-extracted chromosomes can only be surmised until techniques are available for studying it in situ.

Improved isolation techniques resulting in fewer contaminating non-chromosomal proteins in preparations of metaphase chromosomes and histone-depleted chromosomes provide an indication of the minimum protein set necessary for maintaining the chromosome core or scaffold. Using a chromosome isolation medium containing chelators and the polyamines spermine and spermidine as structural stabilizers, Lewis and Laemmli (1982) have purified HeLa metaphase chromosomes through multiple centrifugation steps, and have concluded that following histone dissociation, these chromosomes contain two major non-histone polypeptides, Sc1 (170 kD) and Sc2 (135 kD), which are required for integrity of the scaffold. Gooderham and Jeppesen (1983) have described a different isolation procedure for Chinese hamster metaphase chromosomes in which morphology is stabilized by isotonic KCl solution, and chromosomes purified by a single velocity centrifugation step through a glycerol concentration gradient. Following extraction of histones with 2 M NaCl, a somewhat more complex set of polypeptides was found retained as a protein core, including vimentin and possibly other intermediate filament and cytoskeletal proteins, whereas some proteins which might be expected in chromosome preparations, such as tubulin and the nuclear lamins, were not detected. (Polypeptides corresponding to Sc1 and Sc2 may have been present but were not identified.) The significance of finding cytoskeletal components co-purifying with metaphase chromosomes and retained in the histone-depleted core (Lewis and Laemmli 1982; Gooderham and Jeppesen 1983) is not clear: they may be merely contaminants, but it is also possible they may play a role in nuclear organization. It is of interest to note that considerable sequence homology has been shown between intermediate filament proteins and nuclear lamins (see Sect. 4.4).

An approach aimed at determining the functional significance of the core has been to investigate the effect on chromosome non-histone proteins of treatments which perturb the normal dehistonized chromosome structure. Lewis and Laemmli (1982) reported a reduction in sedimentation coefficient and loss of Sc1 and Sc2 from HeLa scaffolds treated with β-mercaptoethanol (BME) and certain chelators, such as *o*-phenanthroline, which they have interpreted as indicating a metalloprotein involvement in the organization of the scaffold, and from metal replacement studies have suggested that the metal could be copper. Jeppesen and Morten (1985) also demonstrated a loss of axial organization of residual core proteins in dehistonized Chinese hamster metaphase chromosomes exposed to BME, as evidenced by light microscopy, although no alteration in core polypeptide composition could be detected. Since the effects of BME could be reversed by oxidation of free sulphydryl groups to disulphide bridges prior to dehistonization, these results suggested that inter-protein cross-linking via sulphydryl groups is required for a visible axial core structure.

A second functional approach has been to investigate the possible attachment sites of DNA loops to the non-histone protein core. Razin et al. (1978) have reported that digestion of mouse chromosomes with specific restriction endonucleases leaves a fraction of repetitious DNA associated with the chromosome core. Treatment of dehistonized metaphase chromosomes from Chinese hamster with non-specific nucleases, such as DNase I and micrococcal nuclease, leads to a population of short DNA fragments (around 100–140 bp) which are associated with the non-histone protein core, and which are protected from complete degradation. Analysis of the fragments by reassociation kinetics suggested that they were enriched for repeated sequences (Jeppesen and Bankier 1979). Further studies on the sequences of these fragments have confirmed this conclusion, and have indicated that the fragments are GC-rich, with some short sequence repeats (unpublished observations, P.J.). The direct demonstration in vitro, however, of DNA binding to core or scaffold proteins, for example, by probing membrane immobilized core polypeptides with radioactively labelled, cloned DNA, is hampered by the insolubility of these proteins except in denaturing buffers, when DNA-binding activity is likely to be lost. To date, the evidence for specific DNA/protein interaction in metaphase chromosomes remains circumstantial and inconclusive.

4.2 Immunological Methods

One of the most successful approaches so far in defining the structural significance of the chromosome core has been the use of immunological methods. Antisera, monoclonal antibodies and naturally occurring autoantibodies associated with autoimmune disease have the potential for defining specific subchromosome components, both in situ by light and electron microscopy, and biochemically.

The use of autoimmune sera containing anti-centromere antibodies to investigate kinetochore structure has already been described in Section 3.4. This experimental approach has demonstrated that the kinetochore is retained as a structural entity associated with the protein core or scaffold of dehistonized metaphase chromosomes (Cox et al. 1983; Earnshaw et al. 1984; Jeppesen and Nicol 1986). These results are important because they demonstrate that kinetochore antigen is not redistributed or otherwise artefactually affected as a result of histone dissociation, and lends support to the view that the non-histone protein core in toto represents a real structure. The use of a sensitive anti-kinetochore antibody probe also shows that conventionally stained polypeptide maps of protein cores may be misleading, for Earnshaw et al. (1984) detected polypeptides reacting with anti-kinetochore antibody at 110 kD and 70 kD on Western blots of scaffold proteins, whereas corresponding Coomassie Blue

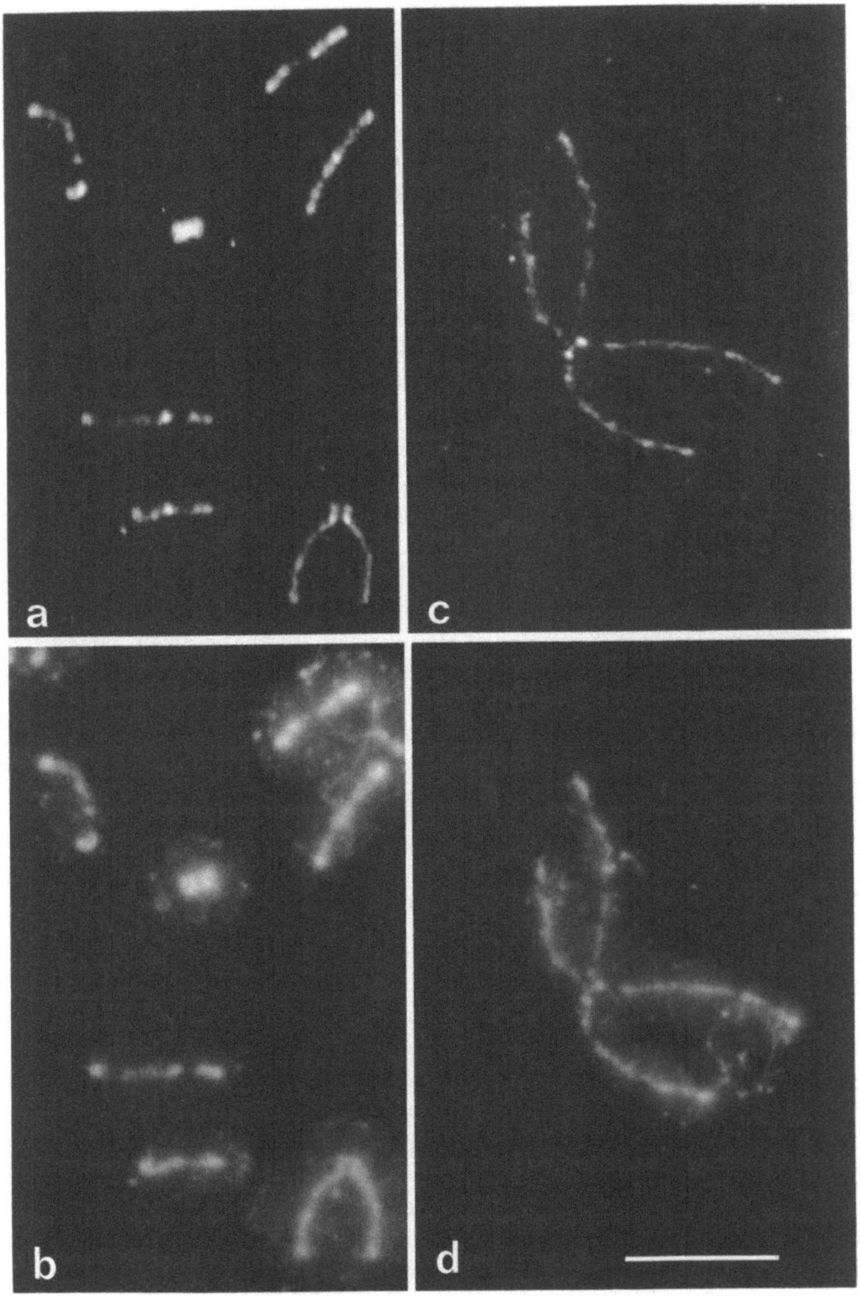

Fig. 1a–d. Organization of protein and DNA in histone-dissociated metaphase chromosomes isolated from Chinese hamster tissue culture cells. **a** Fluorescein isothiocyanate (FITC) indirect immunofluorescence micrograph, demonstrating antibody binding to non-histone protein axial cores, using human autoimmune serum containing anti-kinetochore and anti-chromosome core activities (Jeppesen and Nicol 1986). **b** Same field of view as a, but photographed using DNA-fluorochrome 4,6-diamidino-2-phenylindole (DAPI) fluorescence to show "halo" of DNA loops surrounding brightly fluorescent axes, which contain DNA collapsed on non-histone protein cores. **c** As a, but illustrating a particularly extended example, to show non-uniform distribution of chromosome core antigen and correspondence in distribution patterns between sister chromatids. **d** DAPI fluorescence of the same field of view shown in c. Bar = 10 μm.

Experimental conditions for Figs. 1 and 2 are contained in Jeppesen and Nicol (1986)

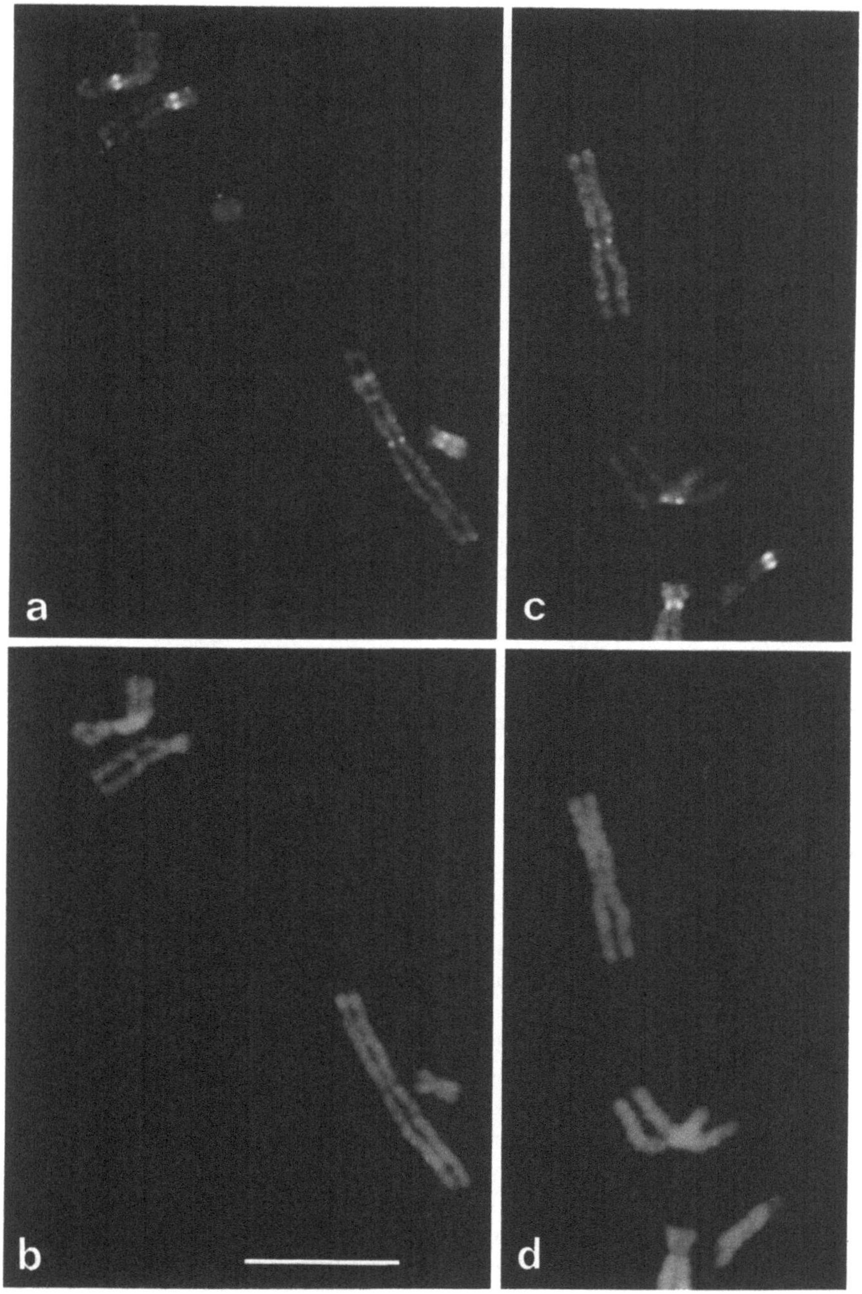

Fig. 2a–d. Distribution of non-histone protein core antigen in native isolated Chinese hamster metaphase chromosomes, using the same human autoimmune serum as in Fig. 1. **a**, c FITC indirect immunofluorescence micrographs illustrating kinetochore binding and "banding-like" non-uniform distribution of chromosome core antigen along chromatid arms. **b**, **d** DAPI fluorescence micrographs of the same fields of view shown in a, c respectively, illustrating DAPI bands. Bar = 10 μm

stained gels show only Sc1 and Sc2 (170 kD and 135 kD respectively). Thus other structurally important polypeptides may also be present, but below the level of detection in stained gels.

Using polyclonal antibodies, Earnshaw et al. (1985) have shown that the scaffold protein Sc1 from chicken corresponds to topoisomerase II, and in situ localization by indirect immunofluorescence and immuno-electron microscopy indicates that topoisomerase II is distributed along the scaffold (Earnshaw and Heck 1985). These studies have also shown that the HeLa chromosome Sc1 is a mixture of at least three co-migrating polypeptides, of which one is topoisomerase II (Earnshaw et al. 1985), again illustrating that one-dimensional SDS/polyacrylamide gel electrophoresis may lead to an underestimate of the complexity of chromosome core composition. The role of topoisomerase II in metaphase chromosome structure is unclear, although during interphase it could be responsible for relaxing the supercoiling of DNA as a prelude to transcription or DNA replication. Base sequences corresponding to topoisomerase II recognition sites have been detected in *Drosophila* interphase nuclear scaffolds (Gasser and Laemmli 1986).

In addition to the more commonly found anti-kinetochore activity, the serum of a patient with the CREST syndrome has also been shown to react with a component of the protein core in dehistonized Chinese hamster metaphase chromosomes (Jeppesen and Nicol 1986). The antigen is non-uniformly distributed along the core (Fig. 1), and whole chromosomes reacted with the serum have a banded appearance (Fig. 2). In vivo the antigen is confined to mitotic chromosomes or the nucleus, demonstrating that it is not associated with metaphase chromosome preparations by an artefact. So far attempts to identify the antigen by Western blotting have failed, no doubt because the antigenicity is destroyed during solubilization of the core polypeptides.

4.3 Loops in Meiotic Chromosomes

Meiotic prophase follows DNA replication in the preceding S phase, and consists of a series of morphologically defined stages in which chromatin progressively condenses and becomes distinguishable as individual chromosomes (leptotene), followed by pairing, or synapsis, in which homologous autosomes and the sex chromosomes align alongside each other at zygotene. During pachytene the bivalents contract further until at diplotene they separate along their length except where held together at points of genetic crossing over, which are evident as chiasmata. Through the use of a silver staining technique, a proteinaceous structure, the synaptonemal complex (SC), can be seen in the light microscope apparently organizing this alignment during zygotene and pachytene (Dresser and Moses 1979), while electron microscopy shows the structure to consist of two lateral elements connected to a finer central element by transverse fibres (reviewed by Bostock and Sumner 1978). How the SC functions in organizing homologous pairing or chiasma formation is not known, nor whether this structure is related to the non-histone protein core of metaphase chromosomes. The kinetochore, which is a component of the metaphase chromosome core (see Sect. 4.2) is also associated with the SC, as demonstrated by immuno-electron microscopy using anti-kinetochore antibodies (A.T. Sumner and R. Speed, pers. commun.), which suggests that the SC and chromosome core might possess common features. Because of the difficulty in obtaining pure preparations of meiotic prophase chromosomes in quantity, biochemical characterization of the SC has proved difficult. It was reported that antibodies reacting with the cytoskeletal contractile proteins actin and myosin bind to the SC (Martino et al. 1980), but this observation has not been confirmed by others (Spyropoulos and Moens 1984). As with metaphase chromosomes, cytoskeletal proteins may be involved in meiotic chromosome structural organization or contraction, but it is hard to distinguish this possibility from the alternative explanation that the presence of these ubiquitous cellular proteins is a result of contamination during preparation.

Following diplotene, and further chromosome contraction (diakinesis), the nuclear membrane disappears at metaphase I with the bivalents still joined at the chiasmata. Finally, at anaphase I, the two sets of genetically recombined homologues separate. The remaining stages of meiosis, which may occur with or without an intervening reformation of the nuclear membrane, resemble mitosis, with each DNA-replicated chromosome appearing at metaphase II as two sister chromatids joined at the centromeres, which separate at anaphase II. The net result of meiotic division is four haploid sets of non-replicated chromosomes. Apart from homologous pairing in prophase, chromosome condensation appears similar in both meiosis and mitosis, and the morphology of chromosomes at meiotic metaphase II closely resembles that observed at mitotic metaphase. The positions of centromeres, NORs and relative arm lengths of the chromatids are the same, and where banding patterns for meiotic chromosomes have been obtained, these too correspond to their mitotic counterparts (Caspersson et al. 1971), indicating that the underlying organization is the same. In this context, the organization of the large "lampbrush" diplotene chromosomes of certain amphibian oocytes is particularly interesting, as it provides convincing support for the loop model of chromosome organization.

Lampbrush chromosomes can be directly observed in unfixed nuclei where they appear for most of their length as regions of condensed chromatin (chromomeres) but, at intervals, pairs of fibres loop out from the axis, at which points the loops seem to be continuous with a pair of axial fibres (reviewed by MacGregor 1980). It has long been recognized that the loops represent single DNA double-helices actively transcribing RNA, and it is this dense coating of ribonuclear protein which enables them to be seen in the light microscope. The positions of loops change with time in a predictable manner, and can be correlated with genetic map positions (Callan and Lloyd 1960; Mancino et al. 1972; Barsacch-Pilone et al. 1974). This is interpreted as different regions of DNA, possibly corresponding to single genes or groups of related genes, being activated at different stages, the untranscribed loops being tightly condensed at the chromomeres. Electron microscope examination of spread meiotic chromosomes confirms this looped structure, and further, through relaxing the tightly condensed chromomeres, indicates that all the chromatin is organized as loops, and suggests that the stems of loops are associated with the lateral elements of the SC (Rattner et al. 1981).

4.4 Nuclei and Looped Domains

Superficially, interphase chromatin lacks the high degree of order and constancy that characterizes mitotic chromosome structure. However, there is evidence that this distinction may be more apparent than real. A dramatic demonstration of the fundamental continuity of chromosome organization throughout the cell cycle is the premature condensation of chromosomes observed when interphase and metaphase cells are fused: within minutes of fusion, the nuclear envelope of the interphase partner disappears and the interphase chromatin condenses into structures which resemble mitotic chromosomes, the exact morphology depending on the stage of the cell cycle immediately prior to fusion (Johnson and Rao 1970). This rapid change from the decondensed interphase state to mitotic-like condensed chromatin strongly suggests that no fundamental reorganization of chromosome structure takes place during the transition. Prematurely condensed chromosomes (PCCs) are even capable of giving banding patterns which correspond closely to those obtained in normal metaphase chromosomes (Unakul et al. 1973), demonstrating further that the linear organization of interphase and mitotic chromosomes is similar.

There is evidence from several sources that chromatin in the interphase nucleus, like that in metaphase chromosomes, is organized as looped domains attached to a protein substructure. Loop sizes inferred from monitoring the relaxation by controlled enzymatic nicking of supercoiled DNA retained in nuclei depleted of histones (Cook and Brazell 1975; Benyajati and Worcel 1976), from direct observation in the fluorescence and electron microscopes (Warren and Cook 1978; McCready et al. 1979) and from measuring the maximum size of chromatin fragments released from nuclei by partial enzymatic cleavage (Igo-Kemenes and Zachau 1978), all lie in the approximate range 50–200 kb, and are therefore comparable to loop sizes measured in metaphase chromosomes (Paulson and Laemmli 1977). DNA loops or domains appear to be attached to the protein substructure at specific locations. Cook and Brazell (1980) found that while DNA sequences corresponding to β- and γ-globin genes are released from dehistonized nuclei by nucleolytic cleavage, α-globin gene sequences are more resistant to release, from which they deduced that the latter gene is closely linked to an attachment site. Mirkovitch et al. (1984) have also demonstrated that specific DNA restriction fragments are found associated with the residual protein in histone-depleted *Drosophila* nuclei, and have further deduced that an attachment site forms part of the histone gene-cluster repeat unit in this genome.

While the evidence for DNA loops in interphase chromosomes indicates an underlying similarity with chromatin organization at mitosis, experimental evidence that the same or similar structural proteins might be involved is less clear. The major non-histone protein constituents of the interphase nucleus are the three lamin polypeptides, A, B and C (70, 67 and 60 kD respectively), which comprise the nuclear envelope or lamina (Gerace and Blobel 1980). Recently, sequence similarities have been detected between lamin polypeptides A and C, and intermediate filament cytoskeletal proteins (Aebi et al. 1986; Fisher et al. 1986; McKeon et al. 1986), which could indicate that the cytoskeleton is continuous throughout the nucleus and cytoplasm. It has been suggested that interphase chromosomes are organized, at least partially, by the lamina, through multiple sites of attachment at the nuclear envelope (reviewed by Comings 1980). At the macroscopic level, a number of subchromosomal structures, such as telomeres, pre-kinetochores and inactive, condensed heterochromatin, appear to be associated with the nuclear periphery, although nucleoli and dispersed chromatin seem to be internal. However, it seems clear that the lamin polypeptides themselves are not involved in mitotic chromosome organization: when the nuclear membrane disappears at mitotic prophase, the lamina also dissociates, lamins A and C solubilizing throughout the cytoplasm, while lamin B is found associated with membrane vesicles (Gerace and Blobel 1980). In confirmation, the lamin polypeptides are not detected in isolated metaphase chromosomes (Gooderham and Jeppesen 1983).

During telophase, the lamina and nuclear membrane reform around the segregated sets of divided chromosomes. Assembly of the nuclear envelope in vitro has been studied by electron microscopy, using Chinese hamster tissue culture cell extracts and metaphase chromosomes (Burke and Gerace 1986). All three lamins cooperatively form the lamina directly on the surface of the still-condensed chromosomes, and the membrane vesicles associated with lamin B coalesce to form the new nuclear membrane in contact with the lamina. Selective removal of individual lamins with specific antibodies was shown to block reassembly. Similar packaging has been observed with non-eukaryotic DNA and *Xenopus* oocytes (Forbes et al. 1983), which suggests that interaction with the lamina is not DNA sequence-specific.

In addition to the lamins, a number of other proteins are found associated with nuclei depleted of histones, and having the bulk of nuclear DNA removed by nuclease digestion. Several methods have been described for the preparation of these structures, which are variously termed nuclear matrices (Berezney and Coffey 1974), nuclear cages (Cook and Brazell 1980) and nuclear scaffolds (Lebkowski and Laemmli 1982). It is probable that all of these structures contain the same structural elements responsible for organizing interphase chromosomes, but differences in preparative procedures no doubt lead to different levels of degradation and contamination with non-structural molecules, making it difficult to distinguish what these basic elements are, and whether they are shared in common with metaphase chromosome organization. Lebkowski and Laemmli (1982) have postulated two types of DNA/protein interactions organizing interphase chromosomes: one associated with the lamina, and another with a protein substructure which is sensitive to β-mercaptoethanol and/or metal chelation. By analogy with similar properties of histone-depleted metaphase chromosomes (Lewis and Laemmli 1982), these workers suggested that the non-lamina interaction is continuous with mitotic chromosome organization, although the common protein components were not defined.

Unlike the inactive condensed chromatin of metaphase chromosomes, DNA present in interphase chromatin undergoes replication during S phase, and expressed genes are transcribed throughout interphase. An important consideration, therefore, is how these activities relate to the loop model of interphase chromosome structure. There are indications that DNA replication units, or replicons, if not identical, are closely related to DNA loops or domains (Buongiorno-Nardelli et al. 1982; Cook and Lang 1984). Based largely on the observed association of sites of replication and transcription with the non-extracted residue retained following histone depletion and nuclease digestion of nuclei, it has been suggested that these activities take place on a nuclear substructure which may, or may not, be related to that responsible for organizing the looped chromatin domains (McCready et al. 1980; Robinson et al. 1983; Smith et al. 1984). The transient nature of sites of replication and transcription, especially the cell cycle and tissue specific dependency of the latter (Robinson et al. 1983), compared with the constant organization evident in metaphase chromosomes, would implicate a different set of attachment sites involved in synthetic mechanisms.

Unfortunately, this important issue has been confused by equating the substructure postulated to organize replication and transcription with the "nuclear matrix", the fibrillar network evident on electron microscope examination of some preparations of nuclease-digested, histone-depleted nuclei (reviewed by Agutter and Richardson 1980). This network is composed primarily of ribonuclear protein (RNP), as evidenced by its sensitivity to RNase digestion (Kaufmann et al. 1981). However, there are a number of grounds for questioning the existence of the fibrillar matrix in vivo. Importantly, matrix-like structures are not observed in non-extracted nuclei, and the possibility of artefacts arising from the redistribution of nuclear RNA and protein upon treatments to extract nuclei cannot easily be discounted. A fibrillar matrix can only be isolated from actively transcribing nuclei (LaFond and Woodcock 1983), which is taken as evidence for its intimate involvement with transcription. However, this observation is equally consistent with an artefactual origin, since active nuclei are rich in transcribed hnRNA which might predispose them to the aggregation of RNP filaments. At present, therefore, until the origin of the nuclear matrix is further clarified, observations indicating its possible association with synthetic activity, particularly RNA transcription, must be interpreted with caution.

Whether the nuclear matrix is real or artefact, it is not relevant to at least one level of chromosome organization. This is clearly demonstrated in transcriptionally inactive and non-dividing cells, such as avian erythrocytes, which, although unable to furnish a fibrillar nuclear matrix (LaFond and Wooodcock 1983), can still be induced to form PCCs on fusion with metaphase cells (Johnson et al. 1970), indicating that normal chromosome organization is retained in these cells.

While the role of a nuclear matrix awaits clarification, a number of observations regarding gene expression are consistent with the loop model of interphase chromosome organization. Conformational changes take place in interphase chromatin in the vicinity of a transcribed gene. These changes appear to be part of a two-stage process, involved in determining which genes are transcribed. Tissue and stage-specific nuclease-hypersensitive sites appear near the coding sequences in precursors of cells which express the gene and are conserved through mitoses. These are augmented by a second set of changes which are first observed around the onset of transcription, and which include the appearance of more hypersensitive sites and a general increase in nuclease sensitivity throughout large regions flanking the expressed gene (reviewed by Eissenberg et al. 1985). The latter changes are consistent with a model in which a condensed loop of chromatin, bound at its ends by protein, is opened out through processes including nicking by topoisomerase II, thus making it accessible to transcription complexes. This relaxation of a tightly coiled chromatin loop during transcription is analogous to the observations made in lampbrush chromosomes (see Sect. 4.3). The pre-transcriptional changes are not yet understood. Heritable alterations in methylation status also occur in sequences flanking expressed genes, but their role remains to be clarified (reviewed by Bird 1984, 1986; Keshet et al. 1986). There is evidence that methylation is involved in the phenomenon of X-inactivation in mammalian cells, although it may be secondary to some other event (Bird 1986).

5 Conclusions

In virtually all eukaryotic systems that have been studied, evidence has been obtained indicating that chromatin is organized as a series of domains or loops rigidly attached at fixed points to a non-histone protein nuclear substructure. The sizes of looped domains estimated for all stages of the cell/life cycle (i.e. interphase, mitosis and meiosis) are comparable and, although it is not proved, there seems no reason at present to dismiss the likelihood that at least some of the attachment sites are the same, thus ensuring a continuity of chromosome organization. A model for eukaryotic chromosome structure can therefore be proposed, in which the transitions from one stage of the cell cycle to another are limited to spatial reorganization of the attachment sites, together with changes in the degree of packing of the loops. The latter parameter can be directly related to local states of transcriptional activity, and is explained in this model by the relaxation and/or uncoiling of individual chromatin looped domains being transcribed. A similar mechanism may also be involved in replication. Sites of RNA transcription and DNA replication may also be transiently attached to the same or a different protein substructure, but this point requires further clarification. The elements of this model are established with reasonable certainty, the details await elucidation. In particular, future research will determine the DNA sequences involved in loop attachment, characterize the proteins and other molecules which comprise the nuclear substructure and construct the spatial distribution of these components in vivo.

The evidence on which this hypothesis is founded derives to a large extent from the study of histone-depleted and nuclease-digested chromosomes and nuclei, and is subject to the criticism, justified in part, that these structures do not exist in vivo, and may suffer considerable perturbation of their original organization during the course of preparation. However, provided the possibility of artefact is considered in interpreting results, especially what is seen, or believed to be seen, in the light and electron microscopes, the validity of this approach should not be underestimated. In reducing the complexities of nuclear structure to a simplified subset of interactions, valuable insight into possible mechanisms of nuclear organization has been obtained. Further, the fractionation of chromosomes into a number of components has enabled the preparation of specific molecular probes, for example antibodies and cloned nucleic acid sequences, with which these ideas may soon be tested under conditions much more closely related to those that obtain in vivo.

Acknowledgments. We thank Prof. H.J. Evans for his encouragement and for reading the manuscript. We are grateful to Dr. J.R. Paulson for a preprint of his review "Scaffolding and radial loops: the structural organization of metaphase chromosomes" to be published in: Adolph KW (ed) *Chromosome and chromatin structure*. CRC Press, Boca Raton, Florida (in press).

References

Adolph KW, Cheng SM, Paulson JR, Laemmli UK (1977) Isolation of a protein scaffold from mitotic HeLa cell chromosomes. Proc Natl Acad Sci USA 74:4937–4941

Aebi U, Cohn J, Buhle L, Gerace L (1986) The nuclear lamina is a meshwork of intermediate-type filaments. Nature (London) 323:560–564

Agutter PS, Richardson JCW (1980) Nuclear non-chromatin proteinaceous structures: their role in the organization and function of the interphase nucleus. J Cell Sci 44:395–435

Baroudy BM, Venkatesan S, Moss B (1982) Incompletely base-paired flip-flop terminal loops link the two DNA strands of the *Vaccinia* virus genome into one uninterrupted polynucleotide chain. Cell 28:315–324

Barsacch-Pilone G, Nardi I, Batistoni R, Andronico F, Beccari E (1974) Chromosome location of the genes for 23 S, 18 S and 5 S ribosomal RNA in *Triturus marmoratus (Amphibia Urodela)*. Chromosoma 49:135–153

Bateman AJ (1975) Simplification of palindromic telomere theory. Nature (London) 253:379

Benyajati C, Worcel A (1976) Isolation, characterization and structure of the folded interphase genome of *Drosophila melanogaster*. Cell 9:393–407

Benz RD, Burki HJ (1978) The distribution of moderately repeated DNA sequences among Chinese hamster chromosomes. Exp Cell Res 112:155–165

Berezney R, Coffey DS (1974) Identification of a nuclear protein matrix. Biochem Biophys Res Commun 60:1410–1417

Bird AP (1984) DNA methylation – how important in gene control? Nature (London) 307:503–504

Bird AP (1986) CpG-rich islands and the function of DNA methylation. Nature (London) 321: 209–213

Blackburn EH, Szostak JW (1984) The molecular structure of centromeres and telomeres. Annu Rev Biochem 53:163–194

Bostock CJ, Sumner AT (1978) The eukaryotic chromosome. Elsevier/North-Holland Biomedical Press, Amsterdam New York

Bower DJ, Jeppesen PGN (1986) Characterization of a polypeptide associated with coated vesicles and the cytoskeleton which is recognized by a CREST serum. Exp Cell Res 167:166–176

Brenner S, Pepper D, Berns MW, Tan E, Brinkley BR (1981) Kinetochore structure, duplication and distribution in mammalian cells: analysis by human autoantibodies from scleroderma patients. J Cell Biol 91:95–102

Britten RJ, Davidson EH (1969) Gene regulation for higher cells: a theory. Science 165:349–357

Britten RJ, Kohne DE (1968) Repeated sequences in DNA. Science 161:529–540

Brutlag DL (1980) Molecular arrangement and evolution of heterochromatic DNA. Annu Rev Genet 14:121–144

Buongiorno-Nardelli M, Micheli G, Carri MT, Marilley M (1982) A relationship between replicon size and supercoiled loop domains in the eukaryotic genome. Nature (London) 298:100–102

Burke B, Gerace L (1986) A cell free system to study reassembly of the nuclear envelope at the end of mitosis. Cell 44:639–652

Callan HG, Lloyd L (1960) Lampbrush chromosomes of crested newts *Triturus cristatus (Laurenti)*. Philos Trans R Soc London Ser B 243:135–219

Caspersson T, Hulten M, Lindsten J, Zech L (1971) Identification of chromosome bivalents in human male meiosis by quinacrine mustard fluorescence analysis. Hereditas 67:147–149

Clarke L, Carbon J (1985) The structure and function of yeast centromeres. Annu Rev Genet 19: 29–56

Comings DE (1980) Arrangement of chromatin in the nucleus. Human Genet 53:131–143

Comings DE, Okada TA (1971) Fine structure of the kinetochore in the Indian muntjac. Exp Cell Res 67:97–110

Cook PR, Brazell IA (1975) Supercoils in human DNA. J Cell Sci 19:261–279

Cook PR, Brazell IA (1980) Mapping sequences in loops of nuclear DNA by their progressive detachment from the nuclear cage. Nucleic Acids Res 8:2895–2907

Cook PR, Lang J (1984) The spatial organization of sequences involved in initiation and termination of eukaryotic DNA replication. Nucleic Acids Res 12:1069–1075
Cooke HJ, Smith BA (1986) Variability at the telomeres of the human X/Y pseudo autosomal region. Cold Spring Harbor Symp Quant Biol 51:213–220
Cooke HJ, Brown WRA, Rappold GA (1985) Hypervariable telomeric sequences from the human sex chromosomes are pseudoautosomal. Nature (London) 317:687–692
Cox JV, Schenk EA, Olmsted JB (1983) Human anti-centromere antibodies: distribution, characterization of antigens, and effect on microtubule organization. Cell 35:331–339
Doolittle WF, Sapienza C (1980) Selfish genes, the phenotype paradigm and genome evolution. Nature (London) 284:601–603
Dounce AL, Chanda SK, Townes PL (1973) The structure of higher eukaryotic chromosomes. J Theor Biol 42:275–285
Dresser ME, Moses MJ (1979) Silver staining of synaptonemal complexes in surface spreads for light and electron microscopy. Exp Cell Res 121:416–419
Drew HR, Travers AA (1985) DNA bending and its relation to nucleosome positioning. J Mol Biol 186:773–790
Earnshaw WC, Heck MMS (1985) Localization of topoisomerase II in mitotic chromosomes. J Cell Biol 100:1716–1725
Earnshaw WC, Rothfield N (1985) Identification of a family of human centromere proteins using autoimmune sera from patients with scleroderma. Chromosoma 91:313–321
Earnshaw WC, Halligan N, Cooke C, Rothfield N (1984) The kinetochore is part of the metaphase chromosome scaffold. J Cell Biol 98:352–357
Earnshaw WC, Halligan B, Cooke CA, Heck MMS, Lin LF (1985) Topoisomerase II is a structural component of mitotic chromosome scaffolds. J Cell Biol 100:1707–1715
Eissenberg JC, Cartwright IL, Thomas GH, Elgin SCR (1985) Selected topics in chromatin structure. Annu Rev Genet 19:485–536
Evans HJ, Buckland RA, Pardue ML (1974) Location of genes coding for 18 S and 28 S ribosomal RNA in the human genome. Chromosoma 48:405–426
Felsenfeld G (1978) Chromatin. Nature (London) 271:115–122
Finch JT, Klug A (1976) Solenoidal model for superstructure in chromatin. Proc Natl Acad Sci USA 73:1897–1901
Fisher DZ, Chaudhury N, Blobel G (1986) cDNA sequencing of nuclear lamins A and C reveals primary and secondary structural homology to intermediate filament proteins. Proc Natl Acad Sci USA 83:6450–6454
Forbes DJ, Kirschner MW, Newport JW (1983) Spontaneous formation of nucleus-like structures around bacteriophage DNA microinjected into *Xenopus* eggs. Cell 34:13–23
Gall JG (1981) Chromosome structure and the C-value paradox. J Cell Biol 91:3s–14s
Gasser SM, Laemmli UK (1986) Cohabitation of scaffold binding regions with upstream enhancer elements of three developmentally regulated genes of *D. melanogaster*. Cell 46:521–530
Gerace L, Blobel G (1980) The nuclear envelope lamina is reversibly depolymerized during mitosis. Cell 19:277–287
Gooderham K, Jeppesen PGN (1983) Chinese hamster metaphase chromosomes isolated under physiological conditions: a partial characterization of associated non-histone proteins and protein cores. Exp Cell Res 144:1–14
Goodpasture C, Bloom SE (1975) Visualization of nucleolar organizer regions in mammalian chromosomes using silver staining. Chromosoma 53:37–50
Goodwin GH, Walker JM, Johns EW (1978) The high mobility group HMG) non-histone chromosomal proteins. In: Busch H (ed) The cell nucleus, vol 6: Chromatin, part C. Academic Press, London New York, pp 181–219
Greider CW, Blackburn EH (1985) Identification of a specific telomere terminal transferase activity in *Tetrahymena* extracts. Cell 43:405–413
Guldner HH, Lakomek H-J, Bautz FA (1984) Human anti-centromere sera recognize a 19.5 kD non-histone chromosomal protein from HeLa cells. Clin Exp Immunol 58:13–20
Hadlaczky G, Sumner AT, Ross A (1981) Protein-depleted chromosomes. Chromosoma 81:537–567

Hamada H, Petrino MG, Kakunaga T (1982) A novel repeated element with Z-DNA-forming potential is widely found in evolutionarily diverse eukaryotic genomes. Proc Natl Acad Sci USA 79:6465–6469

Hatch FT, Mazrimas JA (1970) Satellite DNAs in the kangaroo rat. Biochem Biophys Acta 224(1): 291–294

Henderson AS, Warburton D, Atwood KC (1972) Localization of ribosomal DNA in the human chromosome complement. Proc Natl Acad Sci USA 69:3394–3398

Henikoff S, Keene MA, Fechtel K, Fristrom JW (1986) Gene within a gene: nested *Drosophila* genes encode unrelated proteins on opposite DNA strands. Cell 44:33–42

Honjo T (1983) Immunoglobulin genes. Annu Rev Immunol 1:499–528

Hsu TC, Brinkley BR, Arrighi FE (1967) The structure and behavior of the nucleolus organizer in mammalian cells. Chromosoma 23:137–153

Hughes-Schrader S, Ris H (1941) The diffuse spindle attachment of coccids, verified by the mitotic behaviour of induced chromosome fragments. J Exp Zool 87:429–456

Igo-Kemenes T, Zachau HG (1978) Domains in chromatin structure. Cold Spring Harbor Symp Quant Biol 42:109–118

Jeppesen PGN, Bankier AT (1979) A partial characterization of DNA fragments protected from nuclease degradation in histone-depleted metaphase chromosomes of the Chinese hamster. Nucleic Acids Res 7:49–67

Jeppesen PGN, Morten H (1985) Effects of sulphydryl reagents on the structure of dehistonized metaphase chromosomes. J Cell Sci 73:245–260

Jeppesen PGN, Nicol L (1986) Non-kinetochore directed autoantibodies in scleroderma/CREST: identification of an activity recognizing a metaphase chromosome core non-histone protein. Mol Biol Med 3:369–384

Jeppesen PGN, Bankier AT, Sanders L (1978) Non-histone proteins and the structure of metaphase chromosomes. Exp Cell Res 115:293–302

Johnson RT, Rao PN (1970) Mammalian cell fusion: induction of premature chromosome condensation in interphase nuclei. Nature (London) 226:717–722

Johnson RT, Rao PN, Hughes HD (1970) Mammalian cell fusion III: a HeLa cell inducer of premature chromosome condensation active in cells from a variety of animal species. J Cell Physiol 76:151–157

Jokelainen PT (1967) The ultrastructure and spatial organization of the metaphase kinetochore in mitotic rat cells. J Ultrastruct Res 19:19–44

Kaufmann SH, Coffey DS, Shaper JH (1981) Considerations in the isolation of rat liver nuclear matrix, nuclear envelope, and pore complex lamina. Exp Cell Res 132:105–123

Kavenoff R, Klotz LC, Zimm BH (1974) On the nature of chromosome-sized DNA molecules. Cold Spring Harbor Symp quant Biol 38:1–8

Keshet I, Lieman-Hurwitz J, Cedar H (1986) DNA methylation affects the formation of active chromatin. Cell 44:535–543

Kit S (1961) Equilibrium sedimentation in density gradients of DNA preparations from animal tissues. J Mol Biol 3:711–716

LaFond RE, Woodcock CLF (1983) Status of the nuclear matrix in mature and embryonic chick erythrocyte nuclei. Exp Cell Res 147:31–39

Langmore JP, Paulson JR (1983) Low angle X-ray diffraction studies of chromatin structure *in vivo* and in isolated nuclei and metaphase chromosomes. J Cell Biol 96:1120–1131

Lebkowski JS, Laemmli UK (1982) Non-histone proteins and long-range organization of HeLa interphase DNA. J Mol Biol 156:325–344

Lewis CD, Laemmli UK (1982) Higher order metaphase chromosome structure: evidence for metalloprotein interactions. Cell 29:171–181

Lonberg N, Gilbert W (1985) Intron/exon structure of the chicken pyruvate kinase gene. Cell 40: 81–90

Long EO, Dawid IB (1980) Repeated genes in eukaryotes. Annu Rev Biochem 49:727–764

Lustig AJ, Petes TD (1986) Identification of yeast mutants with altered telomere structure. Proc Natl Acad Sci USA 83:1398–1402

MacGregor HC (1980) Recent developments in the study of lampbrush chromosomes. Heredity 44(1):3–35

Mancino G, Nardi I, Ragghianti M (1972) Structural correspondence between nucleolus and sphere-organizing regions of the lampbrush chromosomes and secondary constrictions of the mitotic chromosomes. Experientia 28:586–588

Maniatis T, Fritsch EF, Laner J, Lawn RM (1980) The molecular genetics of human hemoglobins. Annu Rev Genet 14:145–178

Martino CD, Capanna E, Nicotra MR, Natali PG (1980) Immunochemical localization of contractile proteins in mammalian meiotic chromosomes. Cell Tissue Res 213:159–178

Matsui S, Fuke M, Chai L, Sandberg AA, Elassouli S (1986) N-band proteins of nucleolar organizers: chromosomal mapping, subnucleolar location and rDNA binding. Chromosoma 93:231–242

Maxson R, Cohn R, Kedes L, Mohun T (1983) Expression and organization of histone genes. Annu Rev Genet 17:239–277

McClintock B (1950) The origin and behavior of mutable loci in maize. Proc Natl Acad Sci USA 36:344–355

McCready SJ, Akrigg A, Cook PR (1979) Electron microscopy of intact nuclear DNA from human cells. J Cell Sci 39:53–62

McCready SJ, Godwin J, Mason DW, Brazell IA, Cook PR (1980) DNA is replicated at the nuclear cage. J Cell Sci 46:365–386

McKeon FD, Kirschner MW, Caput D (1986) Homologies in both primary and secondary structure between nuclear envelope and intermediate filament proteins. Nature (London) 319:463–468

Miller DA, Dev VG, Tantravahi R, Miller OJ (1976) Suppression of human nucleolus organizer activity in mouse-human somatic hybrid cells. Exp Cell Res 101:235–243

Mirkovitch J, Mirault M-E, Laemmli UK (1984) Organization of the higher-order chromatin loop: specific DNA attachment sites on nuclear scaffold. Cell 39:223–232

Mitchell AR, Gosden JR, Ryder OA (1981) Satellite DNA relationships in man and the primates. Nucleic Acids Res 9:3235–3249

Moroi Y, Peebles C, Fritzler MJ, Steigerwald J, Tan EM (1980) Autoantibody to centromere (kinetochore) in scleroderma sera. Proc Natl Acad Sci USA 77:1627–1631

Moroi Y, Hartman AL, Nakane PK, Tan EM (1981) Distribution of kinetochore (centromere) antigen in mammalian cell nuclei. J Cell Biol 90:254–259

Nakaseko Y, Adachi Y, Funahashi S, Niwa O, Yanagida M (1986) Chromosome walking shows a highly homologous repetitive sequence present in all centromere regions of fission yeast. EMBO J 5:1011–1021

Naora H, Deacon NJ (1982) Relationship between total size of exons and introns in protein-coding genes of higher eukaryotes. Proc Natl Acad Sci USA 79:6196–6200

Nordheim A, Pardue ML, Lafer EM, Moeller A, Stollar BD, Rich A (1981) Antibodies to left-handed Z-DNA bind to interband regions of *Drosophila* polytene chromosomes. Nature (London) 294:417–422

Okada TA, Comings DE (1980) A search for protein cores in chromosomes: is the scaffold an artifact? Am J Human Genet 32:814–832

Orgel LE, Crick FHC (1980) Selfish DNA: the ultimate parasite. Nature (London) 284:604–607

Padgett RA, Grabowski PJ, Konarska MM, Seiler S, Sharp PA (1986) Splicing of messenger RNA pecursors. Annu Rev Biochem 55:1119–1150

Palmer DK, Margolis RL (1985) Kinetochore components recognized by human autoantibodies are present on mononucleosomes. Mol Cell Biol 5:173–186

Paulson JR, Laemmli UK (1977) The structure of histone-depleted metaphase chromosomes. Cell 12:817–828

Paulson KE, Deka N, Schmid CW, Misra R, Schindler CW, Rush MG, Kadyk L, Leinwand L (1985) A transposon-like element in human DNA. Nature (London) 316:359–361

Petes TD, Newlon CS, Byers B, Fangman WL (1974) Yeast chromosomal DNA: size, structure and replication. Cold Spring Harbor Symp Quant Biol 38:9–16

Prescott DM (1970) The structure and replication of eukaryotic chromosomes. In: Prescott DM, Goldstein L, McConkey E (eds) Advances in cell biology, vol 1. Appleton-Century-Crofts, New York, pp 57–117

Rattner JB (1986) Organization within the mammalian kinetochore. Chromosoma 93:515–520
Rattner JB, Goldsmith M, Hamkalo BA (1981) Chromosome organization during male meiosis in *Bombyx mori.* Chromosoma 82:341–351
Razin SV, Mantieva VL, Georgiev GP (1978) DNA adjacent to attachment points of deoxyribonucleoprotein fibril to chromosomal axial structure is enriched in reiterated base sequences. Nucleic Acids Res 5:4737–4751
Resnick MA, Martin P (1976) The repair of double-strand breaks in the nuclear DNA of *Saccharomyces cerevisiae* and its genetic control. Mol Gen Genet 143:119–129
Rieder CL (1982) The formation, structure and composition of the mammalian kinetochore and kinetochore fiber. Int Rev Cytol 79:1–58
Ris H, Witt PL (1981) Structure of the mammalian kinetochore. Chromosoma 82:153–170
Robinson SI, Small D, Idzerda R, McKnight GS, Vogelstein B (1983) The association of transcriptionally active genes with the nuclear matrix of the chicken oviduct. Nucleic Acids Res 11: 5113–5130
Rogers J (1983) CACA sequences – the ends and the means? Nature (London) 305:101–102
Roos U-P (1973) Light and electron microscopy of rat kangaroo cells in mitosis I: formation and beakdown of the mitotic apparatus. Chromosoma 41:195–220
Scheer U, Rose KM (1984) Localization of RNA polymerase I in interphase cells and mitotic chromosomes by light and electron immunocytochemistry. Proc Natl Acad Sci USA 81:1431–1435
Schmid CW, Jelinek WR (1982) The Alu family of dispersed repetitive sequences. Science 216: 1065–1070
Singer MF (1982) SINEs and LINEs: highly repeated short and long interspersed sequences in mammalian genomes. Cell 28:433–434
Singer MF, Skowronski J (1985) Making sense out of LINEs: long interspersed repeat sequences in mammalian genomes. Trends Biochem Sci 10:119–122
Smith HC, Puvion E, Buchholtz LA, Berezney R (1984) Spatial distribution of DNA loop attachment and replicational sites in the nuclear matrix. J Cell Biol 99:1794–1802
Southern EM (1970) Base sequence and evolution of guinea pig α-satellite DNA. Nature (London) 227:794–798
Spowart G, Forster P, Dunn N, Cohen BB (1985) Clinical and biochemical studies on anti-kinetochore antibody in patients with rheumatic diseases: a diagnostic marker for CREST. Disease Markers 3:103–112
Spyropoulos B, Moens PB (1984) The synaptonemal complex: does it have contractile proteins? Can J Genet Cytol 26:776–781
Stone EM, Rothblum KN, Schwartz RJ (1985) Intron-dependent evolution of chicken glyceraldehyde phosphate dehydrogenase gene. Nature (London) 313:498–500
Sugino A, Hirose S, Okazaki R (1972) RNA-linked nascent DNA fragments in *Escherichia coli.* Proc Natl Acad Sci USA 69:1863–1867
Sutherland GR, Hecht F (1986) Fragile sites in human chromosomes. Oxford Univ Press, New York
Szostak JW, Blackburn EH (1982) Cloning yeast telomeres on linear plasmid vectors. Cell 29: 245–255
Thoma F, Simpson RT (1985) Local protein-DNA interactions may determine nucleosome positions on yeast plasmids. Nature (London) 315:250–252
Unakul W, Johnson RT, Rao PN, Hsu TC (1973) Giemsa banding in prematurely condensed chromosomes obtained by cell fusion. Nature New Biol 242:106–107
Valdivia MM, Brinkley BR (1985) Fractionation and initial characterization of the kinetochores from mammalian metaphase chromosomes. J Cell Biol 101:1124–1134
Walker PMB (1971) "Repetitive" DNA in higher organisms. Prog Biophys Mol Biol 23:147–190
Walter P, Blobel G (1982) Signal recognition particle contains a 7 S RNA essential for protein translocation across the endoplasmic reticulum. Nature (London) 299:691–698
Waring M, Britten RJ (1966) Nucleotide sequence repetition: a rapidly reassociating fraction of mouse DNA. Science 154:791–794

Warren AC, Cook PR (1978) Supercoiling of DNA and nuclear conformation during the cell cycle. J Cell Sci 30:211–226
Weiner AM, Deininger PL, Efstratiadis A (1986) Nonviral retroposons: genes, pseudogenes, and transposable elements generated by the reverse flow of genetic information. Annu Rev Biochem 55:631–661
Wilson GN (1982) The structure and organization of human ribosomal genes. In: Busch H, Rothblum L (eds) The cell nucleus, vol 10: rDNA, part A. Academic Press, London New York, pp 287–318
Yunis JJ, Sawyer JR (1980) The striking resemblance of high-resolution G-banded chromosomes of man and chimpanzee. Science 208:1145–1148

2 The Magic of Cytogenetic Technology

G. P. HOLMQUIST[1] and M. A. MOTARA[2]

1 Introduction

A cytogeneticist's universal substrate is acid-alcohol fixed chromosomes. Biochemists may have understood native chromatin somewhat, but not acid-alcohol fixed chromatin. Our substrate is understood primarily through banding techniques that have been stochastically developed over the last 15 years. Why is this work mostly magic? For example, quinacrine was devised as, and initially deemed a GC-specific fluorochrome. It was later shown to demarcate AT-rich DNA, but not AT-rich mouse satellite. As a first approximation, the fluorochromes quinacrine (Q), bisbenzimide (H 33258), Hoechst (H), or daunomycin stain chromosomes like they stain DNA. Q and H are AT-specific and their G-bands are bright; daunomycin is GC-specific and the R-bands are bright (Comings and Drets 1976). However, DNA in fixed chromosomes often does not behave like free DNA. For example, acridine orange stains single-stranded DNA red. However, when used to stain chromosomes, this does not always hold true. Giemsa stains free DNA, but not DNA in trypsinized R-bands. DNA in fixed chromosomes is bound to protein which makes it inaccessible to many probes. Fixation probably puts it in a different supercoiled state and changes the way bound fluorescent dyes transfer their absorbed energy. These changes are the magic of fixed chromosomes and the purpose of this chapter is to explain some of this magic.

2 Spreading of Chromosomes on Slides

The glass surface is important for obtaining good chromosome spreads. The prominent chemical feature of such a surface is

1 Developmental Biology, Beckman Research Institute, City of Hope, Duarte, CA 91010, USA
2 Department of Zoology, University of Durban-Westville, Private Bag X54001, Durban 4000, South Africa

Cytogenetics. Ed. by G. Obe and A. Basler

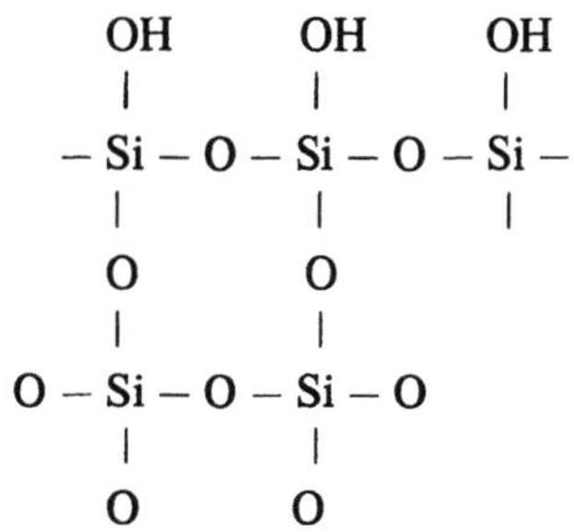

The free-hydroxyl groups makes the surface hydrophilic and slightly negatively charged. If one siliconizes the glass slide or it becomes oily, the chromosomes will spread poorly. Modifying the surface to charge it positively (Van Prooijen-Knegt et al. 1982) has been reported to improve spreading but we did not find this particular method worth the effort. Soaking slides in 80% ethanol:20% hydrochloric acid for 30 min and rinsing in water before storing them in deionized water in a refrigerator has proved very satisfactory.

3 Hypotonic Treatment and Fixation

Some aspects of fixation in 3:1 methanol-acetic acid are understood. Towards metaphase, factors accumulate in the cell which cause chromatin to condense. If a metaphase cell is fused with an interphase cell, these factors cause the interphase chromatin to condense prematurely. At prometaphase, the nuclear membrane breaks down and, if colcemid is present to depolymerize the spindle, the chromosomes are free to swim around. Hypotonic media makes the swimming pool bigger and fixation freezes it. Hypotonic swelling is an active process. The sodium-potassium pump in the membrane uses ATP to maintain Na^+ out- and K^+ inside. In a hypotonic environment, the cell actively absorbs water to equilibrate its salt balance. Poison the pump and the cell does not swell. After prolonged exposure in hypotonic medium the cell almost returns to its original size. Maybe someday a membrane physiologist will show us how to manipulate the ionic pump in order to fix large swollen cells which would result in the perfect chromosome spreads. Hypotonic swelling also frees chromosomes of a ribonucleoprotein complex which usually surrounds them. This makes them band better.

Methanol-acetic acid fixation is a two-step process. As one adds the first few drops of fixative the pH of the cell drops to about 2 and most acidic groups like $-COO^-$ become titrated to $-COOH$ which changes the net charge of macromolecules and permanently denatures (fixes) them. The second step of fixation is dehydration under acid conditions. Replacing hydration water with methanol again (fixes) alters the configuration of proteins and DNA. The energy required to remove a certain percent of water during dehydration increases exponentially as more water is removed. This is a general phenomenon which applies either to air-drying cellulose or dehydrating cells. The free water is easily replaced. Loosely bound water is more difficult to remove and the most tightly bound water may be almost impossible to remove. In general, it takes more energy to dehydrate from 99% dryness to 99.5% dryness than it does to remove the first 99% of water. Very dry cells are necessary for good spreading. To prove this, add a drop of water to the final fix. Many complaints of "my cells will not spread" can be attributed to water from the air getting into the methanol bottle. Use fresh fixative for the final fixation, because on standing the fixative absorbs water, and water is formed during esterification.

By dropping the "dry cells" onto clean, chilled wet slides they burst open and spread out nicely. This is not all magic. Remember all that energy it took to dehydrate the cells: Now we get it back as a change in the free energy of mixing. The energy of rehydration is released, which spreads the cells. To visualize this energy of mixing, try layering water over freshly prepared fixative and note the ensuing turbulence. Another experiment is to drain a colcemid-hypotonic treated cover slip with amniotic cells and, while observing a few metaphases through an inverted scope, add fresh fixative. The turbulence will knock most of the metaphases off the glass. The few metaphase cells which remain will have their chromosomes clustered near the side of the cell which is first fixed. Clustering hinders the spreading of chromosomes; hence, it is important to add the initial few drops of fixative slowly.

Now that the chromosomes are spread and "dry", stain them with a protein stain such as 0.2% fast green, pH 5. The entire metaphase set of the very best chromosome spreads will be seen to be totally surrounded by and imbedded in a green halo of cytoplasm. Any staining one does must penetrate this halo. If the halo is broken, some cytoplasm along with chromosomes may be lost. An unbroken, fast green halo may be used to help eliminate preparation artefacts for the assessment of aneuploidy.

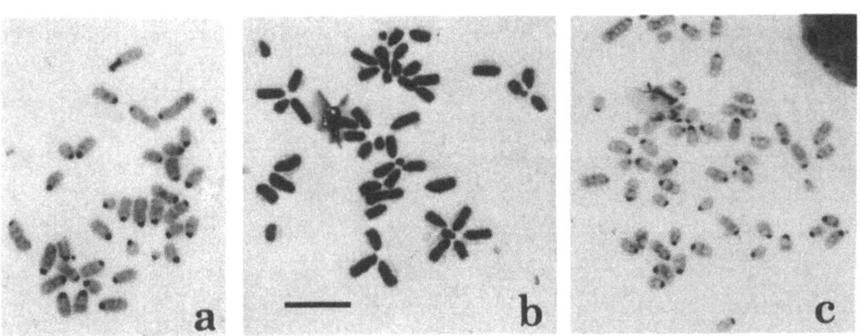

Fig. 1. a Methanol-acetic acid-fixed mouse chromosomes were C-banded as follows: 0.2 N HCl, 24°C, 15 min; 0.01 N NaOH, 0.3 M NaCl, 24°C, 2 min; 0.25 M NaPi, pH 8.3, 65°C, 18 h; and Giemsa stained. Here, the acid treatment depurinated one thymidine residue every 150 base pairs. b Treated as above except that a $NaBH_4$ treatment was inserted after the HCl treatment. $NaBH_4$ reduces the =O, aldehyde, which was created by acid depurination, to an alcohol, –OH ; this prevents subsequent strand breakage by elimination at the carbon beta to the aldehyde carbon. Hence, the molecular weight of the DNA was too high for it to be solubilized from the chromosome during the final hot salt treatment. c Treated as in b but with an additional acid step, thus the sequence of treatments was HCl-$NaBH_4$-HCl-NaOH, 65°C. The additional HCl treatment created more apurinic sites with aldehyde sugars. Strand breakage by beta-elimination occurred at these new sites and the resulting low molecular weight DNA was solubilized from the chromosome during the 65°C treatment. The extent of depurination, aldehyde reduction and beta-elimination was determined from sedimentation rates of DNA extracted from the microscope slides (Holmquist 1979). Bar = 10 μm

For many techniques, the slides must be aged properly. Two weeks at room temperature, 3 days at 55°C or 20 min at 95°C produce the same results. Ageing effects can be prevented by storing slides at –20°C in dry nitrogen. The most important effect of ageing may be the oxidation of the protein sulphydryl groups (Evans 1978).

4 Banding of Chromosomes

4.1 C-Banding

C-banding is partially understood. The procedure removes DNA from euchromatin at a faster rate than it does from heterochromatin. Why this occurs is unknown but why DNA is removed is known (Holmquist 1979) (Fig. 1). The DNA is broken into fragments which are small enough to diffuse out of their fixed, proteinaceous entrapment. In the acid step, the N-glycoside of purines becomes protonated and this bond then breaks, thus releasing the purine base. For C-banding, depurination is stopped when about one purine per 300 bases is removed. Instead of acid, the antibiotic bleomycin at pH 7 can be used to depurinate the DNA (Fig. 2). The banding is better but this requires more work. Next, the slides are put in alkali which denatures the DNA. Strangely, in situ denatured DNA hardly re-anneals. The slide is then placed in hot salt

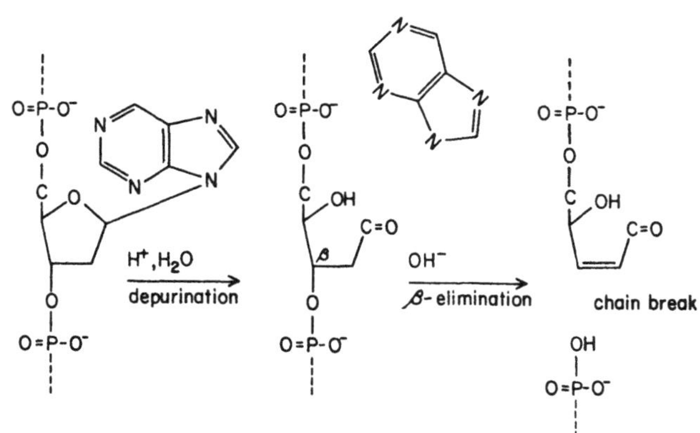

where the depurinated sites undergo beta-elimination. The aldehyde activates the beta-phosphodiester which breaks the DNA into 300-base-long fragments. Being single-stranded, these fragments are free to diffuse out of the chromosome. For some reason, one or more of the processes occurs faster in euchromatin than in heterochromatin (C-bands). Proper C-banding requires that one adjusts the reactions so that the euchromatic DNA is mostly removed while some of the C-band DNA remains.

The first in situ hybridization method was this C-band technique with ^{3}H-RNA in the hot salt. C-banding was an in situ spin-off. Fortunately, the first in situ probes were to C-band DNA.

Depurination generates an aldehyde which can be detected by aldehyde-specific reagents such as the colourless Schiff reagent which, in the Feulgen procedure, reacts with aldehydes and becomes red. The acid hydrolyzes the chromosomes at elevated

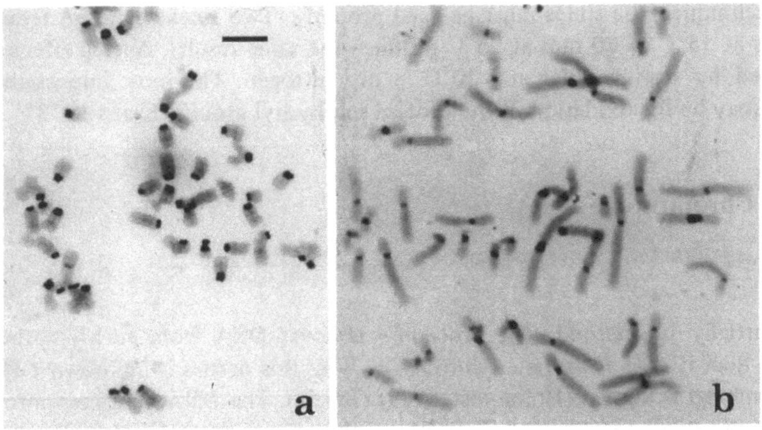

Fig. 2a,b. Mouse (**a**) and human (**b**) chromosomes C-banded by the standard treatment except that depurination was caused by the antibiotic bleomycin, pH 7, 4°C, instead of by HCl

temperatures for various times before Schiff stains the aldehydes. The strong acid (1 N HCl) used also beta-eliminates the DNA (Holmquist 1979). The Feulgen banding techniques utilize the fact that hydrolysis curves for R-, G- and C-bands are different.

4.2 DAPI Fluorescence

Fluorochromes bind to DNA and fluoresce. Some, like Hoechst or DAPI, have a selective affinity for DNA of a certain base composition (Comings 1975b; Comings and Drets 1976). Ohters, like quinacrine, bind all DNA equally but fluoresce more when bound to AT-rich DNA (Evans 1978; Sumner 1982). Dry and re-mount a quinacrine banded chromosome in immersion oil and note that it fluoresces homogeneously (Evans 1978). Quinacrine was originally bound everywhere but fluoresced only in certain bands. Fluorochromes adsorb light energy and have a few picoseconds to decide how they want to release it. Fluorescence is only one option. Consider the fluorochrome DAPI. It has a planar, two-membered ring of conjugated double bonds

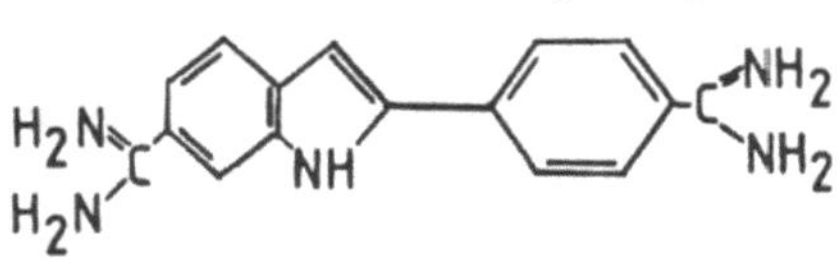

DAPI

along with an attached benzene ring. The rings absorb 360 nm light forcing the double-membered ring's pi bonds into an excited state. Generally, in solution, the excited pi bonds give off their energy to rotation of the benzene ring. The rotations just heat up the water. However, when DAPI's benzene ring cannot rotate, DAPI cannot readily release energy through bond rotations or bond vibrations. It gives off the energy as light. When DAPI binds DNA, its benzene cannot spin, thus it fluoresces.

4.3 Hoechst Fluorescence

Let us review Hoechst fluorescence. The electron clouds of the planar pi bonds absorb 360 nm light, raising them to an excited, unstable state. They return to a lower energy state by giving up some of the extra energy by the easiest means possible. The three common routes are bond vibrations, energy transfer to another molecule and, if all else fails, by re-emitting light. The route used is very sensitive to the local environment and this is poorly understood. Hoechst, like DAPI, gives up its absorbed energy to internal bond vibrations unless bound to something like DNA. Here, it gives most, but not all, of its energy to a re-emitted photon. The re-emitted photon always has a lower energy than the absorbed one, so fluorochromes always re-emit a longer wavelength than they absorb. For some reason, bromodeoxyuridine (BrdUrd)-substituted DNA can absorb the Hoechst's energy. The resulting bond vibrations often cause breaks in the BrdUrd-substituted DNA. The fluorescence plus Giemsa technique takes advantage of these breaks to selectively dissolve the shattered DNA from the BrdUrd-substituted chromatid. Hot salt is used to accomplish this as in the C-band technique. Thus, cytologists have learned to utilize both the fluorescence pathway and the bond vibration pathway to visualize in situ fluorochrome cytochemistry.

4.4 Hoechst-Chromomycin-A_3 Energy Transfer System

Energy transfer is the basis of life on earth. Two photons of sunlight must be absorbed by chlorophyll and the energy transferred to an acceptor site for water photolysis. Energy transfer between fluorochromes requires three factors: (1) the acceptor molecule must be able to absorb the exact energy which the donor is donating. (2) The plane of the acceptor molecule must be parallel to the plane of the donor. (3) The acceptor must be less than about 50 Å away. Consider the Hoechst-chromomycin-A_3 energy transfer system. Hoechst absorbs UV light and emits blue light. Chromomycin absorbs blue light and emits red light. Thus, if a chromomycin molecule is close to and properly oriented relative to an excited Hoechst molecule, the Hoechst pi-bond energy can be transferred to the chromomycin. The excited chromomycin molecule can now fluoresce red when the system is stimulated with UV light even though the chromomycin alone cannot absorb UV light. Now Hoechst has an AT specificity and chromomycin has a GC specificity. They can be used together to demonstrate G- and R-bands. In G-bands, mostly Hoechst is bound. Since it cannot find many chromomycin molecules to transfer energy to, G-bands fluoresce blue with UV illumination. R-bands, however, are GC-rich and bind more chromomycin than Hoechst. When the Hoechst in R-bands absorbs UV light, there is energy transfer to the chromomycin which fluoresces red. Thus, this double-staining method can cause G-bands to appear blue and R-bands to appear red.

4.5 Distamycin-DAPI Energy Transfer System

The distamycin-DAPI energy transfer system is somewhat different. Here, both compounds are AT-specific and distamycin does not fluoresce. For some reason, excited

DAPI molecules cannot transfer energy to distamycin in the heterochromatin of chromosomes 1, 9, 15, 16 and Y, thus these regions alone fluoresce. The senior author once tried to stain DNA by this technique. When male DNA is restricted with Hae III, one sees molecular weight bands on the gel which contain Y-chromosome DNA which maps to the distamycin-DAPI bright regions of Y-heterochromatin. This band in the gel does not stand out when the DAPI-stained gel is quenched with distamycin. Thus, DNA in a gel does not behave like DNA in a chromosome. Magic again! Now the distamycin-DAPI bright regions coincide exactly with those regions which bind anti-5-methyl cytosine antibodies. Also, methyl cytidine-rich DNA converts to Z-form DNA with alcohol dehydration. In the Z-form, the helix pitch appears to preclude energy transfer between bound dyes. Thus, this staining technique may be revealing a Z-form DNA. This is, of course, gross speculation. It is an example of a hunch which can be tested and, if correct, remove a bit more magic from cytogenetic technology.

4.6 G- and R-Bands

G-banding is not well understood. Trypsin, chymotrypsin, urea, detergents and many other protein denaturation agents cause chromatid swelling and a selective loss of R-band basophylia from properly aged slides. No DNA and only some protein is lost during the denaturation treatment (Comings and Avelino 1975; Evans 1978). The R-bands have as much DNA as the G-bands but their DNA just does not stain. How does DNA lose its basophylia (base-loving) character?

DNA is a polyanion with a strong affinity for cationic dyes. One mol DNA can bind 1 mol methylene blue (a Giemsa cation). When methylene blue chloride ($MB^+ Cl^-$) is added to the sodium salt of DNA, the methylene blue displaces the Na^+ and the methylene blue salt of DNA precipitates. In fixed chromatin, 20% of the DNA's potential binding sites are involved in protein interaction and are inaccessible to the dye. When the fixed chromatin is denatured as per G-banding, an additional 30% of the sites become inaccessible (Comings and Avelino 1975). The loss of basophylia is caused by rearrangements of proteins so that they cover up DNA phosphates which could potentially bind methylene blue. A model of G-banding is proposed below: (1) Trypsin nicks and denatures the protein. (2) The protein rearranges so that its positively charged groups titrate the negative DNA phosphates. (3) During staining, the methylene blue competes with proteins for the G-band DNA, but the R-band proteins are bound too tightly to be displaced in R-bands. This model makes a number of tested predictions. (1) During progressive trypsin or protein denaturation treatment,

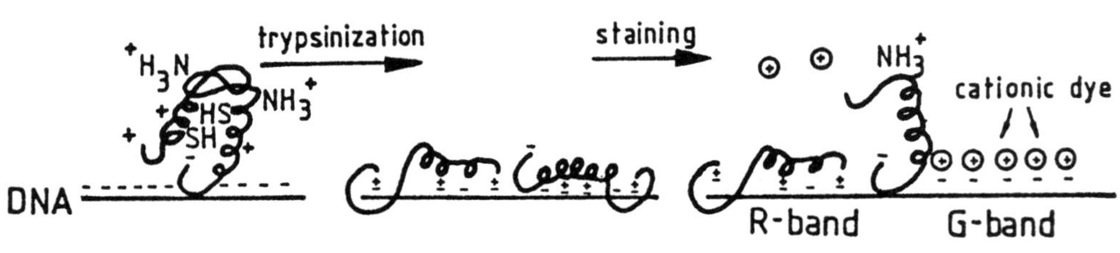

both DNA phosphates and protein amines should become progressively inaccessible. (2) After denaturation, very weak DNA binding dyes should be unable to displace any proteins, while very strong DNA-binding dyes should displace both R- and G-band proteins. Since most previous reviews and published experiments on banding mechanisms have stressed DNA accessibility, we briefly stress protein accessibility.

The ϵ-amines of lysine and N-terminal groups, when available, react with the fluorescent reagents dansyl chloride (Latt and Sober 1967), fluorodinitrobenzene (Clark and Felsenfeld 1971) and fluorescamine (Stein et al. 1974), or bind methyl orange (Clark and Felsenfeld 1971). The ϵ-amines of polylysine, when involved in ionic bonds to DNA phosphodiesters, are not available to bind methyl orange (Hatchard and Parker 1956; Itzhaki and Cooper 1973). We tested the availability of polylysine to react with fluorescamine when polylysine is bound to DNA. Various amounts of DNA were added to similar polylysine solutions; then fluorescamine was added. Fluorescamine reacts with amines to produce a stable fluorescent product, and quickly reacts (half-life about 10 s) with water to produce a non-fluorescent product (Stein et al. 1974). In this experiment, fluorescence decreased with added DNA up to the point of molar equivalence. This was interpreted to mean that the lysine amines involved in ionic bonds to DNA are unavailable to react with fluorescamine, and the same should be true of other amine reactants.

Chromosomes can be dansylated so that they fluoresce brightly (Fig. 3, a1) (Klein and Walker 1970). If treated with fluorodinitrobenzene before dansylation in ethanol, chromosomes do not fluoresce at all (Utakoji and Matsukuma 1974), showing that dansyl chloride treatment of chromosomes dansylates only ϵ-amines and N-terminal amines, which are available to fluorodinitrobenzene. Chromosomes do not fluoresce if dansylated after treatments such as trypsinization, which would induce a bandable state (Fig. 3, a2). This was interpreted to indicate protein removal (Matsukuma and Utakoji 1976), but other experiments show that little or no protein is removed from fixed chromosomes during the limited trypsin digestion necessary for optimal G-banding (Comings et al. 1973). If the trypsinized chromosomes were Giemsa stained and then destained before dansylation, the G-bands became weakly fluorescent (Fig. 3, a3) (Matsukuma and Utakoji 1976). Giemsa staining rendered protein amines in G-bands accessible. To prove that protein amines were present but inaccessible in trypsinized chromosomes, we treated them with either pancreatic DNase or 1 N HCl (warm) to solubilize the DNA before dansylation. In both cases, even though some or much protein was probably removed along with the DNA, the nuclease-treated or acid-trated chromosomes fluoresced more brightly than the controls, which were chromosomes without DNA removal. This shows that after trypsin denaturation, the chromosome's amine groups were present, but inaccessible.

The second prediction from the model involves the staining pattern of trypsinized chromosomes using dyes of various DNA affinities. Clark and Felsenfeld (1974) found staphylococcal nuclease, polylysine and manganese (Mn^{2+}) to be mutually competitive probes for accessible DNA in unfixed chromatin. All of them gave almost identical accessibility values for a given chromatin sample. Low-salt, isolated chromatin was about 50% accessible, which is identical to the accessibility obtained using azure A titration (Klein and Walker 1970), and similar to the 60% value obtained by equilibrium dialysis against methylene blue (Comings and Avelino 1975). Azure A is a component

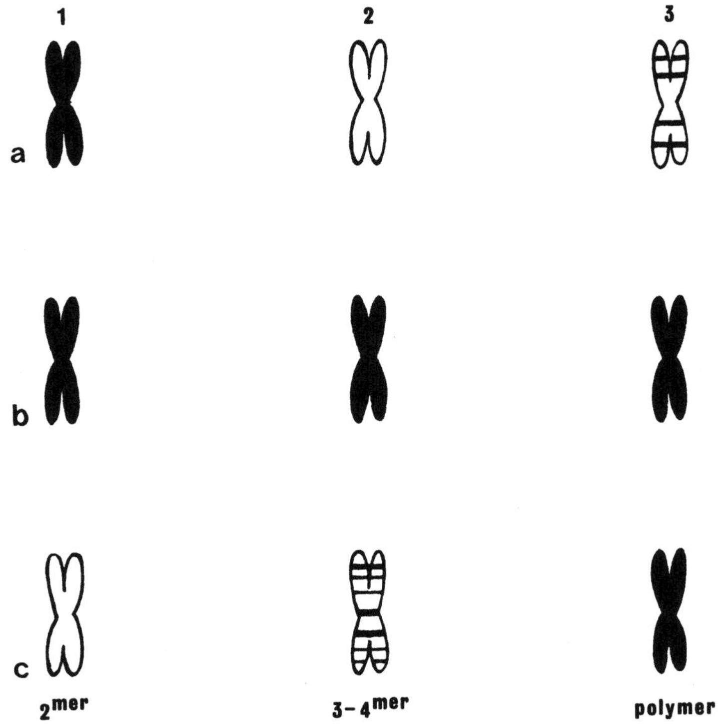

Fig. 3a–c. Dansylated chromosomes with *dark areas* representing bright fluorescent staining. a Results as described by Matsukama and Utakoji (1976). Chromosomes were dansylated after they were: *1* fixed and dried on slides; *2* fixed, dried and trypsinized for optimal G-bandability; *3* fixed, dried, trypsinized, Giemsa banded and destained. The authors interpreted these results as loss of proteins during trypsinization rather than increased inaccessibility of existing ϵ-amines. b Fixed, dried chromosomes. c Fixed, dried and trypsizized chromosomes stained as follows with 1.0 mg ml^{-1} dye in 10 mM Na-phosphate buffer, pH 6.8, and mounted in the same buffer. The dyes used were the dimer to tetramer of α-dansyl-oligolysine (*leftmost columns*) and polylysine (*right column*) with the degree of polymerization = 60 and with 6% (Latt and Sober 1967) of its ϵ-amines dansylated

of Giemsa stain, a spontaneous oxidation product of methylene blue, and stains chromosomes by the same mechanism as methylene blue (Comings 1975a). Because polylysine and methylene blue seem to measure the same accessible DNA in unfixed chromatin, one might expect them to do the same in chromosomes. This is not so, and the probable reason is the higher binding affinity for DNA of polylysine. Polylysine binding is strong and irreversible in low salt (Clark and Felsenfeld 1971; Holmquist and Comings 1976), while methylene blue dissociates slowly ($T_{1/2}$ = 30–60 min, unpublished results). When an unfixed chromatin solution is titrated with polylysine, it initially binds to accessible DNA until all exposed phosphodiesters are titrated. Then the chromatin precipitates, leaving no free polylysine in the supernatant. If more polylysine is added, some binds to the chromatin with concomitant displacement of the proteins that bind chromatin more weakly (Axel et al. 1973). The same phenomenon seems to occur in fixed chromosomes (Fig. 3, a3). Polylysine can be lightly dansylated without appreciably altering its affinity for DNA. When this probe is used in excess to stain fixed chromosomes, they fluoresce uniformly (Fig. 3, b3) (Latt and Gerald 1973). However, this probe also stains bandable chromosomes uniformly (Fig. 3, c3), showing that interband chromatin, which is not accessible to methylene blue, is accessible to polylysine, the higher affinity probe. The measured DNA accessibility is thus a function of the binding constant of the probe. In this case, polylysine, having a greater affinity for DNA than some of the proteins covering it, can displace these proteins and bind to additionally accessible DNA.

The affinity of oligolysine for DNA increases with the length of the oligomer (Latt and Sober 1967). We knew from previous work that if one dips a slide in a polylysine solution, the chromosomes will not subsequently stain with Giemsa, but dipping the slide in a lysine solution has little effect on subsequent staining (Holmquist and Comings 1976). If one makes a series of oligolysines with a fluorescent light bulb at one end, these would represent a series of stains with various affinities for DNA, which could test the protein displacement prediction. All the stains in the α-dansyl lysine (n) series, except dansyl lysine (n = 1), stained non-trypsinized chromosomes (Fig. 3b). The dimer would stain only a halolike outline of trypsinized chromosomes (Fig. 3, c1). The 3^{mer} and 4^{mer} at high concentrations banded trypsinized chromosomes (Fig. 3, c2; Fig. 4). Partially dansylated polylysine homogeneously stained the trypsinized chromosomes (Fig. 3, c3).

There are several other reasons to believe that the probes, Giemsa or methylene blue, also alter the chromatin structure by removing or displacing proteins. 2 μg of accessible DNA can bind 1 μg dye (Comings and Avelino 1975). This is a lot of extra bulk to fit into tightly packed chromatin without altering something. Overstained banded chromosomes do not return to their original condition after destaining. Giemsa staining alters the electron microscopic morphology of the chromosomes (Barnett et al. 1974). Examination of Comings and Avelino's data (1975), concerning equilibrium dialysis of DNA or fixed chromatin against methylene blue, reveals that DNA saturates at 2×10^5 M methylene blue, but chromatin does not saturate until 7×10^5 M. This suggests that higher concentrations are necessary to displace weakly bound protein. Crossen (1973) induced the G-banded state with 0.07 N NaOH for 2 min followed by 20 mM phosphate buffer for 2 h at 60°C. When stained with cresyl violet, these chromosomes did not appear banded, but if they were stained with Giemsa, destained with 50% acetic acid, and re-stained with cresyl violet, they banded as if they were stained with Giemsa. In addition, when bandable chromosomes are treated with formaldehyde just before Giemsa staining, very poor banding is obtained (Crossen 1973). Since formaldehyde forms inter- and intramolecular protein cross-links (Pearse 1961), and cross-links proteins to DNA (Georgiev and Ilyin 1969), this would prevent displacement of protein by Giemsa and inhibit G-banding if displacement is a necessary part of the mechanism. All of these data indicate that Giemsa staining plays an active role in the induction of G-bands by removing or displacing proteins in fixed chromosomes.

Giemsa stain is made by partially demethylating methylene blue, precipitating the product as an eosine salt, adding a little more methylene blue chloride and dissolving the powder in glycerol and methanol (Comings 1975a; Marshall 1978). Methylene

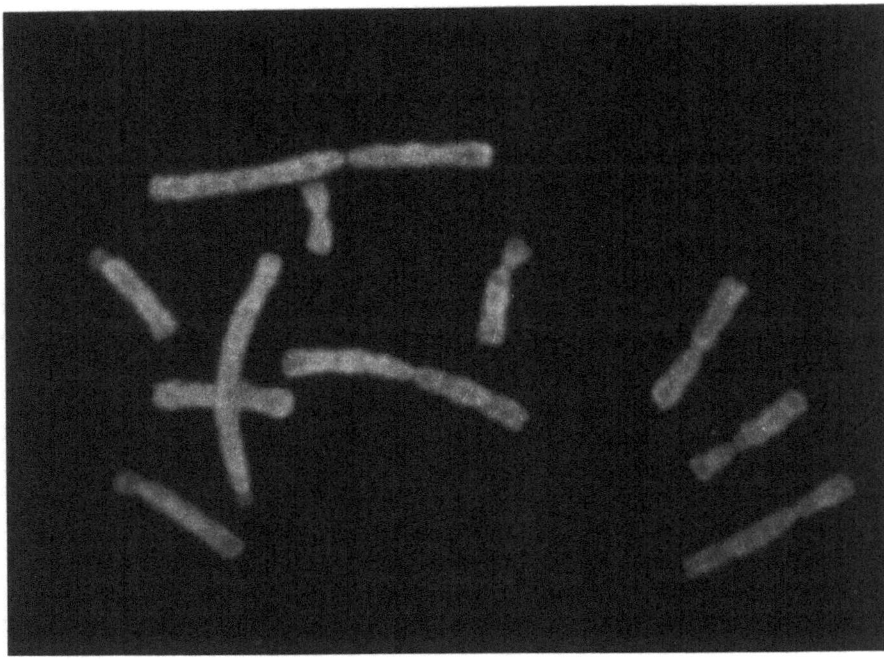

Fig. 4. Chinese hamster chromosomes, trypsinized as for optimal G-banding, were stained with 1 mg ml^{-1} α-dansyl-(lysine)$_3$. The banding of euchromatin was equivalent to poor Q-banding. Lower dye concentrations effected a haloed appearance with visible, but ghostlike cross-bands in te positions of the major G-bands. These disappeared at concentrations below 0.2 mg ml^{-1}. At concentrations above 3 mg ml^{-1}, the trypsinized chromosomes stained homogeneously. The tetrameric and pentameric stains gave similar results except that they revealed optimal banding at lower concentration ranges

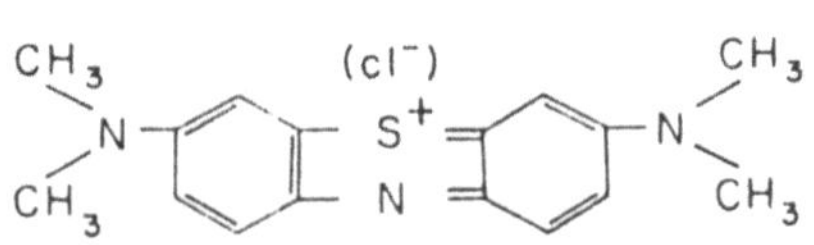

blue, shown above, has four methyl groups, which spontaneously oxidize in warm alkali. Azure B lacks one methyl, azure A two methyls and azure C three methyls, whereas thionine is unmethylated. These various thiozine dyes have different shades of blue and the demethylation process is called polychroming.

In 1880, Ehrlich mixed acid and basic dyes and found that the mixtures could differentially stain what he called acidophilic, neutrophilic or basophylic structures (Marshall 1978). A combination of oppositely charged dyes produces an effect which was not some average of the two dyes used alone. This magic is yet to be understood. Romanowski, in 1891, found that methylene blue and eosine, a yellow anionic dye related to mercurichrome, was a good combination. However, the stain mixture had to be used immediately because it precipitated very rapidly. May and Grünwald, in 1902, found that they could dissolve this precipitate in ethanol and use it as a blood stain. Leishman and Wright soon thereafter substituted polychromed methylene blue chloride for methylene blue chloride in the May-Grünwald stain. These dye mixtures varied greatly from batch to batch. A good stain batch became a prized possession. Giemsa and Lillie improved the reproducibility of the blood stain by controlling the ratio of its polychromed components. A polychromed eosinate powder is now dissolved in glycerol/methanol to improve its stability, a dash of methylene violet is added for contrast and some azure chloride is added to prevent immedaite precipitation of the stain when it is diluted with water. When one reads the contents on the label of a Giemsa stain bottle, one usually encounters azure I (= A) and azure II (= B). Azure I is polychromed methylene blue chloride and azure II is azure I plus methylene blue. Azure I eosinate is just the precipiated eosine salt. These formulations were originally an industrial secret and many of the lower azures still cannot be obtained except by HPLC fractionation of polychromed methylene blue. However, the ratio of the different azures in a polychromed batch of dye can be assessed chromatographically and different batches combined give a consistent product (Dean et al. 1977).

Giemsa stain usually varies from batch to batch. The powder is quite stable but the polychroming process occurs at a measurable rate in liquid Giemsa (Dean et al. 1977). Therefore, after purchasing liquid Giemsa, keep it refrigerated. When one adds Giemsa dye to water or buffer, one is making a supersaturated dye solution which starts to precipitate or come out of solution as a surface film at different rates for different components.

Giemsa dye is designed specifically to stain blood smears and the design has taken many years. No one has yet designed a stain specifically for G-banding. The higher azure chlorides do reveal marginal banding (Comings and Avelino 1975), but the eosine salt is absolutely necessary for good, sharp bands (Evans 1978). Of the pure compounds tested, the senior author has found azure B eosinate to be the best, but a good Giemsa lot is still unsurpassed.

4.7 Biochemical Characteristics of G- and R-Bands

We separated G-band DNA from R-band DNA and characterized these (Holmquist et al. 1982; Goldman et al. 1984). The separation is quite simple, but first one must know that the S-phase is bimodal. Neglecting facultative X's and heterochromatin, R-bands are synthesized in the first half of S-phase, then DNA synthesis slows down or may even come to a halt. Thereafter, G-bands are synthesized in the last half of S-phase. Methotrexate seems to block cells near the R-G transition so that when one releases the methotrexate-arrested cells with BrdUrd instead of thymidine, the BrdUrd is incorporated into G-bands and one gets a burst of metaphases 5.5 h later. The BrdUrd-substituted DNA is denser than normal DNA, thus the BrdUrd G-band DNA can be separated from the R-band DNA in CsCl gradients. The two DNA fractions are different. G-band DNA is about 3.3% richer in AT than the R-band DNA, as expected from AT-specific fluorochromes producing a bright G-band pattern (Comings and Drets 1976). The CCGG sequences in G-bands have their external C more highly methylated than the R-band sequences; MspI cuts R-band DNA more than G-band

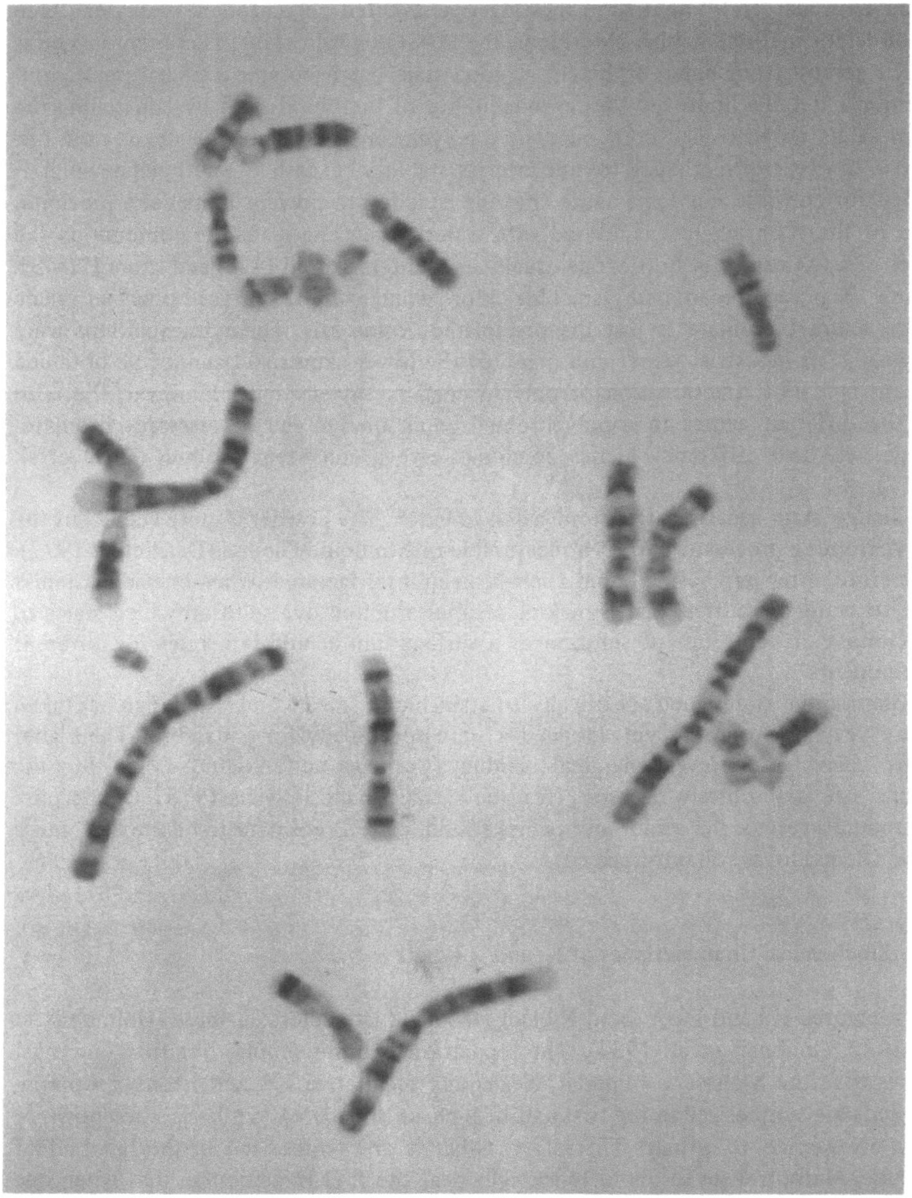

Fig. 5. Trypsin G-banded chromosomes from a Chinese hamster V 79 fibroblast cell line

DNA. Unexpectedly, we found from Rot curves, DNase I digestion kinetics and Southern blots, that G-bands contain about as many genes as R-bands. The base composition difference and its concomitant DNA density difference allowed Bernardi et al. (1985) to fractionate the DNA according to base composition in CsCl gradients.

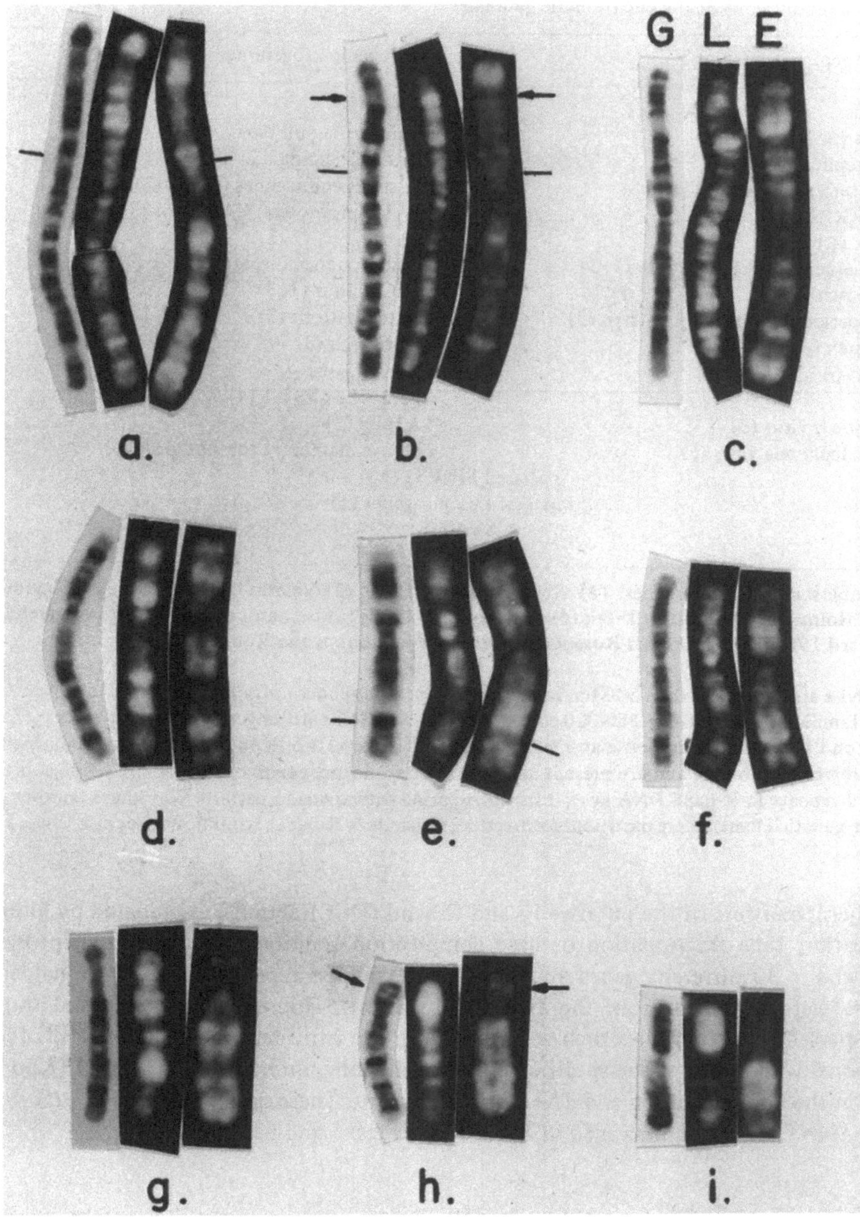

Fig. 6a–i. G-banding and replication banding of Chinese hamster chromosomes. The *leftmost* chromosome of each triplet was trypsin G-banded. The remaining chromosomes were from cells labelled with BrdUrd in the first (*center* chromosome) or last (*rightmost* chromosome) half of S-phase. These were then stained with acridine arange to reveal bands of late and early replicating DNA respectively. For most of the euchromatic bands except those labelled with *arrows*, G-bands replicate late and R-bands replicate early (Holmquist et al. 1982)

Table 1. Subdivisions of the euchromatic genome[a]

Housekeeping subgenome	Ontogenic subgenome
R-bands	G-bands
Quinacrine bright bands	Quinacrine dull bands
Early replicating	Late replicating
Constitutive genes (all) (1)	Tissue-specific genes (most) (1)
GGGCGGG promoters (2)	TATA promoters, CAAT boxes (2)
dG+dC rich (1, 3)	dA+dT rich (1, 3)
Slow sequence divergence rate (1, 3)	Fast sequence divergence rate (1, 3)
CCGG rich (1, 3)	CCGG poor (1, 3)
Short period interspersion pattern (3)	Mixed pattern (3)
SINE rich (3)	LINE rich (3)
Alu (3–6)	Kpn I (3–6)
Bl (3)	Bam HI or MIF-1 (4, 5, 7)
Hepatitis B virus (5)	A36Fc (3)
Bovine leukemia virus (5)	Mouse mammary tumour virus (5)
– – Shared SINES (1, 3) – –	
Processed pseudogenes (1)	
SV40 (8)	

[a] Numbers indicate references: (1) Goldman et al. 1984; (2) Valerio et al. 1985; (3) Holmquist 1987, Holmquist and Caston 1986; (4) Soriano et al. 1983; (5) Bernardi et al. 1985; (6) Manuelidis and Ward 1984; (7) Vizard and Rosenberg 1984; (8) Marchionni and Roufa 1981.

SINEs are short, less than 500 bp, interspersed repeat sequences. *Alu* in man and Bl in mouse are SINE families present at over 100,000 copies per genome. LINEs are longer, interspersed repeats. The common LINE families in mouse and man are called L1, Bam HI or Kpn I respectively. Their members are up to 7 kb long and are present at over 20,000 copies per genome. The paucity of long, interspersed repeats in R-band DNA gives it its short-period interspersion pattern. Why these families of mobile genetic elements are partitioned into either G-bands or R-bands is unknown (see also Chap. 1).

Southern transfers of the putative R- and G-band DNA fractions, as prepared by either replication time fractionation or base composition fractionation, were then probed with about 30 different genes and several interspersed repeat sequences. Coincident results were obtained from the two fractionations. Housekeeping genes and short interspersed repeat families such as *Alu* were in the early replicating and dG+dC rich fractions, while most tissue-specific genes and the long, interspersed repeat L1 (Kpn I) were in the late replicating and dA+dT rich fractions. These data were reviewed (Holmquist 1987) and are summarized in Table 1 (see Figs. 5 and 6).

4.8 Evolution of the Banded Pattern

The banded prepattern has a function associated with its structure. The prepattern coincides with an alternating distribution of constitutively active genes in R-bands versus tissue-specific genes in G-bands. Also, mammalian G-band DNA differs from R-band DNA in many other properties (Table 1). Did all these distinctive properties evolve simulateously or did a few more essential, distinctive properties evolve first?

Bony fish (Wiberg 1983; Delaney and Bloom 1984) and frogs (Stock and Mengden 1975; Schempp and Schmid 1981; Stock 1984) show good G-banding and replication banding of their chromosomes. Holostean fish, but not sharks and rays, show reasonable G-banding of their chromosomes (Clark 1985). Perhaps a cytogeneticist with super hands will someday G-band shark chromosomes, but until then, the limited data implies that G-bandability evolved with bony fish. A band in mammalian chromosomes reflects a cluster of replicons, all of which initiate DNA synthesis at about the same time (Hand 1978; Holmquist et al. 1982). This was first detected by fiber autoradiography of replicating mammalian DNA (Huberman and Riggs 1968). *Drosophila* chromosomes do not show G- or replication bands, and fiber autoradiography of their cultured cells does not reveal clusters of replicons as are characteristic of mammalian cells (Steinemann 1981). Thus, replication banding is not a characteristic of all eukaryotic chromosomes and probably evolved in chordates. Most interestingly, fish (Kligerman and Bloom 1977; Phillips et al. 1985) and frog chromosomes (Schmid 1980; Schempp and Schmid 1981) do not band with quinacrine or other base composition specific fluorochromes. Molecular analysis of frog DNA also shows that it does not have the base composition variation in its DNA which could produce dA+dT rich G-bands (Thiery et al. 1976; Bernardi et al. 1985). Thus, the base compositional differences between R- and G-band DNA evolved much later than the initial banded prepattern.

In conclusion, a banded chromosomal prepattern evolved probably in an early chordate. The essence of the prepattern involved differences in replication time and differences in protein-DNA associations. These differences are visualized by replication banding or G-banding similar to the way one develops film to visualize the prepattern of the metallic silver, latent image in the silver halide crystals of a photographic emulsion. As the prepattern evolved further, additional properties such as base composition came to distinguish G-bands from R-bands and the old properties probably became more pronounced. Visualization of the banded prepattern of higher vertebrates is amenable to a larger variety of banding procedures and the standard procedures work better with chromosomes from higher vertebrates than from lower vertebrates. The great challenge remaining for the magic-fingered cytotaxonomist of today is to push back each and every banding procedure to lower and lower taxa so as to understand from extant species just when the various properties of the banded prepattern evolved.

References

Axel R, Cedar H, Felsenfeld G (1973) Chromatin template activity and chromatin structure. Cold Spring Harbor Symp Quant Biol 38:773–783

Barnett RI, Mackinnon EA, Romero-Sierra G (1974) The lability of acid alcohol fixed human chromosomes. Cytobios 11:115–122

Bernardi G, Olofsson B, Filipski J, Zerial M, Salinas J, Cuny G, Meunier-Rotival M, Rodier M (1985) The mosaic genome of warm blooded vertebrates. Science 228:953–958

Clark B (1985) Chromosomal evolution of holostean fishes. Ph.D. Thesis, Univ S Fla

Clark RJ, Felsenfeld G (1971) Structure of chromatin. Nature New Biol 229:101–106

Clark RJ, Felsenfeld G (1974) Chemical probes of chromatin structure. Biochemistry 13:3620–3628

Comings DE (1975a) Mechanisms of chromosome banding: IV Optical properties of Giemsa dyes. Chromosoma 50:89–110
Comings DE (1975b) Mechanisms of chromosome banding: VIII Hoechst 33258-DNA interaction. Chromosoma 52:229–243
Comings DE, Avelino E (1975) Mechanisms of chromosome banding: VII Interaction of methylene blue with DNA and chromatin. Chromosoma 51:365–379
Comings DE, Drets ME (1976) Mechanisms of chromosome banding: IX Are variations in DNA base composition adequate to account for Quinacrine, Hoechst 33258 and Daunomycin banding? Chromosoma 56:199–211
Comings DE, Avelino E, Okada TA, Wyandt HE (1973) The mechanism of C- and G-banding of chromosomes. Exp Cell Res 77:469–493
Comings DE, Harris D, Okada TA, Holmquist G (1977) Nuclear proteins IV: Deficiency of non-histone proteins in *Drosophila virilis* and mouse heterochromatin. Exp Cell Res 105:349–365
Crossen PE (1973) Factors influencing Giemsa band formation of human chromosomes. Histochemie 35:51–52
Dean WW, Stastny M, Lubrano GJ (1977) The degradation of Romanowski-type blood stains in methanol. Stain Technol 52:35–46
Delaney ME, Bloom SE (1984) Replication banding patterns in the chromosomes of the rainbow trout. J Hered 75:431–434
Evans HJ (1978) Some facts and fancies relating to chromosome structure in man. In: Harris H, Hirschhorn K (eds) Adv Hum Genet 8:347–438
Georgiev GP, Ilyin YV (1969) Heterogeneity of deoxynucleoprotein particles as evidenced by ultracentrifugation in cesium density gradient. J Mol Biol 41:299–303
Goldman MA, Holmquist GP, Gray MC, Caston LA, Nag A (1984) Replication timing of mammalian genes and middle repetitive sequences. Science 224:686–692
Hand R (1978) Eukaryotic DNA: Organization of the genome for replication. Cell 15:317–325
Hatchard CG, Parker CA (1956) A new sensitive chemical actinometer II: Potassium firrioxalate as a standard chemical actinometer. Proc R Soc London Ser A 235:518–536
Holmquist GP (1979) The mechanism of C-banding: Depurination and β-elimination. Chromosoma 72:203–224
Holmquist GP (1987) DNA sequences in G-bands and R-bands. In: Adolph KW (ed) Chromosome and chromatin structure. CRC Press, Fla
Holmquist GP, Caston LA (1986) Replication time of interspersed repetitive DNA sequences in hamsters. Biochim Biophys Acta 868:164–177
Holmquist GP, Comings DE (1976) Histones and G-banding of chromosomes. Science 193:599–602
Holmquist GP, Gray M, Porter T, Jordon J (1982) Characterization of Giemsa dark- and light-band DNA. Cell 31:121–129
Huberman JA, Riggs AD (1968) On the mechanism of DNA replication in mammalian chromosomes. J Mol Biol 32:327
Itzhaki RF, Cooper HK (1973) Similarity of chromatin from different tissues. J Mol Biol 75: 119–128
Klein F, Walker IO (1970) The partial dissociation of nucleohistone by salts. Eur J Biochem 14: 345–350
Kligerman AD, Bloom SE (1977) Distribution of F-bodies, heterochromatin, and nucleolar organizers in the genome of the central mudminnow, *Umbra limi*. Cytogenet Cell Genet 18:182–196
Latt SA, Gerald PS (1973) Staining of human metaphase chromosomes with fluorescent conjugates of polylysine. Exp Cell Res 81:401–406
Latt SA, Sober HA (1967) Protein-nucleic acid interactions II: Oligopeptidepolyribonucleotide binding sites. Biochemistry 6:3293–3306
Manuelidis L, Ward DC (1984) Chromosomal and nuclear distribution of the Hind III 1.9-kilobase human DNA repeat segment. Chromosoma 19:28–38
Marchionni MA, Roufa DJ (1981) Replication of viral DNA sequences intergrated within the chromatin of SV40-transformed Chinese hamster lung cells. Cell 26:245–258

Marshall PN (1978) Romanowski-type stains in haemotology. Histochem J 10:1–29

Matsukuma S, Utakoji T (1976) Uneven extraction of protein in Chinese hamster chromosomes during G-staining procedures. Exp Cell Res 97:297–303

Pearse AGE (1961) Histochemistry. Churchill, London

Phillips RB, Zajicek KD, Utter FM (1985) Q-bands, chromosomal polymorphisms in chinook salmon. Copeia 2:273–278

Schempp W, Schmid M (1981) Chromosome banding in amphibia IV: BrdU-replication patterns in Anura and demonstration of XX/XY sex chromosomes in *Rana esculenta*. Chromosoma 83:697–710

Schmid M (1980) Chromosome banding in amphibia IV: Differentiation of GC- and AT-rich chromosome regions in Anura. Chromosoma 77:83–104

Soriano P, Meunier-Rotival M, Bernardi G (1983) The distribution of interspered repeats is non-uniform and conserved in the mouse and human genomes. Proc Natl Acad Sci USA 80:1816–1820

Stein S, Böhlen P, Udenfriend S (1974) Studies on the kinetics of reaction and hydrolysis of fluorescamine. Arch Biochem Biophys 163:400–403

Steinemann M (1981) Chromosomal replication in *Drosophila virilis* III: Organisation of active origins in the highly polytene salivary gland cells. Chromosoma 82:289–307

Stock AD (1984) The occurrence of G-bands in the mitotic chromosomes of the amphibian *Xenopus muelleri*. Genetica 64:225–228

Stock AD, Mengden GA (1975) Chromosome banding pattern conservatism in birds and non-homology of chromosome banding patterns between birds, turtles, snakes and amphibians. Chromosoma 50:69–77

Sumner AT (1982) The nature and mechanisms of chromosome banding. Cancer Genet Cytogenet 6:59–87

Thiery JP, Maxaya G, Bernardi G (1976) An analysis of eukaryotic genomes by density gradient centrifugation. J Mol Biol 108:219–235

Utakoji T, Matsukuma S (1974) Fluorescent staining of L-cell chromosomes and chromocenters with 1-dimethylaminopathalene-5-sulfonyl chloride and G-banding. Exp Cell Res 87:111–119

Valerio D, Duyvesteyn MGC, Dekker MM, Weeda G, Berkvens TM, Voorn L van der, Ormondt H van, Eb AJ van der (1985) Adenosine deaminase: characterization and expression of a gene with a remarkable promoter. EMBO J 4:437–447

Van Prooijen-Knegt AV, Rapp AK, Burg MJM van der, Vrolijk J, Ploeg M van der (1982) Spreading and staining of human metaphase chromosomes on aminoalkylsilane-treated glass slides. Histochem J 14:333–344

Vizard DL, Rosenberg NL (1984) Temporal replication of an interspersed repeated sequence of mouse DNA. Biochem Biophys Acta 782:402–407

Wiberg UH (1983) Sex determination in the European eel (*Aqguilla anguilla*, L.). Cytogenet Cell Genet 36:589–598

3 Automated Cytogenetics with Modern Computerized Scanning Microscope Photometer Systems

M. E. DRETS and F. J. MONTEVERDE[1]

1 Microphotometry

The aim of this chapter is a brief introduction to some methodological and instrumental aspects involved in microphotometry which may be of interest to the general cytogeneticist. Some application examples and developments using a computerized scanning microscope photometer system are also described, stressing upon the potentialities of this technology, particularly in human cytogenetics.

Microphotometry, a branch of microscopy and photometry, allows one to carry out photometric measurements on minute biological specimens or on microscope slides. The cytogeneticist uses this technique to make precise quantitative measurements of nuclei and chromosomes and other cell components to investigate their structure and molecular composition.

Early in this century, some researchers were interested in applying the microscope to the chemical analysis of cells and tissues. In 1904, the great German optician, Köhler, presented the first clear evidence that UV light is absorbed by a cell component. Using microscope lenses that allow the transmission of UV light, Köhler (1904) took photomicrographs of *Salamander maculosa* nuclei at two different wavelengths (275 and 280 mμ) and discovered that the nuclear components absorbed the UV light differentially. This observation was practically forgotten until Caspersson (1936) confirmed Köhler's discovery. Since then, microphotometry has greatly evolved from a very simple apparatus to the complex microscope instruments for quantitative microphotometry available at present. In this respect, Swift (1966) published an informative review on the historical and methodological developments occurring in this area.

As an introduction to this kind of instrumentation, it is convenient to mention the main features of a classic colorimeter such as the one devised by Duboscq before describing a modern microphotometer for automated cytogenetics. This apparatus is composed of three parts, namely: (1) the light source: (2) two cuvettes, one filled with the solvent (reference) and the other with the substance to be measured dissolved in the same type of solvent (sample); and (3) a light-integrating prism system which directs both light beams to the observer's eye. By analogy, a modern microscope photometer follows similar basic principles (Fig. 1).

1 Division of Human Cytogenetics and Quantitative Microscopy, Instituto de Investigaciones Biológicas Clemente Estable, Avda. Italia 3318, Montevideo, Uruguay

Cytogenetics. Ed. by G. Obe and A. Basler

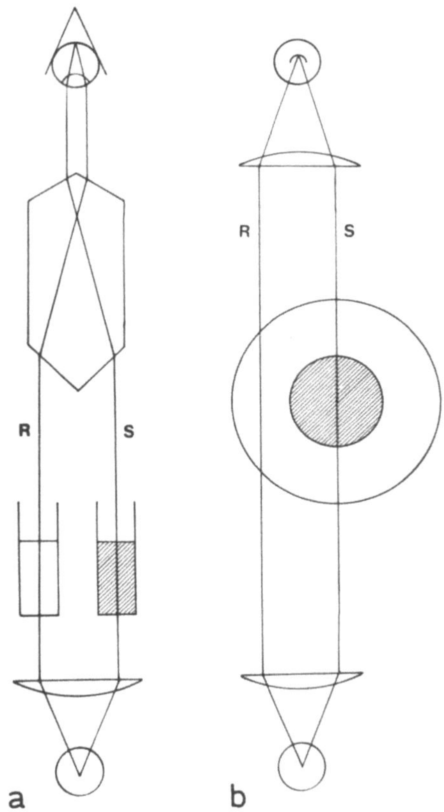

Fig. 1a,b. Analogies between a classic colorimeter (Duboscq) (**a**) and a cytophotometer (**b**). *R* and *S* denote reference and sample beams. A cell with a stained nucleus is diagrammatically depicted in **b**

When a light beam of a given intensity (Io) passes through the sample cuvette an absorption of the light occurs depending on the concentration of the absorbing molecules and the path length. The fraction of light remaining after loss by absorption, reflection, scattering and diffraction is called the transmittance of the sample and their values range between 0.0 (total absorption) and 1.0 (total transmission). These values are usually expressed as a percentage. The Lambert-Beer law usually used in microphotometry explains the conversion of transmission into absorbance. However, the Lambert-Beer law is not always exact in microscopy because of the chemical reactions or interactions with the incident light and because of the overlapping of absorption bands of chemical compounds or dyes present in the specimen to be measured.

The Swedish researcher T. Caspersson, using Köhler's UV microscope system, reported in a classic monograph (1936), that the nuclear components presented an absorption peak at 250 mμ but not the cytoplasm. Further instrumental developments made by Caspersson (1950) followed similar analytical principles found in Duboscq's colorimeter, i.e. the light source was split into two beams: one used as the reference or background beam and the other one as the sample or reference beam. The Zeiss Co. (Oberkochen) manufactured a registering microphotometer (Model UMSP I) largely based on Caspersson's instrument (Trapp 1966). However, the instrument was expensive and somewhat difficult to use, which limited it to a few laboratories.

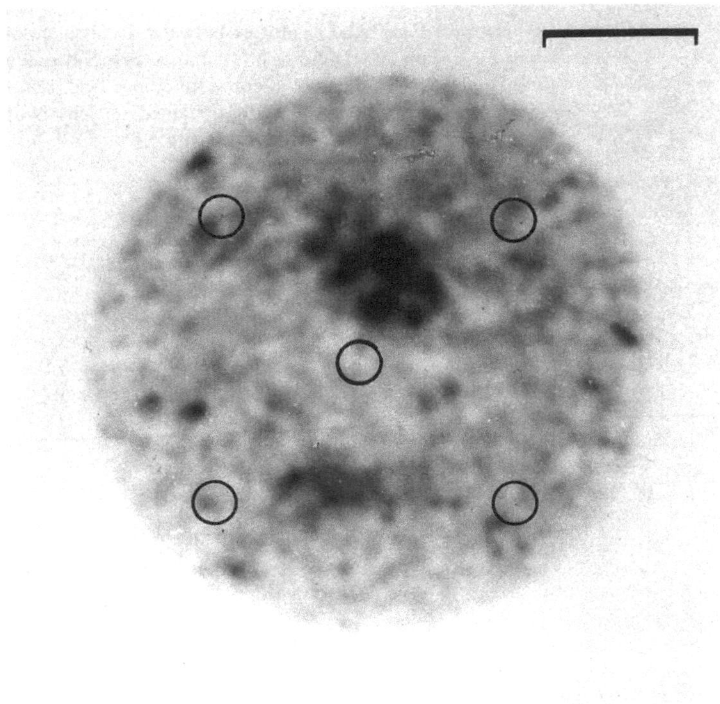

Fig. 2. Errors introduced by measuring a few areas in an interphase nucleus. *Circles* denote the areas selected by the user and covered by the photometer diaphragm. Bar = 5 μm

A more practical solution was reported by Pollister (1952), who devised a single-beam cytophotometer for the visible range of the spectra which was much easier to use. This instrument popularized the use of photometry at the cellular level in small cytogenetics laboratories. The disadvantage of Pollister's apparatus was that each photometer measurement of the sample had to be followed by another one of the reference beam; this meant that the cytologist had to move the microscope stage manually many times to estimate the extinction coefficient of the specimen to be measured. This problem was in part technically overcome in the epoch by measuring the sample and the reference beams by means of an adjustable mirror system which allowed one to explore the specimen without moving the microscope stage (Drets 1961).

One of the greatest problems in microphotometry is the lack of homogeneity usually found in biological samples. In our simple colorimeter, we assumed that the sample cuvette contained the absorbing material uniformly dispersed into the solvent. Obviously, this ideal situation never occurs in biological materials. Cytogeneticists usually find nuclear chromatin organized in many different ways, depending on the cell type, amount of heterochromatin, condensation and degree of ploidy. If, for instance, the DNA content of a nucleus is measured in only five different predefined nuclear areas, it is obvious that a considerable number of errors will be intro-

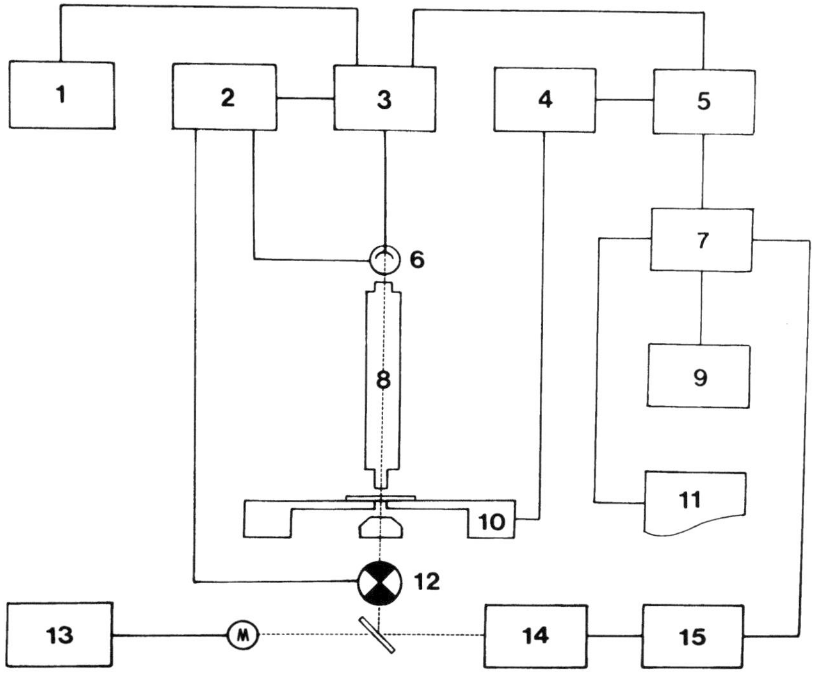

Fig. 3. Basic components of a computerized scanning microscope photometer. *1* Analogue recorder; *2* stabilized power supply and amplifier; *3* digital measuring unit; *4* stage control unit; *5* interface; *6* photomultiplier; *7* computer; *8* microscope system; *9* graphics display terminal; *10* microscope stage with stepping motors; *11* printer; *12* light modulator; *13* stabilized light power unit and illuminating lamp; *14* monochromator; *15* wavelength drive control and interface

duced because of the differential condensation of the chromatin (Fig. 2). These errors, due to the irregular distribution of the components in biological samples, are known as distributional errors. The reader is referred to the works of Ornstein (1952), Pätau (1952) and Piller (1977) for further information on this topic.

The only practical method to eliminate the distributional error introduced by cellular inhomogeneities is thus complete object scanning. This method can additionally provide valuable information on the morphology and chemical composition of cell components since the sample is scanned completely and a high number of measurements carried out.

A modern scanning microscope photometer is basically composed of four major parts: (1) microscope with scanning stage; (2) stabilized light source; (3) measuring electronic device; and (4) computer. These components are usually associated "on line" permitting the specimen to be scanned and measured under computer control, generally through interactive software (Fig. 3). Obviously, it is absolutely necessary to adjust in a very careful way the instrumental setup for a proper and precise operation. For a discussion on adjustment procedures, see Zimmer (1973) and Ruch and Leemann (1973).

Microscope scanning stages have built-in electrical stepping motors allowing one to move the specimen in steps of 0.25, 0.5 or 10 μm in the X- and Y-directions. The

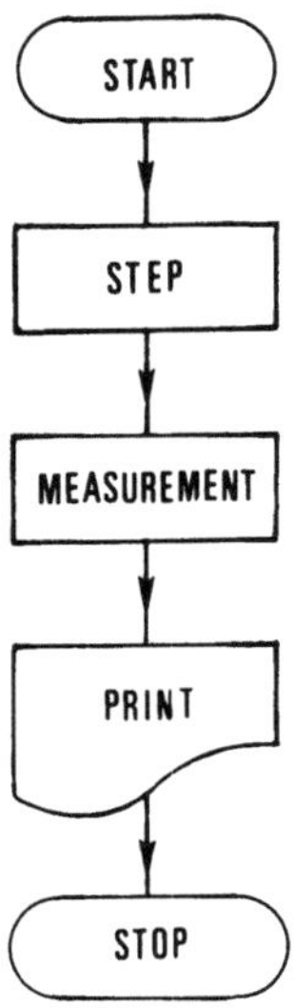

Fig. 4. Simple flowchart showing step, measurement and print commands

scanning stages can be operated by means of electronic devices containing pre-programmed hardware. However, it is much more advisable for the general cytogeneticist to write his own computer software to cover specific purposes for obtaining more scanning versatility and analytical possibilities.

The present great development of computers is making this technology more and more affordable to cytogenetics research, broadening the scope of their application in quantitative analytical problems. To improve the computer power and the performance of the scanning microscope photometer presently available at our Division, a digital PDP LSI 11/23 microcomputer and a color graphics terminal from Tektronix Model 4107 have been combined with the Zeiss MP01 system using simple electronics and programming procedures. Information on this installation can be obtained from the authors which may be taken as an example to combine other micros with the Zeiss MP01 system.

The development of computer software for analytical quantitative microscopy is not a complicated task. A simple computer flowchart for microphotometry, including stage step and photometer measurement orders, is illustrated in Fig. 4. This figure shows an example of the basic logic applicable for programming and scanning biological components or other microscope objects as well. Adequate combination of these commands may allow the user to develop application software to cover a great variety of photometric analyses and scanning patterns.

Values of measurements of microscope objects such as, for instance, nuclei or chromosomes obtained by means of digital measuring devices or through computer software, can be displayed in the graphics color terminal or printed by the computer plotter. Manipulation of data allows the transformation into arbitrary series of printing characters or pre-defined image areas (pixels), each one representing a pre-defined object density. The use of color graphics terminals are adding an extra dimension to the analysis of microscope objects in automated cytogenetics since they allow the

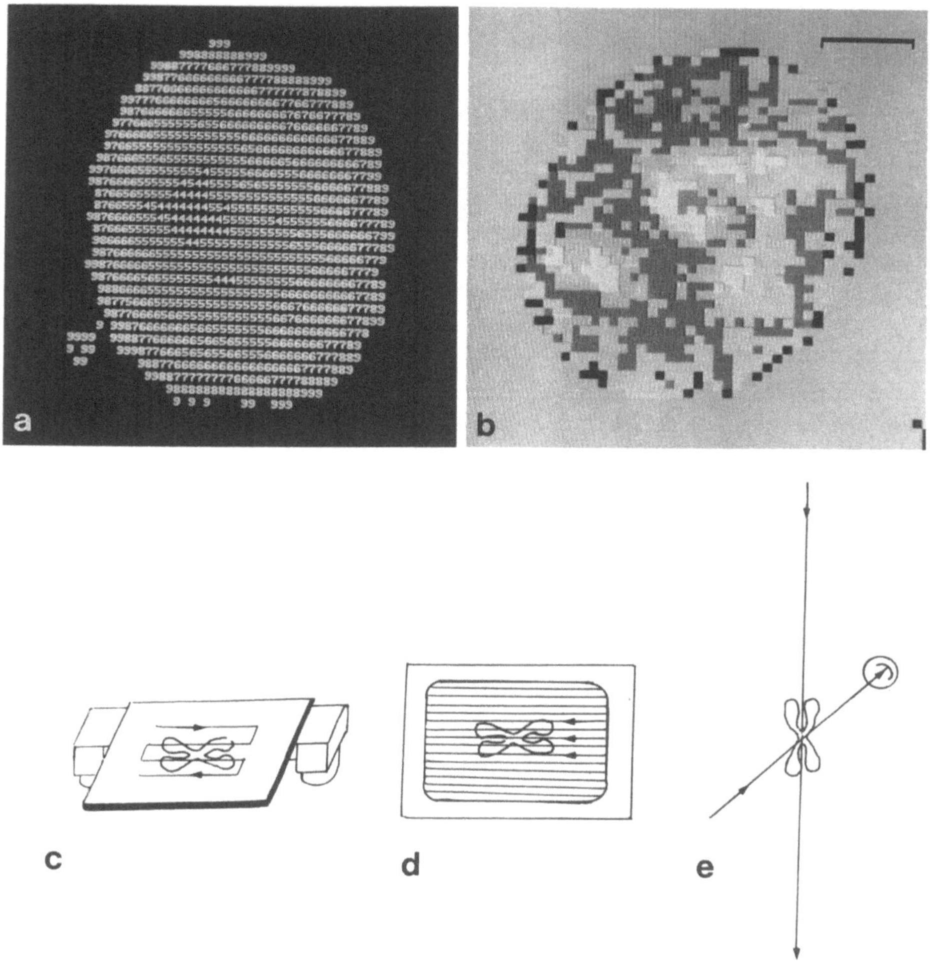

Fig. 5a–e. Computer display of sorted and transformed data after scanning two different phytohemagglutinin-stimulated human lymphocytes. **a** Arbitrary units; **b** image pixels. **c–e** Scanning of chromosomes through different quantitative analytical systems: **c** electromechanical stepping microscope stage; **d** TV image analysis; **e** chromosome flow system. *Vertical arrow* in **e**: direction of chromosome flow. *Perpendicular arrow:* direction of laser beam. In all cases, chromosomes are scanned linearly. Bar = 5 μm

assignment of different colors to characters or image pixels. The result is a displayed image representing different densities of the original microscope image. Figure 5a,b illustrates two examples of this type of image transformation. This procedure is closely related to image analysis including TV image analysis used, for instance, for automated karyotyping. Appropriate computer software additionally allows one to make real-time analyses or to store important parameters such as extinction coefficient, distributional error, total area scanned and total object area detected. These data are used to carry out further statistical analyses of microscope objects. The volume published by Agrawala (1977), a collection of papers on pattern recognition, is a useful introduction to those interested in the analysis of cell images, data manipulation and other related problems.

For systematic studies of microscope specimens it is more convenient to store all measurements into peripheral magnetic devices such as floppy diskettes or hard disks which enable further computer processing or image analysis. The latest models of hard disks of Winchester type offer large storing capacity and faster access making them highly recommendable in microphotometry.

Photomultipliers installed in photometer heads highly amplify the radiant flux impinging on the photocathode because of the property of their dinodes to emit secondary electrons, which collect a considerable amount of energy in comparison with the incident number of photoelectrons. This property of photomultipliers allows the measurement of objects transmitting, emitting or reflecting very low levels of energy both in the visible and the UV spectra which is particularly useful in quantitative fluorescence microscopy.

Chromosome scanning by means of stepping microscope stages is, in several respects, comparable to the ones carried out through TV or flow systems since chromosomes are also linearly scanned by electron or laser beams respectively (Fig. 5c–e).

There is practically no area where these scanning microphotometer systems cannot be successfully applied in cytogenetics. Although a comprehensive review of all possible applications of the present microscope photometers is beyond the scope of this brief chapter, it is of interest to refer to some of the problems usually found in cytogenetic research.

2 Autoradiography

One of the most time-consuming and tedious tasks in cytology is the visual counting of silver grains on autoradiophotographic prints. With automated photometer systems silver grains can be counted conveniently and faster. Thus, it is possible to measure a considerably larger number of microscope objects, i.e. labelled nuclei or chromosomes, directly on slides in a short time using automated microscope systems.

In transmission microscopy each time that the microscope stage moves the specimen one step and finds a grain filling the plane of the photometer-measuring diaphragm a count is recorded (Fig. 6). Transmitted images present the problem that the underlying absorbing structures disturb silver grain images. For this reason, it is preferable to use reflected polarized systems since this method reduces any disturbing background significantly. In this case, silver grains are seen as brilliant spots because of their high reflecting properties (Dörmer and Thiel 1976; Piller 1977). Here, each brilliant spot is counted as an individual grain.

The number of scattered grains (background) is essentially variable from slide to slide and from area to area in the same slide. Therefore, background grains must also be counted and taken into account when writing computer software, in order to estimate radioactive labelling of the sample under measurement. Grain clusters pose an extra problem since they introduce an important source of error especially in heavily labelled slides because the machine obviously cannot distinguish individual grains in

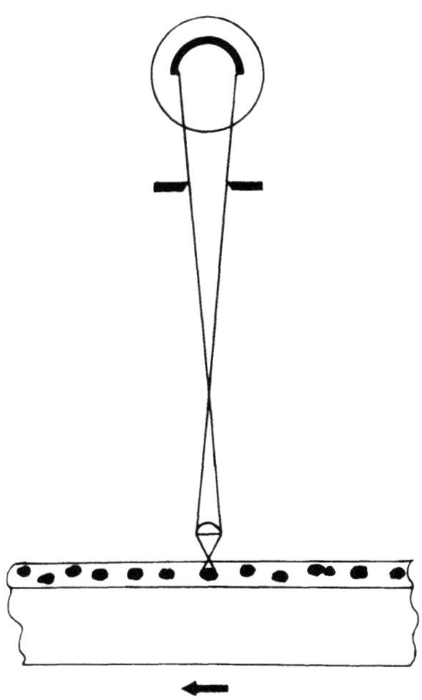

Fig. 6. Autoradiographic silver grain image filling the photometer-measuring diaphragm plane. Each time that a silver grain fills the diaphragm a count is recorded

large accumulations. It is thus necessary to avoid heavily labelled samples to obtain precise results.

The total number of silver grains recorded and the total labelled cell area detected allow one to estimate the mean grain content per area unit. With this basis a computer program (Graincount Program) was written which enabled us to estimate the DNA loss induced in human chromosomes by photo-oxidation in a short time (Drets et al. 1978) proving that the system of automatic grain counting was reliable, rapid and useful (Fig. 7).

3 Nucleic Acid Measurements in Nuclei and Chromosomes

One of the most extensive applications of scanning microscope photometers is at present the accurate measurement of nuclei acids in cell spreads or in histological sections. This technique has proved to be extraordinarily useful for quantitative studies, i.e. DNA content of different cell lines including cancer cells, cell cycle, nuclear DNA distribution, and in evolution problems.

Feulgen stain has been classically used for quantitative determinations of DNA because of its high staining specificity. Feulgen-based measurements have also been used to detect aberrations and polymorphisms in human chromosomes (Groen and Van der Ploeg 1979). Moreover, the Feulgen method permitted the measurement of the DNA content of a variety of cells and animals such as, for instance, mammals, reptiles and birds (Atkin et al. 1965). Relative measurements can be converted into absolute DNA values usually expressed in picograms. To do this, samples are compared with DNA values of genomes determined by chemical means such as in *Xenopus laevis* or chicken.

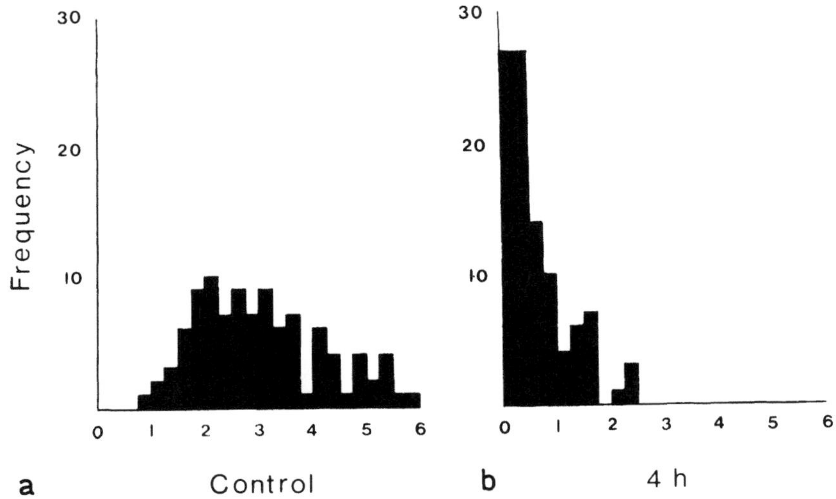

Fig. 7a,b. Histograms of labelled, normal human lymphocyte nuclei (**a**) and nuclei exposed to light for 4 h (**b**), scanned and silver grains counted. An intense DNA loss was induced as detected by automatic silver grain counting. (See text)

Although the Feulgen method is a simple and rapid cytological procedure, it has the great disadvantage that it bleaches with time (Kasten 1962). This instability hinders comparative measurements after some time which, in practice, means that the use of Feulgen stain in quantitative cytology is limited. For many quantitative applications, Gallocyanin Chrome Alum stain is thus the preferred cytological procedure for quantitative DNA determinations because this stain is stable with time and also reacts stoichiometrically with DNA (Sandritter et al. 1966).

In experimental series or evolution studies where a high number of samples and measurements are involved, a practical method is to use the two-wavelength method to measure them manually (Mendelsohn 1966; Pätau 1952). This method or the automated scanning of the sample allows the estimation of the relative variations of the DNA nuclear content with respect to the control expressed in percentage. DNA measurements are reliable in a given diploid tissue or cell line on the basis that the DNA content is constant from cell to cell. Degrees of ploidy are detected as multiple numbers of the basic diploid DNA value as expected (Swift 1950). If a computer is available on line, it is convenient to call a histogram subroutine to facilitate interpretation and to better the presentation of the results obtained (Fig. 8).

In the last decade, cytofluorometry has become very popular for the quantitative measurement of RNA, DNA and other cell components because of the considerable number of fluorochromes presently available and the advances occurring in fluorescence microscope systems. Scanning fluorometry of fluorochrome-stained nuclei or chromosomes is better carried out using reflection microscope systems since they do not introduce measurement errors due to fluorescing background as it usually occurs with transmission fluorescence microscopes. A very important field of fluorochrome application is nowadays in cell-flow techniques but a discussion of this modern method

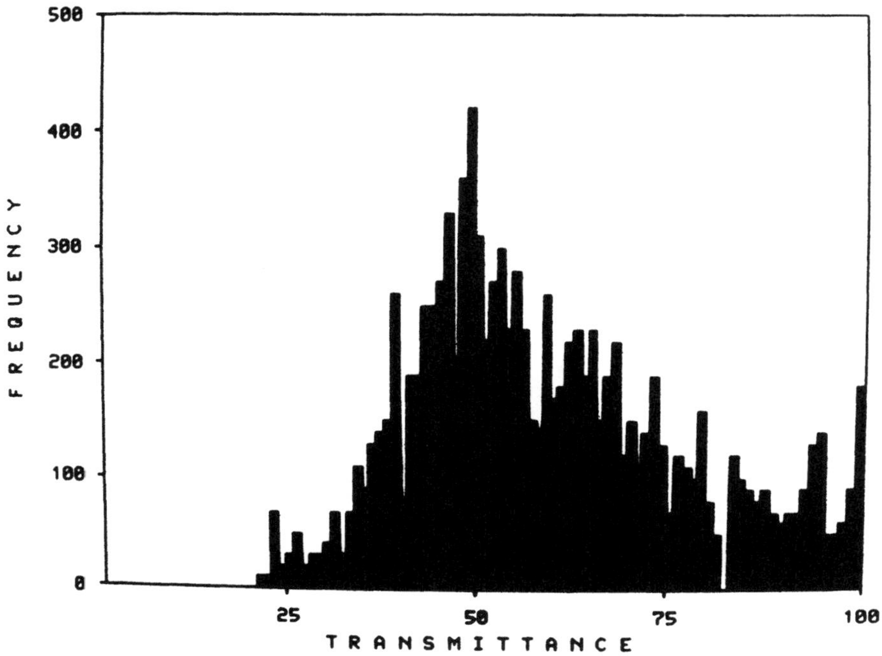

Fig. 8. Histogram displayed by the computer terminal of sorted measurements obtained by scanning a normal human Gallocyanin Chrome Alum stained lymphocyte nucleus showing chromatin distribution

is far beyond the scope of this brief chapter and the reader is referred to Melamed et al. (1979) for more information on the topic. Also, fine articles on flow-system applications are often published in *Cytometry*. Recently, a new and exciting field of automated, quantitative measurement is fluorescence hybridization. Pinkel et al. (1986) have shown that hybrid translocation chromosomes fluoresce differentially according to the species involved, making this technique particularly useful for automation.

4 Analysis of Banded Chromosomes

Giemsa and other reported methods for chromosome banding afforded the cytogeneticist a new, powerful tool for a much more precise description of the normal and pathological human chromosome structure. Complete maps of the human karyotype were thus published (Drets and Shaw 1971). International conferences on human chromosome nomenclature (ISCN 1981 and previous conferences) followed this type of diagrammatic band representation but chromosome arms were divided into different regions. Although this division was informative and gave a practical method of band localization, it was still a system largely based on direct microscope observations and not an actual quantitative estimation on the size and localization of chromosome bands.

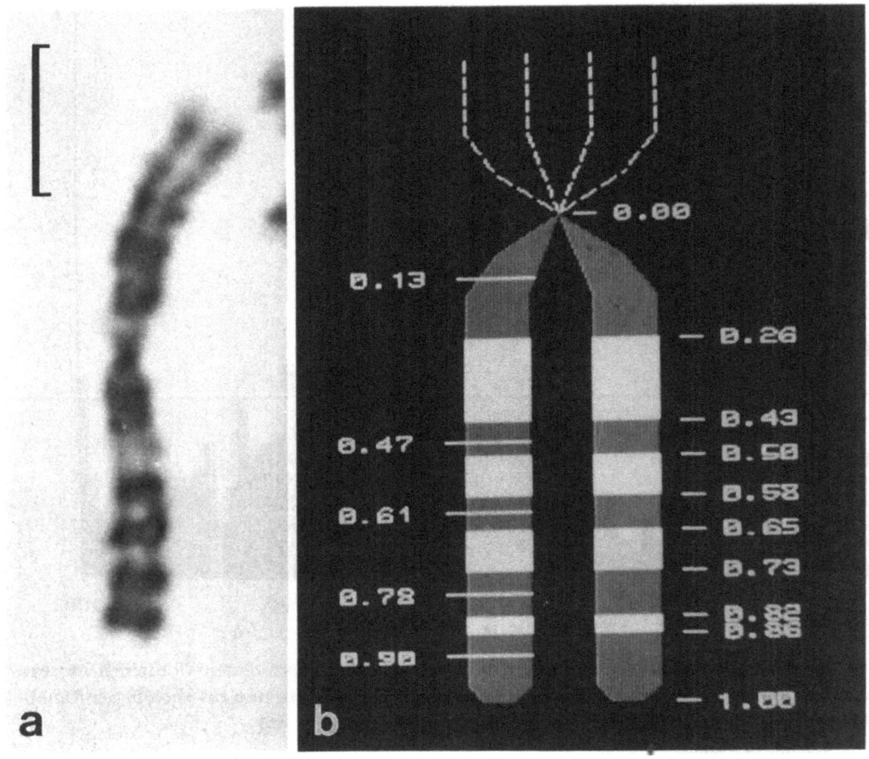

Fig. 9. a G-banded human chromosome No. 1. **b** Computer graphics display of quantitative data obtained by one single scanning of the long arm of the chromosome exemplified in **a**. *Left* chromatid shows the relative position of band densitometer peaks as detected by the Bandscan program. Relative positions of band-interband junctions are seen on the *right* chromatid. In both chromatids, bands and lines were exactly displayed according to the relative positions detected. Bar = 2.5 μm

The problem of quantitative band localization has not been solved as yet. The numerous changing parameters found in the usual metaphase spreads such as division stage, degree of chromosome contraction, chromosome bending or overlapping pose serious difficulties for developing a reliable system of quantitative image analysis using electromechanical microscope scanning stages or TV scanning systems.

In our laboratory, the problem of band localization was studied using C-banded Ys' and Giemsa-banded No. 1 human chromosomes as analytical models. These chromosomes were scanned using a semi-automatic analogue recording microphotometer built in our laboratory. Densitometer tracings thus obtained were measured and quantitative maps of the relative localization of C- and G-bands of the Y- and No. 1 chromosomes were drawn (Drets and Seuanez 1973).

A computer program for the Wang programmable calculator Model 720 C combined with the MP01 Zeiss microphotometer was subsequently written on this basis (Bandscan program). This program allowed the detection of the relative position of

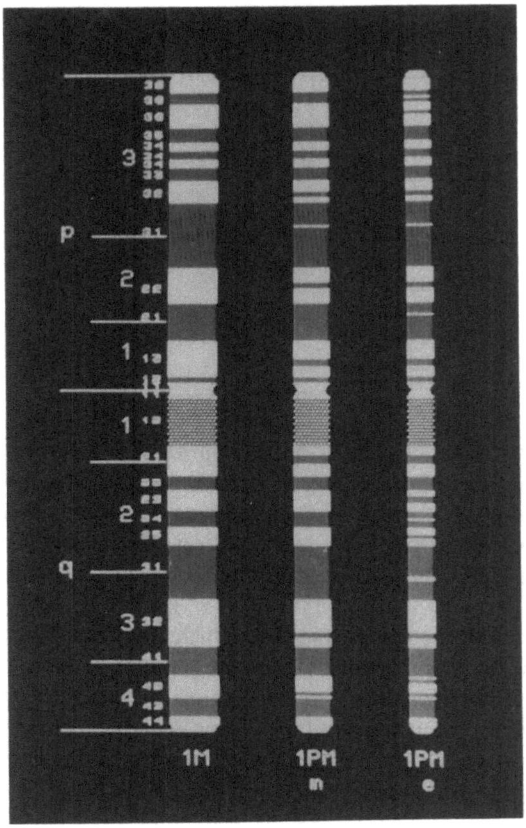

Fig. 10. Computer-generated diagrams showing G-banding patterns of a normal human chromosome No. 1 at three different degrees of condensation. 1M = metaphase; 1PM m = mid-prometaphase; 1PM e = early prometaphase

several characteristic landmarks of the human chromosome No.1, thus confirming data previously estimated from analogue densitometer tracings (Drets 1978). The major feature of this program was the possibility of rejecting interactively minor bands or suspected chromosome artefacts which facilitated the cytogenetic interpretation of band localization.

The Bandscan program was re-written (Fortran) for our present computer system and our previous results on band localization on the long arm of chromosome No. 1 were confirmed preliminarily through this method. Figure 9 presents the results obtained by scanning the long arm of a G-banded chromosome No. 1 (Fig. 9a). Densitometer peaks (left chromatid) and the relative positions (right chromatid) of band-interband junctions were detected and displayed by the system using two subprograms (Fig. 9b).

Additionally, the user can compare on the screen the data thus obtained with analogue computer-generated densitometer tracings or with diagrammatic representations of human chromosome-banding patterns (Fig. 10). The program that generates these chromosomes in the computer graphics terminal was written following the

Table 1. Expected and detected relative values of band-inter-band junctions found after ten scannings of a photograph of the model illustrated in Fig. 11b

Expected	Detected	
	Mean	SD
0.1250	0.1245	0.0003
0.2500	0.2476	0.0002
0.3750	0.3742	0.0003
0.5000	0.4962	0.0004
0.6250	0.6248	0.0004
0.7500	0.7496	0.0004
0.8750	0.8744	0.0003

diagrammatic standards published by the ISCN (1981) on human banding patterns. The system is capable of displaying the whole human karyotype or individual chromosomes, single or combined, in three degrees of condensation. We have found it useful to display these diagrams when comparing quantitative data obtained after scanning a chromosome arm. Moreover, they can also be used as a reference on chromosome structure.

Due to the fact that the information obtained is relative and that the user can check and "see" data from the same chromosome arm through different graphic representations, the final result is highly informative. However, it is important to point out that to obtain consistent, quantitative and reproducible results on band localization with automated scanning systems, it is necessary to standardize the cytological and photographic procedures followed, which may differ from one laboratory to another. We believe that these methods of band localization could provide a precise way of studying banding patterns which may prove useful in experimental and clinical cytogenetics, particularly in those cases where band identification is uncertain.

5 Sister Chromatid Exchange Localization

Closely related to banding-pattern localization is the problem of the detection of the position of sister chromatid exchanges (SCE). It is thus of utmost interest for the cytogeneticist to have access to a precise system capable of detecting them and to compare both relative measurements. Computer analysis techniques to automate the detection of exchanges in their relative positions are the logical way to attack this important area of application in cytogenetics. Computer localization through SCE scanning has already been explored using fluorochrome stains (Zack et al. 1976, 1977). However, Giemsa-SCE staining is by far the best cytological method for a precise SCE localization since it gives higher contrast than the usual fluorochrome stains.

Several authors (Carrano and Wolff 1975; Crossen et al. 1977; Hoo and Parslow 1979; Kato 1979) have been interested in the localization of SCEs in relation to the type of chromatin or band involved. However, no quantitative data on the exchange localization along the chromatids were reported to enable comparisons with specific

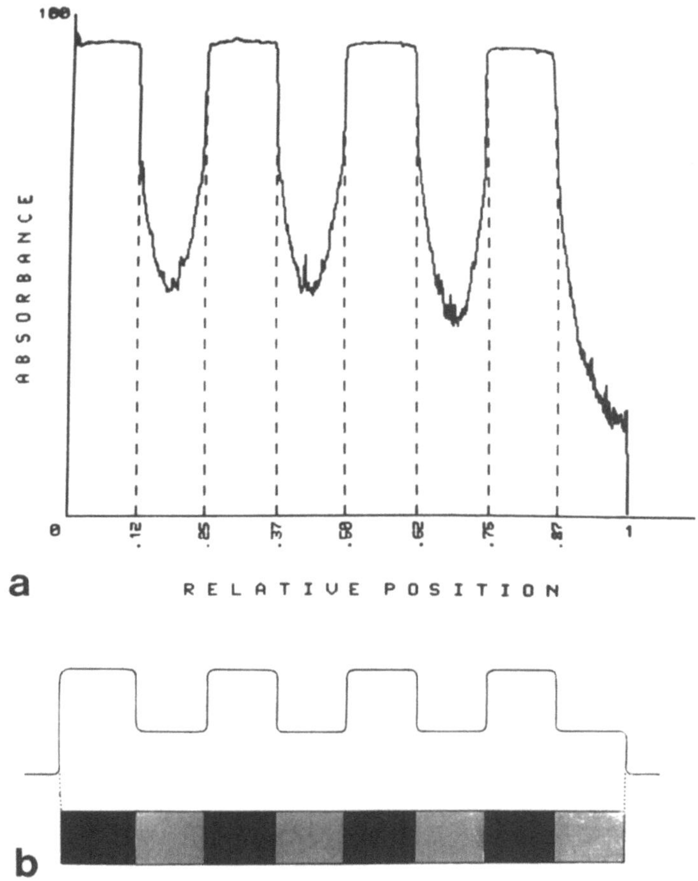

Fig. 11. a Computer-generated analogue curve and relative positions of band-interband junctions obtained after scanning the photograph of a drawing of alternating black-gray squares. b Model and expected curve. The model was used to develop a computer program for SCE detection

bands. To study this problem, a computer program (SCE Program) for the quantitative localization of SCE in human chromosomes was developed. A drawing of alternating black and gray squares was used for testing the accuracy and reliability of the computer program. The model simulated the real situation of Giemsa-stained chromatids exhibiting exchanges where black squares represented unifilarly, and gray squares bifilarly, BrdU-substituted chromatid segments. This model also served to adjust the instrumental setup since it was difficult to determine beforehand the degree of contrast obtained in each case. Table 1 gives mean values obtained after scanning the model ten times, thus showing good agreement between expected and detected values. Figure 11 illustrates a densitometer analogue curve (Fig. 11a) displayed by the graphics terminal of our system and obtained by scanning the test model (Fig. 11b) giving the

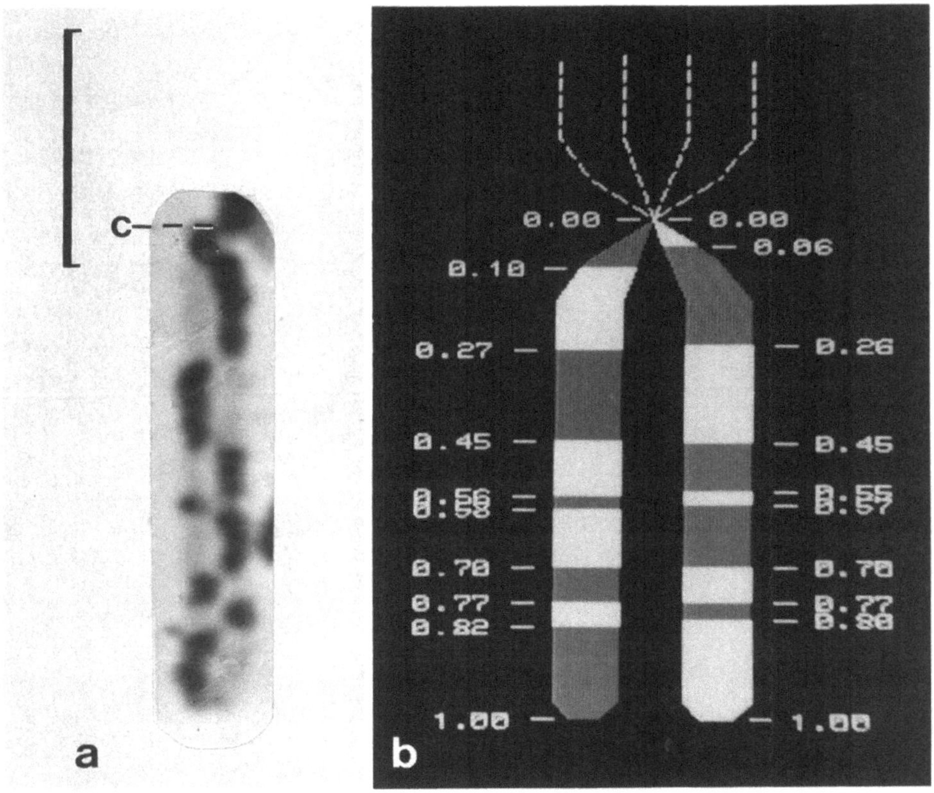

Fig. 12. a Chromosome long arm from Chinese hamster ovary cell incubated in BrdU and UV irradiated exhibiting several sister chromatid exchanges (Courtesy of Dr. N.O. Bianchi). b Relative band-interband junction positions as detected by the SCE computer program after scanning individually each chromatid of the chromosome illustrated in a. Agreement for several SCE values was found. c indicates the centromere area. Bar = 5 μm

detected relative junction values. An example of the potentiality of this technique is presented in Fig. 12 which shows a chromosome cutout carrying exchanges (Fig. 12a) and the graphics image and relative positions of the band-interband junction exchanges as detected by the system (Fig. 12b).

Since the SCE localization is also relative, the points of exchanges can be conveniently compared with the position of Giemsa and R-banding patterns using the Bandscan program developed to detect bands along the chromosome arms as mentioned above. The use of both computer programs can thus advance useful information to the problem of the band-interband junction exchange localization in relation to band positions and characteristic chromosome regions where SCEs are produced. A complete report on these programs will be published elsewhere.

6 Final Comments

In this short chapter some analytical applications of modern scanning microphotometer systems useful for automated cytogenetics were briefly reviewed.

Technical progresses in microscopy, cytological procedures and computation are occurring rapidly. At present, it is foreseeable that in the near future, the cytogeneticist will be able to investigate nuclei and chromosomes in a much more detailed and precise way than ever before (Ploem 1986). Improvements in areas such as, for instance, chromosome flow analysis, automated karyotyping, detection and classification of neoplastic cells and other cell types and quantitative chromosome-banding patterns and sister chromatid exchange localization will undoubtedly become powerful methods which will stress the importance of being informed about the latest technical advances for the successful use and application of microphotometry in the many research problems found in cytogenetics.

References

Agrawala AK (1977) Machine recognition of patterns. IEEE, New York, 463 pp

Atkin NB, Mattinson G, Beçak W, Ohno S (1965) The comparative DNA content of 19 species of placental mammals, reptiles and birds. Chromosoma 17:1–10

Carrano AV, Wolff S (1975) Distribution of sister chromatid exchanges in the euchromatin and heterochromatin of the Indian Muntjac. Chromosoma 53:361–369

Caspersson T (1936) Über den chemischen Aufbau der Strukturen des Zellkernes. Skand Arch Physiol 73 Suppl 8:1–151

Caspersson T (1950) A universal ultramicrospectrograph for the optical range. Exp Cell Res 1: 595–598

Crossen PE, Drets ME, Arrighi FE, Johnston DA (1977) Analysis of the frequency and distribution of sister chromatid exchanges in cultured human lymphocytes. Human Genet 35:345–352

Dörmer P, Thiel E (1976) Methods of quantitative autoradiography using incident light microphotometry. J Histochem Cytochem 24:145–151

Drets ME (1961) Ratio recorder for cytophotometry based on an image discriminator apparatus. Mikroskopie 16:341–348

Drets ME (1978) Bandscan – A computer program for on-line linear scanning of human banded chromosomes. Comput Progr Biomed 8:283–294

Drets ME, Seuanez H (1974) Quantitation of heterogeneous human heterochromatin: microdensitometric analysis of C- and G-bands. In: Coutinho EM, Fuchs F (eds) Physiology and genetics of reproduction, part A. Plenum, New York, pp 29–52

Drets ME, Shaw MW (1971) Specific banding patterns of human chromosomes. Proc Natl Acad Sci USA 68:2073–2077

Drets ME, Comings DE, Folle GA (1978) Mechanisms of chromosome banding. X. Chromosome and nuclear changes induced by photo-oxidation and their relation to R-banding with anti-C antibodies. Chromosoma 69:101–112

Groen FCA. Van der Ploeg M (1979) DNA cytophotometry of human chromosomes. J Histochem Cytochem 27:436–440

Hoo JJ, Parslow MI (1979) Relation between the SCE points and the DNA replication bands. Chromosoma 73:67–74

ISCN (1981) An international system for human cytogenetic nomenclature. High-resolution banding (1981) Rep Standing Committee on Human Cytogenetic Nomenclature. Cytogen Cell Genet 31:1–23

Kasten FH, Kiefer G, Sandritter W (1962) Bleaching of Feulgen stained nuclei and alteration of absorption curve after continuous exposure to visible light in a cytophotometer. J Histochem Cytochem 10:547–555

Kato H (1979) Preferential occurrence of sister chromatid exchanges at heterochromatin-euchromatin junctions in the Wallaby and Hamster chromosomes. Chromosoma 74:307–316

Köhler A (1904) Mikrophotographische Untersuchungen mit ultraviolettem Licht. Z Wiss Mikrosk 21:129–165

Melamed MR, Mullaney PF, Mendelsohn M (1979) Flow cytometry and sorting. John Wiley & Sons, New York, 716 pp

Mendelsohn ML (1966) Absorption cytophotometry: Comparative methodology for heterogeneous objects, and the two-wavelength method. In: Wied GL (ed) Introduction to quantitative cytochemistry. Academic Press, London New York, pp 201–237

Ornstein L (1952) The distributional error in microspectrophotometry, Lab Invest 1:250–262

Pätau K (1952) Absorption microphotometry of irregular shaped objects. Chromosoma 5:341–362

Piller H (1977) Microscope photometry. Springer, Berlin Heidelberg New York, 253 pp

Pinkel D, Straume T, Gray JW (1986) Cytogenetic analysis using quantitative, high-sensitivity, fluorescence hybridization. Proc Natl Acad Sci USA 83:2934–2938

Ploem JS (1986) New instrumentation for sensitive image analysis of fluorescence in cells and tissues. Applications of fluorescence in the biomedical sciences. Alan R Liss, New York, pp 289–300

Pollister AW (1952) Photomultiplier apparatus for microphotometry of cells. Lab Invest 1:106–114

Ruch F, Leemann U (1973) Cytofluorometry. In: Neuhoff V (ed) Micromethods in molecular biology. Springer, Berlin Heidelberg New York, pp 329–346

Sandritter W, Kiefer G, Rick W (1966) Gallocyanin chrome alum. In: Wied GL (ed) Introduction to quantitative cytochemistry. Academic Press, London New York, pp 295–326

Swift H (1950) The constancy of desoxyribose nucleic acid in plant nuclei. Proc Natl Acad Sci USA 36:643–654

Swift H (1966) Analytical microscopy of biological materials. A brief history. In: Wied GL (ed) Introduction to quantitative cytochemistry. Academic Press, London New York, pp 1–39

Trapp L (1966) Instrumentation for recording microspectrophotometry. In: Wied GE (ed) Introduction to quantitative cytochemistry. Academic Press, London New York, pp 427–435

Zack GW, Spriet JA, Latt SA, Grandlund GH, Young IT (1976) Automatic detection and localization of sister chromatid exchanges. J Histochem Cytochem 24:168–177

Zack GW, Rogers WE, Latt SA (1977) Automatic measurement of sister chromatid exchange frequency. J Histochem Cytochem 25:741–753

Zimmer HG (1973) Microphotometry. In: Neuhoff V (ed) Micromethods in molecular biology. Springer, Berlin Heidelberg New York, pp 297–328

4 The Chromosomes of Man: Evolutionary Considerations

H. N. Seuánez[1]

1 Chromosomes, Genes and Evolution

The human chromosome complement, as any other component of man, is a product of a transmutational process that took place at different stages of mammalian radiation. If we envisage evolution as a continual event where change may either be gradual or abrupt, occurring under different degrees of adaptative pressures or perhaps by random non-selective drift, tentative pathways of chromosome phylogenies can be proposed to account for the extant karyotypic diversity within mammals. Chromosomes are discrete organelles with definite morphological attributes and gene content. As structural entities inside the nucleus they represent a particular arrangement of chromatin that appears visible during cell division or, exceptionally, in the interphase nuclei of some larval tissues, like the salivary glands of dipteran flies. As vehicles of inheritance, they represent separate genomic compartments where syntenic loci are precisely aligned with respect to one another.

An analysis of chromosome phylogenies would be incomplete if we overlooked the fact that evolution has also been operative at other levels of biological complexity. This is because macrostructural characteristics might change so drastically and suddenly in the evolutionary history of mammals as if speciation occurred in a non-continuous, abrupt process of saltatory divergence. This observation led to the proposition that species emerged following a punctuated model of phyletic radiation rather than by a gradual process of continuous change (Gould 1980). However, at the opposite side of the spectrum of biological complexity, the evolution of our structural and enzymatic proteins appears to take place at constant rates so that fixations of new mutations occur stochastically, following the model of a molecular, evolutionary clock (Wilson et al. 1977). Two important corollaries arise from this finding. Firstly, that most mutations ought to be non-adaptative, or presumably neutral, if they were to be fixed at each "tick" of the molecular clock, independently from the selective pressures that each phylad is bound to bear. And secondly, that molecular metrics may be used as reasonably accurate timekeepers, whenever a clade can be calibrated at a single point of dichotomy.

1 Genetics Section, Instituto Nacional do Cancer and Department of Genetics, Universidade Federal do Rio de Janeiro, Brazil

Cytogenetics. Ed. by G. Obe and A. Basler

The remarkable paradox between macrostructural and molecular evolutionary rates challenged the reasonable expectation that change at the entire, or organismic, level was the simple consequence of variation at the particulate elements of the biological universe. This led to the postulation that regulatory, but not structural, genes must account for drastic organismal disparities independently from the degree of biochemical identity that any two species might share (King and Wilson 1975). This situation is especially relevant to chromosomes because they are components of intermediate complexity between macromolecules, on the one side, and whole cells or tissues, on the other. Moreover, chromosomes have *both* morphological and genetic attributes. They are organelles with gross structural characteristics such as size, arm ratio and presence or absence of secondary constriction regions, and with more subtle macromolecular particularities that account for their banding and DNA replication patterns and for their highly specific, sometimes regional, affinity to dyes. Moreover, they represent separate genomic enclaves of linkage associations with crossover exchanges normally limited to homologues within each chromosome pair. But both their morphological and genetic characteristics are bound to bear the pressure of biological instability; either by selective or random events the mammalian chromosome complement has been extensively shuffled. Whether such rearrangements preceeded or succeeded the budding of the profuse, yet intricate, offshoots of the evolutionary tree is still an open question. However, once an emerging group has acquired a peculiar chromosome constitution, its reproductive isolation becomes more readily guaranteed with respect to other karyotypically distinct groups. Since normal meiosis requires homologue recognition, pairing, crossover between homologue chromatids and chromosome segregation, karyodimorphic hybrids are generally precluded from producing balanced gametes.

The study of human chromosome phylogeny can be based on different approaches. One is the simple comparison of the morphological attributes of chromosomes, especially of their banding patterns with different dyes. This has allowed for the recognition of presumed interspecific homologues, or *homoeologous,* chromosomes when the chromosome complement of man is compared to those of other mammalian species. Tentative chromosome phylogenies have been proposed within the primates (de Grouchy et al. 1978; Dutrillaux 1979a; Seuánez 1979, 1984); these studies show that, with the exception of a few groups such as gibbons or the night monkeys, the primate order is chromosomally conserved as to allow for the recognition of tentative homoeologies from prosimians to man. A strikingly different situation occurs in the rodent order where karyotypic shuffling has been so intense as to preclude the recognition of human chromosome homoeologies by simple comparisons of banded karyotypes. Yet other methods, such as comparative gene assignment has allowed for the identification of homoeologous syntenic associations among different orders (Nash and O'Brien 1982; O'Brien et al. 1985), thus resulting in a more comprehensive overview of the genome evolution in mammals. In this paper, I will comment on how these findings have helped us to understand chromosome evolution, and how some logical assumptions might even be challenged by the recently emerging evidence.

2 Humans and the Great Apes are very Similar

Humans and great apes are biochemically so similar as sibling species of the same genera (Bruce and Ayala 1978). The average protein of the chimpanzee, for example, has been found to share 99% of the amino acid homology with that of man (King and Wilson 1975). This extraordinary resemblance has been more recently confirmed by gel "maps" using double dimension electrophoresis (Goldman et al. 1987). Immunological distances between these species prompted Goodman (1975) to include the great apes and man into a single family (Hominidae), comprising two different subfamilies: Ponginae (with one genus, *Pongo*) and Homininae (with three genera: *Pan, Gorilla* and *Homo*). Nuclear DNA studies (Benveniste and Todaro 1976; Benveniste 1985) have also demonstrated the extraordinary, whole-genomic, similarity between man and our closest living relatives. This finding has been further confirmed by detailed analyses of (1) specific gene families, such as ribosomal DNA (Arnheim et al. 1980; Wilson et al. 1984); (2) isolated gene sequences, such as globins (Barrie et al. 1981); and (3) repeated DNA of retorviral origin (Benveniste 1985). Furthermore, the analysis of extranuclear DNA in human and great ape mitochondria supplies conclusive evidence of the close similarity among hominoid primates (Ferris et al. 1981a,b; Brown et al. 1982).

Not surprisingly, humans and great apes are chromosomally very similar (Fig. 1) despite the fact that the diploid number is 46 in our species and 48 in the chimpanzee, the pygmy chimpanzee, the gorilla and the orangutan. Morphological comparisons allow the recognition of presumed homoeologues between species so that each human chromosome has a recognizable counterpart (or arm counterpart in the case of the human chromosome 2) in the other species. Since the karyotypes of these species are similar but not identical, it follows that we may *derive* them from one another by presumed chromosome rearrangements. Comparative studies have shown that the most common type of presumed rearrangement within this group is the pericentric inversion (Turleau et al. 1972; Dutrillaux 1979a; Seuánez 1979, 1984; Stanyon and Chiarelli 1982; Yunis and Prakash 1982). However, other types of less common rearrangements have also been postulated, such as a fusion of non-homologous acrocentrics (or subtelocentrics). Such an event must have occurred in the homozygous condition and only in the human lineage, after its divergence from the hominoid common stock, originating human chromosome 2. It must be stated, however, that the great majority of the rearrangements herewith postulated must have resulted in a considerable shuffling of our common ancestor's chromosome complement, producing significant relocations of euchromatic material. Consequently, it might be questioned whether their occurrence was compatible with normal fitness in the individuals where they presumably appeared because, in man, most rearrangements involving significant amounts of euchromatic material are deleterious, resulting in clear pathological conditions. Interestingly, pericentric inversions involving considerably large euchromatic regions have been found to be present at polymorphic frequencies in the two subspecies of orangutan (Seuánez et al. 1976), while another smaller inversion has contributed to the subspecific divergence between the Bornean and the Sumatran variety of this species (Seuánez et al. 1979). These findings are good evidence that despite karyotypic similarities between two species (such as man and the orangutan) their chromosome complement is not identical with respect to how much shuffling they are

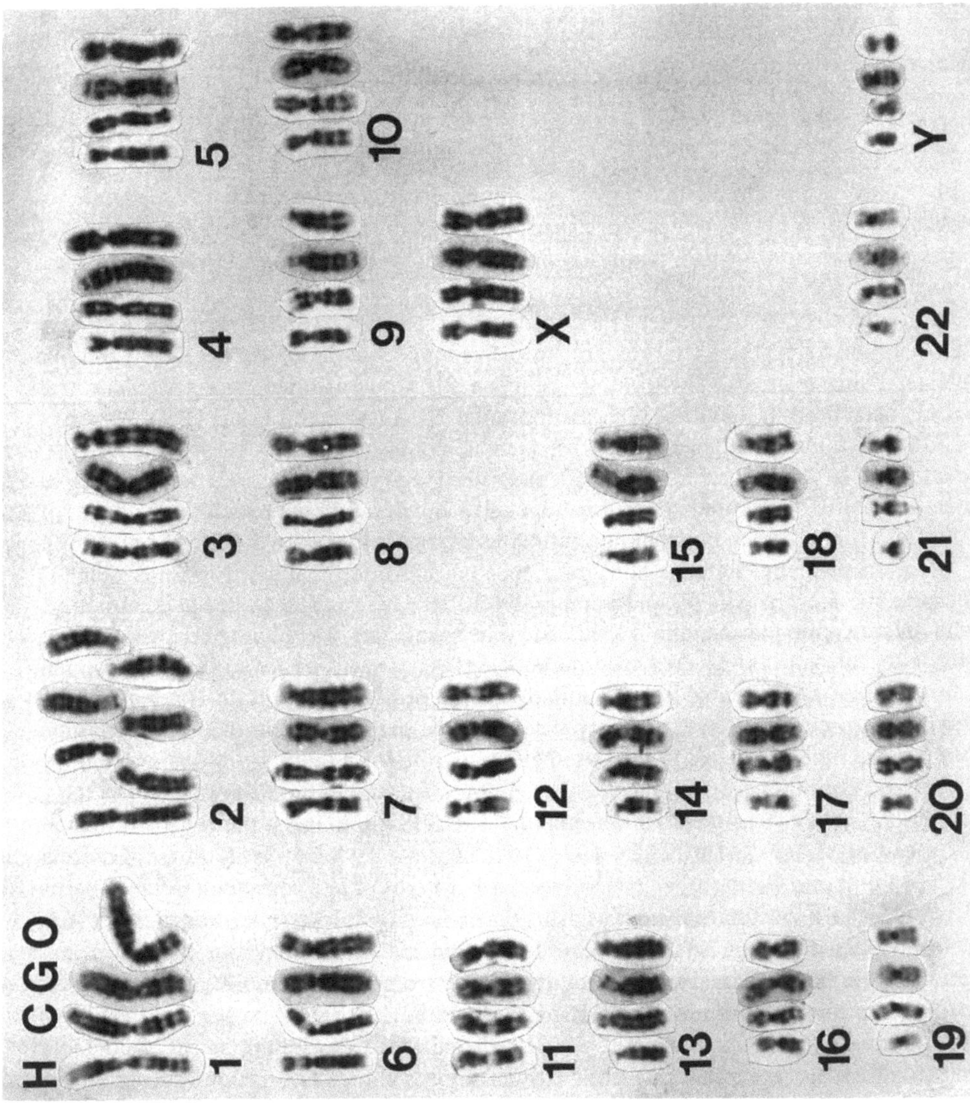

Fig. 1. G-band pattern of the chromosomes of man (*H*), chimpanzee (*C*), gorilla (*G*) and orangutan (*O*). The great ape chromosomes have been matched to their human homoeologues according to the criterion of the Stockholm Conference (1977). See Table 1 for chromosome nomenclature and syntenic homoeologies between species

Table 1. Chromosomal positions of homologous loci in man and the great apes[a] (O'Brien et al. 1985)

Human chromosome	Marker	Chimpanzee chromosome	Gorilla chromosome	Orangutan chromosome
1p	*PGM1*	1	1	1q
1p	*PGD*	1	1	1q
1p	*ENO1*	1	1	1q
1p	*AK2*	--	--	1q
1p	*FUCA*	--	--	--
1q	*PEPC*	1	1	1
1q	*FH*	--	1	1p
2q	*IDH1*	12 (13)	12 (11)	12 (11)
2p	*MDH1*	13 (12)	11 (12)	11 (12)
2p	*ACP1*	13 (12)	--	11 (12)
3	*GPX1*	2	2	(2)
4p	*PGM2*	3	3	3
5	*HEXB*	(4)	(4)	4
6p	*GLO*	5	--	--
6p	*MHC*	5	5	5
6q	*PGM3*	5	5	5
6q	*SOD2*	5	5	5
6q	*ME1*	5	5	5
7q	*GUSB*	6	6	(10)
8p	*GSR*	7	7	6
9q	*AK1*	11	13	(13)
9p	*ACO1*	11	13	--
9p	*AK3*	11	--	--
10p	*GOT1*	8	8	7
11p	*LDHA*	9	9	8
11p	*ACP2*	9	--	--
12p	*GAPD*	10	10	9
12p	*TPI1*	10	10	9
12p	*LDHB*	10	10	9
12q	*PEPB*	10	10	9
13q	*ESD*	(14)	14	(14)
14q	*NP*	15	18	(15)

[a] The great ape cytological homoeologue (Stockholm Conference 1977) and the syntenic homoeologues are the same in every case except human chromosomes, 2p, 2q and 20. In these cases and in those cases when no homoeologues have been mapped, the cytological homoeologue is in parentheses.

Table 1 (continued)

Human chromosome	Marker	Chimpanzee chromosome	Gorilla chromosome	Orangutan chromosome
15q	*MPI*	16	--	(16)
15q	*PKM2*	16	15	--
15q	*HEXA*	16	15	--
17q	*TK*	19	19	19
17q	*GALK*	19	19	--
18q	*PEPA*	17	16	(17)
19	*GPI*	20	20	20
20p	*ITPA*	15 (21)	18 (21)	(21)
21q	*SOD1*	22	22	22
X	*G6PD*	X	X	X
X	*HPRT*	X	--	--
X	*GLA*	X	X	X
X	*PGK*	X	X	--

capable of bearing in compatibility with a fitness threshold. Thus, it might well be possible that our ancestral chromosome complement was more plastic and, therefore, more tolerant to such rearrangements than the chromosome complement of modern man.

Chromosome similarities between man and the great apes, at the morphological level, are generally confirmed by comparative gene mapping (Table 2). This concordance is only valid for the syntenic homoeologies of *structural* genes, such as enzymes (for example, see O'Brien 1984), but not for moderately repeated sequences, like ribosomal DNA (Henderson et al. 1977), or for highly repetitive sequences, like human-homologous satellite DNAs (Gosden et al. 1977; Mitchell et al. 1977; Seuánez 1979). This is because the evolution of these reiterated DNA sequences seems to be independent of chromosome phylogeny in the primates, regardless of the morphological similarities that might persist between species. Moreover, the good agreement between morphological and syntenic homoeologies is not complete. The arm homoeologues of human chromosome 2 in the apes are not coincident; human 2p contains loci that are located in the 2q-homoeologue in the other species and vice versa (Sun et al. 1978a,b; Table 1). Moreover, the human 20 homoeologue in the great apes does not contain the inosine-triphosphatase locus (ITPA), in spite of the obvious morphological similarity between this human chromosome and its great ape counterparts (Fig. 1). In fact, ITPA in the great apes has been found to be syntenic with the nucleoside phosphorylase (NP) locus that in man is assigned to chromosome 14. Thus, a punctual gene transposition, yet unnoticed at the cytological level, has been responsible for the emergence of a new syntenic association in the chimpanzee and the gorilla, similar to that of human 14 + 20. While in the chimpanzee these loci are assigned to chromosome 15 (the morphological homoeologue of chromosome 14 in

Table 2. Comparative gene mapping: man, Old World and New World monkeys, and the mouse lemur (O'Brien et al. 1985)

Human chromosome	Human loci	Gibbon chromosome	Rhesus chromosome	Baboon chromosome	African green monkey chromosome	Capuchin monkey chromosome	Owl monkey (*Aotus* K-VI) chromosome	Mouse lemur chromosome
1p	*PGM1*	5	1	1	1	15	12	3
1p	*PGD*	24	1	1	1	15	12	3
1p	*ENO1*	24	1	1	1	15	12	3
1p	*AK2*	--	--	--	--	--	12	--
1p	*FUCA*	3	--	--	1	--	--	3
1q	*PEPC*	--	--	1	6	--	--	U4
1q	*FH*	--	1	--	6	U2	6	--
1q	*GUKI*	5	1	--	--	U2	--	U5
2p	*MDH1*	--	15	--	--	4	2	4
2p	*ACP1*	19	--	--	--	4	--	4
2q	*IDH1*	--	9	12	--	14	16	--
2	*UGP2*	--	--	--	--	4	2	4
3	*GPX1*	4	3	--	5	--	--	1
3p	*ACY1*	--	--	--	--	18	--	--
4p	*PEPS*	--	--	5	--	--	--	--
4p	*PGM2*	--	6	5	--	2	14	--
5	*HEXB*	--	5	--	--	--	--	--
6p	*GLO*	--	--	4	--	--	9	6
6p	*HLA*	--	2	--	--	--	9	--
6q	*PGM3*	17	2	--	--	3	9	6
6q	*SOD2*	3	2	4	--	--	9	--
6q	*MEI*	--	--	4	--	3	9	6
7q	*GUSB*	--	2	3	--	1	--	--
7	*MDH2*	--	2	--	--	--	4	--
8p	*GSR*	--	8	--	10	--	--	--
9q	*AK1*	8	--	--	U1	--	--	10
9p	*ACO1*	8	--	--	U1	--	15	--
9p	*AK3*	--	--	--	--	12	--	10
10q	*GOT1*	3	--	--	--	--	--	15
11p	*LDHA*	15	11	14	12	16	19	5
11p	*ACP2*	15	11	--	--	16	--	--
12p	*GAPD*	--	12	--	13	--	--	7
12p	*TPI1*	U1	12	--	13	10	10	7
12p	*LDHB*	U1	12	11	13	10	10	7

U = unassigned.

Table 2 (continued)

Human chromosome	Human loci	Gibbon chromosome	Rhesus chromosome	Baboon chromosome	African green monkey chromosome	Capuchin monkey chromosome	Owl monkey (*Aotus* K-VI) chromosome	Mouse lemur chromosome
12q	*PEPB*	--	12	--	13	10	--	7
12	*CS*	--	12	--	13	--	--	7
12	*ENO2*	U1	--	--	--	--	--	7
14q	*NP*	U2	7	7	U2	--	11	2
14q	*CKBB*	U2	--	7	U3	--	--	2
15q	*MPI*	6	7	7	U2	U1	11	2
15q	*PKM2*	6	7	7	U2	U1	11	2
15q	*HEXA*	6	7	--	--	--	--	2
15q	*IDH2*	--	--	7	--	--	--	--
15q	*SORD*	6	--	7	--	--	11	2
15q	*B2M*	--	--	--	--	--	11	--
16p	*PGP*	--	--	--	--	1	--	--
19	*GPI*	--	19	--	25	8	25	U3
19	*PEPD*	--	--	--	--	--	--	U3
20q	*ADA*	--	--	10	--	--	--	--
20p	*ITPA*	--	13	10	--	--	--	--
21q	*SOD1*	--	--	3	--	9	--	--
22	*NAGA*	--	13	--	--	--	--	--
X	*G6PD*	--	X	--	X	--	X	X
X	*HPRT*	X	--	--	--	--	--	X
X	*GLA*	X	X	--	X	--	X	X
X	*PGK*	X	--	--	X	--	X	X

man), they are located in chromosome 18 in the gorilla (the morphological similarities between human chromosome 14 and gorilla 18 are far less evident; see Fig. 1). Furthermore, a presumptive reciprocal translocation has been proposed by analyzing high resolution banded chromosomes, suggesting that the gorilla has two chromosome pairs, one similar to human 5q–17q and another to human 5p–17p (Yunis and Prakash 1982). Though the pattern of chromosome banding might give partial support to this presumptive rearrangement, comparative gene assignment of the thymidine kinase locus (TK) has ruled out the possibility that such rearrangement has ever occurred (see Seuánez 1984).

The identification of presumptive chromosome homoeologies by morphological comparisons within the hominoids is, however, limited to man and the great apes. The lesser apes (gibbons and siamangs) comprise a karyologically shuffled group within

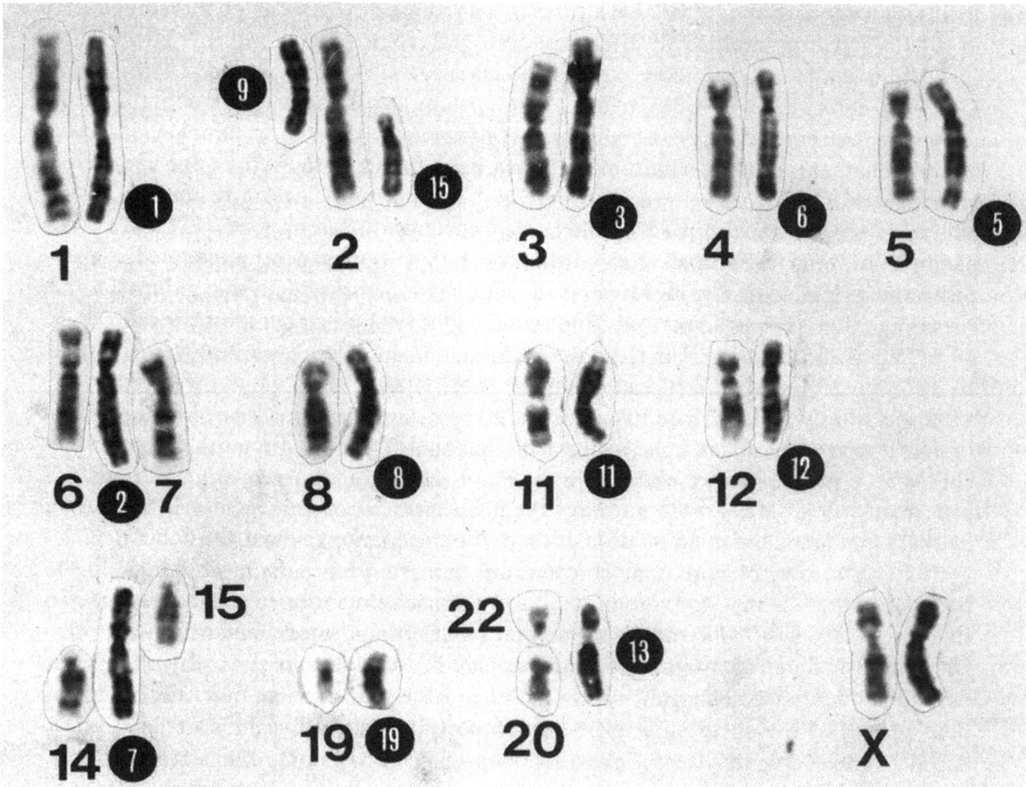

Fig. 2. G-band pattern of human and rhesus chromosomes showing syntenic homoeologies (see Table 2). Human chromosomes are indicated by black numbers; rhesus chromosomes by white numbers inside black circles. Since rhesus and baboon share thè same karyotype this figure should also correspond to human-baboon chromosome homoeologies. [The human 6 + 7 linkage association, however, has been questioned in the baboon (see Creau-Goldberg et al. 1983)]

few karyotypic similarities are evident (Couturier et al. 1982), even between species that might accidentally crossbreed (Myers and Shafer 1979). Although some presumptive homoeologies between man and *Hylobates concolor* were initially proposed (Dutrillaux 1979a), practically none of them were confirmed by comparative gene assignment, showing that even some evolutionary, conserved syntenic associations (such as those of human 1p and 6q) were disrupted in this species (Turleau et al. 1983; see Table 2).

3 Chromosome Phylogenies can be Traced in Old World and New World Monkeys as well as in Prosimians

Contrary to what is observed between the lesser apes and the larger hominoids, chromosome homoeologies, both morphological and syntenic, arè evident between man and some Old World monkeys (rhesus, baboon and African green monkey; Finaz et al. 1977; Estop et al. 1979; Fig. 2) as well with some New World monkey species (capuchins and wooly monkey; see Dutrillaux 1979b; Dutrillaux et al. 1980; Creau-Goldberg et al. 1981, 1982). In fact, chromosome evolution in the cercopithecoids shows an extreme example of absolute conservation; the rhesus monkey (*Macaca mulatta*) and the baboon (*Papio papio*) have been found to share the same karyotype (Finaz et al. 1978) in spite of the fact that the morphological attributes of these allopatric species have allowed for their inclusion into different genera. This is a clear example of how organismal and chromosome evolution can be uncoupled. Cercopithecoid primates and man, however, do not share identical syntenic groups; in the rhesus monkey, for example, we find associations of human syntenic groups 6 + 7, 14 + 15 and 20 + 22. Surprisingly, in the baboon, human 6 and 7 markers were assigned to different chromosomes (Creau-Goldberg et al. 1982, 1983), while the human 14 + 15 association was confirmed. The human 20 + 22 association observed in the rhesus monkey has not yet been confirmed in the baboon (Table 2). In another cercopithecoid, the African green monkey, as in the Platyrrhine species *Cebus capucinus* and the prosimian species *Microcebus murinus*, there is a dissociation for the human 1p and 1q markers that are maintained in association in the rhesus monkey and the baboon.

It is, therefore, obvious that chromosome evolution has been more drastic in the lesser apes than in the larger hominoids and the above mentioned cercopithecoids, so that these latter are still conserving several relict chromosome similarities with man. On the principle of parsimony, that explains an evolutionary pathway by the minimum number of changing events, it is logical to assume that these relict similarities, proper of the ancestral primate karyotype, drastically vanished in the lesser apes. An alternative, yet less logical explanation, would be to assume that extensive chromosome rearrangements, similar to those observed between Old World monkeys and lesser apes, occurred before hominoid radiation and that a second set of rearrangements, in the human-great ape common stock, partially restored their relict chromosome homoeologies with the Old World monkeys. This second pathway would not only disregard the principle of parsimony, but it would overlook additional findings of particular relevance. One is the fact that at the morphological level, there is an evident karyotypic similarity between man and the Platyrrhine species *Cebus capucinus* (Dutrillaux 1979b) and also with the prosimian *Microcebus murinus* (Dutrillaux 1979a), a primate species phylogenetically further apart from man than Old and New World monkeys. Since several human *Cebus capucinus–Microcebus murinus* homoeologies have been confirmed by comparative gene assignment (Creau-Goldberg et al. 1981; Cochet et al. 1982) these findings point to a remarkable chromosome conservation in the primate order with the exception of gibbons and owl monkeys (*Aotus*, Platyrrhini), the karyotypes of which have suffered extensive shuffling as shown by both the morphological study of chromosomes and gene assignment. In *Aotus trivirgatus*, a platyrrhine group comprising some 11 distinct karyomorphic populations (see Ma et al. 1976; Ma 1984), syntenic associations of human 1p + 19, 2p + 12, 11 + 12 groups have been observed, as well as dissociation of some human 12q loci that are kept together in the primates and other mammals (see Table 3). It is reasonable to propose that these shuffled karyotypes are more recent, and that numerous rearrangements probably occurred after this group branched off from the Platyrrhine stock.

Table 3. Linkage and syntenic relationship in chromosomes of owl monkeys (with karyotypes I, III, V, VI, VII) homologous to man[a] (O'Brien 1984)

	Human	*Aotus* V (46)	*Aotus* III (53)	*Aotus* I (54)	*Aotus* VI (49/50)	*Aotus* VII (51/50)
ENO1	1	--	10	12	12	12
PGD	1	2	--	12	12	--
AK2	1	2	10	12	12	--
PGM1	1	2	--	12	12	12
FH	1	8	4	--	6	--
MDH1	2	1	14	2	2	2
UGP2	2	--	14	2	2	--
IDH1	2	15	16	--	16	--
PGM2	4	--	--	--	14	--
HLA	6	10	7	--	9	9
GLO1	6	10	7	--	9	--
MEI	6	10	7	11	9	9
SOD2	6	10	7	--	9	--
PGM3	6	--	7	--	9	--
MDH2	7	5	3	4	4	4
ACON1	9	--	--	--	15	--
LDHA	11	4	2	15	19	19
LDHB	12	1	13	2	10	10
TPI	12	--	2	--	10	--
NP	14	11	12	16	11	11
B2M	15	11	12	--	11	--
MPI	15	11	12	16	11	--
PKM2	15	11	12	16	11	11
SORD	15	11	12	16	11	11
GPI	19	2	--	9	25	--
PGK	X	X	--	--	X	--
GLA	X	--	--	--	X	--
G6PD	X	--	--	--	X	--

[a] The numbers refer to chromosomes to which the loci are assigned. *Aotus* chromosome nomenclature follows Ma et al. [Lab. Anim. Sci. 26:1022 (1976)].

4 Human Syntenic Associations can be Traced Outside the Primate Order Within the Mammalian Radiation

The availability of different techniques for gene assignment, especially those based on the analysis of somatic cell hybrids (O'Brien et al. 1985) allow for the comparison of the human gene chart with that of other mammals (see Table 4 and 5). It is, therefore, evident that many human syntenic associations have been maintained in the mammalian stock for long periods of phyletic radiation, as shown by the remarkable human-cat homoeologies that have been maintained during the last 80 million years (O'Brien and Nash 1982; Berman et al. 1986). In the domestic cat, however, some human-homologous syntenic associations are evident, such as 1p–2q, 2p–20p–20q, 14q–15q and 8p–21q, whereas two human chromosome 10 markers are dissociated (see Fig. 3).

Table 4. Homologous gene assignments in mouse and man (O'Brien et al. 1985)

Human chromosome	Human locus	Mouse locus	Mouse chromosome
1p	*PGD*	*Pgd*	4
1p	*GDH*	*Gpd-1*	4
1p	*ENO1*	*Eno-1*	4
1p	*FUCA1*	*Fuca*	4
1p	*AK2*	*Ak-2*	4
1p	*PGM1*	*Pgm-2*	4
1p	*AMY1*	*Amy-1*	3
1p	*AMY2*	*Amy-2*	3
1p	*NGF*	*Ngf*	3
1q	*PEPC*	*Pep-3*	1
2	*ACP1*	*Acp-1*	12
2p	*IGK*	*Igk*	6
2q	*IDH1*	*Idh-1*	1
3p	*ACY1*	*Acy*	9
3p	*GLB1*	*Bgl*	9
3q	*SST*	*Sst*	16
4	*PGM2*	*Pgm-1*	5
4	*PEPS*	*Pep-7*	5
4q	*ALB*	*Alb-1*	5
4q	*AFP*	*Afp*	5
6p	*HLA*	*H-2*	17
6p	*GLO1*	*Glo-1*	17
6p	*BF*	*Bf*	17
6q	*SOD2*	*Sod-2*	17
6q	*PGM3*	*Pgm-3*	9
6q	*ME1*	*Mod-1*	9
6q	*CGA*	*Tsha*	4
6q	*MYB*	*Myb*	10
7p	*BLVR*	*Blvr*	2
7	*GUSB*	*Gus*	5
7	*MDH2*	*Mor-1*	5
7	*ASL*	*Asl*	5
7	*PSP*	*Psp*	5
7q	*TRV1*	*Trv-1*	6
7q	*CPA*	*Cpa*	6
8p	*GSR*	*Gr-1*	8
8q	*MOS*	*Mos*	4
8q	*MYC*	*Myc*	15
9p	*ACO1*	*Aco-1*	4
9p	*GALT*	*Galt*	4
9q	*AK1*	*Ak-1*	2
9q	*ABL*	*Abl*	2
10	*PP*	*Pyp*	10
10q	*HK1*	*Hk-1*	10
10	*LIPA*	*Lipa*	19
10q	*GOT1*	*Got1*	19
10q	*ADK*	*Adk*	14
11p	*INS*	*Ins-1*	7
11p	*HBB*	*Hbb*	7
11p	*LDHA*	*Ldh-1*	7
11p	*HRAS1*	*Hras1*	7
11p	*ACP2*	*Acp-2*	2
11p	*CAT*	*Cs-1*	2
11q	*APOA1*	*Alp-1*	9
11q	*UPS*	*Ups*	9
11q	*ESA4*	*Es-17*	9
12p	*GAPD*	*Gapd*	6
12p	*TPI1*	*Tpi-1*	6
12p	*KRAS2*	*Kras-2*	6
12p	*LDHB*	*Ldh-2*	6
12q	*CS*	*Cs*	10
12q	*PEPB*	*Pep-2*	10
12q	*IFG*	*Ifg*	10
12	*ELA1*	*Ela-1*	15
12	*INT-1*	*Int-1*	15
13	*ESD*	*Es-10*	14
14q	*NP*	*Np*	14
14q	*IGH*	*Igh*	12
14q	*PI*	*Pre-1*	12
15	*SORD*	*Sdh-1*	2
15q	*B2M*	*B2m*	2
15q	*MPI*	*Mpi*	9
15q	*PKM2*	*Pk-3*	9
15q	*IDH2*	*Idh-2*	7
15q	*FES*	*Fes*	7
16p	*HBA*	*Hba*	11
16	*GOT2*	*Got-2*	8
16	*CTRB*	*Ctrb*	8
16q	*APRT*	*Aprt*	8
17p	*MYH-1*	*Myh*	11
17p	*MYH-2*	*Myh*	11
17p	*MYH-3*	*Myh*	11
17q	*TK1*	*Tk-1*	11

Table 4 (continued)

Human chromosome	Human locus	Mouse locus	Mouse chromosome	Human chromosome	Human locus	Mouse locus	Mouse chromosome
17q	*GALK*	*Glk*	11	22q	*ARSA*	*Arsa*	15
17	*ERBAI*	*Erba*	11	22q	*DIAI*	*Dia-1*	15
				22q	*SIS*	*Sis*	15
18q	*PEPA*	*Pep-1*	18	22	*IGLC*	*Igl*	16
				Xq	*G6PD*	*G6pd*	X
19	*GPI*	*Gpi-1*	7	Xq	*HPRT*	*Hprt*	X
19	*PEPD*	*Pep-4*	7	Xq	*GLA*	*Ags*	X
19	*LHB*	*Lhb*	7	Xq	*PGK*	*Pgk-1*	X
				X	*TFM*	*Tfm*	X
20p	*ITPA*	*Itp*	2	X	*PYK*	*Phk*	X
20q	*ADA*	*Ada*	2	Xp	*MDD*	*Mdx*	X
20	*SRC*	*Src*	2	Xp	*STS*	*Sts*	X
				X	*OTC*	*Spf*	X
21q	*SODI*	*Sod-1*	16	X	*HPDR*	*Hyp*	X
21q	*IFRC*	*Ifrc*	16				
21	*PRGS*	*Prgs*	16	Y	*HYA*	*H-Y*	Y

Table 5. Homologous gene assignments in mammalian species with preliminary gene maps (O'Brien et al. 1985; Berman et al. 1986)

Human chromosome	Human locus	Mouse locus	Cat (*Felis catus*)	Rabbit (*Oryctolagus cuniculus*)	Dog (*Canis familiaris*)	Cattle (*Bos taurus*)	Sheep (*Ovis ovis*)
1p	*PGD*	*Pgd*	C1	--	U1	U1	U1
1p	*ENO1*	*Eno-1*	--	U2	U1	U1	U1
1p	*PGM1*	*Pgm-2*	C1	U3	U2	U6	U1
1q	*PEPC*	*Pep-3*	--	--	U4	U13	--
1q	*GUK1*	*Guk-1*	--	15	U3	U19	--
2p	*ACP1*	*Acp-1*	A3	--	U5	--	--
2p	*MDH1*	*Mor-2*	A3	U5	U6	--	--
2q	*IDH1*	*Idh-1*	C1	--	U7	U17	--
3p	*ACY1*	*Acy*	--	9	U8	U12	--
3	*GPX1*	*Gpx*	--	9	U8	--	--
4	*PGM2*	*Pgm-1*	--	--	U9	U15	--
4	*IL2*	*Il-2*	B1	--	--	--	--
4	*PEPS*	*Pep-7*	B1	--	--	--	--
6p	*HLA*	*H-2*	--	--	--	--	U6
6p	*GLO1*	*Glo-1*	B2	--	--	--	--
6q	*SOD2*	*Sod-2*	B2	--	U10	U2	U8
6q	*PGM3*	*Pgm-3*	B2	--	U10	U2	--

Table 5 (continued)

Human chromosome	Human locus	Mouse locus	Cat (*Felis catus*)	Rabbit (*Oryctolagus cuniculus*)	Dog (*Canis familiaris*)	Cattle (*Bos taurus*)	Sheep (*Ovis ovis*)
6q	*MEI*	*Mod-1*	B2	--	U10	U2	--
7	*GUSB*	*Gus*	E3	--	--	--	--
7	*MDH2*	*Mor-1*	--	15	U11	U8	--
8p	*GSR*	*Gr-1*	C2	19	--	U14	--
9p	*AK3*	--	--	--	U12	--	--
9q	*AK1*	*Ak-1*	U5	--	--	U16	--
10	*PP*	*Pyp*	D4	--	U13	--	--
10q	*HK1*	*Hk-1*	D2	--	--	--	--
11p	*LDHA*	*Ldh-1*	D1	1	U14	U7	--
11p	*ACP2*	*Acp-2*	D1	1	U15	--	--
11p	*CAT*	*Cs-1*	--	--	--	U20	--
11q	*ESA4*	*Es-17*	--	--	U16	--	--
12p	*GAPD*	*Gapd*	B4	4	U17	U3	--
12p	*TPI1*	*Tpi-1*	B4	4	U17	U3	U2
12p	*LDHB*	*Ldh-2*	B4	4	U17	U3	U2
12q	*PEPB*	*Pep-2*	B4	17	--	U3	U2
13	*ESD*	*Es-10*	A1	--	U18	--	--
14q	*NP*	*Np*	B3	17	U19	U5	U3
15q	*HEXA*	--	B3	--	--	--	--
15q	*MPI*	*Mpi-1*	B3	--	U20	U4	U9
15q	*PKM2*	*Pk-3*	B3	--	U20	U5	U4
18q	*PEPA*	*Pep-1*	U4	--	--	--	--
19	*GPI*	*Gpi-1*	U2	U6	U22	U9	--
20p	*ITPA*	*Itp*	A3	17	U23	U11	--
20q	*ADA*	*Ada*	A3	--	--	U11	--
20	*SRC*	*Src*	--	--	--	--	--
21q	*SODI*	*Sod-1*	C2	--	U24	U10	U10
21q	*IFRC*	*Ifrc*	--	--	--	U10	--
21	*PRGS*	*Prgs*	--	--	--	--	U10
21	*PAIS*	--	--	--	--	--	U10
22	*ACO2*	--	--	--	U25	--	--
22	*IDUA*	--	D4	--	--	--	--
Xq	*G6Pd*	*G6pd*	X	X	X	X	X
Xq	*HPRT*	*Hprt*	X	--	X	X	X
Xq	*GLA*	*Ags*	X	--	X	X	X
Xq	*PGK*	*Pgk-1*	--	--	--	X	X

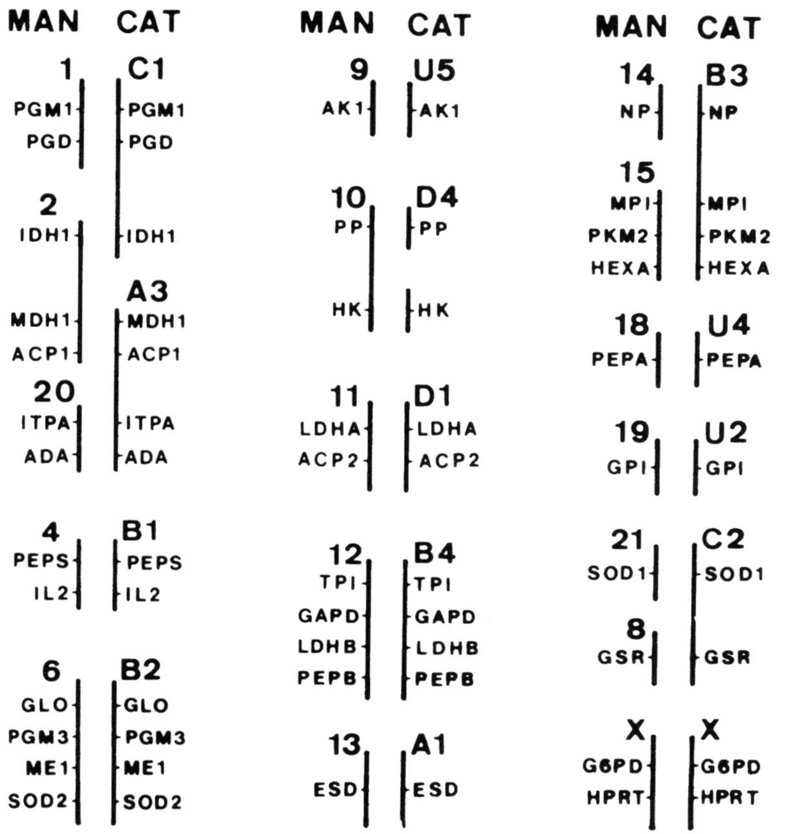

Fig. 3. Linkage group associations in man and the domestic cat (see Berman et al. 1986). Human loci are ordered according to the human gene chart

By comparing the distribution of homologous loci in man and the domestic cat these authors concluded that such a similarity could have occurred at a random probability of 1.4×10^{-10}. Moreover, this approach allowed for the identification of homoeologous chromosome regions between man and the domestic cat, comprising approximately 20% of our chromosome complement (Nash and O'Brien 1982).

A comparison between the primate-murine gene chart shows a remarkable conservation of several linkage associations although humans and mice are karyotypically very distinct. A minimum of 178 ± 39 chromosome rearrangements are estimated as having occurred between these species since their separation from the common mammalian stock (Nadeau and Taylor 1984). This explains why some nine murine linkage groups have been disrupted in man, and, conversely, why eight human groups have been disrupted in the murine genome (O'Brien and Nash 1982). However, seven of these eight human groups, in the mouse, have conserved their (human) arm-syntenic associations, as though breakage had occurred at the centromere region, while only human chromosomes 1, 11 and 15 appeared as having been split in the arms. A more detailed comparison (see O'Brien et al. 1985) shows that some murine chromosomes carry contiguously arranged clusters of human-homologous loci. For example, the distal end of murine chromosome 4 contains six loci that are homologous to those found in human 1p. Moreover, the proximal segment of the same murine chromosome contains a pair of contiguous loci homologous to those of human 9p.

A further comparison between the human and other mammalian species (see Table 5) shows that several linkage associations are evolutionarily conserved. Three human 6q markers, for example, are kept together in cat, dog and cattle. The human 12 markers are an even better example of evolutionary conservation in several mammalian species except for the rabbit. In this species, three human 12 markers have been found to be syntenic but dissociated from a fourth marker, peptidase-B (PEPB) that was found to be in association with NP and ITPA (Soulie and de Grouchy 1982). It is, therefore, clear that in the rabbit, there is an association of three loci, each of which is in a separate human chromosome (12, 14 and 20). Human chromosome 21 markers show a clear evolutionary conservation in the species listed in Table 5 as well as in the mouse. In this species, as in man, the Ifrc–Sod-1–Prgs cluster (interferon receptor-soluble superoxide dismutase-1-phosphorybosylglycinamide synthetase) is contained in one chromosome (number 16) that is almost empty of other genes (Polani and Adinolfi 1980). This murine chromosome contains genes that control stature, cerebellar size and motor coordination, and may contain a mutant gene involved in the integration of a murine leukemia virus in AKR/N mouse strains. Moreover, trisomic 16 mice have been reported with a 10% survival rate and a phenotype not too different from the normal (Miyapara and Gropp, pers. commun.). Since stature, motor coordination and cerebellar size are affected in Down's syndrome, the parallel between human 21 and murine 16 is remarkable (Polani and Adinolfi 1980). Interestingly, a similar clinical condition to human Down's syndrome has been observed in chimpanzee and orangutan trisomic for chromosome 22, the homoeologous counterpart for human 21 (McClure et al. 1969; Andrle et al. 1979).

Finally, the best example of evolutionary conservation is that of the mammalian X-chromosome that has been characterized by Ohno (1973) as an evolutionary, frozen accident. The gene cluster formed by the glucose-6-phosphate dehydrogenase (G6PD), phosphoglycerate-kinase (PGK), hypoxanthine-phospho-ribosyl-transferase (HPRT) and alfa-galactosidase (GLA) loci was found to be present in rabbit, sheep, cattle, pig, cat, hamster, mouse, rat, lemur, chimpanzee and man (O'Brien 1984). In other mammals, where only three of these loci have been mapped, they have always been assigned to the X-chromosome. Yet, this syntenic conservtion has not precluded internal rearrangements accounting for the extant differences at the morphological level between the X-chromosomes of different mammalian species. In the mouse, where the X-chromosome is acrocentric, the order of X-linked genes is not precisely identical with that of the human X, a submetacentric chromosome; while in the murine X the order is centromere–Hprt–Ags (Francke and Taggart 1980), the homologous human loci are differently aligned along the X-long arm (centromere–GLA–HPRT). But leaving these minor differences aside, the mammalian X-chromosome is strikingly conserved, even outside the eutherian stock, as it is evident in the kangaroo (a marsupial).

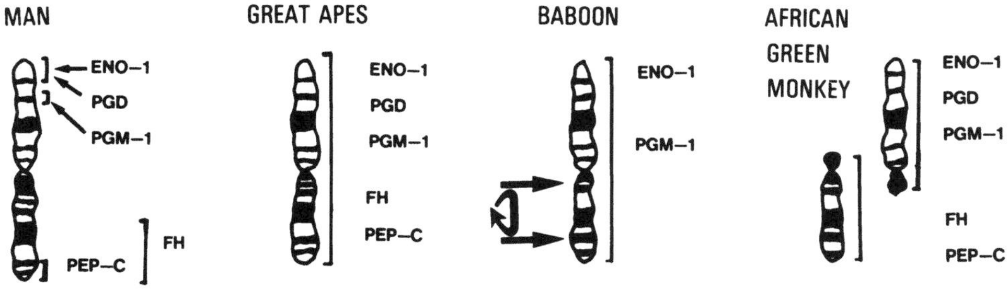

Fig. 4. Comparative gene assignment of chromosome 1 in man and its homoeologues in the great apes, rhesus-baboon and African green monkey (see Table 2 for more complete data). Ape and rhesus-baboon chromosomes are inverted. Straight arrows indicate presumptive breaks and the curved arrow indicates the paracentric inversion between cercopithecoids and hominoids (see Finaz et al. 1977; Seuánez 1979)

5 The Principle of Parsimony Challenged in Chromosome Evolution as Inferred from Comparative Gene Assignment

So far we have accepted the principle of parsimony as the most logical postulate for the understanding of chromosome change. We will now show two special cases where this principle is untenable and, therefore, questionable as a universal proposition. The first example will deal with the evolution of the human 1p–1q linkage association and the second with the NP–ITPA association in mammals.

As shown in Fig. 4, the human 1p–1q linkage association [enolase-1 (ENO1), phosphogluconate-dehydrogenase (PGD), phosphogluconate mutase (PGM1), alfa-fucosidase (FUCA) and peptidase-C (PEPC)] is maintained in the great apes and the rhesus-baboon but not in the African green monkey (Finaz et al. 1977; Garver et al. 1977). A morphological study of human chromosome 1 homoeologues in these species shows that human chromosome 1 is unique in having a proximal, secondary constriction in its long arm. The absence of such a constriction in the great ape-rhesus-baboon chromosome 1 changes the arm ratio, so that human 1p is homoeologous to 1q in these species, and vice versa. Moreover, the rhesus-baboon 1 can be derived by a simple fusion of the African green monkey arm-homoeologues, while a paracentric inversion is needed at the human 1q homoeologous arm to derive the great ape 1 from the rhesus-baboon 1. In view of these findings, Finaz et al. (1977) proposed three alternative pathways of chromosome evolution. The first hypothesis (Fig. 5) postulates that the 1p–1q dissociation was already existing in the ancestral catarrhine, similarly to that found in the African green monkey. This would require the occurrence of one paracentric inversion and a fusion in the hominoid branch and a second fusion (of the same type, involving the same chromosomes) in the rhesus-baboon lineage, within the cercopithecoids. Alternatively, if the ancestral catarrhine chromosome 1 was similar to that of the rhesus-baboon (Fig. 6) fusions are no longer required; only one paracentric inversion (in the hominoid lineage) and one fission (in the African green monkey lineage) would explain the origin of human chromosome 1. Finally, a third hypothesis postulates that the ancestral catarrhine chromosome was similar to that of the great apes (Fig. 7). This would require that all rearrangements (a paracentric inversion and a

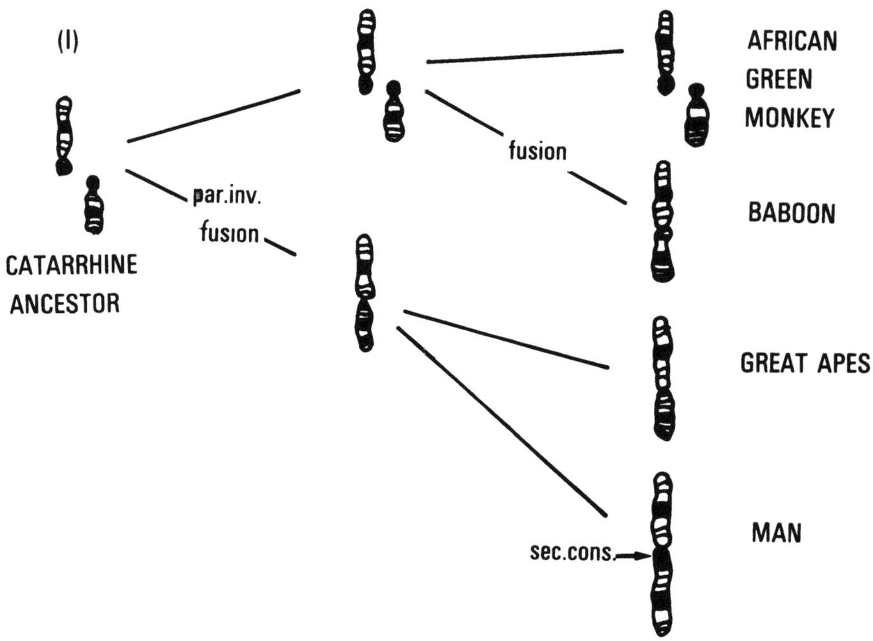

Fig. 5. First hypothesis: derivation of chromosome 1 from a common catarrhine ancestor with arm homoeologues similar to those in the African green monkey. sec. cons. = secondary constriction; par. inv. = paracentric inversion (Finaz et al. 1977; Seuánez 1979)

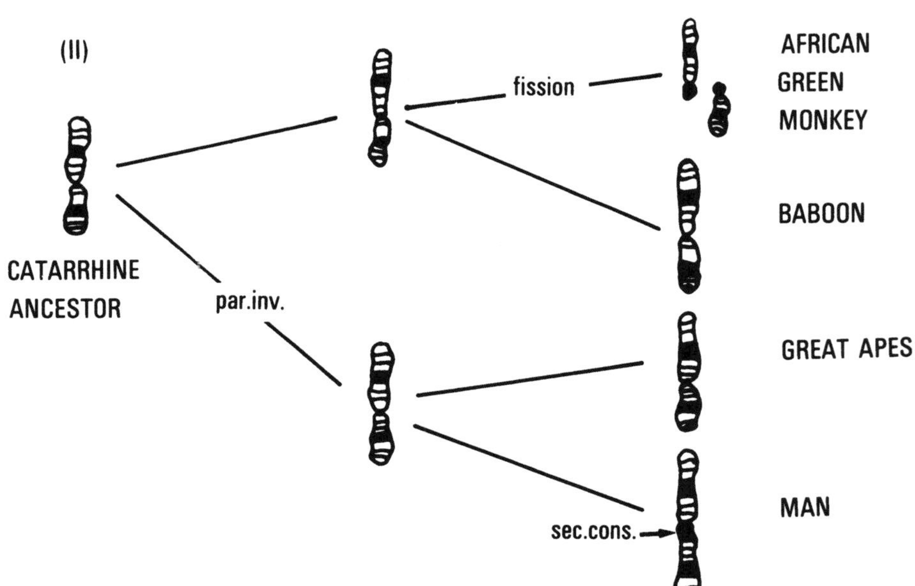

Fig. 6. Second hypothesis: derivation of chromosome 1 from a catarrhine ancestor with a homoeologous chromosome similar to that of the rehsus-baboon (Finaz et al. 1977; Seuánez 1979)

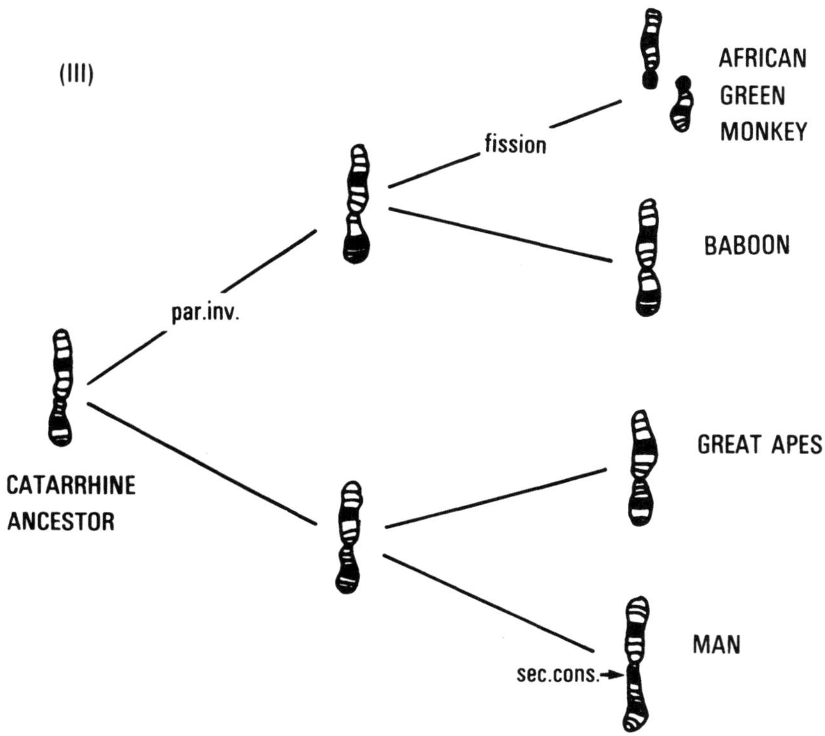

Fig. 7. Third hypothesis: derivation of chromosome 1 from a catarrhine ancestor with a homoeologous chromosome similar to that of the great apes (Finaz et al. 1977; Seuánez 1979)

fission) place within cercopithecoid radiation. It must be noted that the two latter hypotheses postulate *two* rearrangements, while the first hypothesis requires *three,* in which the same kind of fusion has to occur twice and independently in two lineages. It is, therefore, clear that the first hypothesis is less parsimonious than the other two.

Comparative gene assignment in *Cebus capucinus* and *Microcebus murinus* proves that human 1p and 1q markers are dissociated in these species (Creau-Goldberg et al. 1981; Cochet et al. 1982; see Table 2). Since the divergence of platyrrhines (like *Cebus capucinus*) from the common primate stock anteceded the emergence of cercopithecoids and hominoids, the dissociation of human 1p–1q markers must have been already present in the common catarrhine ancestor of Old World monkeys and hominoids. Moreover, the fact that such dissociation is still present in a prosimian (*Microcebus murinus*) strongly suggests that it might represent the ancestral situation in the primate order. Moreover, a study of homologous gene assignments in mammals shows that the human 1p–1q dissociation is present in mouse, dog and cattle, with additional 1p dissociations in dog, cattle and rabbit, 1q dissociations in dog and cattle, and an 1p–2q association in the domestic cat (O'Brien and Nash 1982). These data suggest that the human 1p–1q association is probably non-existent outside the primate order. However, if we wish to explain its emergence in the primates, we must conclude that the most likely hypothesis is *not* the most parsimonious one because

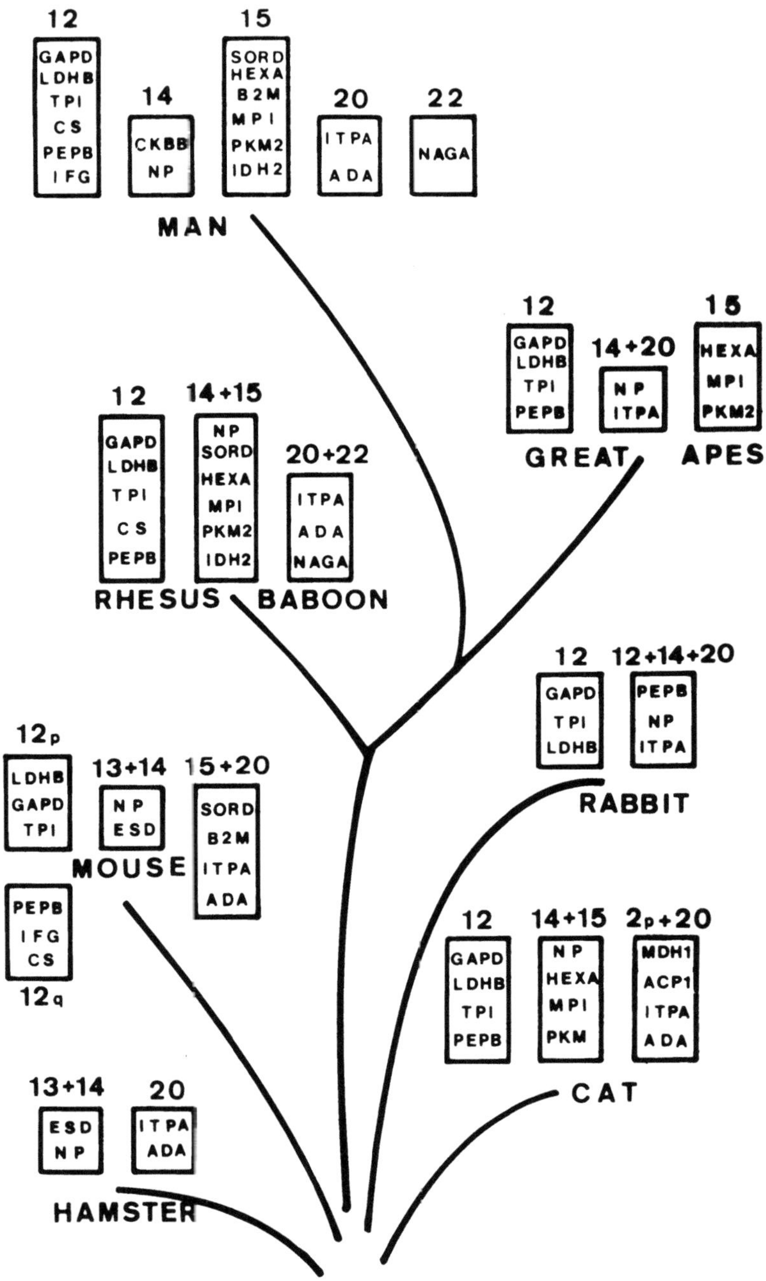

Fig. 8. Phylogeny of human syntenic associations 12, 14, 15, 20 and 22 showing the association of ITPA with ADA in mouse, hamster, cat, rhesus-baboon and man, and the association of ITPA and NP in the rabbit and the great apes. Numbers above boxes indicate human or human-homoeologous syntenic groups. Acronyms inside boxes identify gene markers (see O'Brien 1984)

it requires a minimum of three rearrangements, one of which occurs twice and independently.

The second case where the principle of parsimony does not seem to apply is the NP-ITPA (nucleoside phosphorylase/inosine triphosphatase) association in the great apes and the rabbit. In man, NP (a 14q marker) is dissociated from ITPA (a 20p marker) that is syntenic with ADA (adenosine deaminase, in human chromosome 20q). In the great apes, however, NP-ITPA are syntenic (Table 1) while in rhesus-baboon these loci are dissociated (Table 2 and Fig. 8). NP is syntenic with some human 15q markers in these two species, but not with ITPA [in rhesus, ITPA is syntenic with NAGA (n-acetyl-alfa-D-galactosaminidase), thus representing a human syntenic association 20 + 22; data on ADA are not yet available]. Moreover, in the baboon, ITPA and ADA are syntenic, as in man and other mammals like mouse, Chinese hamster, domestic cat and cattle, while NP is dissociated from ITPA in these species. Surprisingly, NP and ITPA are syntenic in the rabbit (there is no data for ADA). What would be the most parsimonious approach to this problem?

It seems as if the NP-ITPA association, though exceptional, was already present before the emergence of the primate order; the rule being NP/ITPA dissociation and ITPA-ADA association. The same rule appears to have been maintained in the cercopithecoids (rhesus and baboon) where NP and ITPA are dissociated from one another. Quite surprisingly, another NP-ITPA association is found again in the great apes, while NP-ITPA are again dissociated in man in which ITPA is *again* syntenic with ADA. If man were the ultimate offshoot of the hominoid primates it is reasonable to assume that NP and ITPA became, for a second time, associated at the emergence of the hominoid stock. A strict parsimonious approach would lead us to include the great apes and the rabbit in one group, aside from man, the cercopithecoids and the other mammals, where NP is dissociated from ITPA, and ITPA is syntenic with ADA. This possibility is, obviously, unacceptable so that we must admit that the ITPA locus has clearly been transposed *twice and independently* (in the lagomorph and hominoid lineages) to become syntenic with the *same* locus (NP) to be later relocated (by a *third* transposition) to its *original* association with ADA, exclusively in the human lineage after divergence from the hominoid stock. Obviously, this is not a parsimonius pathway, though it seems to be the most logical one.

6 Epilogue

Karylogical comparisons and comparative gene assignment have clearly demonstrated a remarkable evolutionary conservation, both at the morphological and syntenic levels, within mammals. These studies have allowed us to trace the phylogenies of several human chromosomes and linkage associations. Several conclusions are, therefore, evident from these studies.

1. The mammalian chromosome complement has been drastically shuffled since the time this class appeared on earth. Chromosome rearrangements have occurred at different rates in each mammalian order; an extreme example of highly rearranged karyotypes can be found among the rodents while other orders, such as Primates, are more conserved. However, rates of chromosome change are also variable within

orders; the lesser apes and night monkeys show high rates of chromosome rearrangement in contrast to other groups like the great apes and man. An extreme case of chromosome conservation, as in rhesus-baboon, is illuminating in showing that speciation might occur in the absence of chromosome rearrangement, because organismal and chromosome evolution are obviously uncoupled. Chromosome evolution in mammals is, therefore, punctuated and not gradual because long periods of karyotypic invariance might suddenly alternate with others where shuffling becomes prominent. We should not, therefore, assume that chromosomal differences might be straightforwardly correlated to genetic distances between mammalian groups.

2. There is a general coincidence between the morphological (banding) attributes of chromosomes and their gene content within the primate order. Thus, homoeologous chromosomes are expected to contain similar linkage associations when two or more species are compared, though exceptions to this rule have been reported. This coincidence, however, applies to structural genes but not to repetitive sequences (ribosomal and satellite DNAs) the evolution of which seems to be independent from chromosome change. The identification of human chromosome homoeologies might be extended to other mammals outside the primate order, as it is the case in the domestic cat. This approach, however, must be cautious and based on the comparative gene assignment of those species whose karyotypes are compared.

3. In the absence of apparent karyotypic similarities, as between man and mouse, comparative gene assignment is illuminating in showing regions of continuous homologous loci that have been maintained intact in the mammalian genome. The extraordinary resemblance between humans and other mammals with respect to their gene distribution strongly suggests that the association of several loci is likely to be adaptative.

4. Finally, a detailed analysis of chromosome and linkage association phylogenies has shown that the principle of parsimony cannot be universally accepted for the tracing of evolutionary pathways. It is clearly evident that some specific types of chromosome rearrangements appear to have occurred more frequently than expected under a random assumption. The same observation is valid for the relocation of genes, like the ITPA locus, that seems to be specifically associated either with NP or ADA in mammalian radiation. Whether the frequent, yet independent, occurrence of these events is adaptative or selectively neutral is still an open question.

Acknowledgements. This work was supported by the following grants: CNPq 40.2404/82, Fundação José Bonifacio, CEPG-UFRJ and FINEP 4394073400.

References

Andrle M, Fiedler W, Rett A, Ambros P, Schweizer D (1979) A case of trisomy 22 in *Pongo pygmaneus*. Cytogenet Cell Genet 24:1–6

Arnheim N, Krystal M, Schmickel R, Wilson G, Ryder O, Zimmer E (1980) Molecular evidence for genetic exchanges among ribosomal genes on nonhomologous chromosomes in man and apes. Proc Natl Acad Sci USA 77:7323–7327

Barrie PA, Jeffreys AJ, Scott AF (1981) Evolution of the alfa-globin gene cluster in man and the primates. J Mol Biol 149:319–336

Benveniste RE (1985) The contributions of retrovirus to the study of mammalian evolution. In: MacIntyre RJ (ed) Molecular evolutionary genetics. Plenum, New York London, pp 359–417

Benveniste RE, Todaro G (1976) Evolution of type C viral genes: Evidence for an Asian origin of man. Nature (London) 261:101–108

Berman EJ, Nash WG, Seuánez HN, O'Brien SJ (1986) Chromosomal mapping of enzyme loci in the domestic cat: GSR to C2, ADA and ITPA to A3, and LDHA-ACP2 to Di. Cytogenet Cell Genet 41:114–120

Brown WM, Prager EM, Wang A, Wilson AC (1982) Mitochondrial DNA sequences of primates: Tempo and mode of evolution. J Mol Evol 18:225–239

Bruce EJ, Ayala FJ (1978) Human and apes are genetically very similar. Nature (London) 276: 264–265

Cochet C, Creau-Goldberg N, Turleau C, Grouchy J de (1982) Gene mapping of *Microcebus murinus* (Lemuridae): A comparison with man and *Cebus capucinus* (Cebidae). Cytogenet Cell Genet 33:213–221

Couturier J, Dutrillaux B, Turleau C, Grouchy J de (1982) Comparaisons chromosomiques chez quatre especes ou sous-especes des gibbons. Ann Genet 25:5–10

Creau-Goldberg N, Cochet C, Turleau C, Grouchy J de (1981) Comparative gene mapping of man and *Cebus capucinus:* A study of 23 enzymatic markers. Cytogenet Cell Genet 31:228–239

Creau-Goldberg N, Turleau C, Cochet C, Grouchy J de (1982) Comparative gene mapping of the baboon (*Papio papio*) and man. Ann Genet 25:14–18

Creau-Boldberg N, Turleau C, Cochet C, Grouchy J de (1983) New gene assignments in the baboon and new chromosome homologies with man. Ann Genet 26:75–78

Dutrillaux B (1979a) Chromosomal evolution in the primates: Tentative phylogeny from *Microcebus murinus* (Prosimian) to man. Human Genet 48:251–314

Dutrillaux B (1979b) Very large analogy of chromosome banding between *Cebus capucinus* (Platyrrhini) and man. Cytogenet Cell Genet 24:84–94

Dutrillaux B, Couturier J, Fosse AM (1980) The use of high resolution banding in comparative cytogenetics: Comparison between man and *Lagothrix lagothrica cana* (Cebidae). Cytogenet Cell Genet 27:45–51

Estop A, Garver JJ, Meera Khan P, Pearson PL (1979) Rhesus-human chromosome homologies via cytogenetic and mapping studies. Cytogenet Cell Genet 25:150–151

Ferris SD, Brown WM, Davidson WS, Wilson AC (1981a) Extensive polymorphism in the mithocondrial DNA of apes. Proc Natl Acad Sci USA 78:6319–6323

Ferris SD, Wilson AC, Brown WM (1981b) Evolutionary tree for apes and humans based on cleavage maps of mithocondrial DNA. Proc Natl Acad Sci USA 78:2432–2436

Finaz C, Van Cong N, Cochet C, Frézal J, Grouchy J de (1977) Histoire naturelle du chromosome 1 chez les primates. Ann Genet 20:85–92

Finaz C, Cochet C, Grouchy J de (1978) Identité des caryotypes de *Papio papio* et *Macaca mulatta* en bandes R, G, C et Ag-NOR. Ann Genet 21:149–151

Francke U, Taggart RT (1980) Comparative gene mapping: Order of the loci on the X chromosome is different in mice and humans. Proc Natl Acad Sci USA 77:3595–3599

Garver JJ, Estop A, Pearson PL, Dijksman TM, Wijnen LMM, Meera Kahn P (1977) Comparative gene mapping in the Pongidae and Cercopithecoidea. In: Chapelle A de la, Sorsa M (eds) Chromosomes today, vol 6. Elsevier/North-Holland Biomedical Press, Amsterdam New York, pp 191–199

Goldman D, Rathna Giri P, O'Brien SJ (1987) Molecular phylogeny of the hominoid primates as indicated by two-dimensional protein electrophoresis. Proc Natl Acad Sci USA 84:3307–3311

Goodman M (1975) Protein sequence and immunological specificity: Their role in phylogenetic studies in the primates. In: Luckett WP, Szalay JS (eds) Phylogeny of the primates. Plenum, New York, pp 219–248

Gosden JR, Mitchell AR, Seuánez HN, Gosden C (1977) The distribution of sequences complimentary to satellite I, II and IV DNAs in the chromosomes of the chimpanzee (*Pan troglodytes*), the gorilla (*Gorilla gorilla*), and the orangutan (*Pongo pygmaeus*). Chromosoma 63:253–271

Gould SJ (1980) Is a new and general theory of evolution emerging? Paleobiology 3:115–151

Grouchy J de, Turleau C, Finaz C (1978) Chromosome phylogeny of the primates. Annu Rev Genet 12:289–328

Henderson AS, Warburton D, Megraw-Ripley S, Atwood KC (1977) The chromosomal location of rDNA in selected lower primates. Cytogenet Cell Genet 19:281–302

King MC, Wilson AC (1975) Evolution at two levels in human and chimpanzee. Science 188: 107–116

Ma NSF (1984) Linkage and syntenic relationship in chromosomes of owl monkeys (with karyotypes I, III, V, VI, VII) homologous to man. In: O'Brien SJ (ed) Genetic maps, vol 3. Cold Spring Harbor Press, New York, pp 410–413

Ma NSF, Jones TC, Miller AC, Morgan LM, Adams EA (1976) Chromosome polymorphisms and banding in the owl monkey (*Aotus*). Lab Anim Sci 26:1022–1036

McClure H, Belden KH, Pieper WA (1969) Autosomal trisomy in a chimpanzee: Resemblance to Down's syndrome. Science 65:1010–1012

Mitchell AR, Seuánez HN, Lawrie S, Martin DE, Gosden JR (1977) The location of DNA homologous to satellite III DNA in the chromosomes of the chimpanzee (*Pan troglodytes*), gorilla (*Gorilla gorilla*) and orangutan (*Pongo pygmaeus*). Chromosoma 61:345–358

Myers RH, Shafer DA (1979) Hybrid ape offspring of a mating of gibbon and siamang. Science 205:308–310

Nadeau JH, Taylor BA (1984) Lengths of chromosomal segments conserved since the divergence of man and mouse. Proc Natl Acad Sci USA 81:814–818

Nash WG, O'Brien SJ (1982) Conserved regions of homologous G-banded chromosomes between orders in mammalian evolution: Carnivores and primates. Proc Natl Acad Sci USA 79:6631–6635

O'Brien SJ (ed) (1984) Genetic maps, vol 3. Cold Spring Harbor Press, New York

O'Brien SJ, Nash WG (1982) Genetic mapping in mammals: Chromosome map of the domestic cat. Science 216:257–265

O'Brien SJ, Seuánez HN, Womak JE (19785) On the evolution of genome organization in mammals. In: MacIntyre RJ (ed) Molecular evolutionary genetics. Plenum Press, New York London, pp 519–589

Ohno S (1973) Ancient linkage groups and frozen accidents. Nature (London) 244:259–262

Polani PE, Adinolfi M (1980) Chromosome 21 of man, 22 of the great apes and 16 of the mouse. Dev Med Child Neurol 22:223–224

Seuánez HN (1979) The phylogeny of human chromosomes. Springer, Berlin Heidelberg New York

Seuánez HN (1984) Evolutionary aspects of human chromosomes. In: Roodyn DB (ed) Subcellular biochemistry, vol 10. Plenum, New York London, pp 455–537

Seuánez HN, Fletcher J, Evans HJ, Martin DE (1976) A polymorphic structural rearrangement in two populations of orangutan. Cytogenet Cell Genet 17:327–337

Seuánez HN, Evans HJ, Martin DE, Fletcher J (1979) An inversion in chromosome 2 that distinguishes between Bornean and Sumatran orangutans. Cytogenet Cell Genet 23:137–140

Soulie J, Grouchy J de (1982) Of rabbit and man: Comparative gene mapping. Human Genet 60: 172–175

Stanyon R, Chiarelli B (1982) Phylogeny of the Hominoidea: The chromosome evidence. J Human Evol 11:493–504

Stockholm Conference (1977) An international system for human cytogenetic nomenclature. 1978 I.S.C.N. Cytogenet Cell Genet 21:313–409

Sun NC, Sun CRY, Ho T (1978a) Chimpanzee chromosome 12 is homologous to human chromosome arm 2q. Cytogenet Cell Genet 22:594–597

Sun NC, Sun CRY, Ho T (1978b) Chimpanzee chromosome 13 is homologous to human chromosome 2p. Cytogenet Cell Genet 22:598–601

Turleau C, Grouchy J de, Klein M (1972) Phylogenie chromosomique de l'homme et des primates hominiens (*Pan troglodytes, Gorilla gorilla* et *Pongo pygmaeus*). Essai de reconstitution du caryotype de l'ancetre commun. Ann Genet 15:225–240

Turleau C, Creau-Goldberg N, Cochet C, Grouchy J de (1983) Gene mapping of the gibbon. Its position in primate evolution. Human Genet 64:65–72

Wilson AC, Carlson SS, White TJ (1977) Biochemical evolution. Annu Rev Biochim 46:573–639

Wilson GN, Knoller M, Szura L, Schmickel RD (1984) Individual and evolutionary variation of primate and ribosomal DNA transcription initiation regions. Mol Biol Evol 1:221–237

Yunis JJ, Prakash O (1982) The origin of man; A chromosomal pictorial legacy. Science 215: 1525–1529

5 Chromosome Evolution of Cervidae: Karyotypic and Molecular Aspects

H. NEITZEL[1]

1 Introduction

The role of chromosomal rearrangements in mammalian evolution is the subject of considerable controversy, since the average rate of karyotypic changes has been much faster in mammals than in other vertebrates. The fact that this rapid chromosomal evolution is paralleled by rapid anatomical diversification in mammals supports the idea that a high level of chromosomal rearrangements may be responsible for differentiation in the speciation process (Bender and Chu 1963; Wilson et al. 1974).

We focused our attention, therefore, to the family Cervidae which has a very high degree of karyotypic evolution. Although cervids are a comparatively young mammalian family with only 32 recent species (Thenius and Hofer 1960; Haltenorth 1963), no other taxa are found among mammals with a similarly extreme extent of chromosomal diversification. Diploid chromosome numbers vary from 2n=6♀/7♂ in the Indian muntjac (*Muntiacus muntjac*) (Wurster and Benirschke 1970) to 2n=80 in the Sibirian roe (*Capreolus pygargus*).

By applying various improved banding techniques the phylogeny of chromosomal rearrangements can be largely resolved in this family. In addition to the reconstruction of the course of chromosomal evolution, aspects of molecular and anatomical diversification are discussed, since only within the general context of organic and molecular change can the role of chromosomal rearrangements in phylogeny be understood.

2 Course of Karyotypic Evolution in Cervidae

2.1 Reconstruction of the Ancestral Karyotype of Cervidae

Based on comparison of the G-banded karyotypes of 11 cervid species, the hypothetical ancestral karyotype of the common ancestor of Cervidae is assumed to be 2n=70, NF=70 (Neitzel 1982). Data concerning diploid chromosome numbers, NF values (NF = fundamental number = number of chromosome arms) and general chromosome

1 Institut für Humangenetik, Freie Universität Berlin, Heubnerweg 6, D–1000 Berlin 19

Cytogenetics. Ed. by G. Obe and A. Basler

Table 1. Data on diploid chromosome numbers, NF values and general chromosome morphology in different cervid species studied in this report and previously by other authors

Family	Subfamily	Genus	Species	2n	NF	Autosomes m/sm	Autosomes a	Gonosomes X	Gonosomes Y	B-chromosomes	References
Cervidae	Hydropotinae	*Hydropotes*	*H. inermis*	70	70	–	68	a	a	–	Hsu and Benirschke (1973)
	Muntiacinae	*Muntiacus*	*M. muntjac*	6♀/7♂	12	4	–	sm	sm	–	Wurster and Benirschke (1970)
			M. feae	13♀	16	1	10	1m	–	–	Soma et al. (1983)
			M. reevesi	46	46	–	44	a	a	–	Hsu and Benirschke (1968)
	Odocoilinae	*Capreolus*	*C. capreolus*	70	72	–	68	sm	a	–	Gustavsson and Sundt (1968)
			C. pygargus	74–80	76–80	–	68	sm	a	6–10	Present report
		Odocoileus	*O. virginianus*	70	74	2	66	sm	m	–	Wurster and Benirschke (1967)
			O. hemionus	70	74	2	66	sm	m	–	Hsu and Benirschke (1967)
			O. bezoarticus	68	74	4	62	sm	sm	–	Present report
			O. dichotomus	66	74	6	58	sm	sm	–	Present report
		Mazama	*M. americana*	68	74						Taylor et al. (1969)
				50	72	20	28	sm	m	–	Jorge and Benirschke (1977)
				57	61	2	48	sm	–	5	Present report
			M. gouazoubira	70	70	–	68	a	a	–	Present report
			M. pudu	70	74	2	66	m	a	–	Koulischer et al. (1972)
			M. mephistopheles	69	74	3	64	sm	a	–	Sperling (unpubl.)
	Rangiferinae	*Rangifer*	*R. tarandus*	70	74	2	66	m	a	–	Amrud and Nes (1966)
	Alcinae	*Alces*	*A. alces*	70	74	2	66	sm	a	–	Hsu and Benirschke (1969)
				68	74	4	62	sm	a	–	Gustavsson and Sundt (1968)
	Cervinae	*Elaphurus*	*E. davidianus*	68	70	2	64	a	–	–	Hsu and Benirschke (1971)
		Cervus	*C. axis*	66	70	4	60	a	a	–	Hsu Benirschke (1974)
			C. porcinus	68	70	2	64	a	a	–	Present report
			C. nippon	64–68	70	2–4	60–66	a	sm	–	Gustavsson and Sundt (1969)
			C. dama	68	70	2	64	a	sm	–	Gustavsson and Sundt (1969)
			C. duvauceli	56	70	14	40	a	m	–	Chandra et al. (1967)
			C. eldi thamin	58	70	12	44	a	–	–	Present report
			C. unicolor	64	70	6	56	a	m	–	Chandra et al. (1967)
				58	70	12	44	a	a	–	Benirschke (1969)
				60	70	10	48	a	a	–	Present report
			C. timorensis	60	70	10	48	a	a	–	Present report
			C. elaphus	68	70	2	64	a	m	–	Gustavsson and Sundt (1968)

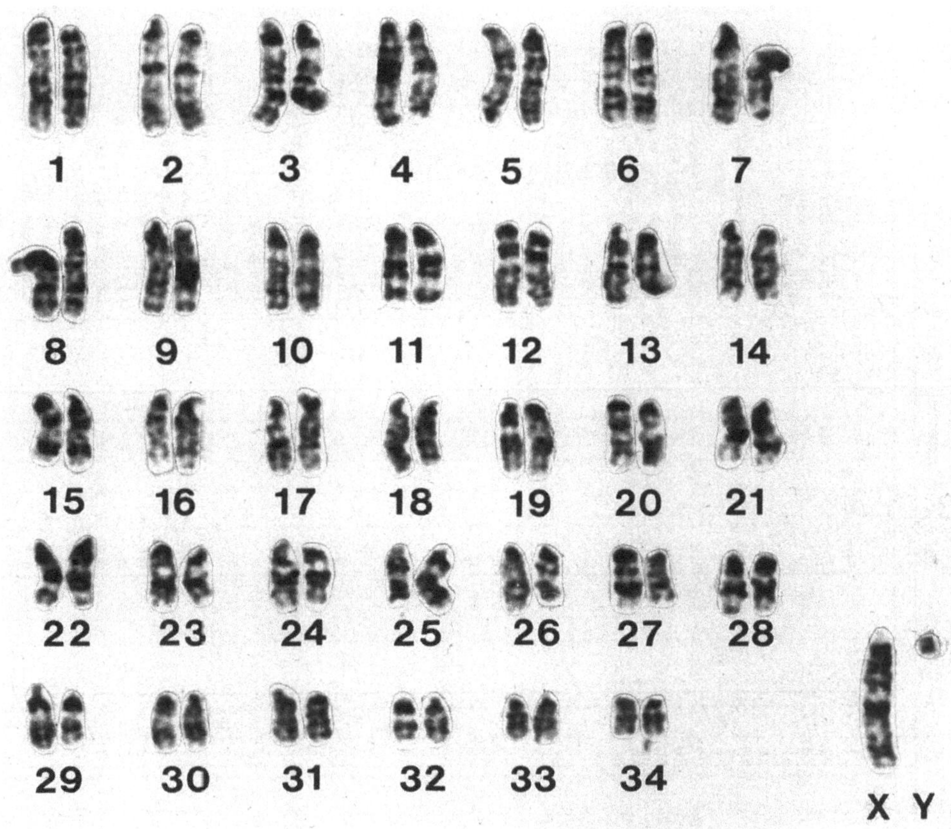

Fig. 1. G-banded karyotype of a male *Mazama gouazoubira.* This karyotype with 2n=70, NF=70 represents the hypothetical ancestral karyotype of the family Cervidae

morphology of the species studied in the present report and previously by other authors are summarized in Table 1. The reconstruction of the ancestral karyotype is confirmed by the fact that this karyotype is retained in two, distantly related, recent species, namely *Mazama gouazoubira* (Figs. 1 and 2) and *Hydropotes inermis,* which belong to different subfamilies. Since no obvious differences are evident between the karyotypes of these two species after G-, C- and AgNOR staining, the retention of the ancestral karyotype for more than 25 million years must be considered in the two species which separated early in evolution and accumulated a high degree of anatomical divergence (Thenius and Hofer 1960).

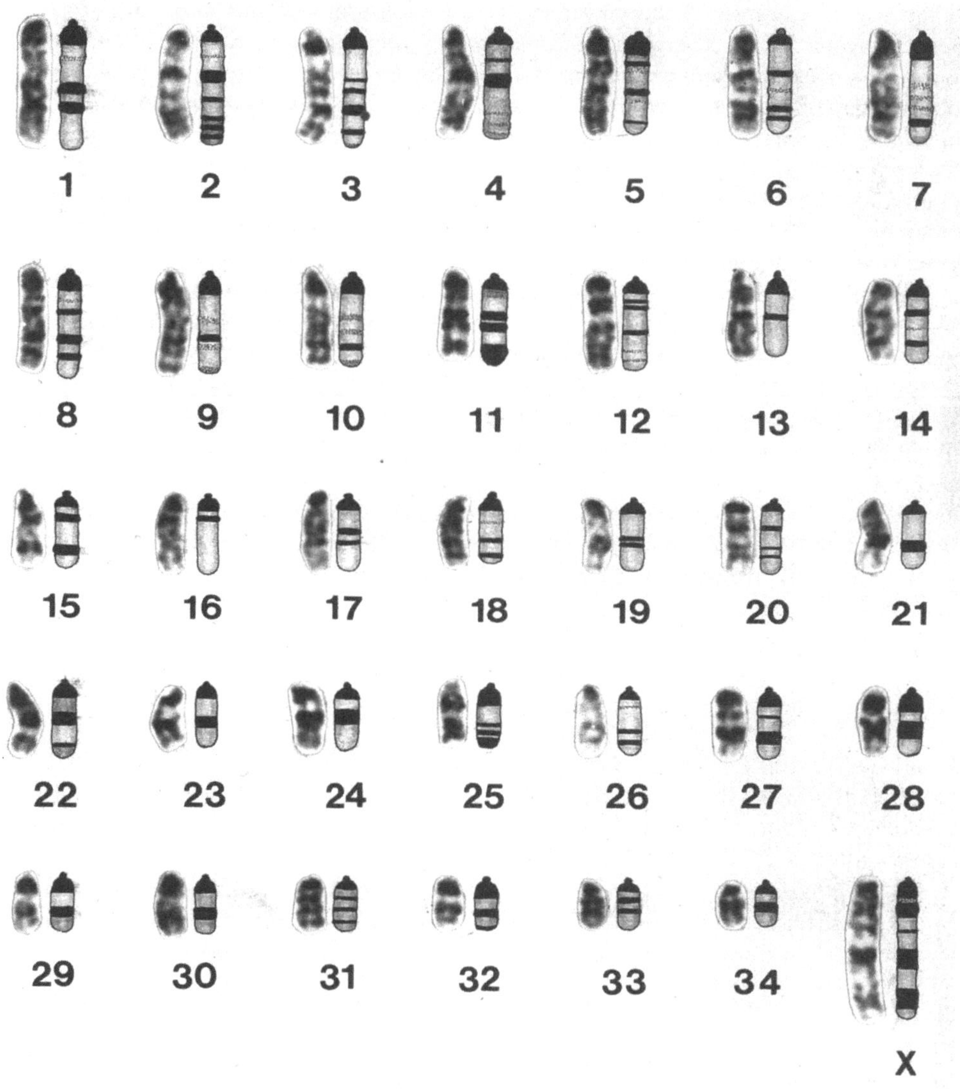

Fig. 2. Schematic representation of the G-banding pattern of the hypothetical karyotype with 2n=70, NF=70

2.2 Chromosomal Rearrangements in the Phylogeny of Different Cervid Subfamilies

2.2.1 Odocoilinae

The karyotypes of the genera *Capreolus* and *Odocoileus* first diverged from the ancestral karyotype by a pericentric inversion of the X-chromosome. Assuming a monophyletic origin of the rearranged X-chromosome, the common origin of *Capreolus* and *Odocoileus* is evident, while the genus *Mazama*, having the original acrocentric X-chromosome type, must have emerged earlier from this lineage. Further karyotypic diversification within the genus *Odocoileus* is due to a pericentric inversion of autosome No. 5, increasing the fundamental number to 74, followed by Robertsonian translocations in the karyotypes of *Odocoileus bezoarticus* and *Odocoileus dichotomus*, decreasing the diploid numbers to 2n=68 and 2n=66 respectively. The Robertsonian translocation chromosome t 9;11 (Fig. 2), found in both sepcies, indicates clearly their close phylogenetic relationship which was previously controversial (Frädrich 1981).

Despite the retention of the ancestral karyotype in *Mazama gouazoubira* extensive and complex rearrangements took place in the course of evolution in other *Mazama* species. Karyotypes with 2n=68, NF=74 (Taylor et al. 1969) and 2n=49/50, NF=74 (Jorge and Benirschke 1977) have been reported for *Mazama americana*. A female *Mazama americana* originating from Paraguay had a karyotype with 2n=52 plus four to five B-chromosomes, NF=56 (Figs. 4 and 13). This karyotype can be derived from the ancestral one by a pericentric inversion of the X-chromosome, an X-autosome translocation, one Robertsonian translocation and several tandem fusions (Neitzel, in prep.). Interbreeding between this female and a male of *Mazama gouazoubira* resulted in a viable female hybrid with 2n=61 plus two B-chromosomes which was assumed to be infertile (Fig. 13).

2.2.2 *Cervinae*

In the lineage of the subfamily Cervinae, only Robertsonian translocations contributed to differentiation of the karyotypes, which can already be expected from the strikingly constant NF value of 70. The Robertsonian translocation involving the original acrocentric autosomes 16 and 24 characterizes the karyotypes of all *Cervus* species and that of *Elaphurus davidianus*, indicating a close taxonomic relationship between both genera irrespective of the phenotypic peculiarities of *Elaphurus*. This rearrangement must have taken place in a common ancestor more than 7 million years ago.

The karyotypes of *Cervus duvauceli, Cervus eldi thamin* and *Cervus unicolor* differ from the other *Cervus* species studied by an increasing number of Robertsonian translocations (Fig. 3) (Neitzel 1982).

2.2.3 *Muntiacinae*

Among cervid species chromosomal divergence is extreme within the genus *Muntiacus*, however, the morphological similarity is strikingly retained. In the Indian muntjac (*Muntiacus muntjac*) the karyotype is 2n=6♀/7♂ (Figs. 5 and 6; Wurster and Benirschke 1970), whereas the diploid number is 2n=13 in *Muntiacus feae* (Soma et al. 1983) and 2n=46 in *Muntiacus reevesi* (Fig. 5; Hsu and Benirschke 1968). Since there can be no doubt about the monophyletic origin of Cervidae, the karyotypes of the muntjacs must have differentiated from the ancestral one with 2n=70, NF=70. However, because of the high degree of chromosomal divergence only a few homologies with the

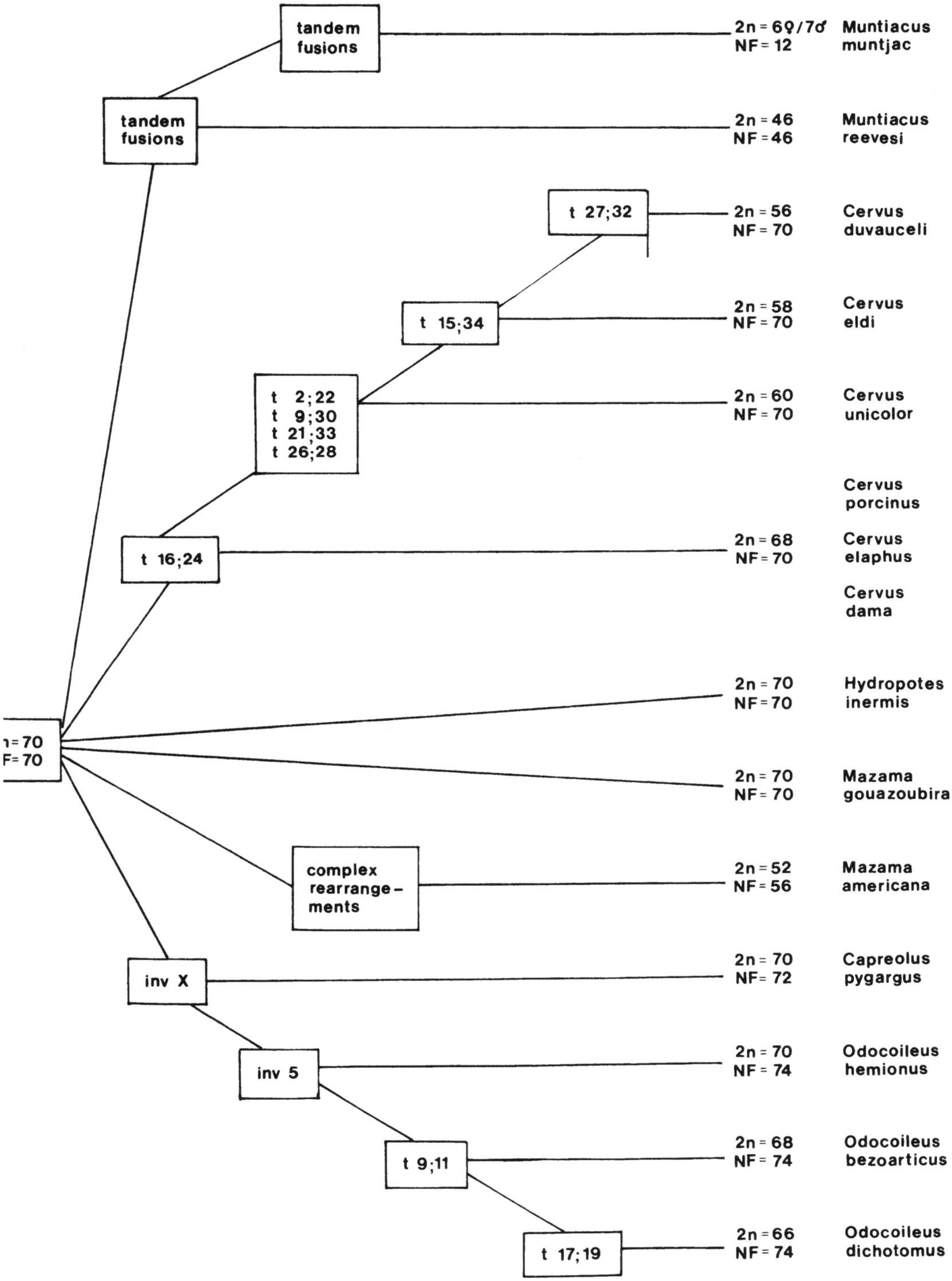

Fig. 3. Course of karyotypic evolution in the different subfamilies of Cervidae

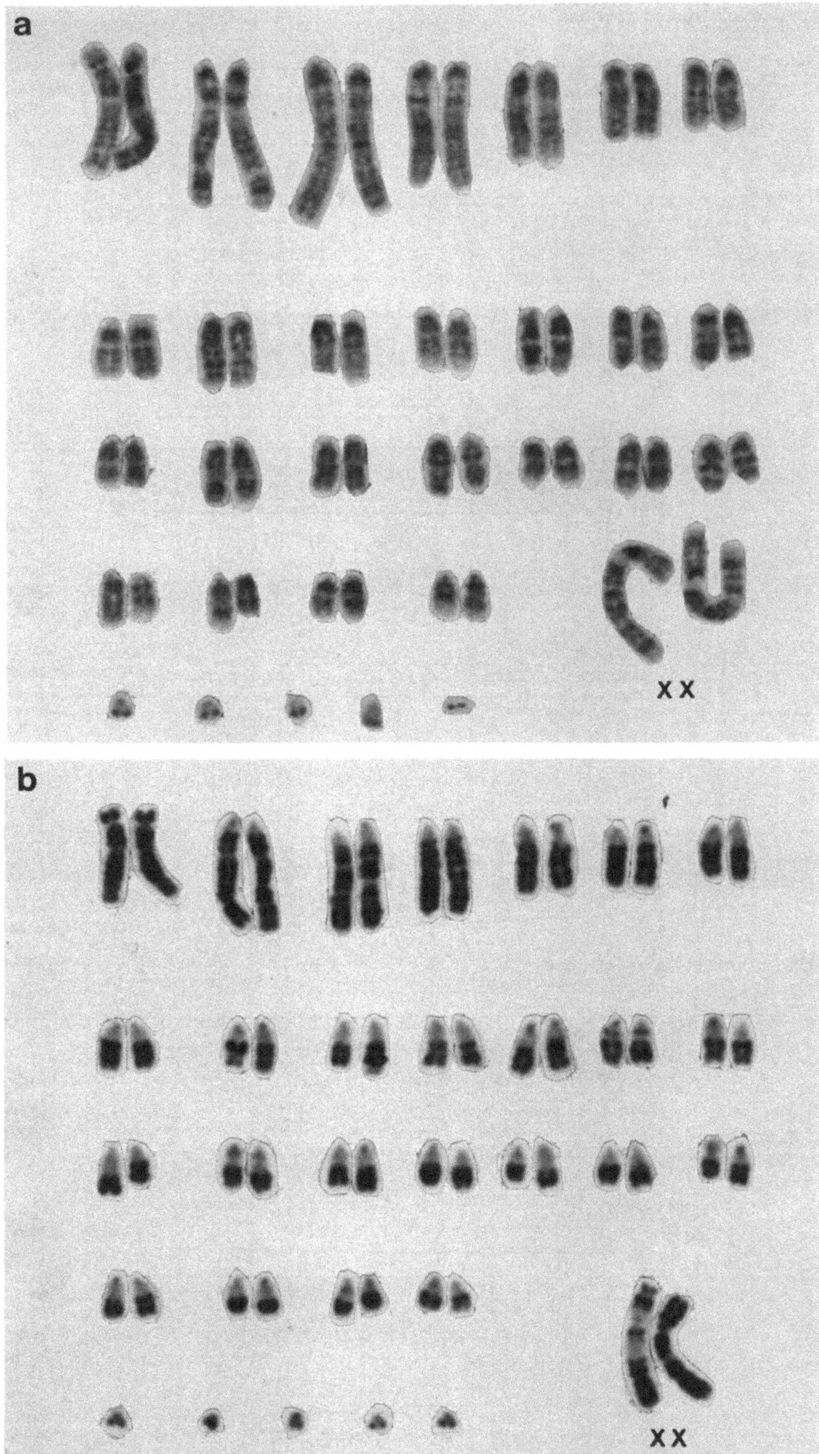
a
X X
b
X X

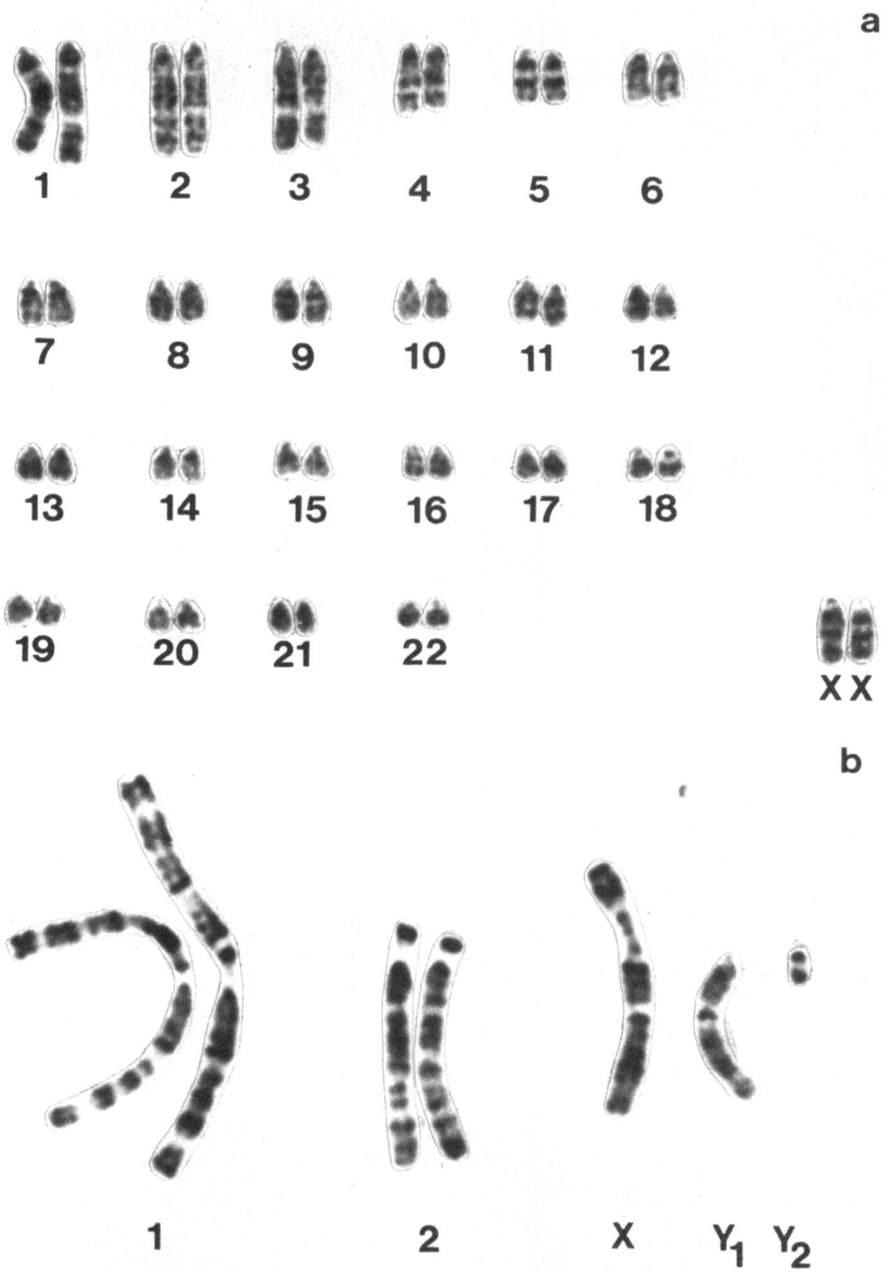

Fig. 5a,b. G-banded karyotypes of *Muntiacus reevesi* (**a**) with 2n=46, and *Muntiacus muntjac* with 2n=7 (**b**)

Fig. 4a,b. Karyotype of a female *Mazama americana*. This specimen has a basic karyotype with 2n=52, NF=56 plus four to five B-chromosomes. **a** G-banding; **b** late replication pattern demonstrating the X-autosome translocation

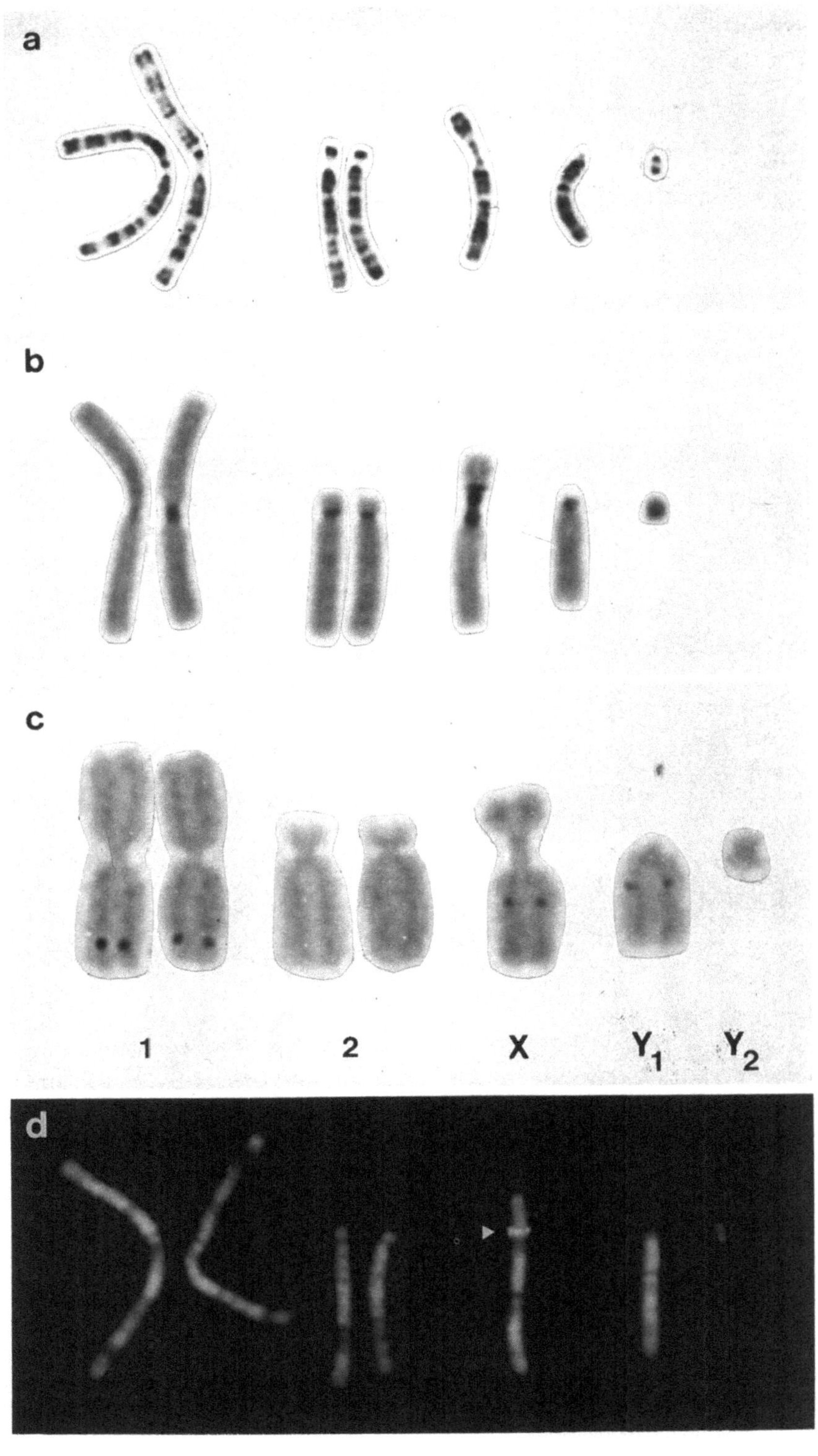
a
b
c
1
2
X
Y_1
Y_2
d

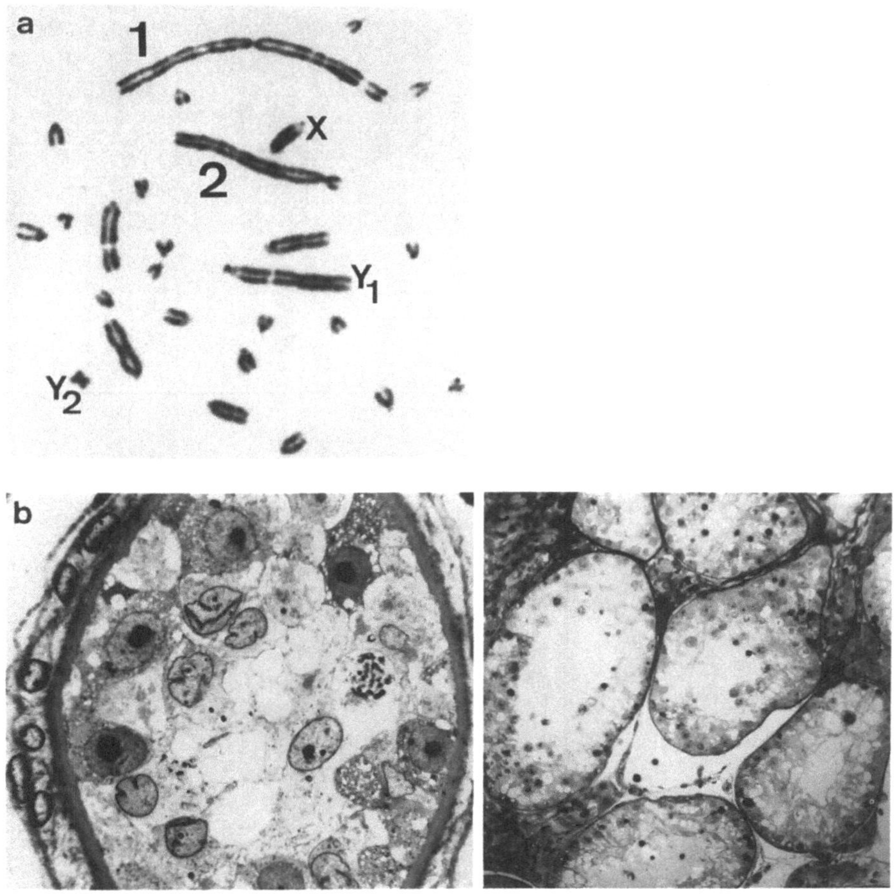

Fig. 7a–c. Metaphase of a male hybride between *Muntiacus muntjac* and *Muntiacus reevesi* with a diploid number of 2n=27 (**a**). **b**,**c** Histological preparations of testicle tubuli

ancestral karyotype can be demonstrated, while between the complements of both muntjac species a high degree of homology is found (Shi Liming et al. 1980; Neitzel 1982).

The most striking peculiarity in the diversification process from the ancestral karyotype to that of *Muntiacus reevesi* as well as to the further evolved *Muntiacus muntjac* is the seemingly exclusive occurrence of tandem fusions. Since this type of rearrangement should result in a high frequency of unbalanced gametes in the heterozygotes, this mechanism is only realized in the chromosomal evolution of very few taxa.

Fig. 6a–d. Karyotype of a male *Muntiacus muntjac* 2n=7. **a** G-banding; **b** distribution of constitutive heterochromatin as revealed by C-banding; **c** AgNOR staining demonstrating active nucleolus organizer regions; **d** DAPI staining showing a bright fluorescence in the constitutive heterochromatin of the short arm of the X-chromosome

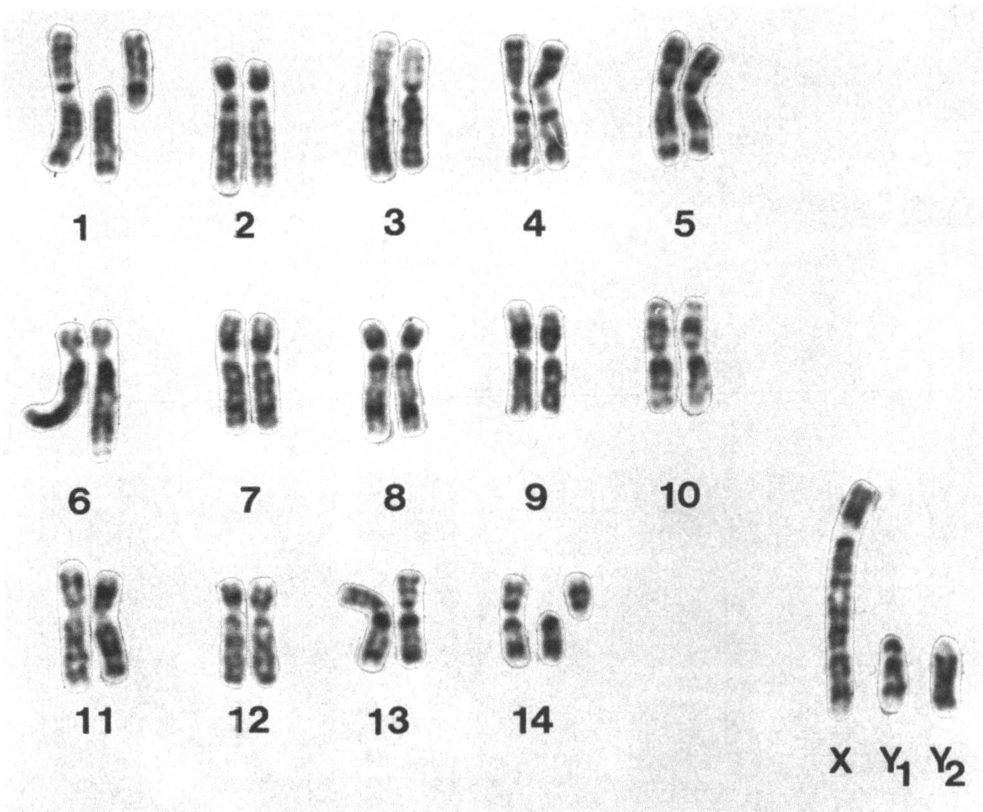

Fig. 8. G-banded karyotype of a male *Antilope cervicapra* with 2n=33, NF=60. The specimen has two heterozygote Robertsonian translocations and an X-autosome translocation

Despite the karyotypic differences between Indian (*Muntiacus muntjac*) and Chinese muntjac (*Muntiacus reevesi*), both species can be interbred, yielding viable F1 hybrids. A male hybrid from a cross between a female Chinese and a male Indian muntjac born in the Berlin Zoo showed the expected karyotype with 2n=27 (Fig. 7). Histological and cytogenetic examination of testicle preparations showed the absence of synapsis at pachytene. In the tubuli, spermatogonia, spermatocytes and single spermatids were found while mature spermatozoa were absent (Fig. 7). These findings are consistent with the data obtained previously in another F1 hybrid (Shi Liming and Pathak 1981).

However, from the F1 hybrid in the Berlin Zoo a number of spermatids were recovered from the epididymis and these had a constant DNA value of 2c. This can be interpreted to mean that spermatocytes progress through meiosis I up to pachytene. This is characterized by missing synapsis, consequently, no separation of homologous elements can be achieved, thus resulting in the degeneration of a considerable number

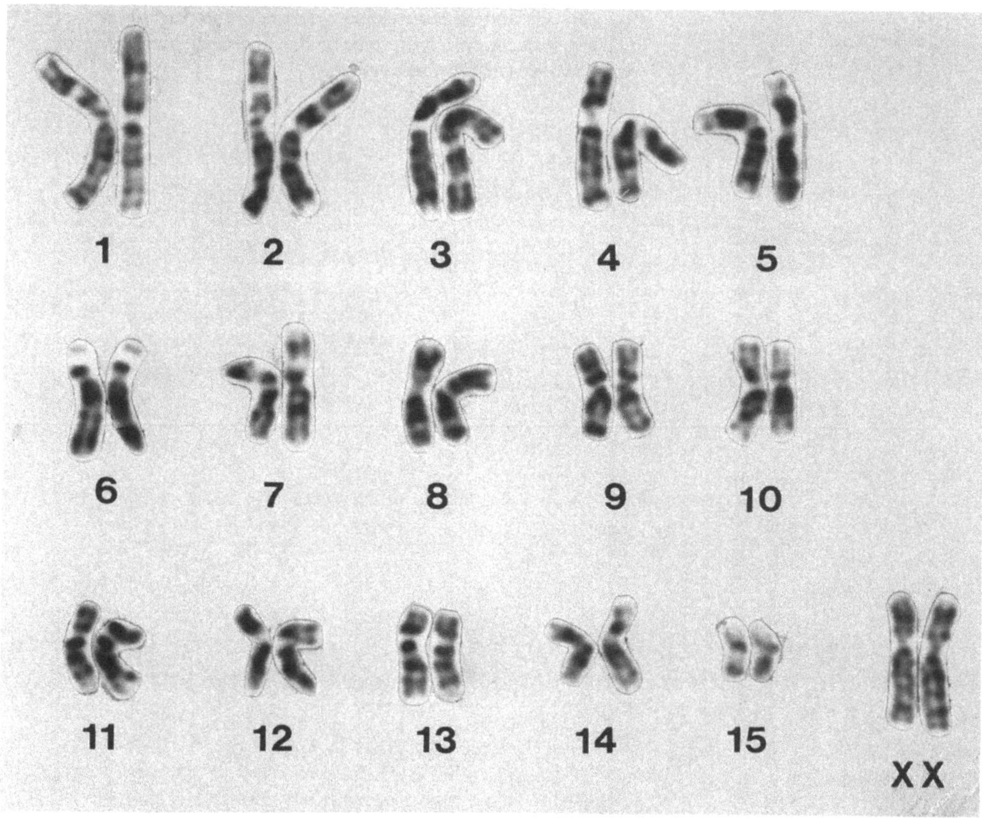

Fig. 9. G-banded karyotype of a female *Tragulus javanicus* with 2n=32, NF=64

of spermatocytes. However, a small number seems to be able to undergo meiosis II, even without completion of meiosis I, leading to diploid spermatids (Neitzel and Hoehn, in prep.).

3 Comparison of the Cervid Karyotypes with the Related Tragulidae and Bovidae

In order to obtain information on the degree of karyotypic diversification between the three related artiodactyl families Cervidae, *Bovidae* and *Tragulidae,* we compared the hypothetical ancestral karyotype with those of *Antilope cervicapra* (Bovidae) 2n=33, NF=60 (Fig. 8) and *Tragulus javanicus* (Tragulidae) 2n=32, NF=64 (Fig. 9). However, because of the high degree of chromosomal divergence only few G-band homologies can be demonstrated between the cervid karyotype and *Antilope cervicapra,* while no homologies are evident with the karyotype of *Tragulus javanicus* (Fig. 10; Neitzel 1982).

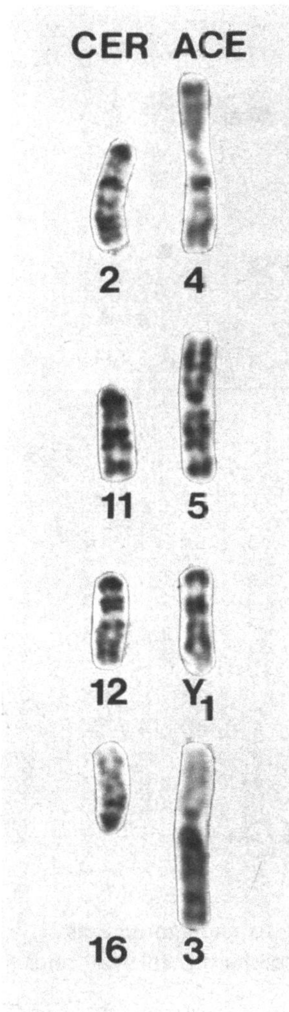

Fig. 10. G-banding homologies between the karyotypes of *Mazama gouazoubira* representing the ancestral karyotype of Cervidae and *Antilope cervicapra*

4 Single-Copy DNA Divergence

Assuming that nucleotide exchanges in genomes were random events during evolution, occurring with a rate of 0.1–0.3% per million years (typical for single-copy DNA), the time of divergence can be calculated.

The extent of nuclear single-copy DNA divergence between related species can be estimated by measuring the thermal stability of interspecific DNA-DNA hybrids. According to Bonner et al. (1973), depression of 1°C in the melting point reflects 1% nucleotide mismatch. The data for different cervid species are given in Table 2 (Schmidtke et al. 1981; Schmidtke and Neitzel, unpubl.).

Table 2. Single-copy divergence in different cervid species

Species			Nucleotide mismatch
Muntiacus muntjac	–	*Muntiacus reevesi*	2.2%
Muntiacus muntjac	–	*Capreolus pygargus*	5.5%
Muntiacus muntjac	–	*Mazama gouazoubira*	5.5%
Capreolus pygargus	–	*Mazama gouazoubira*	4.5%

The calculated divergence periods for different species are not inconsistent with the palaeontological data, indicating the relatively recent evolutionary origin of cervids which makes the high rate of chromosomal divergence in this family even more surprising.

5 Phylogenetic Alterations in the Amount and Distribution of Constitutive Heterochromatin, and in Satellite DNA Evolution

5.1 Amount of Constitutive Heterochromatin

Comparative studies have shown that constitutive heterochromatin as revealed by C-banding is a highly variable component in the genomes of different mammals, both with respect to its distribution and quantity. With the exception of the two muntjac species, within the family *Cervidae* relatively high amounts of constitutive heterochromatin are found. As demonstrated for other mammalian taxa (Pathak et al. 1973; Hatch et al. 1976; Deaven et al. 1977; Miklos and John 1979), there is a strong correlation between the amount of constitutive heterochromatin in the genome of a given species and the nuclear DNA content. Compared to *Mazama gouazoubira,* representing the hypothetical ancestral karyotype, Chinese and Indian muntjac show a reduction in relative DNA content of 17% and 20% respectively. An increase of 7% in *Odocoileus bezoarticus* can be explained by additional heterochromatic blocks in the X-chromosomes. The increase of 15% in *Capreolus pygargus* is probably due to the presence of ten supernumerary chromosomes in the specimen studied. Altogether in the cervid family, there are differences in relative DNA content of 34% which probably mainly result from the addition or elimination of constitutive heterochromatin during karyotypic evolution (Neitzel 1982).

5.2 Chromosomal Distribution of Constitutive Heterochromatin

The distribution of constitutive heterochromatin in the genomes of Cervidae is obviously not random, but, as in the majority of mammals (Miklos and John 1979), preferentially in the centromeric regions. It is striking that most acrocentric elements in the genomes have prominent blocks of heterochromatin, while numerous bi-armed chromosomes show partial or even complete reduction of centromeric heterochromatin.

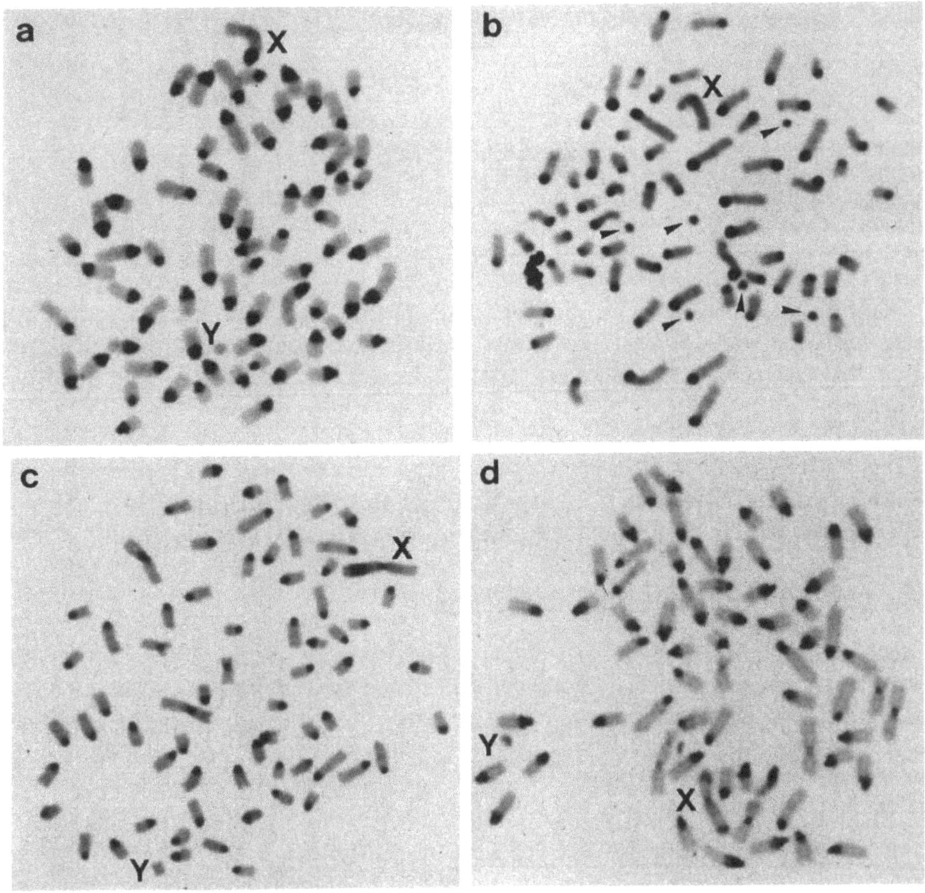

Fig. 11a–d. Distribution of constitutive heterochromatin as revealed by C-banding in the subfamily Odocoilinae. **a** *Mazama gouazoubira* 2n=70; **b** *Capreolus pygargus* 2n=70 and six B-chromosomes (*arrowheads*); **c** *Odocoileus bezoarticus* 2n=68; **d** *Odocoileus dichotomus* 2n=66

From data in Cervidae it is evident that in bi-armed chromosomes, which were rearranged early in the course of karyotypic evolution, centromeric heterochromatin is eliminated. Bi-armed elements which are of very recent evolutionary origin still exhibit a considerable amount of constitutive heterochromatin. Since this is also found in other mammals (Zech et al. 1971; Evans et al. 1973), it implies that chromosomal rearrangements leading to bi-armed chromosomes can be realized without a visible loss of heterochromatin. After fixation of such rearrangements in a population the amount of heterochromatic material tends to be decreased. Since the majority of acrocentric elements exhibit centromeric heterochromatin it can be concluded that centromeric heterochromatin probably makes an essential contribution to the structural organiza-

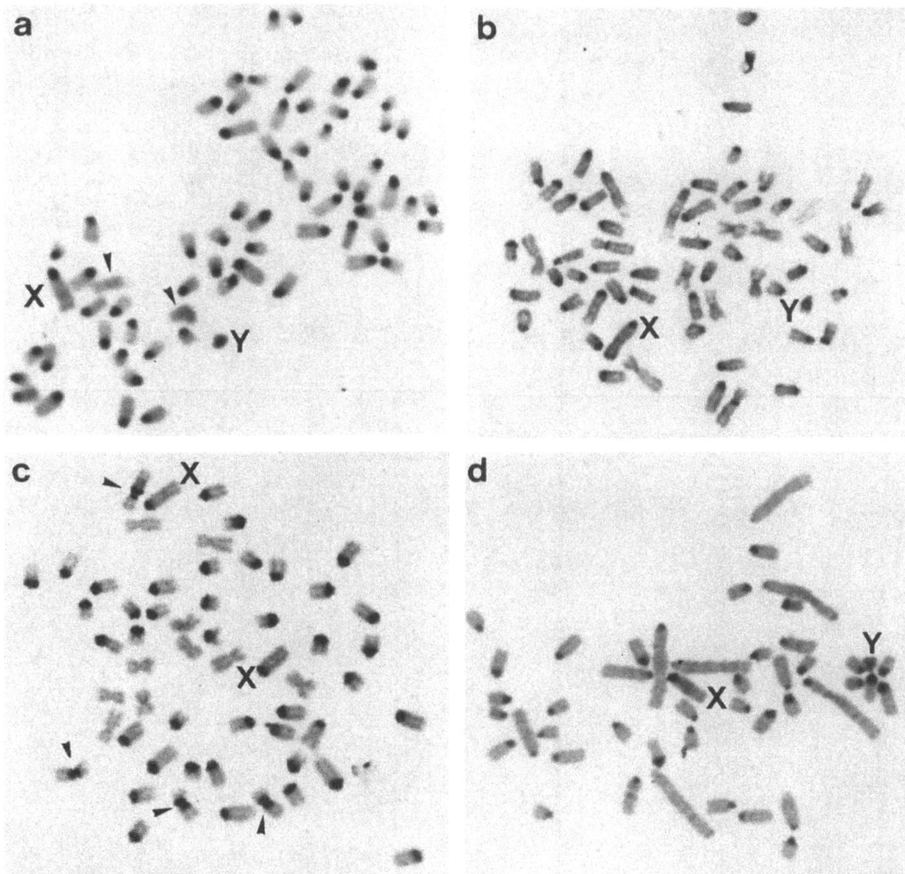

Fig. 12a–d. Distribution of constitutive heterochromatin as revealed by C-banding in the plesiometacarpal genera *Cervus* and *Muntiacus*. **a** *Cervus porcinus* 2n=68; **b** *Cervus unicolor* 2n=60; c *Cervus duvauceli* 2n=56; d *Muntiacus reveesi* 2n=46

tion of acrocentric chromosomes. It seems that this function is forfeited in bi-armed chromosomes where heterochromatin tends to decrease (Figs. 11–13).

5.3 Satellite DNA Evolution

In order to obtain more information about the molecular organization and possible role of constitutive heterochromatin, a cloned satellite DNA sequence derived from *Muntiacus muntjac* DNA (Bogenberger et al. 1982) was used to compare the presence and organization of this satellite in related species.

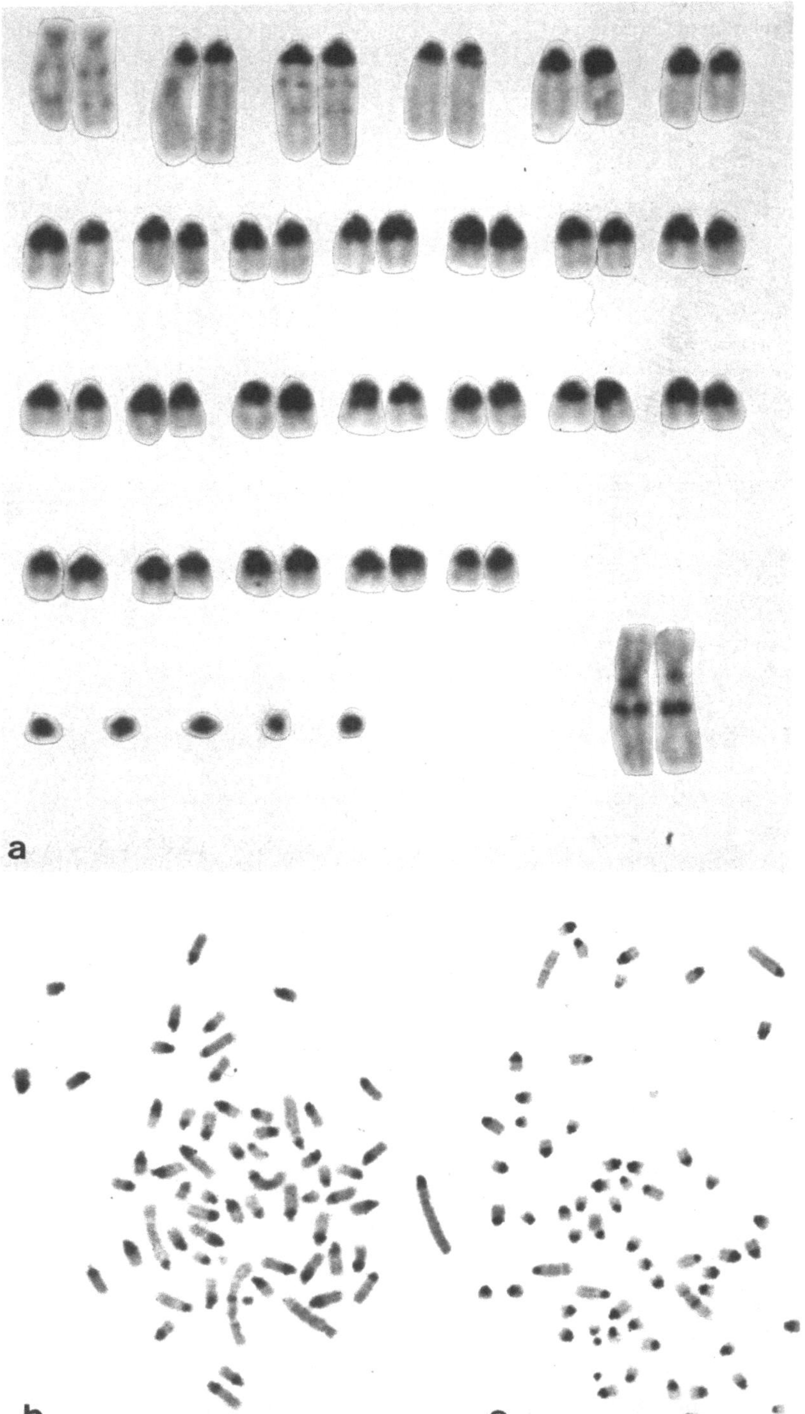

Fig. 13. **a** C-banding in a female *Mazama americana* with a basic karyotype with 2n=52 and five additional B-chromosomes. **b,c** C-banded metaphases of a hybrid between the female *Mazama americana* and a male *Mazama gouazoubira* with one and two B-chromosomes respectively

In the genome of the Indian muntjac the tandemly arranged satellite DNA IA is localized in the heterochromatin on both sides of the centromere of the X-chromosome. However, a cross-hybridizing repetitive DNA component, IB, is not clustered, but rather interspersed in the muntjac genome (Benedum et al. 1986).

The tandemly arranged satellite IA is present in the genomes of all cervid species, and in *Bos taurus*, but not in *Tragulus javanicus*, suggesting that these sequences were generated after the separation of Cervidae and Bovidae from the Tragulidae. This is in contrast to the palaeontological data since tragulids and cervids were thought to be more closely related.

Studies on the organization of this satellite reveal three different repeat lengths: 807, 1000 and 1400 bp. The relative proportion of satellite IA sequences present in any one of the three registers is different in various species.

In *Bos taurus* the majority of the satellite IA sequences are organized in a repeat unit of 1400 bp, while in *Plesiometacarpalia* and *Telemetacarpalia* of Cervidae, amplification units of 807 and 1000 respectively, predominate. This indicates that the satellite diverged by several mutation and amplification steps during evolution (Neitzel et al., in prep.).

The chromosomal locations of the satellite IA sequences were determined by in situ hybridization (Fig. 14). In the subfamilies Cervinae and Odocoilinae satellite IA is present in the centromeric heterochromatin of nearly all chromosomes, but absent in those heterochromatic regions which are assumed to have evolved more recently, for example, the additional heterochromatin in the X-chromosomes of the two South American *Odocoileus* species. Furthermore, complete elimination is found in the evolutionary, ancient, bi-armed chromosomes which lack constitutive heterochromatin. In the course of karyotypic evolution of both *Muntiacus* species the reduction of chromosome number by tandem fusions is paralleled by a loss of satellite IA sequences at most locations (Neitzel et al., in prep.).

In conclusion, these data demonstrate that despite the evolutionary conservation of this satellite DNA in bovids and cervids, the sequences diverged by several mutation and amplification steps. Different amplification mechanisms must have been involved in the generation of the tandemly arranged satellite IA and the interspersed repeat units of the component IB. While in the first case amplification might have occurred by unequal crossing over, in the second case retrotransposition as a mechanism must be considered (see Singer 1982).

During karyotypic evolution the satellite DNA IA tends to be eliminated in rearranged chromosomes, apparently by unequal crossing over.

6 Implications for the Role of Karyotypic Evolution in Cervidae

From our data concerning chromosomal and molecular evolution the following conclusions can be made:

1. Obviously, different mechanisms predominate during the course of karyotypic evolution of different cervid lineages. While in the subfamily Cervinae exclusively Robertsonian translocations led to divergence of karyotypes, in the genera *Mazama* and *Muntiacus* extensive tandem fusions predominate, even though this type of

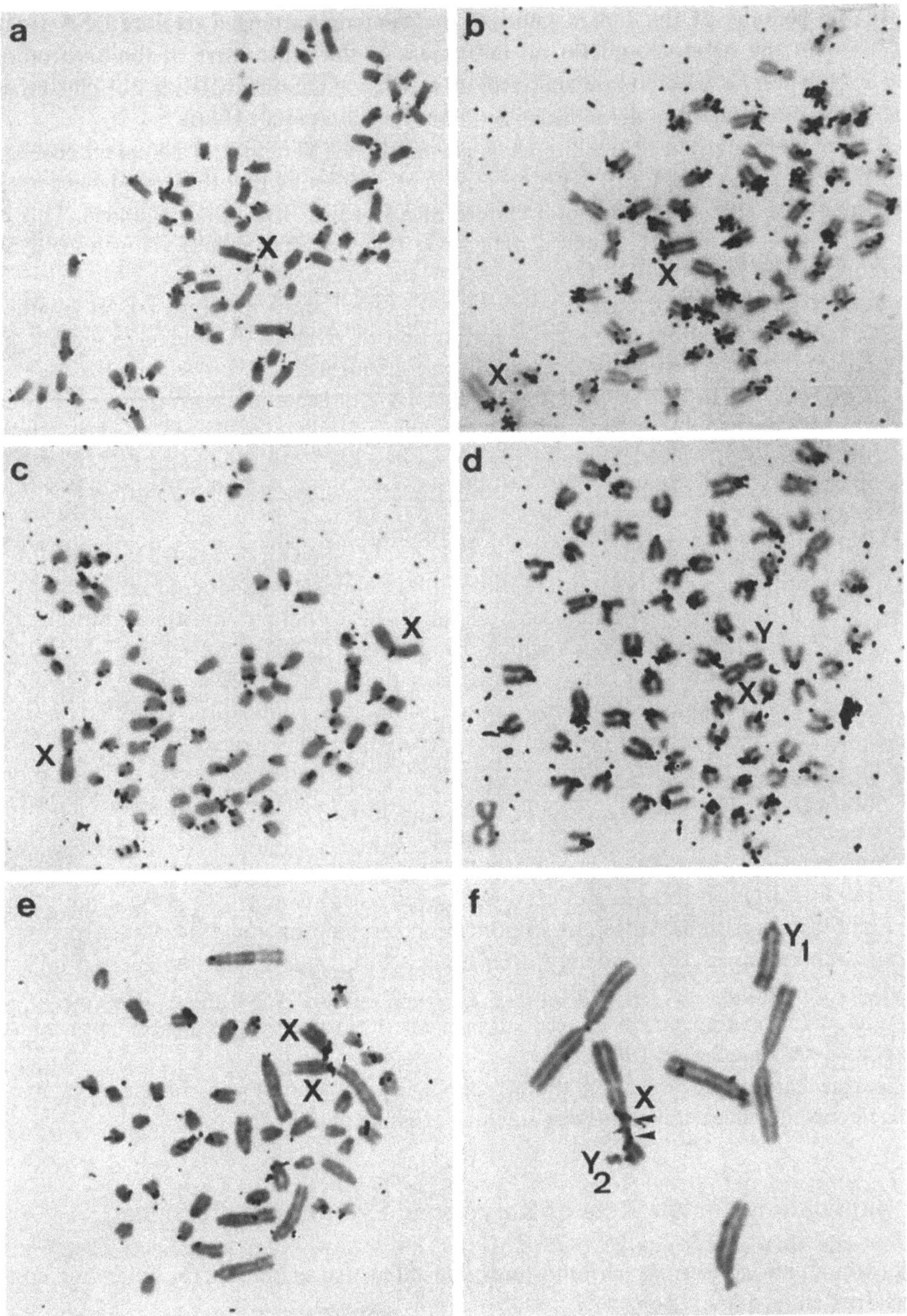

Fig. 14a–f. In situ hybridization with the satellite DNA IA probe demonstrating that this satellite DNA is localized in the centromeric heterochromatin. a *Mazama gouazoubira;* b *Cervus duvauceli;* c *Odocoileus bezoarticus;* d *Odocoileus dichotomus;* e *Muntiacus reevesi;* f *Muntiacus muntjac*

rearrangement should lead to a high frequency of unbalanced gametes in heterozygotes and, therefore, to a significant reduction in the fertility of a population. Presumably it is for this reason that tandem fusions are realized in the karyotypic evolution of only a very few mammalian taxa. With respect to karyotypic evolution of the muntjacs it is speculated that a specific, repetitive sequence was located in the centromeric and telomeric regions of the ancestor genome, promoting repeated chromosome translocations by sequence-specific recognition (Schmidtke et al. 1981).

Based on karyological observations in cell cultures it can be assumed that some of these sequences are still retained in the muntjac genome since in an established cell line of *Muntiacus muntjac* characteristic tandem fusions occurred (Fig. 15). Similarly, in cell lines of Cervinae chromosomal rearrangements of the Robertsonian type have been observed in vitro. These results implicate the existence of an inherent intragenomic capacity for realizing a specific type of chromosomal rearrangement, presumably by recognition of specific DNA sequences.

2. In contrast to the hypothesis of Chu and Bender (1963), no phenotype-karyotype correlation can be demonstrated for the family Cervidae since phenotypically more primitive forms can have either a conserved ancestral or a highly diverged karyotype as demonstrated for the species *Hydropotes inermis* and *Muntiacus muntjac* respectively.

Therefore, karyotypic diversification does not seem to be necessarily paralleled by anatomical divergence as claimed by some authors (Wilson et al. 1974). Chromosomal diversification is obviously due to neutral mutations which occur in a single specimen of a population and which can be subsequently transmitted throughout the population without having any phenotypic effects.

3. The rate of chromosomal evolution within a taxa depends on both the frequency of chromosome mutations in a population, and particularly on the population structure.

There are some indications that the frequency of chromosomal mutations might have been higher in the cource of evolution of Cervidae than in other taxa, since it can be demonstrated that the satellite IA, which is present in the genomes of all Cervidae, contains two internal sequences which were found to be recombination hot spots in several other organisms (Bogenberger et al. 1985; Neitzel et al., in prep.).

The transmission of chromosomal rearrangements depends mainly on the population structure. Within panmictic populations, chromosomal rearrangements should have only a minimal chance of being fixed, especially those which result in a decrease of fertility in the heterozygotes. These rearrangements can be fixed only in populations which exhibit a high degree of inbreeding since only mating within small demes promotes the rapid homozygote fixation of chromosomal rearrangements within a short time.

There is firm evidence that particularly the genera *Muntiacus* and *Mazama*, which have the highest rate of karyotypic evolution in Cervidae, form very small breeding groups as part of their social behavior (Clutton-Brock et al. 1980).

Fixation of the rearranged karyotype can act as an isolation mechanism, allowing the accumulation of further isolating factors, which in turn may promote anatomical diversification.

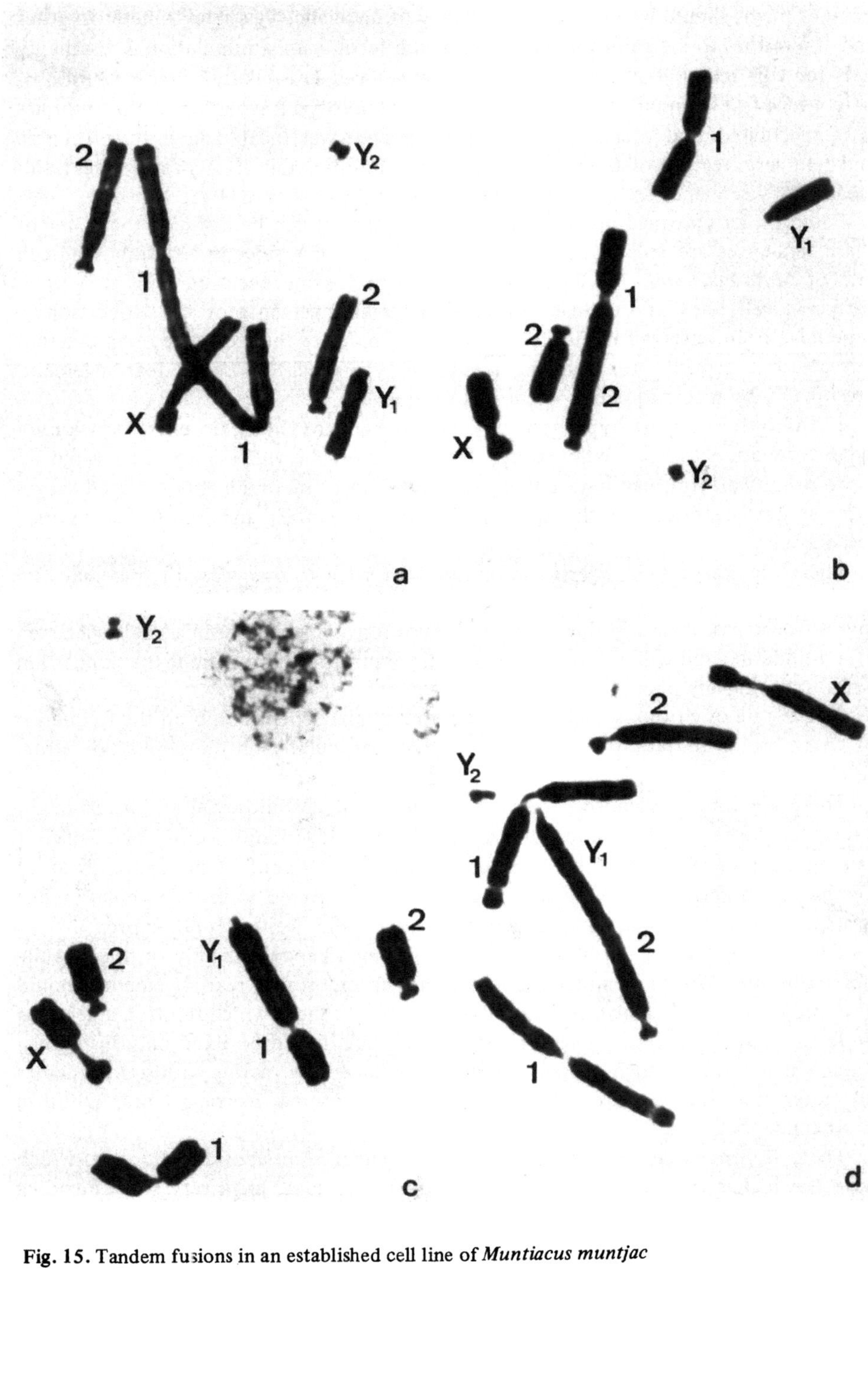

Fig. 15. Tandem fusions in an established cell line of *Muntiacus muntjac*

References

Amrud J, Nes N (1966) The chromosomes of the roe (Capreolus capreolus). Heriditas 56:217–220

Bender MA, Chu EHA (1963) The chromosomes of primates. In: Evolutionary and genetic biology of primates. Academic Press, London New York

Benirschke K (1969) Comparative mammalian cytogenetics. Springer, Berlin Heidelberg New York

Benedum UM, Neitzel H, Sperling K, Bogenberger J, Fittler F (1986) Organization and chromosomal distribution of a novel repetitive DNA component from Muntiacus muntjak vaginalis with a repeat length of more than 40 kb. Chromosoma 94:267–272

Bogenberger J, Schnell H, Fittler F (1982) Characterization of X-chromosome specific satellite DNA of Muntiacus muntjac vaginalis. Chromosoma 87:9–20

Bogenberger J, Neumaier PS, Fittler F (1985) The muntjac satellite IA sequence is composed of 31-bp-pair internal repeats that are highly homologous to the 31-bp-pair subrepeats of the bovine satellite 1.715. Eur J Biochem 148:121–144

Bonner TI, Brenner BJ, Neufeld BR, Britten RJ (1973) Reduction in the rate of DNA reassociation by sequence divergence. J molec Biol 81:123–135

Buckland RA, Evans JH (1978) Cytogenetic aspects of phylogeny in the Bovidae. I. G-banding. Cytogenet Cell Genet 21:64–71

Chandra HS, Hungerford A' Wagner J (1967) Chromosomes of five artiodactyl mammals. Chromosoma 21:211–220

Deaven LL, Vidal-Rioja L, Jett JH, Hsu TC (1977) Chromosomes of Peromyscus (Rodentia, Cricetidae). VI. The genomic size. Cytogenet Cell Genet 19:241–249

Evans JH, Buckland RA, Summer AT (1973) Chromosomes homology and hterochromatin in goat sheep and ox studied by banding techniques. Chromosoma 42:383–402

Frädrich H (1981) Internationales Zuchtbuch für den Pampashirsch, Blastoceros bezoarticus. Bongo 5:73–80

Gustavsson I, Sundt CO (1968) Karyotypes in five species of deer (Alces alces L., Capreolus capreolus L., Cervus elaphus L., Cervus nippon Temm. and Dama dama L.). Heriditas 60:233–247

Gustavsson I, Sundt CO (1969) Three polymorphic chromosome systems of centric fusion type in a population of Manchurian Sika deer (Cervus nippon hortulorum Swinhoe). Chromosoma 28:245–254

Haltenorth TH (1963) Klassifikation der lebenden Säugetiere. Familie Hirsche Cervidae. Handbuch der Zoologie, VIII. De Gruyter, Berlin, pp 38–60

Hatch FT, Bodner AJ, Mazrimas JA, Moore DH (1976) Satellite DNA and cytogenetic evolution: DNA quantity, satellite DNA and karyotypic variations in kangaroo rats (genus Dipodomys). Chromosoma 58:155–165

Hsu TC, Benirschke K (1967–1974) Mammalian chromosomes atlas. Springer, Berlin Heidelberg New York

Jorge W, Benirschke K (1977) Centromeric heterochromatin and G-banding of the Red Brocket Deer, Mazama americana temama (Cervoidea, Artiodactyla) with a probable No-Robertsonian translocation. Cytologia 42:711–721

Koulischer L, Tyskens J, Mortelmanns J (1972) Mammalian cytogenetics. The chromosomes of Cervus canadensis, Elapharus davidianus, Cervus nippon (Temminch) and Pudu pudu. Acta Zool Pathol Antverp 56:25–29

Mayr E (1958) Behavior and evolution. Yale Univ Press, New Haven

Miklos GLG, John B (1979) Heterochromatin and satellite DNA in man: Properties and prospects. Am J Human Genet 31:264–280

Neitzel H (1982) Karyotypenevolution und deren Bedeutung für den Speciationsprozess der Cerviden (Cervidae; Artiodactyla; Mammalia). Thesis, Freie Univ Berlin

Pathak S, Hsu TC, Arrighi FE (1973) The role of heterochromatin in karyotypic evolution: Chromosomes of Peromyscus. Cytogenet Cell Genet 12:315–326

Schmidtke J, Brennecke H, Schmid M, Neitzel H, Sperling K (1981) Evolution of muntjac DNA. Chromosoma 84:187–193

Shi Liming YE Yingying, Duan Xingsheng (1980) Comparative cytogenetic studies on the red muntjac, Chinese muntjac and their F1 hybrids. Cytogenet Cell Genet 26:22–27

Shi Liming YE, Pathak S (1981) Gametogenesis in a male Indian muntjac x Chinese muntjac hybrid. Cytogenet Cell Genet 30:152–156

Singer M (1982) Highly repeated sequences in mammalian genomes. In: Gourne GH, Danielli JF (eds) Int Rev CYtol 76:67–112

Soma H, Kada H, Mtayoshi K, Suzuki Y, Meckvichal C, Mahannop A, Vatanaromya B (1983) The chromosomes of Muntiacus feae. Cytogenet Cell Genet 35:156–158

Taylor KM, Hungerford DA, Snyder RL (1969) Artiodactyl mammals: their chromosome cytology in relation to patterns of evolution. In: Comparative mammalian cytogenetics. Springer, Berlin Heidelberg New York

Thenius E, Hofer H (1960) Stammesgeschichte der Säugetiere. Springer, Berlin Heidelberg New York

Wilson AC, Sarich VM, Maxson LR (1974) The importance of gene rearrangement in evolution: Evidence from studies on rates of chromosomal. Protein and anatomical evolution. Proc Natl Acad Sci USA 71:3028–3030

Wurster DH, Benirschke K (1970) Indian muntjac Muntiacus muntjac: A deer with a low diploid chromosome number. Science 168:1364–1366

Zech L, Evans EP, Ford CE, Gropp A (1971) Banding patterns in mitotic chromosomes of tobacco mouse. Exp Cel Res 70:263–250

Zimmermann EG, Lee MR (1968) Variation in chromosomes of the cotton rat, Sigmodon hispidus. Chromosoma 24:243–250

6 Cytogenetic Studies in Human Neoplasia

E. GEBHART[1]

1 Historical Background

The roots of tumor cytogenetics date back to the late 19th century, when Arnold (1879) was the first to describe mitotic abnormalities in the cells of tumors, and von Hansemann (1890) postulated that all carcinomas are characterized by asymmetric karyokineses (mitoses). His assumption that the unequal chromatin distribution as a result of asymmetric mitoses might be responsible for the disordered growth of cancer cells was resumed and further developed by Boveri (1914), who formulated a first "chromosome theory of cancer origin". His profound knowledge of normal mitosis and of the high biological importance of its integrity led him to suggest that the cells of malignant tumors should be characterized by an abnormal chromosome constitution, and that any event causing chromosome abnormalities should result in malignancy. In addition, he anticipated the idea of a monoclonal origin of neoplasias when stating: "Typically each tumor takes its origin from one and a single cell" (English translation 1929). It was also due to him that his ideas drew attention to the importance of the genotype of the somatic cell in the origin of malignancy, although it took more than 40 years until his hypothesis could be definitely tested in human neoplasia.

As an indispensable prerequisite the exact chromosome number of man had to be determined first which was not attained before 1956 (Tjio and Levan 1956). Of course, much effort had been devoted to the chromosome constitution of tumor cells just before this date, as reviewed by Sandberg (1980), but no study presented reliable results on the detailed human tumor karyotype at that time.

In spite of the fact that the first years of more sophisticated cytogenetic research on human neoplasia yielded a nearly inextricable thicket of different karyotypic changes, it was now possible to establish the importance of selection for the evolution of tumor cells (Levan 1956; Hauschka 1961). A first milestone in the development of human tumor cytogenetics, however, was set by Nowell and Hungerford (1960) who discovered the first specific chromosomal anomaly in a human neoplastic disease (chronic myeloic leukemia, CML), which was called "Philadelphia chromosome" (Ph') after the place of its detection. But even these detections would not have brought tumor cytogenetics to its present importance if modern banding techniques for chromosome identification had

1 Institut für Humangenetik der Universität, Schwabachanlage 10, D–8520 Erlangen, FRG

Cytogenetics. Ed. by G. Obe and A. Basler

not been introduced about 10 years later. Their utilization for human tumor cytogenetics induced an explosive development starting with the recognition of the Ph' chromosome as a product of translocation between the long arms of chromosomes 22 and 9 (Rowley 1973). The consequential detection of a series of other specific translocations, deletions, inversions and cytogenetic equivalents of gene amplification has not ended yet. Sandberg's voluminous collection of past results of human neoplasia cytogenetics (1980) and Mitelman's catalog of chromosome aberrations in cancer (2nd edn. 1985), though marking a certain completion of the data of classical tumor cytogenetics, are not its definite keystones. On the contrary, new methods are now applied and since they are based on the insights of classical neoplasia cytogenetics, they now opened the door to the molecular understanding of the specific chromosomal changes in human neoplastic cells. Seventy years after Boveri had formulated his hypothesis on the importance of mutations for the origin of malignancy in the genome of somatic cells, it could now be shown that he was right in a deeper sense than he could ever have dreamed.

2 Present State of Classical Cytogenetics of Human Neoplasia

2.1 The Chromosomes of Human Cancer and Leukemia

It was not by chance that the first specific chromosome abnormality of a human neoplastic disease was found in a leukemia. Hematologic neoplasias were a favorite object of cytogenetic studies because of the availability of the target cell material as a single cell suspension and the rather simple techniques of chromosome preparation from this material. Stimulated by the detection of the Ph' chromosome, vast literature on the cytogenetics of human leukemias has accumulated so far (Sandberg 1980; Rowley 1984; Mitelman 1985). The most important collections of pertinent data were presented at the "International workshops on chromosomes in leukemia" (1979–1984). Human solid tumors, on the other hand, were less amenable to cytogenetic analysis due to diverse technical reasons, discussed in detail in the pertinent literature (e.g. Trent 1984, 1986).

Since the fundamental detection of translocations t(9;22) in chronic myeloid leukemia and t(8;14) in Burkitt lymphoma and other malignant lymphomas a series of more or less specific chromosomal changes were described in the various types of leukemia and lymphoma (e.g. Rowley 1984); the most characteristic ones are compiled in Table 1. The majority of leukemias are characterized by the occurrence of less chromosomal changes per cell than solid tumors, translocations in leukemic cells often being of the "balanced" type. The nonrandomness of involvement of chromosomes and chromosomal sections in marker formation is striking (Mitelman 1985).

In the first years of cytogenetic research on human leukemias the prevailing impression was that about 50% of all cases did not show any cytogenetic change at all. This point, however, had to be corrected as banding techniques were improved to "high resolution banding" (Yunis 1981) and new techniques of cytodiagnosis increased the possibilities of a selective and reliable analysis of the neoplastic cells. Utilizing these facilities, several authors suggested that, for instance, up to 100% of the acute nonlymphocytic leukemias (ANLL) actually present detectable cytogenetic changes

Table 1. Characteristic chromosome changes in human leukemias and lymphomas and their clinical implications

Specific chromosome change	Characteristic for[a] (percentage of cases)	Subtype (FAB)	Clinical implications	References[b]
1. Translocations				
t(9;22) (q34;q11) = Ph'	CML (95%)		Only change in chronic phase: good prognosis	1
	AML (?)	M 1		10
	ALL (17.5% adult;	L 2	Poor prognosis	2
	5.8% child.)	L 1		
t(8;14) (q24;q32)	Burkitt lymphoma; large			10
	cell immunobl. lymphoma;			10
	small cell non-Burkitt l.			10
	ALL (5.8% adult; 3.8% child.)	L 3	Poor response to therapy, short survival	2
der 14 q+	ALL (6.9% adult; 1.9% child.)	L 2	"Intermediate" prognosis	2
t(5;11) (q21;q23)	ALL (5.5%)	L 1	Poor response to therapy; short survival	2
t(8;21) (q22;q22)	ANLL (12.5%)	M 2	Remission rate and survival "good" (incombination with missing sex chromosome "poor")	3
t(15;17) (q22;q11.2)	APL (70%)			
	ANLL (12%)	M 3	In optimally treated cases good progn.; long survival	3
t(6;9) (q21.2;q34)	ANLL	M 1, M 2		10
t(9;11) (q22;q23)	ANLL	M2,4,5a		10
t(11;14) (p13;q13)	ALL	L 1 (T-cell)		10
t(11;14) (q12;q32)	CLL (rare)	B-cell		6
	Small cell lymphocyt. lymph.	B-cell		10
t(1;19) (q23;p13)	ALL	L1 (pre-B)		10
t(14;18) (q32.3;q21.3)	Follicular lymphomas	Div. types		10
der 11 q (23)	ANLL (8%)	M 5	Short median duration of remission	3
inv (16) (p13;q22)	ANLL (4.7%)	M 4 (AMMoL)	Good prognosis (?)	3
		M 2		
i (17q)	CML (11.5%)		Blastic crisis	4

Table 1 (continued)

Specific chromosome change	Characteristic for[a] (percentage of cases	Subtype (FAB)	Clinical implications	References[b]
2. Deletions and Monosomies				
del (6) (q21–25)	ALL			10
	Diffuse large cell lymphoma			10
	CLL	T-cell		10
der 6 q-	ALL (1.7% adult; 6.4% child.)	L 1, L 2	Good prognosis	2
del (9) (q13–q22.1)	ANLL	M 2		10
-5/5q-	ANLL (10%)	M 2	Poor prognosis (short survival); frequent in secondary ANLL	3
	Preleukemia (16%)			7, 8
-7/7q-	ANLL (14%)	(M 2)	Long survival	3
	Preleukemia (19%)			7, 8
del (7) (p11.2–p22)	ANLL	M 2		10
3. Trisomies				
+ 8	CML (37,5%)		Preceding blastic crisis	4
	ANLL (10%)	M 4	"Intermediate" prognosis	3
	Preleukemia (17%)			7, 8
+ 12	CLL (40%)	B-cell		9
	Small cell lymphoc. lymphoma	B-cell		10
+ 21	ANLL (3.5%)		Long median duration of remission; rel. long survival	3

[a] Abbreviations: *ALL* acute lymphocytic leukemia; *AML* acute myelocytic leukemia; *ANLL* acute nonlymphocytic leukemias; *CLL* chronic lymphocytic leukemia; *CML* chronic myelocytic leukemia; *AMMoL* acute myelomonocytic leukemia.

[b] References: *1* Sandberg (1980); *2* 3rd Int Worksh on Chromosomes in Leukemia (1981); Le Beau and Rowley (1984); *3* 4th Int Worksh on Chromosomes in Leukemia (1984); *4* Alimena et al. (1982); *5* Zech et al. (1984); *6* San Roman et al. (1982); Ueshima et al. (1984); *7* 2nd Int Worksh on Chromosomes in Leukemia (1980); *8* Van den Berghe (1984); *9* Gahrton et al. (1985); *10* Yunis (1985).

(Yunis 1985). Thus, cytogenetics of human leukemias and lymphomas has now reached a level of development which not only allows reliable diagnostic application, but also presents valuable support to prognosis as well as to the control of progression and regression of the diseases (see Sect. 3). In addition, it was the cytogenetic data on the specificity of chromosomal translocations in certain neoplastic diseases of the hematopoietic system (Table 1) which formed one of the most promising fields of modern cancer research, i.e. the investigation of the role of oncogenes in human cancer development (see Sect. 4).

Interestingly, the first solid tumor in which a specific chromosomal change was detected (i.e. a missing or deleted chromosome 22) was a benign tumor: meningeoma (Zang and Singer 1967; Mark et al. 1972). In contrast to the "minor" chromosomal changes of this tumor, human malignant carcinomas, in general, are characterized by a large variety of numerical and structural chromosome abnormalities. The difficulties of obtaining suitable cell suspensions from solid carcinomas as well as of culturing these cells and preparing them for cytogenetic investigation, so far has hindered an extensive and specific cytogenetic definition of characteristic markers comparable to those in leukemias. Only during the last years has a more definite picture of the cytogenetic peculiarities of human solid carcinomas become available (Table 2).

One of the most striking cytogenetic features, once more, was the unexpectedly nonrandom involvement of rather few chromosomes of the human karyotype in marker formation in human malignant tumors. On the involved chromosomes only specific bands are the favorite sites of interchange and deletion. Figure 1 shows the chromosome most frequently involved in marker formation in human solid carcinomas; i.e. chromosome 1 (Brito-Babapulle and Atkin 1981; Atkin 1986; Gebhart et al. 1986b). As reviewed by Rowley (1978) the same chromosome also plays an important role in the spectrum of anomalies found in leukemias and lymphomas. Of particular interest is the implication of the pericentric bands (p13–q12), including the centromer (which is the region of isochromosome formation), the distal region (p31–p36) of the short arm and the bands q21–q23 on the long arm.

While, in general, the karyotypic anomalies in cells of advanced stages of neoplasia appear confounding by number and variety, Mitelman (1986) paid particular attention to those cases characterized by one single abnormality only. Among 5345 reviewed cases he found 318 different aberrations as the only change from the normal diploid karyotype. Seventy-seven abnormalities were selected that were identical in at least two neoplasms of the same or related entity. The identified 161 breakpoints were clustered to a total of only one-fourth of the available bands of the standard human karyotype. At least one of these breakpoints was found to be involved in 96% of leukemias, lymphomas and solid tumors with complex karyotypic changes, indicating that only a limited number of breakpoints are regularly involved in chromosomal aberrations of human neoplasia. In addition, he pointed out that breakpoints of diseases affecting related cell types seem to cluster to certain chromosomal regions (for details see Mitelman 1986).

Changes often found in carcinoma cells pertain to isochromosomes, the long arm of chromosome 1 being the most frequently involved chromosome arm (Table 2; Gebhart et al. 1986b). Another characteristic and frequent finding in cells, particularly of solid carcinomas, is the presence of cytogenetic manifestations of gene amplification

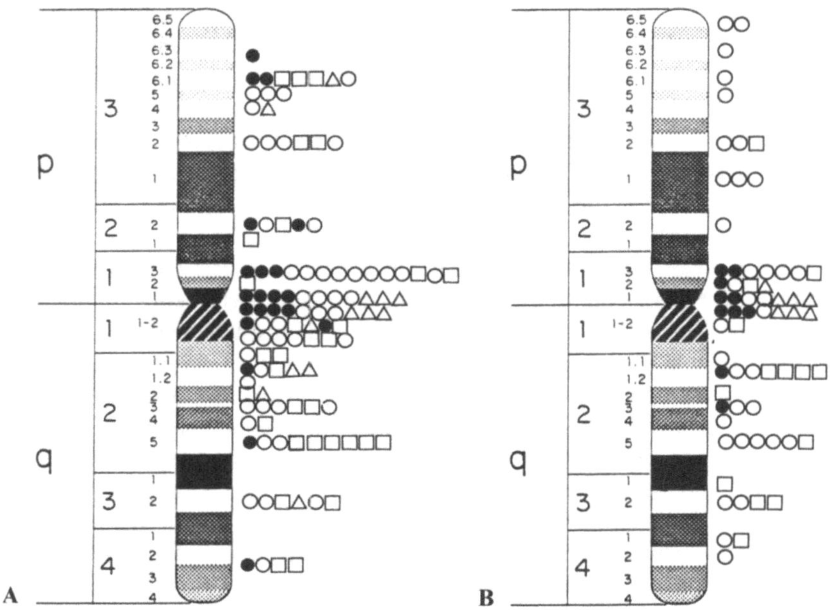

Fig. 1A–C. Distribution of breakpoints of marker formation on chromosome 1 in human solid carcinomas. **A** Clonal markers; **B** nonclonal markers of a study of this group (Gebhart et al. 1986); **C** breakpoints from the literature. *Circles:* breast carcinoma; *squares:* ovarian ca.; *triangles:* cancers of other sites

Table 2. Some characteristic marker chromosomes from human solid tumors. (After Yunis 1985)

Description of anomaly	Mainly observed in (type of carcinoma)
t (1; . .) (q. .; . .)	Breast cancer
t (3;8) (p21;q12)	"Mixed salivary gland tumors"
t (6;14) (q21;q24)	Papillary cystadenocarcinoma of ovary
t (11;22) (q24;p12)	Ewing's sarcoma; neuroepithelioma
del (1) (p31–36)	Neuroblastoma
del (3) (p14–p23)	Small cell lung cancer; rhabdomyosarcoma; ovarian cancer; renal cell carcinoma
del (6) (q21)	Malignant melanoma
del (11) (p13)	Wilm's tumor (with aniridia)
del (13) (q14)	Retinoblastoma
del (20) (p12.2)	Multiple endocrine cancer
– 22 / del (22q)	Meningeoma
i (1p)	Breast carcinoma; uterus carcinoma; bladder cancer; lung ca.
i (1q)	Breast ca.; colon ca.; bladder ca.; uterus ca.; ovarian ca.; renal cell carcinoma
i (5p)	Breast ca.; bladder ca.; lung ca.; renal cell ca.
i (12p) / 12p-	Testicular tumors
i (17q)	Uterus ca.; colon ca.; breast ca.; ovarian ca.

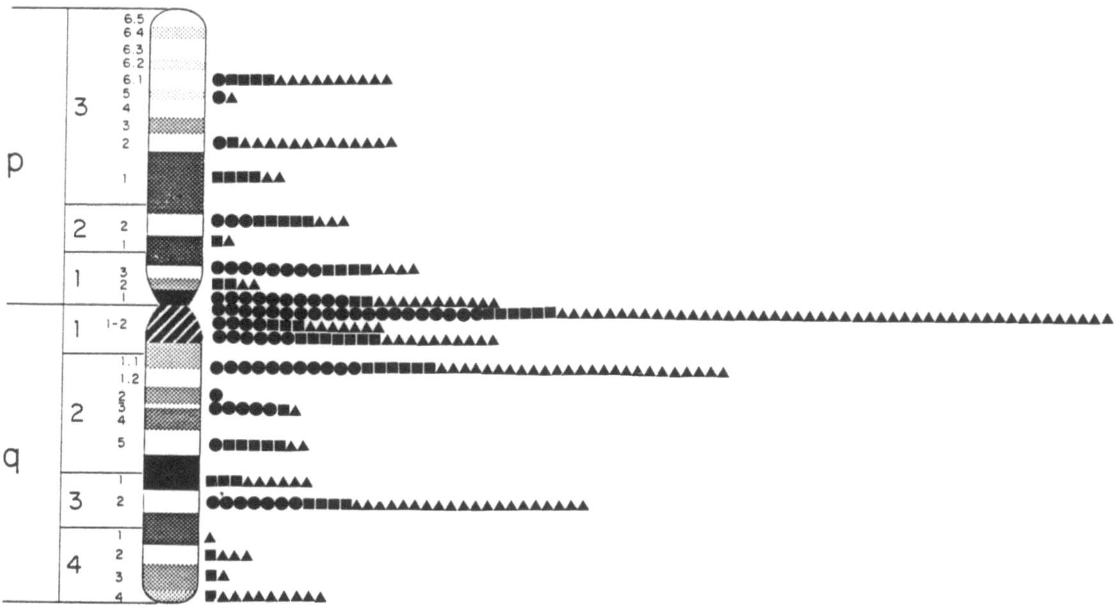

Fig. 1C

which appear as "double minutes" (DM) and/or "homogeneously staining regions" (HSR). While the presence of cytogenetic equivalents of gene amplification originally seemed to be restricted to exceptional solid tumors of neurogenic origin (Biedler and Spengler 1976; Balaban-Malenbaum and Gilbert 1977; Brodeur et al. 1984), DM and HSR have now been found as a frequent marker in carcinomas of different entities (Barker 1982; Gebhart et al. 1984, 1986b), but occasionally also in leukemias (Table 3) (Hartley and Toolis 1980; Cooperman and Klinger 1981; Li 1983; Oguma et al. 1985). It could be shown recently that DM are also preserved in interphase chromatin made visible by the technique of premature chromosome condensation (PCC; Brüderlein and Gebhart 1985).

2.2 Karyotypic Evolution in Neoplastic Cells

Cell populations from human neoplasias are composed of a more or less heterogeneous variety of cells (Owens et al. 1982; Heppner 1984), this fact has, in particular, clinical implications (Schnipper 1986). According to the stem-line concept of tumor progression, this heterogeneity may be explained by the genetic lability of malignant cells and selective mechanisms acting on those cell populations which were originally derived from one single transformed cell (reviewed by Nowell 1976; Yosida 1983). The karyotype of any neoplasia, therefore, is not an invariable feature, but in most

Table 3. Human neoplasia: cytogenetic equivalents of gene amplification

Equivalent	Primary tumors	Metastatic tumors	Cell lines
Double minutes	Breast cancer Ovarian cancer Bladder cancer Gastric cancer "Mixed salivary gland tumor" AML Colon cancer Retinoblastoma Glioma Neublastoma Medullablastoma Rhabdomysarcoma Testicular tumor	Breast cancer Ovarian cancer Cancer of the cervix uteri Gastric cancer Lung cancer Medulloblastoma Neuroblastoma Osterosarcoma Leukemias	Breast cancer Colon cancer Neoplastoma Rhabdomyosarcoma Thyroid cancer Cancer of the cervix uteri Chondrosarcoma
Homogeneously Staining Regions (HSR)	Neuroblastoma Breast cancer Pharyngeal cancer Esophagus cancer Ovarian cancer Melanoma Bladder cancer	Neuroblastoma Breast cancer Ovarian cancer Melanoma	Neuroblastoma Colon cancer Retinoblastoma Melanoma Small cell lung cancer Breast cancer Oral cancer

cases is subject to steady changes. This so-called karyotypic evolution may proceed very slowly, as documented, for instance, by the extremely stable karyotype of a breast tumor cell line recently established by us (ER BT 134; Gebhart et al. 1986), or more or less rapidly as shown by many pertinent reports. Human tumor cell lines may present a good in vitro model for this karyotypic evolution within tumor cell populations as was demonstrated by Göhl (1985) in established human breast tumor cell lines (MDA MB 231 and MDA MB 435; Cailleau et al. 1978). By a subtle combination of cloning procedures he succeeded in reconstructing major routes of this evolution. While chromosome 7 in line 435 and chromosome 9 in line 231 were the first involved in marker formation, chromosome 1 was generally that chromosome mostly involved in "early" as well as in "late" markers in both cell lines.

A very similar in vitro and in vivo pattern of karyotypic changes could be demonstrated in a carcinoma of the cervix uteri most recently (Gebhart et al. 1986). This observation is in contrast to the assumptions of other authors (e.g. Swindell and Ockey 1983; Rodgers et al. 1984), who argued that in vitro evolution might not be representative for in vivo conditions. As a general rule, however, an increasing number and variety of marker chromosomes, sometimes accompanied by the emergence of cytogenetic equivalents of gene amplification (Brodeur et al. 1984; Gebhart et al. 1986a), can be interpreted as signs of an advanced stage of tumor progression (see Nowell 1983 for review).

The genetic lability of transformed cells claimed by this latter author as the motor of tumor cell variability, however, is apparently expressed to a different extent in various tumors. For instance, Gebhart et al. (1986a) studied nonclonal structural

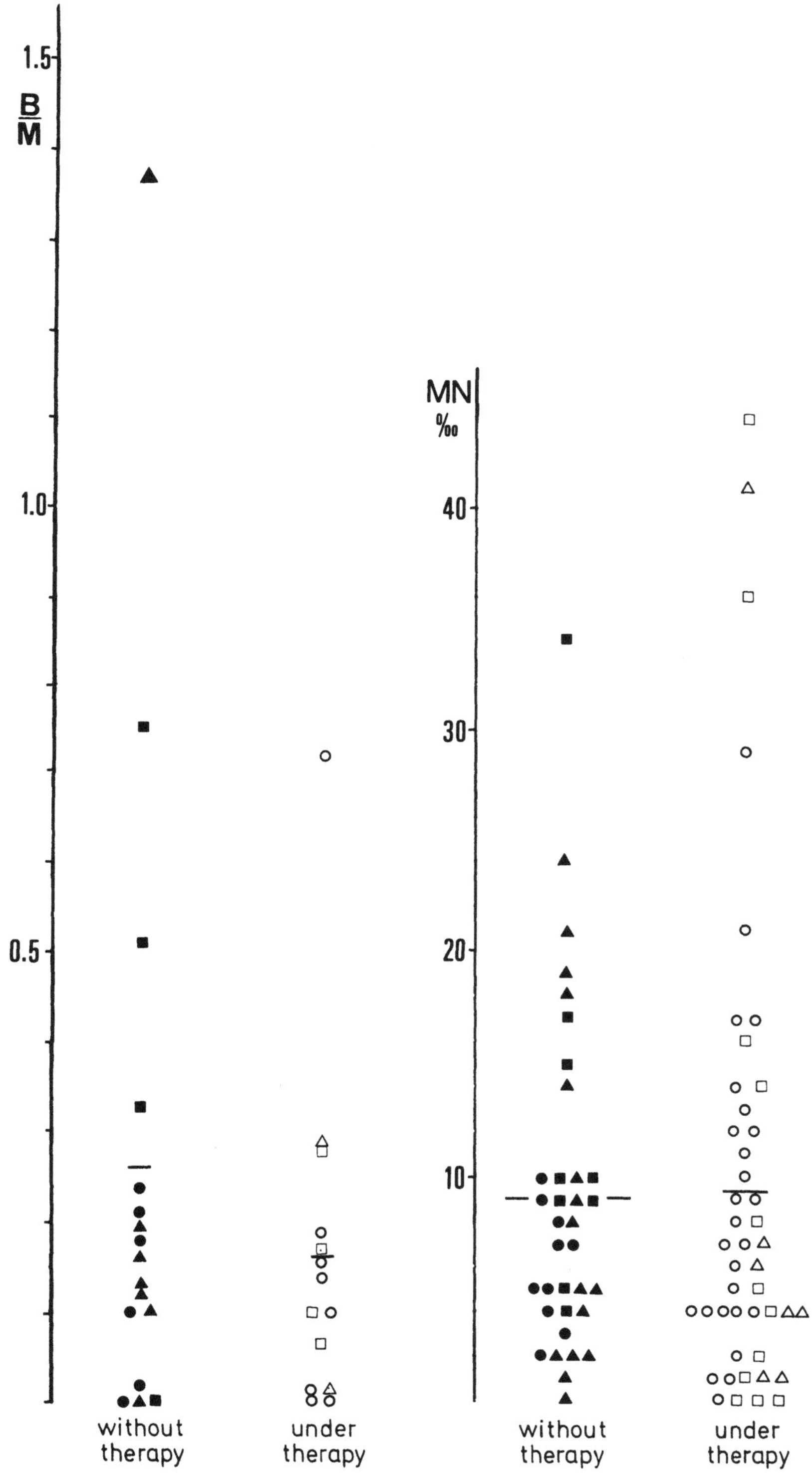

Fig. 2. Breakage rate and frequency (breaks per metaphase $\frac{B}{M}$) of micronuclei (*MN*) in human carcinoma cells. Symbols as in Fig. 1 (Gebhart et al. 1986a)

chromosome damage and micronuclei in metastatic carcinoma cells obtained from cancerous effusions of tumor patients. As shown in Fig. 2 in cases without any therapy as well as in treated cases not only increased, but also baseline levels of chromosomal breakage and micronuclei could be detected. It, therefore, was concluded that the observed chromosomal instability had to be regarded as a facultative rather than an obligatory attitude of malignant cells. As a consequence it could be taken as one of several causes of tumor cell heterogeneity.

Several reports have also been published on SCE in malignant cells as reviewed by Shiraishi and Sandberg (1980) and Gebhart (1981). Once more, increased as well as baseline levels of SCE were reported from different human neoplasias, and even within the same type of tumor this variability of outcome was maintained (e.g. Reisbach et al. 1982).

As a result, besides visible changes in the karyotype of human tumor cells, many additional genetic changes along tumor progression may be held responsible for tumor cell heterogeneity. In particular, amplification of proto-oncogenes (see below) may play an important role in this process since gene amplifications may provide drug resistance to the tumor cells (Schnipper 1986).

2.3 Constitutional Chromosome Instability and Malignancy

As chromosomal instability syndromes (CIS) have so far been subject to a variety of extensive and sophisticated reviews (recently in: German 1983), it remains to note their fundamental importance for our understanding of the close connections between chromosomal mutability and malignant transformation. These autosomal, recessive hereditary diseases (Table 4) are characterized, among other symptoms, by an increased spontaneous and/or inducible frequency of chromosome aberrations and a striking tendency to develop malignant diseases. Differences in their cellular response to clastogenic influences indicate different mechanisms leading to chromosomal instability, but do not affect the common pathway from chromosomal aberration to malignancy. A major role of chromosomal changes was particularly suggested by the detection of clonal marker chromosomes in those syndromes which were highly similar to those found in malignancies common to CIS (reviewed by Chaganti 1983; Ray and German 1983).

There are, however, also some other cytogenetic indicators for the predisposition to malignant diseases. Interstitial deletions involving band 13q14 in retinoblastoma and 11p13 in Wilm's tumor (if associated with aniridia) have been reported as constitutional anomalies in normal cells, but sometimes also in tumor cells of a number of children suffering from these diseases (Riccardi et al. 1978; Francke et al. 1979; Sparkes 1984). More recently, Pathak and Goodacre (1986) reviewed findings that specific constitutional translocations or at least an increased fragility of specific common fragile sites in peripheral lymphocytes can indicate a predisposition to breast, renal cell and colorectal carcinoma of the respective individuals. In this context the high coincidence between "common fragile sites" on human chromosomes (made visible by folate deprivation or DNA polymerase inhibition) and the favorite breakpoints of markers in neoplasias must be mentioned (Sutherland 1979; Yunis and Soreng 1984; Sutherland and Hecht 1985; Le Beau 1986; see also Fig. 3).

Table 4. Spontaneous and induced chromosome damage (breakage) and sister chromatid exchange (SCE) in human chromosome instability syndromes. (According to Heddle et al. 1983, modified)[a]

Syndrome:	Bloom's syndr.		Ataxia tele-angiect.		Fanconi-syndr.		Xeroderma pig-ment.	
Agent	Breakage	SCE	Breakage	SCE	Breakage	SCE	Breakage	SCE
Spontaneous rate	i	i	i	e	i	e	e	e
Induced by UV	e	e	e		i		i	i
Ionizing irrad.	e/i		i		e/i	e/i	e	
N-Lost			e		i	i	e	i
Diepoxybutane	e		i		i	e	e	i
4-NQO	i	i	e	e	i	e	i	i
MMS	i	i	i/e	e	i		e	i
EMS	i	i		e	i/e	e		i/e
MNNG	i	i	i	e	i		i/e	i
ENU		i						i
MNU	i	i						
Cytosine arabin.			d		i		e	
BRDU	e				e			
Streptonigrin			i		i		e	
Bleomycin			i	e	i	e	e	
Dichromate			(i)		i		(i)	
Tallysomycin			i				e/d	
cis-Pt			e		i		i	
Mitomycin C	e/i	e/i	e	e	i		e/i	i
Neocarcinostatin			i		i		e	
Presumed basic defect[b]	Intracellular clastogenic factors; defect in DNA chain elongation		Defects in site-specific processing of DNA; repair defect?		Defect in "cross-link" repair?		Defect in excision repair	

[a] Abbreviations: *i* increased compared to normal cells; *e* equal to normal cells; *d* decreased compared to normal cells.

[b] Recent review of Hanawalt and Sarasin (1986).

3 Clinical Implications of Neoplasia Cytogenetics

The present state of cytogenetics of human neoplasia, at least in certain areas, only allows practical applications of the obtained data for diagnostic purposes. The extensive collection of data at the international level, in addition, yielded important insights into the prognostic meaning of certain karyotypic constitutions of leukemic cells. These data, together with newly developed techniques, also render the control of remission, disease progression and success of therapy possible.

3.1 Diagnostic Application of Cytogenetic Techniques in Human Malignant Diseases

As the vast majority of human malignancies is characterized by karyotypic abnormalities, cytogenetic analysis is a practical means for discerning between normal and malignant cells. This is of particular importance in premalignant states and early malignancies as well as in histologically unclear tissue (Knörr-Gärtner et al. 1977; Fisher and Paulson 1978).

The best known area of diagnostic application of cytogenetic analyses, however, is that of leukemias, in particular of CML (Ph' chromosome), as the karyotypic changes in hematologic diseases not only allow typification and grading (International Workshops on Chromosomes in Leukemia 1978–1984), but also present important opportunities for prognosis with respect to progression and therapy of the respective disease (see Table 1).

With respect to nonhematologic malignancies the presence of structural and/or numerical changes, in any case, defines the malignant stage of the respective cell. Similar to leukemias the diagnostic importance of certain chromosomal configurations may be established as soon as sufficient data is available on specific markers in human solid tumors.

3.2 Cytogenetic Changes and Prognosis in Human Leukemias

As shown in Table 1, a large number of data collected on human leukemias now can associate specific chromosomal changes to features of response to therapies, duration of remission and survival. In addition, a very fascinating cytogenetic means for the control of remission and prediction of relapse in leukemia was contributed by Hittelman and Rao (1978) and Hittelman (1982), who utilized the technique of premature chromosome condensation (PCC) to demonstrate differences in interphase kinetics and in that of the disease. In spite of the rather limited cytogenetic data from human carcinomas, first evidence has been gained for the prognostic importance of DM (Gebhart et al. 1986) and HSR (Brodeur and Seeger 1986) in these malignancies, as both phenomena are apparently associated with an advanced stage and rapid progression of the respective diseases.

Preliminary data (Brüderlein et al. 1986) point to possibilities of a limited application of PCC studies as a prognostic tool in human carcinoma.

3.3 Cytogenetic Control of the Progression of Neoplastic Disease

Changes of the chromosomal pattern within neoplastic cells, in principle, can indicate changes in the disease progression. The best known example, once more, is CML, as it could be shown that a successful therapy was associated with the disappearance of the Ph-positive cell population in this disease (Sandberg 1980; Talpaz et al. 1986), and vice versa, an increase in the percentage of Ph-positive cells indicates the advance from an early stage of the disease to its full expression. In addition, chromosomal abnormalities supervening the Ph' chromosome, announce the forthcoming blastic crisis. In general, a high level of karyotypic abnormality is associated with advanced stages of neoplasia, as is the presence of an increased number of cytogenetic equivalents of gene amplification (DM and HSR). Hittelman's PCC model (1982), if confirmed conclusively, would offer an appreciated chance of control of remission and relapse in leukemias, and thus, of the success or failure of therapeutic efforts. The definition of resistance against antineoplastic therapeutics can be assisted by cytogenetic methods, too. The presence of DM or HSR in neoplastic cells as well as the lack of increased chromosome damage after application of therapeutics to the respective cells in vitro can be taken as indicative for resistance (Goldie and Coldman 1984).

4 Actual Problems of Cytogenetics of Human Neoplasia

Although the definition of "actual" may be debatable and the selection of actual problems subjective, there might be full agreement among all concerned with this field that the development of neoplasia cytogenetics to molecular dimensions is the most important one in recent years. It not only helped to detect proto-ancogenes on human chromosomes, but also helped to understand their normal function and their activation to oncogenes in human neoplasia. Theories on the mechanisms of malignant transformation can now be substantiated by experimental and epidemiological data.

4.1 Chromosomal Localization of Proto-Oncogenes

The highly specific chromosomal changes found in certain human neoplasias (see above) and the detection of transforming genes in viruses led to an intense search for a possible connection between both phenomena. Using techniques of in situ DNA hybridization for the detection of DNA sequences homologous to viral oncogenes, a fascinating close relationship could be determined between chromosomal bands nonrandomly involved in marker formation and loci of proto-oncogenes (i.e. DNA sequences on human chromosomes homolous or very similar to viral oncogenes). A present state map of localization of proto-oncogenes and marker breakpoints is given in Fig. 3 (see also review by Tereba 1985). On grounds of the extreme conservativity of these genes in the phylogenetic tree, their normal function soon was regarded to be a fundamental one. By the time it was found that these genes are expressed for strictly limited time lapses in the course of early development (Slamon and Cline 1984) and cell proliferation by action as growth factors (Stiles 1985; Goustin et al. 1986). Apparently an untimely activation of these genes leads to malignant transformation (Bishop 1982; Land et al. 1983; Weinberg 1984). Table 5 presents a collection of oncogenes found activated in a variety of human malignant diseases.

4.2 Cytogenetics of Oncogene Activation

It has been shown above that the specificity of translocation in the process of marker formation in human neoplasia eventually led to the detection of oncogene localization. In addition, cytogenetic studies were also of basic importance for the understanding of oncogene activation (recently reviewed by Pedersen-Bjergaard et al. 1986).

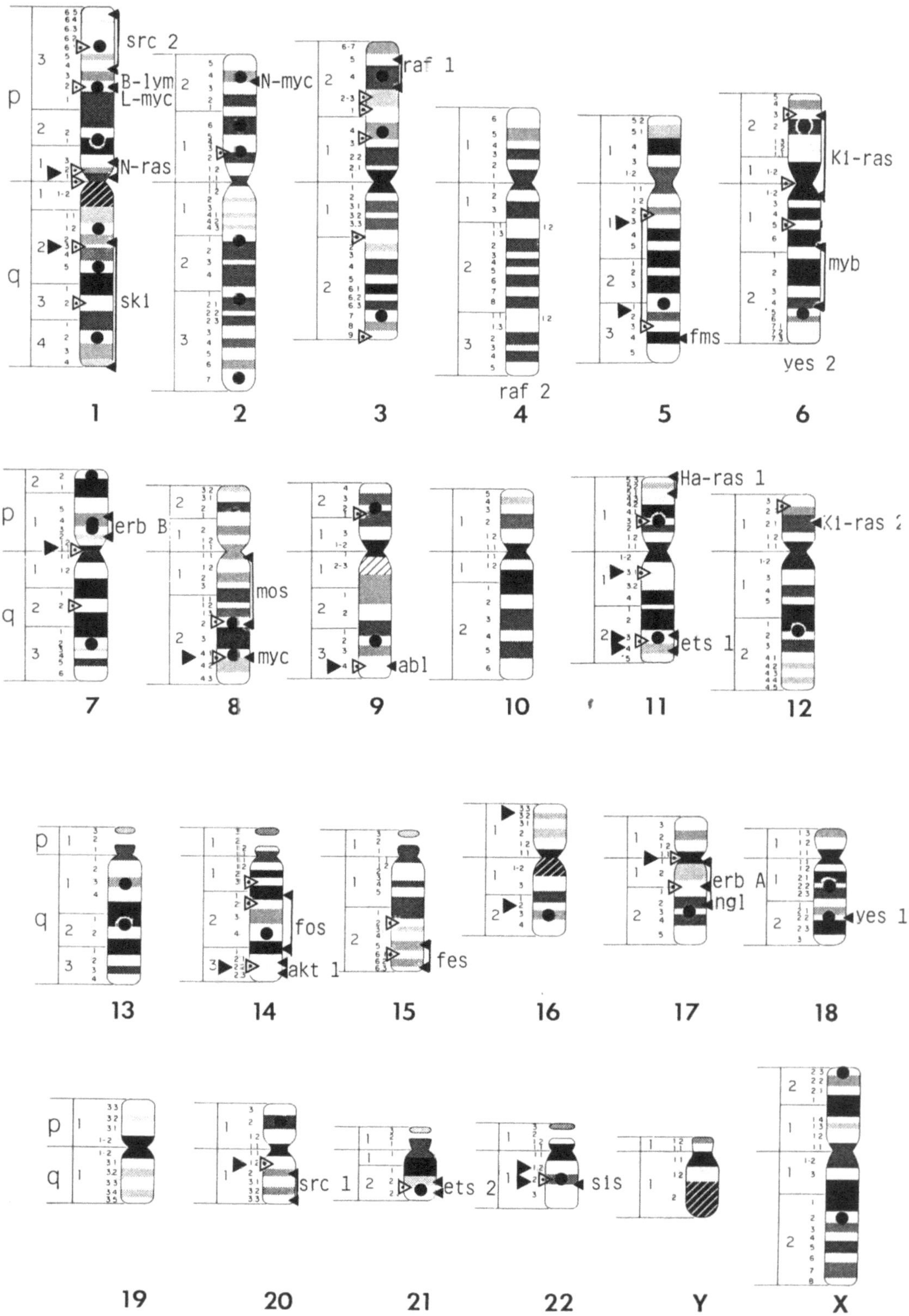

p
q
src 2
B-lym
L-myc
N-ras
ski
N-myc
raf 1
raf 2
fms
Ki-ras
myb
yes 2
1
2
3
4
5
6
erb B
mos
myc
abl
Ha-ras 1
ets 1
Ki-ras 2
7
8
9
10
11
12
fos
akt 1
fes
erb A
ngl
yes 1
13
14
15
16
17
18
src 1
ets 2
sis
19
20
21
22
Y
X

Table 5. Oncogenes found activated in human neoplastic diseases. (According to Blick et al. 1984; McClain 1984; Slamon et al. 1984)

Oncogene	Chromosome localization	Found activated in[a] (type of neoplasia; numbers in parantheses: cases with activation per cases studied)
c-myc	8 q 24	Burkitt-lymphoma (nearly all); AML (12/14); CML (14/14); CML-BT (4/4); CLL (3/3); PBL (1); colorectal adenoca. (9/10); NHLY (3/3); M.Hodgkin (1); sarcomas (2/2); lymphosarcoma (1); renal cell ca. (8/9); ovarian adenoca. (6/6); lung ca. (4/4); breast ca. (4/4); colon ca.; colonic polyps.
N-myc	2 p 23/24	Neuroblastoma
L-myc	1 p 32	Small cell lung cancer
c-abl	9 q 34	CML (nearly all); AML (15/26); ALL (11/15); CML-BT (1/4); CLL (2/6)
c-myb	6 q 21–23	AML (17/17); ALL (5/6); CML (14/14); CML-BT (4/4); breast ca. (2/4); NHLY (1/2); small cell lung ca.; sarcomas
c-Ha-ras 1	11 p 141	AML (14/18); ALL (6/9); CML (2/2); CML-BT (1); CLL (5/5); PBL (2/2); colorectal ca. (8/10); breast ca. (4/4); renal cell ca. (7/9); ovarian ca. (6/6); lung ca. (4/4); sarcomas (2/2); lymphomas (4/4); lymphosarcoma (1); urinary tract tumors; squamous cell ca.; prostatic ca.
c-Ki-ras 1: 2:	6 p 11–12 12 p 121	Ovarian ca. (4/6); colorectal ca. (7/10); renal call ca. (5/9); breast ca. (3/4); lung ca. (4/4); AML (10/11); ALL (10/11); squamous cell ca.; prostatic ca.; (low activity in CML; sarcomas; lymphomas)
N-ras	1 p 11–13	AML; CML; melanoma; neuroblastoma
c-fms	5 q 34	5q-Syndrome; pancreatic ca. (1); lung ca. (2/4); breast ca. (3/4); renal cell ca. (4/9); CML; AMMoL; M.Hodgkin (1); (low activity in cecal adenoca.; ovarian adenoca.; AML)
c-fos	14 q 21–31	Renal cell ca. (7/9); ovarian adenoca. (6/6); colorectal ca. (10/10); lung ca. (4/4); breast ca. (4/4); sarcomas (2/2); AML (6/7); M.Hodgkin (1); NHLY (2/2)
c-fes/fps	15 q 261	AML (12/19); ALL (7/12); CML (2/2); renal cell ca. (2/9); lung ca. (4/4); breast ca. (3/4)
c-src 1: 2:	20 q 12–13 1 q 34–36	ALL (4/11); AML (5/11); CML (1/2); lymphosarcoma (1)
c-erb B	7 pter-22	ALL (5/9); AML (2/6); M.Hodgkin (1)
c-sis	22 q 131	CML-BT (1/5)

[a] For abbreviations see Table 1.

Fig. 3. Preferential breakpoints of marker formation in human neoplasia (*triangles*), common fragile sites (*circles*) and localization sites of proto-oncogenes (*arrows to the right* of the chromosomes) an human chromosomes. *Black triangles* "sole abnormality" markers (see Mitelman 1986); *white triangles* common markers

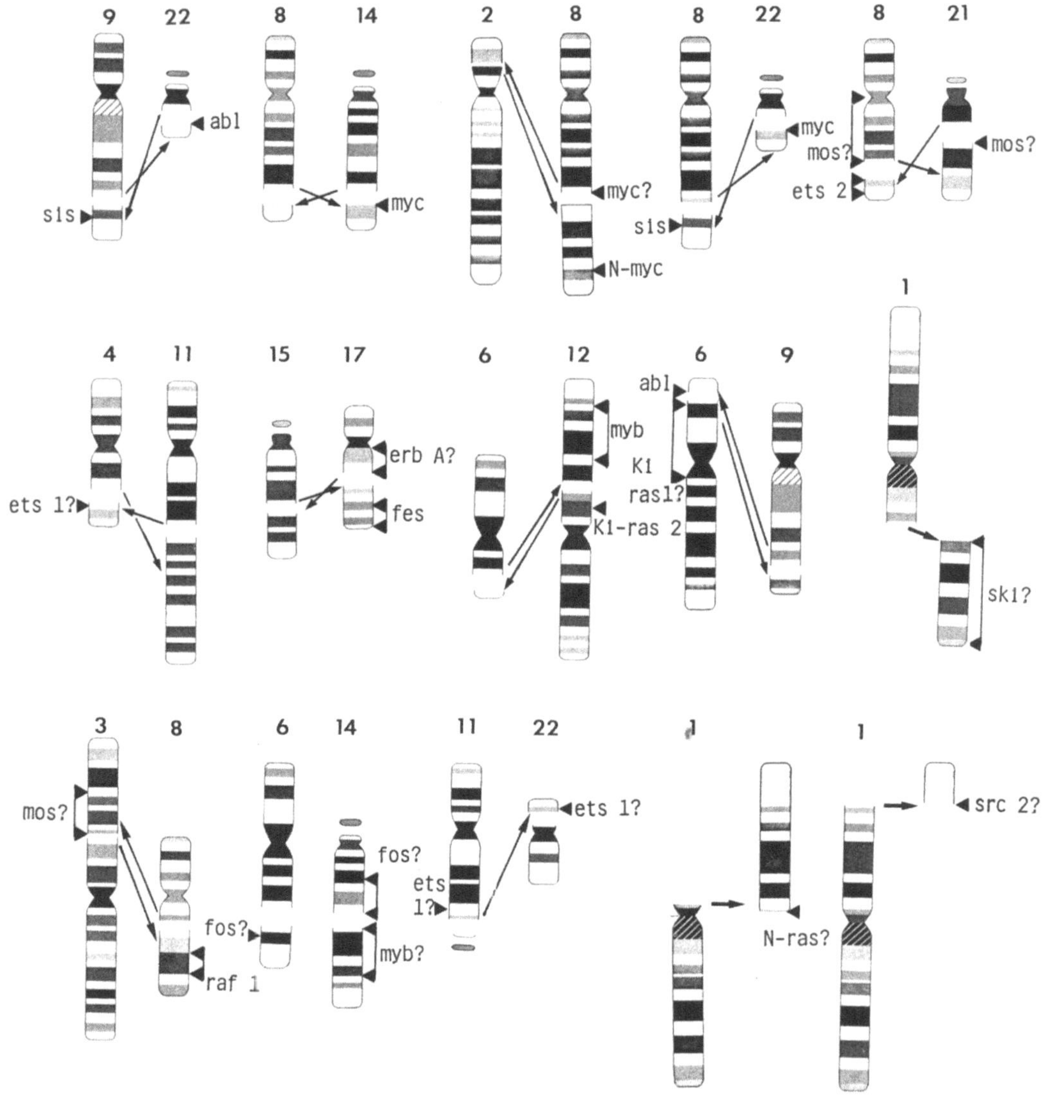

Fig. 4. Examples of translocations in human neoplastic diseases involving localization sites of proto-oncogenes

4.2.1 The Role of Translocation in Oncogene Activation

Once more, the well-known Philadelphia chromosome played a key role when the activation of cellular oncogenes was examined. De Klein et al. (1982) could demonstrate that the oncogene c-abl, which is located in band q34 of chromosome 9, is transferred to the rest of chromosome 22 by "Philadelphia" translocation (reviewed by

Bartram 1985). This transfer from the original region, where it might have been under the control of neighboring genes, into a region of high gene activity (chain of immunoglobulins) was first suggested to be the cause of its activation. More recently it could be shown that the c-abl oncogene is linked to specific sequences (bcr) on chromosome 22, the rearranged sequences transcribing a novel abl/brc hybrid RNA species of about 8 kb (Gale and Canaani 1984; Konopka et al. 1984; Bartram et al. 1986). Another translocation, also impressively demonstrated the role of the transfer of an oncogene from its normal location into the area of highly active genes: the translocation t (8;14) (q24; q32) by which the proto-oncogene c-myc is brought into the chromosomal site on 14q where the genes coding for the heavy chains of immunoglobulins are situated (Croce et al. 1979, 1984; Taub et al. 1982).

An identical recombination was found in plasmocytoma of the mouse (Adams et al. 1983; Klein and Klein 1984). As shown in Fig. 4 a series of other translocations have been suspected to cause oncogene activation, the molecular mechanisms of which, however, are not understood in every case, but sometimes seem to be of a more complex nature than originally assumed (see Tereba 1985 for review).

4.2.2 Oncogene Amplification

As mentioned above, two cytogenetic phenomena have attracted the acute interest of tumor cytogeneticists as well as molecular geneticists, i.e. double minutes and homogeneously staining regions (Fig. 5). Both cytogenetic peculiarities have been shown to be rather frequent in solid carcinomas (Gebhart et al. 1984, 1986) and to be related to clinical features of the malignancies (Brodeur et al. 1984). Soon after their definition as the sites carrying amplified genes (Schimke 1980), several authors detected amplified oncogenes on these regions (Collins and Groudine 1982; Alitalo et al. 1983, Dalla-Favera et al. 1983), suggesting an important role of amplification for the activation of proto-oncogenes. A systematic screening with specific oncogene probes revealed the presence of lower grades of amplification for a series of different human neoplasias (Table 6), which were not present as a detectable cytogenetic manifestation. Thus, these results could support the idea that primary amplifications of oncogenes might play an important role for tumor progression at a very early stage of the disease, while their "visible" cytogenetic manifestations are more likely signs of an advanced stage of neoplasia. Some authors (e.g. Heilbronn et al. 1985) have even discussed an initiating role of gene amplification in the neoplastic process.

4.3 Cytogenetics of the Process of Malignant Transformation

Since the detection of specific marker chromosomes in human neoplasia and also in experimentally induced animal tumors (Kato 1968; Levan 1969; Di Paolo et al. 1971; Ikeuchi and Honda 1971), the role of chromosomal aberrations in malignant transformation has been discussed. This discussion was favored by the findings of the induction of chromosomal changes by well-known carcinogens and the general mutation hypothesis of carcinogenesis.

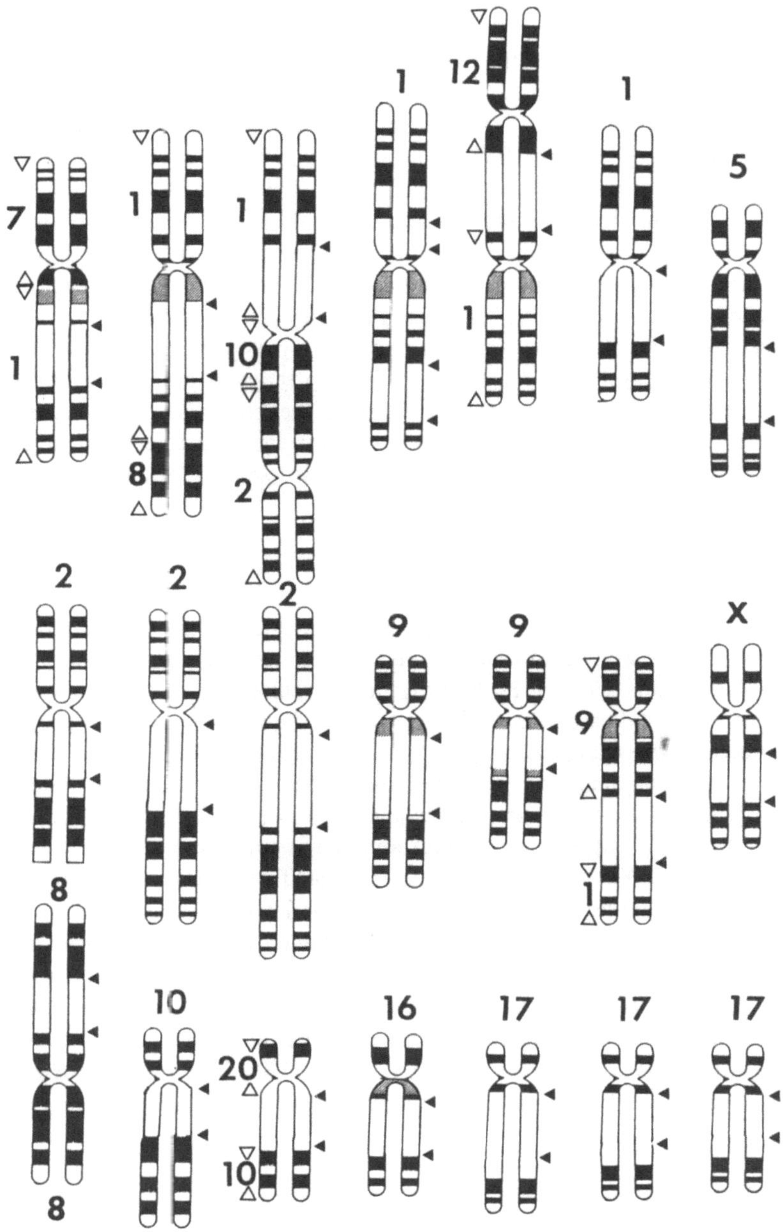

Fig. 5. Graphic definition of homogeneously staining regions (HSR) found in human breast cancer and ovarian tumors (Gebhart et al. 1987). *White arrows* limiting the indicated chromosome; *HSR* white areas on the chromosomes limited by *black arrows*)

Table 6. Amplified oncogenes in human neoplasias (data collected from numerous publications)

Amplified oncogene	Amplification factor	Cytogenetic equivalents[a]	Type of neoplasia amplification was observed in
c-myc	– 10 x	–	Stomach cancer; breast cancer; giant cell lung cancer;
	10 – 30 x	DM / HSR	Neuroendocrine colon ca.; glioblastoma; APL; CML; cervical ca.; gastric adenoca;
	20 – 90 x	DM / HSR	Small cell lung ca.; breast ca.
	+ 100 x	HSR	Neuroblastoma
N-myc	– 10 x	–	Neuroblastoma
	10 – 90 x	?	Retinoblastoma
	+ 100 x	HSR	Neuroblastoma; retinoblastoma; small cell lung ca.
L-myc	10 – 20 x	–	Small cell lung cancer
c-Ha-ras 1	– 10 x	–	Bladder ca.; ca. of cervix uteri
	10 – 30 x	–	Melanoma
c-Ki-ras	10 – 30 x	– / DM	Ovarian cystadenoca.; epidermoid lung ca.; giant cell lung ca.
	30 – 50 x	–	Gastric ca.; lymph node metast. of lung ca.
N-ras	20 x	–	Breast cancer
c-erb B1	– 10 x	–	Breast ca.; glioblastoma; epidermal ca.
	10 – 30 x	–	Glioblastoma; epidermal ca.
	30 – 90 x		Glioblastoma; epidermal ca.
c-erb B2	30 x	–	Adenoca. of salivary gland; other adenoca.
c-abl	4 – 8 x	–	CML
c-myb	– 10 x	–	AML, colon ca.
c-yes 1	– 10 x	–	Gastric ca.
c-ets 1	10 – 30 x	HSR	AMMoL, lymphoma

[a] DM double minutes; *HSR* homogeneously staining regions; + more than

4.3.1 Epidemiological Data

The detection of a specific pattern of chromosome markers in secondary leukemia (4th International Workshop on Chromosomes in Leukemia 1984; Sandberg et al. 1982; Alimena et al. 1982; Rowley 1983) as well as in leukemias of persons exposed to occupational toxicants (Mitelman et al. 1981; Brandt et al. 1983; Mitelman 1984) has attracted attention to the relation between exogenous factors and early chromosomal changes in human neoplasia. By a comparison of a group of 52 occupationally exposed ANLL patients (exposure: chemical solvents, insecticides, petrol products) with 110 ANLL patients with no history of occupational exposure, striking differences were detected (Mitelman et al. 1981): clonal chromosomal aberrations were present in 75% of exposed patients compared with only 32% in the nonexposed group. Of the patients exposed to solvents and insecticides 92% had abnormal chromosomes, whereas only 29% of patients exposed to petrol products showed abnormalities. The incidence of characteristic karyotypic abnormalities, i.e. -5/5q-, -7/7q-, +8, +21, +(8;21) and +(9;22), was decidedly more common in exposed than in nonexposed patients. The results of the 4th International Workshop on chromosomes in leukemia 1982 (publ. 1984), regarding secondary leukemias, also demonstrated the frequent involvement of chromosome 5(-5/5q-) and chromosome 7(-7/7q-), and, in addition to the above pattern, of chromosome 17(-17). Furthermore, Mitelman's findings (1986) in neoplastic cells of single structural abnormalities could point to the role that specific chromosomal changes play in the transformation process (see Fig. 3). Of particular interest were those regions shared by neoplastic disorders originating from different cell types, e.g. 1q21–23; 3p21–25; 8q21–24, 11q22–24; and 22q11–12, as those might be of specific importance in the process of malignant transformation. This role may be stressed by the local coincidence or close neighborhood of oncogene localization sites to those breakpoints (see Fig. 4).

4.3.2 Experimental Data

The first results showing the involvement of specific chromosomal changes in experimentally induced animal tumors were presented by Levan's group (e.g. Kato 1968; Kato and Levan 1969; Levan 1969; Mitelman et al. 1972; Mitelman and Levan 1972), Di Paolo et al. (1971) and Ikeuchi and Honda (1971). Of particular interest were the structurally abnormal "markers" emerging in the induced sarcomas in an obvious association to the agent used for transformation (Ahlström 1974; Levan 1974; Levan and Levan 1975). Later on, specific chromosomal changes could also be detected by others in a large variety of experimentally induced animal neoplasias, e.g. in neurogenic tumors of the rat (Au et al. 1977), in rat erythroblastic leukemia (Maeda et al. 1980), mouse sarcomas (Rachko and Brand 1983), rat leukemia (Sugiyama et al. 1978), tumors of the Indian spiny mouse (Yosida 1981) and mouse lymphomas (Yoshida et al. 1982).

Chromosome 15 of the mouse is apparently that one most frequently involved in trisomies (Chan et al. 1979, 1981; Carbonell et al. 1982; Hagemeijer et al. 1982), but also in translocations (Oikawa et al. 1984) of chemically induced mouse tumors. Many attempts were also made to define specific chromosomal patterns in animal cell lines transformed in vitro (e.g. Pathak et al. 1981; Tyrkus et al. 1982; Oshima and Barrett 1985; Ray et al. 1986), but only a few results are available which relate to the primary chromosomal changes in the transformation process (reviewed by Kakunaga et al. 1983; Dzarlieva-Petrusevska and Fusenig 1985). Even less information has been presented on chromosome changes in experimentally transformed human cells (Namba et al. 1985, 1986; Walen and Arnstein 1986). No characteristic pattern of markers comparable to that described by Levan (1974) in animal tumors has so far been observed in transformation experiments on human cells in vitro, although Walen and Arnstein (1986) reported nonrandom involvement of chromosome 8 in aberrations.

In summary, the role of chromosomal changes in the primary steps of malignant transformation of human cells has not been sufficiently elucidated by experimental studies so far.

4.4 Premature Chromosome Condensation in Cytogenetics of Neoplasia

Cytogenetic studies of human neoplasias are often complicated by the low yield of mitotic cells and the poor quality of their metaphase chromosomes. Therefore, any technique providing additonal advantages for cytogenetic studies of tumor cells should be appreciated. Studies on prematurely condensed chromosomes (PCC) of interphase cells could be a practical supplement to the classical methods as most of the tumor cells are in interphase at the time of observation.

The phenomen of PCC or "prophasing" (Johnson and Rao 1970; Matsui et al. 1972; Rao et al. 1976) can be induced by fusion of the interphase target cells with mitotic "inducer" cells. It changes "homogeneous" interphase chromatin to visible, individual chromosomes. As shown in Section 3.2, Hittelman and Rao (1978) and Hittelman (1982) has demonstrated the use of PCC analyses in predicting relapse in leukemias. This technique, however, not only allows a reliable estimation of interphase kinetics by comparative analyses of the PCC pattern of the concerned cells (Sperling and Rao 1974; Rao et al. 1976), but may also be used for the analysis of specific cytogenetic phenomena in poorly proliferating cells. Brüderlein and Gebhart (1985), for instance, demonstrated the usefulness of this method in verifying the presence of "double minutes" in interphasic tumor cells. Some authors also could attain banding in PCC of normal cells (Unakul et al. 1973; Röhme and Heneen 1982). There are only a few studies on PCC of human solid tumors (Williams et al. 1976; Atkin 1979; Reichmann and Levin 1981; Brüderlein et al. 1986). The latter authors studied PCC in 52 cancerous effusions and found a highly individual pattern of distribution of the various stages of interphase, reflecting the heterogeneity of human solid tumors in an advanced stage. Nevertheless, a variety of clinical, biological and technical factors were examined for their possible influence on these PCC patterns. The duration in culture was one of the influencing factors, as were the time lapse between the first diagnosis and the sampling of the respective effusion, or the nature of cytostatic therapy. In addition, the authors could detect DM in 35 of the 52 studied effusions and achieved an adequate banding quality in 21 of 24 effusions, yielding a sufficient number of well-spread PCC. They concluded that the PCC technique may prove to be an indispensable, additional source of cytogenetic information in cells of human solid tumors, particularly, if the possibilities of molecular cytogenetic studies are taken into consideration. But also spontaneous PCC apparently caused by spontaneous cell fusion in certain neoplasias or by the asynchrony in the cell cycle of micronuclei often found in neoplastic cells was described by several authors as an additional interesting feature of tumor cells (see above).

5 Conclusions

Cytogenetic studies from a variety of viewpoints have contributed important insights in the meaning for malignant transformation of genomic changes in somatic cells: chromosomal changes are an integrated mechanism of malignancy, probably starting with the initiation process and extending to the advanced stages of neoplastic development. Several observations clearly document the nonrandom involvement of specific chromosomal sites in marker formation and demonstrate "major routes" of karyotypic evolution from single, reciprocal translocations or gains and losses of single chromosomes to highly aberrant – simetimes "chaotic" – karyotypes.

Detailed information, so far, is missing with regards to the role of chromosomal changes in the very first steps of initiation and promotion, although a huge body of data on mutagenicity testing points to a rather specific role. This idea is also supported by the reported close interrelations between sites of interchange and localization sites of proto-oncogenes. The high specificity of certain marker chromosomes has not only just proven its practical value for diagnostic and prognostic measures as well as for the choice of therapeutic schedules and the control of their successful application, but also has given support to a global "mutation theory" of malignant transformation. Considering mutations in or near proto-oncogenes as the basic mechanisms of initiation and/or malignant transformation, any mutagenic event in somatic cells – independent of its eventual nature – can be interpreted as potentially carcinogenic. Certainly, the data and insights having contributed so much to this view have also raised a large variety of new and unanswered questions (e.g. Gilbert 1983; Le Beau 1986). The role of chromosomal instability as an indicator of cancer risks, and thus for the definition of high-risk populations, the possibilities of the utilization of cytogenetic techniques for diagnostic and prognostic measures and for therapy control, the interrelationships of fragile sites, specific sites of marker formation, hot spots of clastogenic damage and localization of proto-oncogenes, as well as a number of further problems are far from being clear. The end of tumor cytogenetics has been predicted several times, but every time it revived to even more fruitful heydays and gained even more fascination.

References

Adams JM, Gerondakis S, Webb E, Corcoran LM, Cory S (1983) Cellular myc gene is altered by chromosome translocation to an immunoglobulin locus in murine plasmacytomas and is rearranged similarly in human Burkitt lymphomas. Proc Natl Acad Sci USA 80:1982–1986

Ahlström U (1974) Chromosomes of primary carcinomas induced by 7,12-dimethylbenz(a)-anthracene in the rat. Hereditas 78:235–244

Alimena G, Dallapiccola B, Gastaldi R, Mandelli F, Brandt L, Mitelman F, Nilsson PG (1982) Chromosomal, morphological and clinical correlation in blastic crisis of chronic myeloid leukemia. A study of 69 cases. Scand J Haematol 28:103–117

Alitalo K, Schwab M, Lin CC, Varmus HE, Bishop JM (1983) Homogeneously staining chromosomal regions contain amplified copies of an abundantly expressed cellular oncogene (c-myc) in malignant neuroendocrine cells from a human colon carcinoma. Proc Natl Acad Sci USA 80:1701–1711

Arnold J (1879) Beobachtungen über Kernteilungen in Zellen der Geschwülste. Virchows Arch Pathol Anat 78:279–301

Atkin NB (1979) Premature chromosome condensation in carcinoma of the bladder: presumptive evidence for fusion of normal and malignant cells. Cytogenet Cell Genet 23:217–219

Atkin NB (1986) Chromosome 1 aberrations in cancer. Cancer Genet Cytogenet 21:279–286

Au W, Soukup SW, Mandybur TI (1977) Excess of chromosome No. 4 in ethylnitrosoureau-induced neurogenic tumor lines in the rat. J Natl Cancer Inst 59:1709–1716

Balaban-Malenbaum G, Gilbert F (1977) Double minute chromosomes and the homogeneously staining regions in chromosomes of a human neuroblastoma cell line. Science 198:739–741

Barker PE (1982) Double minutes in human tumor cells. Cancer Genet Cytogenet 5:81–94

Bartram CR (1985) Activation of proto-oncogenes in human leukemias. Blut 51:63–71

Bartram CR, De Klein A, Hagemeijer A, Carbonell F, Kleihauer E, Grosveld G (1986) Additional c-abl/bcr rearrangements in a CML patient exhibiting two Ph' chromosomes during blast crisis. Leukemia Res 10:221–225

Biedler JL, Spengler BA (1976) Metaphase chromosome anomaly: Association with drug resistance and cell-specific products. Science 191:185–187

Bishop JM (1982) Oncogenes. Sci Am 246:68–78

Blick M, Westin E, Gutterman J, Wong-Staal F, Gallo R, McCredie K, Keting M, Murphy E (1984) Oncogene expression in human leukemia. Blood 64:1234–1239

Boveri T (1914) Zur Frage der Entstehung maligner Tumore. Fischer, Jena (Engl Transl 1929: The origin of malignant tumours. Baillière, Tindall & Cox, London)

Brandt L, Mitelman F, Nilsson PG (1983) Chromosome pattern and survival in acute non-lymphocytic leukaemia in relation to age and occupational exposure to potential mutagenic/carcinogenic agents. Scand J Haematol 30:227–231

Brito-Babupulle V, Atkin NB (1981) Break points in chromosome No. 1 abnormalities of 218 human neoplasms. Cancer Genet Cytogenet 4:215–225

Brodeur GM, Seeger RC (1986) Gene amplification in human neuroblastomas: Basic mechanisms and clinical implications. Cancer Genet Cytogenet 20:101–111

Brodeur GM, Seeger RC, Schwab M, Varmus HE, Bishop JM (1984) Amplification of N-myc in untreated human neuroblastomas correlates with advanced disease stage. Science 224:1121–1124

Brüderlein S, Gebhart E (1985) Double minutes in prematurely condensed chromatin of human tumor cells. Cancer Genet Cytogenet 16:145–152

Brüderlein S, Gebhart E, Siebert E, Augustus M (1986) Premature chromosome condensation – studies on human metastatic carcinoma cells. Human Genet 73:44–52

Cailleau R, Olive M, Cruciger QVJ (1978) Long-term human breast carcinoma cell lines of metastatic origin: Preliminary characterization. In Vitro 14:911–915

Carbonell F, Seidel HJ, Saka S, Kreja L (1982) Chromosome changes in butylnitrosourea (BNU)-induced mouse leukemia. Int J Cancer 30:511–516

Chaganti RSK (1983) The significance of chromosome change to neoplastic development. In: German J (ed) Chromosome mutation and neoplasia. Liss, New York, pp 359–396

Chan FPH, Ball JK, Sergovich FR (1979) Trisomy No. 15 in murine thymonas induced by chemical carcinogens, X-irradiation, and an endogenous murine leukemia virus. J Natl Cancer Inst 62:605–610

Chan FPH, Ens B, Frei JV (1981) Cytogenetics of murine thymic lymphomas induced by N-methyl-N-nitrosourea: Difference in incidence of trisomy 15 in thymic lymphomas induced in neonatal and adult mice. Cancer Genet Cytogenet 4:337–344

Collins SJ, Groudine MT (1983) Rearrangement and amplification of c-abl sequences in the human chronic myelogenous leukemia cell line K-562. Proc Natl Acad Sci USA 80:4813–4817

Cooperman BS, Klinger HP (1981) Double minute chromosomes in a case of acute myelogenous leukemia resistant to chemotherapy. Cytogenet Cell Genet 30:25–30

Croce CM, Shander M, Martinis J, Cicurel L, D'Ancona GG, Dolby TW, Koproowski H (1979) Chromosomal location of the genes for human immunoglobulin heavy chains. Proc Nat Acad Sci USA 76:3416

Croce CM, Tsujimoto Y, Erikson J, Nowell P (1984) Biology of disease: Chromosome translocations and B cell neoplasia. Lab Invest 51:258–267

Dalla-Favera R, Wong-Staal F, Gallo R (1982) Oncogene amplification in promyelocytic leukaemia cell line HL-60 and primary leukaemic cells of the same patient. Nature (London) 299:61–63

De Klein A, Van Kessel G, Grosveld G, Bartram CR, Hagemeijer A, Bootsma D, Spurr NK, Heisterkamp N, Groffen J, Stephenson JR (1982) A cellular oncogene is translocated to the Philadelphia chromosome in chronic myelocytic leukemia. Nature (London) 300:756–767

Di Paolo JA, Nelson RL, Donovan PJ (1971) Morphological, oncogenic and karyological characteristics of Syrian hamster embryo cells transformed in vitro by carcinogenic polycyclic hydrocarbons. Cancer Res 31:118–127

Dzarlieva-Petrusevska RT, Fusenig NE (1985) Tumor promoter 12-o-tetradecanoylphorbol-13-acetate (TPA)-induced chromosome aberrations in mouse keratinocyte cell lines: a possible genetic mechanism of tumor promotion. Carcinogenesis 6:1447–1456

Fisher ER, Paulson JD (1978) Karyotypic abnormalities in precursor lesions of human cancer of the breast. Am J Clin Pathol 69:284–288

Francke U, Holmes LB, Atkins L, Riccardi VM (1979) Aniridia-Wilm's tumor association: evidence for specific deletion of 11 p 13. Cytogenet Cell Genet 24:185–192

Gahrton G, Juliusson G, Robert KH, Zech L (1985) Review: Specific chromosomal markers in B- and T-cell chronic lymphocytic leukemia. Tumour Biol 6:1–12

Gale RP, Canaani E (1984) A 8-kilobase abl RNA transscript in chronic myelogenous leukemia. Proc Natl Acad Sci USA 81:5648–5652

Gebhart E (1981) Sister chromatid exchange (SCE) and structural chromosome aberration in mutagenicity testing. Human Genet 58:235–254

Gebhart E, Brüderlein S, Tulusan AH, Maillot K von, Birkmann J (1984) Incidence of double minutes, cytogenetic equivalents of gene amplification, in human carcinoma cells. Int J Cancer 34:369–373

Gebhart E, Augustus M, Brüderlein S (1987) Homogeneously staining regions (HSR) on chromosomes of human solid carcinomas. Biol Zentralbl 106:191–200

Gebhart E, Brüderlein S, Augustus M, Siebert E, Hosari E, Reisbach G, Göhl K, Schmidt W, Feldner J (1986a) Chromosomale Indikatoren der malignen Entartung. In: Kersten W (ed) DFG-Rep SFB 118. Chemie, Weinheim (in press)

Gebhart E, Brüderlein S, Augustus M, Siebert E, Feldner J, Schmidt W (1986b) Cytogenetic studies on human breast carcinomas. Breast Cancer Res Treat 8:125–138

German J (ed) (1983) Chromosome mutation and neoplasia. Liss, New York, 451 pp

Gilbert F (1983) Chromosomes, genes, and cancer: A classification of chromosome abnormalities in cancer. J Natl Cancer Inst 71:1107–1114

Göhl K (1985) Studien zur klonalen Karyotypevolution in etablierten Zellinien aus metastasierten Mammakarzinomen. Thesis, Univ Erlangen

Goldie JH, Goldman AJ (1984) The genetic origin of drug resistance in neoplasms: Implications for systemic therapy. Cancer Res 44:3643–3653

Goustin AS, Leof EB, Shipley GD, Moses HL (1986) Growth factors and cancer. Cancer Res 46: 1015–1029

Hagemeijer A, Smit EME, Govers F, De Both NJ (1982) Trisomy 15 and other nonrandom chromosome changes in Rauscher murine leukemia virus – induced leukemia cell lines. J Natl Cancer Inst 69:945–951

Hanawalt PC, Sarasin A (1986) Cancer-prone hereditary diseases with DNA processing abnormalities. Trends Genet 2:124–129

Hansemann D von (1890) Über asymmetrische Zellteilungen in Epithelkrebsen und deren biologische Bedeutung. Arch Pathol Anat Physiol 119:299–326

Hartley SE, Toolis F (1980) Double minute chromosomes in a case of acute myeloblastic leukemia. Cancer Genet Cytogenet 2:275–280

Hauschka TS (1961) The chromosomes in ontogeny and oncogeny. Cancer Res 21:957–974

Heddle JA, Krepinsky AB, Marshall RR (1983) Cellular sensitivity to mutagens and carcinogens in the chromosome-breakage and other cancer-prone syndromes. In: German J (ed) Chromosome mutation and neoplasia. Liss, New York, pp 203–234

Heilbronn R, Schlehofer JR, Yalkinoglu AO, Hausen H zur (1985) Selective DNA amplification induced by carcinogens (initiators): Evidence for a role of proteases and DNA polymerase alpha. Int J Cancer 36:85–91

Heppner GH (1984) Tumor heterogeneity. Cancer Res 44:2259–2265

Hittelman WN (1982) Premature chromosome condensation in the diagnosis of malignancies. In: Rao PN, Johnson RT, Sperling K (eds) Premature chromosome condensation. Academic Press, London New York, pp 309–358

Hittelman WN, Rao PN (1978) Predicting response or progression of human leukemia by premature chromosome condensation of bone marrow cells. Cancer Res 38:416–423

Ikeuchi T, Honda T (1971) Cytologic studies of tumors. XLVIII. Chromosomes of nine primary rat hepatomas induced by administration of 3'-methyl-4-dimethylamino azobenzene. Cytologia 36: 173–182

1st International Workshop on Chromosomes in Leukemia, 1977 (1978) Chromosomes in acute non-lymphocytic leukemia. Br J Haematol 39:311–316
2nd International Workshop on Chromosomes in Leukemia, 1979 (1980) Chromosomes in preleukemia. Cancer Genet Cytogenet 2:108–113
3rd International Workshop on Chromosomes in Leukemia, 1980 (1981) Clinical significance of chromosomal abnormalities in acute lymphoblastic leukemia. Cancer Genet Cytogenet 4:111–137
4th International Workshop on Chromosomes in Leukemia, 1982 (1984) Clinical significance of chromosomal abnormalities in acute nonlymphoblastic leukemia. Cancer Genet Cytogenet 11:332–350
Johnson RT, Rao PN (1970) Mammalian cell fusion: Induction of premature chromosome condensation in interphase nuclei. Nature (London) 226:717–722
Kakunaga T, Crow JD, Hamada H, Hirakawa T (1983) Mechanisms of neoplastic transformation of human cells. In: Harris CC, Autrup HN (eds) Human carcinogenesis. Academic Press, London New York, pp 371–399
Kato R (1968) Chromosomal studies on carcinogenesis in the Chinese hamster. Berlingska Boktryckeriet, Lund, 15 pp
Kato R, Levan A (1969) Chromosomal aberrations in carcinogenesis. Jpn J Genet 44 Suppl 2:90–91
Klein G, Klein E (1984) Oncogene activation and tumor progression. Carcinogenesis 5:429–435
Knörr-Gärtner H, Schuhmann R, Kraus H, Uebele-Kallhardt B (1977) Comparative cytogenetic and histologic studies on early malignant transformation in mesothelial tumors of the ovary. Human Genet 35:281–297
Konopka JB, Watanabe SM, Witte ON (1984) An alteration of the human c-abl protein in K 562 leukemia cells unmasks associated tyrosine kinase activity. Cell 37:1035–1042
Land H, Parada LF, Weinberg RA (1983) Cellular oncogenes and multistep carcinogenesis. Science 222:771–778
Le Beau MM (1986) Chromosomal fragile sites and cancer-specific rearrangements. Blood 67:849–858
Le Beau MM, Rowley JD (1984) Recurring chromosomal abnormalities in leukemia and lymphoma. Cancer Surv 3:371–394
Levan A (1956) Chromosomes in cancer tissue. Ann NY Acad Sci 63:774–789
Levan A (1969) Chromosome abnormalities and carcinogenesis. In: Lima-de-Faria A (ed) Handbook of molecular cytology. Elsevier/North Holland Biomedical Press, Amsterdam New York, pp 717–731
Levan G (1974) The detailed chromosome constitution of a benzpyrene-induced rat sarcoma. A tentative model for G-banding analysis in solid tumors. Hereditas 78:273–290
Levan G, Levan A (1975) Specific chromosome changes in malignancy: Studies in rat sarcomas induced by two polycyclic hydrocarbons. Hereditas 79:161–198
Li YS (1983) Double minutes in acute myeloid leukemia. Int J Cancer 32:455–459
Maeda S, Uenaka H, Ueda N, Shiraishi N, Sugiyama T (1980) Establishment and chromosome studies of in vitro lines of chemically induced rat erythroblastic leukemia cells. J Natl Cancer Inst 64:539–546
Mark J, Mitelman F, Levan G (1972) On the specificity of the G abnormality in human meningiomas studied by the fluorescence technique. Acta Pathol Microbiol Scand Sect A 80:812–820
Matsui S, Weinfeld H, Sandberg AA (1972) Fate of chromatin of interphase nuclei subjected to "prophasing" in virus-fused cells. J Natl Cancer Inst 49:1621–1630
McClain KL (1984) Expression of oncogenes in human leukemias. Cancer Res 44:5382–5389
Mitelman F (1984) Chromosomal changes in cancer in relation to exposure to carcinogenic agents. In: Berlin A, Draper M, Hemminki K, Vainio H (eds) Monitoring human exposure to carcinogenic and mutagenic agents. IARC Publ No 59:351–360
Mitelman F (1985) Catalog of chromosome aberrations in cancer, 2nd edn. Liss, New York, 707 pp
Mitelman F (1986) Clustering of chromosomal breakpoints in neoplasia. Cancer Genet Cytogenet 19:67–71

Mitelman F, Levan G (1972) The chromosomes of primary 7,12-dimethylbenz(a)anthracene-induced rat sarcomas. Hereditas 71:325–334

Mitelman F, Mark J, Levan G (1972) Chromosomes of six primary sarcomas induced in the Chinese hamster by 7,12-dimethylbenz(a)anthracene. Hereditas 72:311–318

Mitelman F, Nilsson PG, Brandt L, Alimena G, Gastaldi R, Dallapiccola B (1981) Chromosome pattern, occupation, and clinical features in patients with acute nonlymphocytic leukemia. Cancer Genet Cytogenet 4:197–214

Namba M, Nishitani K, Hyodoh F, Fukushima F, Kimoto T (1985) Neoplastic transformation of human diploid fibroblasts (KMST-6) by treatment with ^{60}Co-gamma rays. Int J Cancer 35: 275–280

Namba M, Nishitani K, Fukushima F, Kimoto T, Nose K (1986) Multistep process of neoplastic transformation of normal human fibroblasts by ^{60}Co gamma rays and Harvey sarcoma viruses. Int J Cancer 37:419–423

Nowell PC (1976) The clonal evolution of tumor cell populations. Science 194:23–28

Nowell PC (1983) Tumor progression and clonal evolution: The role of genetic instability. In: German J (ed) Chromosome mutation and neoplasia. Liss, New York, pp 413–432

Nowell PC, Hungerford DA (1960) A minute chromosome in human chronic granulocytic leukaemia. Science 132:1497

Oguma N, Kamada N, Kuramoto A, Tanaka K, Misawa S, Testa JR (1985) Double minute chromosomes in acute myeloblastic leukemia. J Natl Cancer Inst 74:1007–1013

Oikawa T, Kuzumaki N, Yamada T (1984) Specific chromosome translocation in pristane-induced plasmacytomas of NZB mice. J Natl Cancer Inst 72:347–353

Oshimura M, Barrett JC (1985) Double nondisjunction during karyotypic progression of chemically induced syrian hamster cell lines. Cancer Genet Cytogenet 18:131–139

Owens AH, Coffey DS, Baylin SB (1982) Tumor cell heterogeneity: origins and implications. Academic Press, London New York

Pathak S, Goodacre A (1986) Specific chromosome anomalies and predisposition to human breast, renal cell, and colorectal carcinoma. Cancer Genet Cytogenet 19:29–36

Pathak S, Hsu TC, Trentin JJ, Butel JS, Panigrahy B (1981) Nonrandom chromosome abnormalities in transformed syrian hamster cell lines. In: Arrighi FE, Rao PN, Stubblefield E (eds) Genes, chromosomes, and neoplasia. Raven, New York, pp 405–418

Pedersen-Bjergaard J, Andersson P, Philip P (1986) Possible pathogenic significance of specific chromosome abnormalities and activated proto-oncogenes in malignant diseases of man. Scand J Haematol 36:127–137

Rachko D, Brand G (1983) Chromosomal aberrations in foreign body tumorigenesis of mice. Proc Soc Exp Biol Med 172:382–388

Pao PN, Wilson B, Puck TT (1976) Premature chromosome condensation and cell cycle analysis. J Cell Physiol 91:131–142

Ray FA, Bartholdi MF, Kraemer PM, Cram LS (1986) Spontaneous in vitro neoplastic evolution: Recurrent chromosome changes of newly immortalized Chinese hamster cells. Cancer Genet Cytogenet 21:35–51

Ray JH, German J (1983) The cytogenetics of the "chromosome-breakage syndromes". In: German J (ed) Chromosome mutation and neoplasia. Liss, New York, pp 135–167

Reichmann A, Levin B (1981) Premature chromosome condensation in human large bowel cancer. Cancer Genet Cytogenet 3:221–225

Reisbach G, Gebhart E, Cailleau R (1982) Sister chromatid exchanges and proliferation kinetics of human metastatic breast tumor cell lines. Anticancer Res 2:257–260

Riccardi VM, Sujansky E, Smith AC, Francke U (1978) Chromosomal imbalance in the aniridia-Wilms tumor association: 11p interstitial deletion. Pediatrics 61:604–610

Rodgers CS, Hill SM, Hulten MA (1984) Cytogenetic analysis in human breast carcinoma. I. Nine cases in the diploid range investigated using direct preparations. Cancer Genet Cytogenet 13: 95–119

Röhme D, Heneen WK (1982) Banding patterns in prematurely condensed chromosomes and the underlying structure of the chromosome. In: Rao PN, Johnson RT, Sperling K (eds) Premature chromosome condensation. Academic Press, London New York, pp 131–157

Rowley JD (1973) A new consistant chromosomal abnormality in chronic myelogenous leukaemia identified by quinacrine fluorescence and Giemsa staining. Nature (London) 243:290–293

Rowley JD (1977) Mapping of human chromosomal regions related to neoplasia: evidence from chromosomes no. 1 and 17. Proc Natl Acad Sci USA 74:5729–5733

Rowley JD (ed) (1983) Chromosome changes in leukemic cells as indicators of mutagenic exposure. Chromosomes and cancer. Academic Press, London New York, op 139–159

Rowley JD (1984) Biological implications of consistent chromosome rearrangements in leukemia and lymphoma. Cancer Res 44:3159–3168

Sandberg AA (1980) The chromosomes in human cancer and leukemia. Elsevier, New York Amsterdam

Sandberg AA, Abe S, Koalczyk JR, Zegdenidze A, Takeuchi J, Kakati S (1982) Chromosomes and causation of human cancer and leukemia. L. Cytogenetics of leukemias complicating other diseases. Cancer Genet Cytogenet 7:95–136

San Roman C, Ferro MT, Fernandez Ranada JM, Steegman JL (1982) Translocation (11;14) in B-cell lymphoroliferative disorders. Cancer Genet Cytogenet 7:279–286

Satya-Prakash KL, Pathak S, Hsu TC, Olive M, Cailleau R (1981) Cytogenetic analysis on eight human breast tumor cell lines: High frequencies of 1q, 11q and HeLa-like marker chromosomes. Cancer Genet Cytogenet 3:61–73

Schimke RT (1980) Gene amplification and drug resistance. Sci Am 243:60–69

Schnipper LE (1986) Clinical implications of tumor-cell heterogeneity. N Engl J Med 314:1423–1431

Shiraishi Y, Sandberg AA (1980) Sister chromatid exchanges in human chromosomes, including observations in neoplasia. Cancer Genet Cytogenet 1:363–380

Slamon DJ, Cline MJ (1984) Expression of cellular oncogenes during embryonic and fetal development of the mouse. Proc Natl Acad Sci USA 81:7141–7145

Slamon DJ, Kernion JB de, Verma IM, Cline MJ (1984) Expression of cellular oncogenes in human malignancies. Science 224:256–262

Sparkes RS (1984) Cytogenetics of retinoblastoma. Cancer Surv 3:479–496

Sperling K, Rao PN (1974) The phenomenon of premature chromosome condensation: its relevance to basic and applied research. Humangenetik 23:235–258

Stiles CD (1985) The biological role of oncogenes – Insights from platelet-derived growth factor: Rhoads Memorial Award Lecture. Cancer Res 45:5215–5218

Sugiyama T, Uenaka H, Ueda N, Fukuhara S, Maeda S (1978) Reproducible chromosome changes of polycyclic hydrocarbon-induced rat leukemia: incidence and chromosome bandin pattern. J Natl Cancer Inst 60:153–160

Sutherland GR (1979) Heritable fragile sites on human chromosomes. Am J Human Genet 31: 125–148

Sutherland GR, Hecht F (1985) Fragile sites on human chromosomes. Oxford Univ Press, New York

Swindell JA, Ockey CH (1983) Cytogenetic changes during the early stages of liver carcinogenesis in Chinese hamster. An in vivo-in vitro comparison. Cancer Genet Cytogenet 10:23–36

Talpaz M, Kantarjian HM, McCRedie K, Trujillo JM, Keating MJ, Gutterman JU (1986) Hematologic remission and cytogenetic improvement induced by recombinant human interferon $alpha_A$ in chronic myelogenous leukemia. N Engl J Med 314:1065–1069

Taub R, Kirsch I, Morton C, Lenoir G, Swan D, Tronick S, Aaronson S, Leder P (1982) Translocation of the c-myc gene into the immunoglobin heavy chain locus in human Burkitt lymphoma and murine plasmacytoma cells. Proc Natl Acad Sci USA 79:7837–7841

Tereba A (1985) Chromosomal localization of protooncogenes. Int Rev Cytol 95:1–43

Tjio JH, Levan A (1956) The chromosome number of man. Hereditas 42:1–6

Trent JF (1984) Chromosomal alterations in human solid tumors: Implications of the stem cell model to cancer cytogenetics. Cancer Surv 3:395–422

Trent JF (1986) Summary perspectives. 1st Int Worksh on Chromosomes in Solid Tumors. Cancer Genet Cytogenet 19:191–197

Tyrkus M, Diglio CA, Gohle N (1983) Karyotype evolution in a transformed rat cerebral endothelial cell line. Int J Cancer 32:485–490

Ueshima Y, Rowley JD, Variakojis D, Gordon L (1984) Cytogenetic studies in patients with chronic T cell leukemia/lymphoma. Blood 63:1028–1038

Unakul W, Johnson RT, Rao PN, Hsu TC (1973) Giemsa-banding in prematurely condensed chromosomes obtained by cell fusion. Nature (London) 254:245–247

Van den Berghe H (1984) Chromosome analysis of leukemia and preleukemia: Preliminary data on the possible role of environmental agents. In: Eisert WG, Mendelsohn ML (eds) Biological dosimetry. Springer, Berlin Heidelberg New York, pp 61–65

Walen KH, Arnstein P (1986) Induction of tumorigenesis and chromosomal abnormalities in human amniocytes infected with simian virus 40 and Kirsten sarcoma virus. In Vitro Cell Dev Biol 22:57–65

Weinberg RA (1984) Molekulare Grundlagen von Krebs. Spekt Wiss 1:58–71

Williams DM, Scott CD, Beck TM (1976) Premature chromosome condensation in human leukemia. Blood 47:687–693

Yoshida MA, Takagi N, Sasaki M (1982) Influence of strain difference on the karyotypic changes in N.nitroso-N-butylurea-induced mouse lymphomas. Cancer Genet Cytogenet 7:19–31

Yosida TH (1981) Chromosome alteration and the development of tumors. XXIII. Banding karyotype analyses of methylcholanthrene-induced tumors in the Indian spiny mouse, Mus platythrix, with special regard to the anomalies of chromosomes with nucleolar organizer regions. Cancer Genet Cytogenet 3:211–220

Yosida TH (1983) Karyotype evolution and tumor development. Cancer Genet Cytogenet 8:153–179

Yunis JJ (1981) Specific fine chromosomal defects in cancer: An overview. Human Pathol 12: 503–515

Yunis JJ (1985) Genes and chromosomes in human cancer. Progr Med Virol 32:58–71

Yunis JJ, Soreng AL (1984) Constitutive fragile sites and cancer. Science 226:1199–1204

Zang KD, Singer H (1967) Chromosomal constitution of meningiomas. Nature (London) 216: 84–85

Zech L, Gahrton G, Hammarström L, Juliusson G, Mellstedt H, Robert KH, Smith CIE (1984) Inversion of chromosome 14 marks human T-cell chronic lymphocytic leukaemia. Nature (London) 308:858–860

7 Surveillance of Germinal Human Mutations for Effects of Putative Environmental Mutagens and Utilization of a Chromosome Registry in Following Rates of Cytogenetic Disorders

ERNEST B. HOOK[1]

1 Rationale and Definitions

Within the past 2 decades, there has been a great increase in concern about the potential impact of environmental hazards upon human health. Part of this concern applies to effects upon the human genome. Is there a significant impact of environmental substances upon mutation rate? Has the putative recent continuing increase in environmental pollutants had an effect upon human genetic makeup? How may one follow human populations for effects upon mutation rates?

There are of course several different approaches to such questions depending upon the precise question under investigation. An investigator may have identified a population at presumptive high risk because of exposure to some suspected hazard and may wish to follow the population, evaluating reproductive outcomes. Such evaluation of populations selected at risk for raised mutation rates has been done occasionally. Perhaps the best known example has been the study of the populations exposed to radiation at Hiroshima and Nagasaki, although the overall evidence for increased mutation rates in this group is still equivocal at best. Some of the issues in evaluating genetic outcomes (and birth defects) around point sources of pollutants have been discussed in detail elsewhere (Hook 1981a, 1982b; IPCS 1985). These activities depend critically upon the specific details of the episode of concern, the suspect environmental agent, and the particular population; it is relatively difficult to make generalizations about useful approaches for any specific episode.

A different approach involves following a large population, usually an entire political or civil jurisdiction to determine if there is any temporal trend in mutation rates either in the entire population or some subset. It is analogous to notification and analysis of infectious disease rates for detection of epidemics. If one identifies an apparent increase in mutation rate through this activity, one may then undertake an investigation of the possible cause.

The two types of activities are quite different but often confused, so that it is useful to denote them by different terms. I have found it useful heuristically to refer to

1 Bureau of Environmental Epidemiology and Occupational Health, New York State Department of Health, Albany, N.Y. 12237, USA, and
Department of Pediatrics, Albany Medical College, Albany, N.Y. 12208, USA

Cytogenetics. Ed. by G. Obe and A. Basler

following high risk populations as *monitoring* and the evaluation of background rates in the populations as *surveillance* (Hook and Cross 1982). The literature is, I caution, not consistent in usage, and "monitoring" has sometimes been used to denote either activity. In any event in "monitoring", as defined here, the investigator already begins with a subgroup at risk and (usually) an agent of suspicion. In "surveillance" one evaluates the entire population for effect of introduction or increase in unsuspected hazards. It is easier to make generalizations about surveillance than monitoring that are applicable in many different jurisdictions, although obviously what can be achieved will vary with the available resources and data bases. In the discussion that follows I will concentrate on surveillance of human mutations.

2 Somatic and Germinal Mutations

One can of course subdivide the mutations of concern into somatic and germinal. The impact of somatic mutation is primarily, albeit not exclusively, upon occurrence of malignancies, and has no (direct) effect upon the next generation. (This is not to imply all carcinogens are mutagens.) Currently, studies of somatic mutation (and related processes) in human populations involve the investigation of chromosome breakage and rearrangements of sister chromatid exchange, and of alterations of proteins such as HGPRTase in circulating lymphocytes (see Albertini 1985; IPCS 1985 for references). Many extensive investigations worldwide of these events and of the epidemiology of cancer provide evidence pertaining to possible environmental effects upon somatic mutation. But with the exception of high doses of radiation (Brewen et al. 1975; Martin et al. 1986) it has not been proven strictly that agents which influence rates of somatic mutations, measured either directly in lymphocytes or indirectly through following rates of some cancers, necessarily have any effect upon human germinal mutation.[2]

Thus, one cannot use somatic mutations studied experimentally in humans or rates of cancer investigated epidemiologically as a clear guide or index to what is happening to the human genome. Germinal mutations must be studied directly.

In the analysis of germinal human mutation rates, one must distinguish at a minimum three different general categories: specific locus mutations, structural rearrangements of human chromosomes, and numerical chromosome abnormalities.

[2] There is one exception to this generalization about malignancies. Some instances of sporadic bilateral retinoblastoma, Wilm's tumor, and possibly congenital tumors, i.e., those diagnosed in early infancy, are highly likely to result from a germinal mutation, in some cases in addition to a somatic mutation (see e.g. Knudson 1971; Knudson and Strong 1972). These rare types of tumors, which may be indicators of a germinal event, are discussed further below.

3 Surveillance of Chromosome Abnormalities: Background and Need for Prenatal Data

Structural and numerical cytogenetic abnormalities differ markedly in etiologies and epidemiology. Nevertheless, for the most part the optimal methodologies for surveillance of each type are roughly similar. I will discuss first the issue of data collection for both types of abnormalities and then deal separately with subtypes of chromosome abnormality.

Chromosome study of all live births for the purpose of surveillance is economically unfeasible. The average cost of cytogenetic study is about $300 (1986 dollars) and the frequency of abnormality is roughly 5 to 6 per 1,000 live births, of which perhaps two-thirds to three-fourths represent fresh mutations. Thus, the cost of systematic study of the newborn population to detect a mutation would be exorbitant (about $ 50,000 each). This is independent of the cost of specimen acquisition, shipping, and associated data retrieval which would perhaps double the cost. Nevertheless, analogous to surveillance of sentinel phenotypes for specific locus mutations (see below) one could survey and record data on "indicator phenotypes" of chromosome abnormalities in newborns. Moreover, at least in many countries, the occurrence of a suggestive "chromosomal phenotype" in an infant is often followed by cytogenetic study.

Of the syndromes associated with chromosome abnormality probably only Down syndrome is recognized sufficiently frequently in many jurisdictions by the average pediatrician so that ascertainment is likely to be relatively high and consistent. Other phenotypes such as Patau and Edwards syndrome, associated with trisomies 13 and 18, respectively, may result in death before either phenotypic suspicion or recognition (or later cytogenetic confirmation) occurs. The frequent sex chromosome trisomies are not likely to be suscepted phenotypically in infancy and the Turner syndrome phenotype is to nondistinctive and the associated 45,X genotype too rare.

Nevertheless, at an institution where a knowledgeable dysmorphologist examined all newborns, and ordered cytogenetic study on the "suspicious" cases, perhaps about 0.5% of the total live births would be studied cytogenetically and probably about 50% of the infants with chromosome abnormalities could be detected. This would result in almost complete ascertainment of the autosomal trisomies and unbalanced autosomal structural rearrangements at least, but poor if any ascertainment of those with sex chromosomal trisomies and balanced structural rearrangements. Mehes (1973) has used this approach to follow at least the 21, 18, and 13 trisomies in live births in Baranya County in southern Hungary. Of 32,023 newborn infants in the years 1968–1972, 108 (0.3%) were karyotyped because of phenotypic suspicion of a cytogenetic abnormality and 64 reported positive. A main difficulty with this method in a large jurisdiction is that it requires cooperative and knowledgeable individuals at different hospitals who have a *consistent* index of suspicion for infants on whom cytogenetic diagnosis is sought.

Those who analyze birth certificate diagnoses of Down syndrome [which present problems with both accuracy and completeness (Johnson et al. 1985)] or diagnoses in ad hoc congenital malformation data sources in one sense are undertaking surveillance of a mutational indicator phenotype. My own belief is that it is worth using even incomplete data sources that may be available as long as the consistency of diagnosis can be assayed – that is as long as one may make a judgement as to whether any observed fluctuation in frequency is attributable to simply a change in ascertainment or to a real underlying change in population prevalence. Even birth certificates, roughly 35% complete in upstate New York, may be useful in this regard, albeit with some difficulty, although obviously a method which systematically ascertains all structural malformations such as that in use in Atlanta since 1969 (Edmunds et al. 1981; C.D.C. 1985) or in New York State since 1981 (Polan et al. 1985) is preferable.

The major problem with the use of live birth data, however, in almost all "developed" countries today is that the frequencies of indicator cytogenetic phenotypes in live births (as well as cytogenetic abnormalities detected directly in cord bloods or live births) as biased downwards because of selective termination of cytogenetically abnormal pregnancies following prenatal diagnosis. Fluctuations in measured live birth rates even if ascertainment is 100% accurate *and* 100% complete (i.e., there are no falsely positive and no falsely negative cases) will not necessarily reflect fluctuations in other causal factors either "biological" or 'environmental". This is not a major problem for USA data at least before the NIH report on the safety and accuracy of amniocenteses in 1976 (NICHD 1976), but after that time this factor is of increasing concern because of marked increase in prenatal diagnosis.

For this reason data from prenatal cytogenetic diagnoses are necessary in any current surveillance program for chromosome disorders, although it must be remembered that such data have limitations of their own. An ongoing register of all prenatal cytogenetic diagnoses may be used for surveillance purposes, as long as (1) all (or at least representative) results, both normal and abnormal, on women within a particular jurisdiction are reported; (2) the dates of the studies are reported; (3) the reason for each prenatal cytogenetic study is noted; (4) data on maternal age of those studied are available. The latter two requirements enable adjustments for various biases that may arise because of selective use of amniocentesis by women at high risk of a cytogenetic abnormality. An attractive feature of such a registry is that the *cost* of the laboratory effort is essentially free for purposes of surveillance, as the study has been ordered for independent clinical reasons, yet the data are collected and analyzed for other socially useful purposes. (But see also Addendum, p. 165.)

The approaches and results of a particular chromosome registry whose data have been used for surveillance are described further in the next sections. The approaches, methods, and results of the New York State Chromosome Registry which has ascertained systematically, data on prenatal cytogenetic diagnoses since 1977 are reviewed.

4 The New York State Chromosome Registry: Methods and Use for Surveillance

The mechanisms and logistic arrangements of the Registry have been described in detail elsewhere (Hook et al. 1981). Briefly, data are collected from all laboratories with New York State cytogenetic permits and some additional laboratories from near-near-by States. The Registry was founded in 1969 and initially only collected data on detected cytogenetic *abnormalities*. Occasionally, results on an abortus or fetus were reported, but the vast majority of studies were on infants, children, and adults.

With the expansion of prenatal diagnostic services in the mid-1970s two subregistries were established. One designated the "yellow" registry (because amniotic fluid is straw-colored) has collected since January 1, 1977 results of *all* prenatal cytogenetic studies, normal or abnormal. The other, designated the "white" chromosome registry (since the vast bulk of such results are on white blood cells), includes all reported data on *abnormalities* detected in infants, children, adults, as well as the occasional report of the study of abortion or stillbirth. (Abnormalities detected prenatally are, however, reported to the Yellow Registry.)

Cytogenetic studies reported from the White Registry have been found of limited usefulness for surveillance. Chromosome investigation in infants, fetal deaths, children, and adults are highly selective and based on clinical indications only, so that the population studied is highly biased. Following either the frequency of detected abnormalities or even the rate of abnormalities among all those undergoing cytogenetic investigation (if the results of normal studies were also to be reported to the White Registry) provides little data directly pertinent to estimating live birth prevalence rates of these conditions. The one exception is rates in live births of two rare structural chromosome abnormalities: unbalanced Robertsonian translocations producing Down syndrome or Patau syndrome. An indirect method has been used to follow these two cytogenetic events and also for following their mutation rates. This is discussed below in the sections on structural cytogenetic abnormalities.

Data from the White Chromosome Registry nevertheless, are of some pertinence to surveillance because it at least provides a repository for those cytogenetic diagnoses which have been made in live births and older individuals, although of course fluctuation in such reports results from variation in clinicians' decisions to seek cytogenetic studies and the probability of death of cases before study, as well as from variation in live birth prevalence. The White Registry data are of use in other ways, however, for example, as starting points in case-control studies of specific types of cytogenetic abnormalities and for investigations of selected aspects of the natural history and epidemiology of selected cytogenetic conditions, e.g., age of ascertainment, relative frequency of translocations, etc. (Albright and Hook 1980; Hook and Liss 1982).

Nevertheless, it is primarily the data from the Yellow Registry which have been pertinent to surveillance. Because results are reported on all prenatal cytogenetic studies it is possible to derive directly "rates" of abnormalities in the fetuses studied after correction and adjustment for the reasons for amniocentesis. The vast bulk of the data are from results of amniocenteses on fetuses about 16 to 20 weeks of gestational age. Thus, abnormalities associated with embryonic and early fetal lethality are not ascertained through amniocentesis reports. The Registry does collect reports of chorionic villus assays (which are undertaken at 9 to 12 weeks of gestational age) but this is still a very infrequent procedure and was only introduced in New York State to our knowledge in 1985.

Between 1977 and 1984, 78,567 fetuses reported to the New York State Chromosome Registry had prenatal cytogenetic diagnoses upon amniotic fluid for reasons listed in Table 1. About 80% of cases have been studied because of maternal age. This is a risk factor of course for most trisomies and some method must be introduced to adjust for variation in age over time or between populations in undertaking a systematic analysis. Age, we have found, is not a significant risk factor, however, for the struc-

Table 1. General reasons for prenatal cytogenetic study: 1977–1984 New York State Chromosome Registry

Reason for study	Maternal age			Total (%)
	< 35	> 35	Unknown	
A. Known translocation carrier parent[a]	227	44	7	278 (0.4)
B. Putative mutagen exposure[b,c]	365	80	2	447 (0.6)
C. Offspring or relative with chromosome abnormality[a,c]	3,200	645	48	3,893 (5.0)
D. Pregnancy management; suspect fetal pathology[a–c]	1,901	227	68	2,196 (2.8)
E. Offspring (or relative) with malformation and/or abortion[a]	2,029	450	37	2,516 (3.2)
F. Miscellaneous or unknown	648	207	202	1,023 (1.3)
G. Advanced parental age[b,c]	7,971	55,787	223	63,981 (81.4)
H. Incidental; no known cytogenetic risk	3,906	225	68	4,233 (5.4)
Total	20,247	57,665	655	78,567 (100.0)

[a] Definite risk factor for an inherited unbalanced structural abnormality.

[b] Definite risk factor for mutant structural abnormality (in case of age, for extra structural chromosome abnormalities only).

[c] Definite risk factor for numerical abnormality. (In case of mutagen exposure the increased risk is only presumptive; not proven.)

tural rearrangements (except those which are extra or "supernumerary"). Aside from age there are a number of other reasons for prenatal diagnosis, at least some of which are associated with some types of chromosome abnormalities (see Table 1, footnotes). After extensive consideration it was decided that the most useful initial course in surveillance is to undertake two separate types of analyses, one of numerical abnormalities, the other of structural abnormalities. For the numerical abnormalities consideration is limited just to those studied because of advanced age and "incidentally". (Fetuses of parents with translocations or with previous offspring with malformations are excluded in this analysis because of putative, although still unproven, association with numerical chromosome abnormalities.) Those studied incidentally include those in whom amniotic fluid was obtained, for example, in the course of study of inborn error of metabolism, i.e., for a reason other than the concern about a cytogenetic disorder or one statistically associated with a cytogenetic disorder, but on whom cytogenetic study was carried out because the fluid was available. The initial analysis of mutant structural cytogenetic abnormalities is more inclusive, excluding from the surveillance analysis only those studied because of suspect fetal pathology or exposure to a suspect mutagen. The latter reason for study is of intrinsic interest itself and has been considered separately from the surveillance analysis (see below). Fetuses with suspect pathology have been excluded in all analyses because of the known association of these outcomes with autosomal trisomies and unbalanced structural rearrangements, among other abnormalities. (See also Addendum, p. 165.)

5 Surveillance of Trisomies Diagnosed Prenatally

There are six numerical abnormalities on which there are now sufficient data to analyze trends. These are 47,+21, 47,+18, 47,+13, 47,XXY, 47,XXX, and 47,XYY.

Even restriction of analysis of these events to fetuses studied only because of advanced maternal age or "incidentally" still leaves some major questions about rate adjustment. All but 47,XYY exhibit a significant positive association with maternal age (Ferguson-Smith and Yates 1984; Hook et al. 1984). "Crude" rates may not be useful for surveillance because of changing maternal age composition of the population undergoing prenatal cytogenetic diagnosis. For example, the median age of women having this procedure dropped from 36.9 years in 1977 to 35.7 years in 1984. This change of 1.2 years may seem relatively minor but all data I have been able to accumulate on this issue, from a great variety of different sources, suggest that (up to some upper age limit) 47,+21 increases by about 30% per year over age 30 and 47,+18 by about the same amount (Lamson and Hook 1981; Hook et al. 1979, 1984). The expected rate of 47,+21 (at the time of amniocentesis) in a woman aged 36.9 years (the 1977 median age) is about 6.1 per 1,000 and at 35.7 years (the 1984 median age) is 4.5 per 1,000 (Hook et al. 1984). Some adjustment for variation in age is clearly necessary.

A great deal of study has been given to the precise approaches that might be used for such an adjustment. One aspect of the data is that the vast bulk of those studied involves women 35 and over. There is, however, an increasing number under 35 being studied because of concern about risk at younger ages, especially 33 and 34 or just because of "anxiety" without any specific risk factor. For the purposes of this analysis, the data have been subdivided in two age subgroups, those under 35 and those 35 and over. In the older group direct age standardized rates as well as crude rates were calculated. The reference population for this standardization is the published data on women having prenatal cytogenetic diagnosis in all of North America (see Table 2). This reference group was chosen because the distributions are readily available in the literature and it would appear to be most pertinent to at least North American trends. [Distributions from a collaborative European study (Ferguson-Smith and Yates 1984) could also have been used.] For those under 35, there is not as much need to standardize because there is much less variation with maternal age. There are, moreover, several technical reasons why direct age standardization in this age group is not attrac-

Table 2. Frequency and proportion of amniocentesis by maternal age in 56,075 fetuses studied prenatally in North America. Distribution used for direct age standardization of rates in those 35 and over

	Maternal age	Number	Proportion
	35	13,525	0.2412
	36	11,694	0.2085
	37	9,450	0.1685
	38	7,062	0.1259
	39	5,298	0.0945
	40	3,883	0.0692
	41	2,323	0.0414
	42	1,404	0.0250
	43	732	0.0131
	44	433	0.0077
	45	161	0.0029
	46	67	0.0012
	47	29	0.0005
	48	9	0.0002
	49	5	0.0001
Total		56,075	

Data includes results from systematic USA studies by 1983 (Hook et al. 1984).

tive. These include the facts that the absolute numbers affected are very small, that the distribution of ages under 35 of those having amniocentesis is markedly different from the distribution of ages of the entire child-bearing population, and that these factors may result in direct standardized rates yielding a marked distortion of underlying trends.

Consideration has also been given to other methods for adjustment including indirect age standardization or the related derivation of standardized morbidity ratios [i.e., ratios of the observed number (or rate) of cases to the expected number (or rate) (Fleiss 1973)]. This would be straightforward in those 35 and over as there are readily available rate schedules to derive the numbers of expected cases. But for those under 35 there are insufficient data to derive stable expected rates for direct use for this calculation. One would have to rely upon estimates derived from *live birth* studies adjusted for the proportion of spontaneous fetal deaths for each trisomy between the time of amniocentesis and live birth. These alternative approaches are under review and I hope that we will soon have results of both types of analyses for comparison.

In any event, the most significant aspect of the observations in Tables 3 and 4 on numerical abnormalities is that there is no evidence for any temporal increase in rates of the six most common categories of trisomies diagnosed prenatally. The only suggestive trend of any note is in fact to a decrease in age standardized rate of 47,+18 in those 35 and over. 47,+18 (which results in Edwards syndrome) has not been investigated as extensively as 47,+21, but these two abnormalities have almost identical changes with maternal age, both in live births and at amniocentesis, except perhaps at the very upper extremes of age (Hook et al. 1979, 1984; Ferguson-Smith and Yates 1984). It is noteworthy that despite the change in rate for 47,+18 in 1984, and

Table 3. Number and rates (per 1000) of selected trisomies diagnosed prenatally at amniocentesis in women 35 and older studied with no known cytogenetic risk (aside from age) reported to the New York State Chromosome Registry 1977–1984

		1977	1978	1979	1980	1981	1982	1983	1984	Total
47,+21	Number:	24	26	39	55	52	73	92	99	460
	Rate per 1000:	12.4	8.3	8.4	8.6	7.6	7.6	8.1	8.2	8.2
	Standardized rate:	9.3	7.4	8.1	8.8	7.9	8.1	8.6	8.6	8.5
	SE:	1.9	1.5	1.3	1.2	1.1	0.9	0.9	0.9	0.4
47,+18	Number:	8	9	8	15	13	20	18	6	97
	Rate per 1000:	4.1	2.9	1.7	2.3	1.9	2.1	1.6	0.5	1.7
	Standardized rate:	3.2	2.6	1.7	2.4	2.0	2.2	1.7	0.5	1.8
	SE:	1.2	0.9	0.6	0.6	0.6	0.5	0.4	0.2	0.2
47,+13	Number:	0	0	6	5	7	3	5	8	34
	Rate per 1000:	0.0	0.0	1.3	0.8	1.0	0.3	0.4	0.7	0.6
	Standardized rate:	0.0	0.0	1.3	0.8	1.1	0.3	0.5	0.7	0.6
	SE:	0.0	0.0	0.5	0.4	0.4	0.2	0.2	0.3	0.1
47,XXX	Number:	5	3	3	7	6	11	12	11	58
	Rate per 1000:	2.6	1.0	0.6	1.1	0.9	1.1	1.1	0.9	1.0
	Standardized rate:	1.8	0.9	0.7	1.1	0.9	1.1	1.1	0.9	1.0
	SE:	0.8	0.5	0.4	0.4	0.4	0.3	0.3	0.3	0.1
47,XXY	Number:	3	6	3	9	6	12	5	16	60
	Rate per 1000:	1.5	1.9	0.6	1.4	0.9	1.2	0.4	1.3	1.1
	Standardized rate:	1.7	1.6	0.7	1.5	0.9	1.3	0.4	1.3	1.1
	SE:	1.0	0.6	0.4	0.5	0.4	0.4	0.2	0.3	0.1
47,XYY	Number:	1	2	2	3	6	4	6	6	30
	Rate per 1000:	0.5	0.6	0.4	0.5	0.9	0.4	0.5	0.5	0.5
	Standardized rate:	0.4	0.6	0.4	0.5	0.9	0.4	0.5	0.5	0.5
	SE:	0.4	0.4	0.3	0.3	0.4	0.2	0.2	0.2	0.1
Autosomal	Number:	32	35	53	75	72	96	115	113	591
trisomies	Rate per 1000:	16.5	11.1	11.4	11.7	10.5	10.0	10.1	9.4	10.6
(21,18,13)	Standardized rate:	12.5	10.0	11.1	12.0	11.0	10.6	10.9	9.9	10.9
	SE:	2.3	1.7	1.5	1.4	1.3	1.1	1.0	0.9	0.4
47,+X	Number:	8	9	6	16	12	23	17	27	118
(XXY,XXX)	Rate per 1000:	4.1	2.9	1.3	2.5	1.7	2.4	1.5	2.2	2.1
	Standardized rate:	3.5	2.5	1.3	2.6	1.7	2.4	1.5	2.2	2.1
	SE:	1.3	0.8	0.5	0.6	0.5	0.5	0.4	0.4	0.2
Total	Number:	1942	3149	4641	6391	6862	9642	11385	12000	56012
number of	Mean maternal age:	37.9	37.5	37.3	37.2	37.2	37.1	37.0	37.1	37.2
women	(SD):	(2.3)	(2.2)	(2.2)	(2.1)	(2.1)	(2.1)	(2.0)	(2.1)	(2.1)
studied										

Table 4. Number and rates (per 1000) of selected trisomies diagnosed prenatally at amniocentesis in women less than 35 studied with no known cytogenetic risk (aside from age) reported to the New York State Chromosome Registry 1977–1984

		1977	1978	1979	1980	1981	1982	1983	1984	Total
47,+21	Number:	1	0	3	5	5	7	6	9	36
	Rate per 1000:	8.4	0.0	5.1	4.8	3.1	3.1	2.1	2.9	3.0
47,+18	Number:	0	0	0	1	0	3	5	0	9
	Rate per 1000:	0.0	0.0	0.0	1.0	0.0	1.3	1.7	0.0	0.8
47,+13	Number:	0	0	0	0	0	1	2	0	3
	Rate per 1000:	0.0	0.0	0.0	0.0	0.0	0.4	0.7	0.0	0.3
47,XXX	Number:	0	0	0	0	0	0	2	1	3
	Rate per 1000:	0.0	0.0	0.0	0.0	0.0	0.0	0.7	0.3	0.3
47,XXY	Number:	0	1	0	0	2	2	0	6	11
	Rate per 1000:	0.0	3.5	0.0	0.0	1.2	0.9	0.0	1.9	0.9
47,XYY	Number:	1	0	0	2	0	0	1	1	5
	Rate per 1000:	8.4	0.0	0.0	1.9	0.0	0.0	0.3	0.3	0.4
Autosomal trisomies	Number:	1	0	3	6	5	11	13	9	48
(21,18,13)	Rate per 1000:	8.4	0.0	5.1	5.8	3.1	4.9	4.5	2.9	4.0
47,+X	Number:	0	1	0	0	2	2	2	7	14
(XXY, XXX)	Rate per 1000:	0.0	3.5	0.0	0.0	1.2	0.9	0.7	2.2	1.2
Total	Number:	119	282	594	1039	1616	2239	2874	3115	11878
number	Mean maternal age:	31.4	31.7	32.2	32.6	32.8	32.8	32.8	32.8	32.7
of women	(SD):	(3.6)	(3.4)	(3.1)	(2.7)	(2.4)	(2.4)	(2.1)	(2.1)	(2.4)

the suggestion of a temporal decline over the entire 8-year period (in those 35 and over), there is no such decline in 47,+21. While it is possible that there are biological and/or environmental factors that specifically affect rates of nondisjunction for individual chromosomes, specifically 18 (or fetal survival of those with trisomy), and that those affecting 47,+18 have diminished in this time, it appears more likely to me that this trend reflects simply the result of statistical fluctuation. It will be important to evaluate data from subsequent years to determine if this trend persists. (See also Addendum, p. 165.)

6 Surveillance of Structural Cytogenetic Abnormalities Diagnosed Prenatally

Structural cytogenetic abnormalities raise completely different issues in surveillance. First, the main rationale for surveillance is of course to detect putative environmental mutagens. Cases with inherited abnormalities are not usually relevant to this concern.

Almost all numerical abnormalities are (presumably) fresh mutations. But a significant proportion, i.e., 60% of structural abnormalities detected in fetuses are estimated to be inherited from a carrier parent (Hook and Cross 1987). Moreover, this proportion varies by abnormality; it is, for example, about 88% for inversions, but under 10% for rings and other deletions. Such cases are of cource excluded in surveillance analysis. There are, however, some fetuses with structural chromosome abnormalities for whom an inherited origin cannot be ruled out because at least one parent (usually the father) is not studied. Thus, results on structural *mutants* must be presented as "ranges" allowing for the cases of unknown origin. Secondly, structural chromosome abnormalities are collectively less frequent than numerical abnormalities and the many different subtypes may have considerably variable etiological factors, although (almost) all involve some interruption in the structural integrity of a chromosome. (The sole possible exception would be translocations resulting from telomeric fusion of two chromosomes; some dicentric Robertsonian translocations may perhaps represent examples of this process.) Because of relatively small numbers, in surveillance analyses here I considered the three mutually exclusive groups: extrastructurally abnormal chromosomes (e.s.a.c.), all other unbalanced abnormalities, and all balanced abnormalities. The rates of none of these groups, which each comprise heterogeneous subsets, are much greater than 1.0 per 1,000 (Table 5). I have also analyzed separately the mutant Robertsonian translocations (both unbalanced and balanced) as some have suggested that at least in experimental animals these outcomes may differ from other rearrangements in their susceptibility to radiation (Ford et al. 1977). Maternal age, however, is not associated with structural abnormalities except probably for e.s.a.c. (Hook and Cross 1987). For this reason, rates of those studied are presented without subdivision by maternal age category. (Recall that the analysis presented of structural abnormalities excludes only those studied because of exposure to a putative mutagen or because of fetal pathology.)

There are no obvious trends in the rates in Table 5 but clearly some statistical fluctuation. The only apparent outlier in the analysis is e.s.a.c. in 1977 in which the rate was 3/2483 = 1.6 per 1,000 compared to a rate of about 0.4 per 1,000 in remaining years. An analysis of direct age standardized rates was carried out on the e.s.a.c. In those 35 and over, the crude rate was 2 per 1,000 and the standardized rate 1.7 (± 0.9) per 1,000 in 1977, whereas the standardized rate for all years was 0.44 (± 0.09) to 0.55 (± 0.10) per 1,000. No strong inferences are possible about this single year; the high rate resulting from just three cases may well represent just statistical fluctuation. (See also Addendum, p. 165.)

It is emphasized that structural cytogenetic abnormalities reflect completely different etiological mechanisms than do numerical abnormalities. As they are almost all consequences of actual deletion and/or other damage to the chemical composition of the genome, they are analogous to the Mendelian specific locus mutations, i.e., "gene" mutations. Indeed, it may be difficult to distinguish precisely the boundary of the two categories of outcomes because at a certain level of resolution it may be difficult, if not impossible, to draw a definitive boundary between gene mutations and structural chromosome abnormalities. For example, a deletion of a single base pair (or a few base pairs) from a gene resulting in a specific locus mutation may result from the same or similar molecular processes resulting from deletion of several thousand

Table 5. Rates per 1,000 fetuses of de novo structural chromosome abnormalities reported to the New York State Chromosome Registry by year of sample[a,b]

	Year of sample 1977	1978	1979	1980	1981	1982	1983	1984	Total 1977–1984
Unbalanced									
e.s.a.c.[c]	1.6	0.5–0.8	0.0	0.7	0.3–0.5	0.5	0.2–0.3	0.2–0.4	0.4–0.5
Other	0.4	0.8	0.5–0.9	0.0–0.1	0.3–0.6	0.46–0.54	0.4–0.8	0.3–0.7	0.4–0.6
All unbalanced:	2.0	1.3–1.5	0.5–0.9	0.7–0.8	0.6–1.2	1.0 –1.1	0.6–1.0	0.5–1.1	0.7–1.1
Balanced	0.8	0.8–1.3	0.5	0.7–1.0	1.3–1.5	0.8 –1.2	1.0–1.2	1.0–1.3	0.9–1.2
All Robertsonian	0.0	0.3–0.5	0.2–0.3	0.1–0.2	0.5–0.6	0.3	0.3–0.4	0.3–0.5	0.3–0.4
All abnormalities excluding e.s.a.c	1.2	1.5–2.0	1.0 1.4	0.7–1.1	1.6–2.1	1.3 –1.7	1.3–1.9	1.3–2.1	1.3–1.8
All abnormalities	2.8	2.0–2.8	1.0–1.4	1.5–1.8	1.9–2.7	1.8 –2.2	1.5–2.2	1.6–2.4	1.7–2.3
Total studied (Number with maternal age not stated)	2457 (19)	3965 (44)	5874 (22)	8261 (35)	9351 (28)	13,015 (60)	15,605 (90)	16,373 (85)	74,901 (383)
Mean (and (SD) of maternal age:	36.3 (4.4)	36.1 (4.1)	36.0 (3.8)	35.9 (3.6)	35.8 (3.6)	35.8 (3.5)	35.7 (3.3)	35.8 (3.4)	35.8 (3.6)
Median maternal age:	36.9	36.5	36.2	36.1	35.9	35.8	35.8	35.8	35.9

[a] Range in rates is attributable to cases of unknown origin.
[b] Fetuses with reason for study not stated, miscellaneous reasons for study, and those studied because of putative exposure to mutagens or because of suspected fetal pathology are excluded from the analyses.
[c] e.s.a.c. = extrastructurally abnormal chromosomes.

or more base pairs that occurs in a microscopically detectable chromosome deletion. The importance of this point is that an environmental factor resulting in an increase in structural chromosome abnormalities would also be expected to have an effect on at least some specific locus mutations, which are very difficult to follow directly in human populations (see below). Thus, structurally abnormal chromosomes are likely indicators of other types of mutational events as well, and as indicators have significance beyond themselves for surveillance and public health purposes.

7 Surveillance of Robertsonian Translocation: Use of Data on Infants and Children

It is also worth elaborating upon an indirect method of surveillance for one subgroup of structural cytogenetic abnormalities which has been undertaken for some years using cytogenetic data primarily from the White Chromosome Registry on unbalanced translocations. About 5% of the Down syndrome phenotype results from structural chromosome rearrangements caused by a Robertsonian translocation of which about three-fourths are mutant and one-fourth inherited from a carrier parent (Hook 1981c, 1982a). The key point is that there is no known phenotypic difference between mutant translocation Down syndrome cases and those resulting from 47,+21. Thus, for a sporadic case there is no reason (aside from maternal age) for a clinician to suspect one or the other genotype when ordering a cytogenetic evaluation. Therefore, by following the *proportion* of mutant Robertsonian translocations among Down syndrome cases (and stratifying on maternal age) one indirectly has a mechanism for evaluating changes in the mutation rate for unbalanced Robertsonian translocations. For example, a twofold rise in this proportion could be attributable, in principle, either to a doubling in the mutation rate of translocations or roughly a halving in the prevalence of the main genotype 47,+21, resulting in Down syndrome. Since 47,+21 accounts for 95% of the phenotype, a halving in the rate of this karyotype would result in considerable diminishment in the live birth phenotype. This could be detected through relatively crude methods. But if a doubling in the mutation rate for translocations occurred, while it would be expected to have little overall effect upon the rate of the entire Down syndrome phenotype, it would be reflected in an abrupt change in the proportion of translocations among cases studied cytogenetically. A similar method may be used for Robertsonian translocations resulting in Patau syndrome. About 14% of this phenotype is produced by mutant Robertsonian rearrangements, about 7% by inherited rearrangements, and the remainder, about 80% by 47,+13 (Hook 1981c).

This method, originally called the "ratio" method because initially it considered the *ratio* of mutant translocations to 47,+21 instances of Down syndrome, was introduced and first applied to White Registry data in the late 1970s (Hook 1978a). Surprisingly, evidence was found for a rather abrupt increase in mutation rate for translocation trisomy 21 in 1973–1975 compared with four earlier years. (There was also suggestive evidence for an increase in translocation Patau syndrome.) Despite a search for artifacts of reporting or other biases that could have explained this trend, it is clear that there was a true change in the mutation rate in these translocations. There are two

Table 6. Number and percentage of Down syndrome cases with mutant interchange trisomy[a]

Maternal age	Year of birth 1968	1969	1970	1971	1972	1973	1974	1975	1976
< 30									
No.	3–4	5–6	3–4	5	1–3	6–8	10–11	8–10	4–7
%	5.4–7.1	4.9–5.8	3.3–4.3	4.6	0.9–2.6	5.4–7.2	8.2–9.0	7.0–8.8	6.7–7.6
Total DS	56	103	92	108	114	111	122	114	103
> 30									
No.	1–2	1	1	1	3	2	3	2	0
%	1.6–3.1	1.0	0.9	0.8	3.2	2.0	3.1	1.3	0
Total DS	64	101	112	123	95	100	98	150	114
All									
No.	4–6	6–7	4–5	6	4–6	8–10	13–14	10–12	4–7
%	3.3–5.0	2.9–3.4	2.0–2.5	2.6	1.9–2.9	3.8–4.7	5.9–6.4	3.8–4.5	1.8–3.2
Total DS	120	204	204	231	209	211	220	264	217

[a] Total DS are the total number of Down syndrome cases reported to the White Chromosome Registry by March 15. 1985, excluding known out-of-state residents. Cases that did not have year of birth indicated were assigned here to an estimated year of birth. Cases with maternal age not stated were distributed here according to the proportions of those with known maternal age. The rantes in the mutant interchange trisomy reflect instances in which one or both parents were not investigated to determine if the translocation was inherited.

possible explanations. Either there was a real change in biological or environmental factors that affect the measured mutation rate of this condition in these years, or else simply a rare and unusual event has occurred resulting from what may be termed statistical fluctuation. Information from other jurisdictions on this issue were equivocal although none of them provided as much data on this point as the New York Registry itself (Hook 1978b). Because of concern about environmental factors affecting mutation rate, a case-control investigation was undertaken in 1975–1977. This revealed, however, no evidence for any obvious environmental factor in the history of translocation cases as compared to controls.

A great variety of detailed analyses of the data from 1968 to 1977 have already been published (Hook and Albright 1981) and the episode has been commented upon in detail elsewhere (Hook 1978b). Table 6 presents an update of one summary of the data from 1968 through 1984. Because 47,+21 is very sensitive to maternal age but Robertsonian translocation trisomy 21 in balance has little association with age, the analysis is stratified on maternal age to adjust for differential age-related changes in reproduction.

The evidence for a change in mutation rate in 1973 to 1977 was confined to those under 30 years, although because the bulk of the cases occurred at these younger ages, there was also an effect when data on those of all ages were considered. The data from the 7 years after 1977 are equivocal. The rates in 1978–1979 are not notably higher than those before 1973, whereas those for the years 1980, 1982, 1984 are as

Year of birth								
1977	1978	1979	1980	1981	1982	1983	1984	Total 1968–1984
8–9	1–2	0–5	7–9	5–8	6–10	4–11	7–11	83–123
6.7–7.6	1.0–1.9	0.0–4.8	5.4–7.0	4.1–6.6	6.2–10.3	3.7–10.3	6.9–10.8	4.6–6.8
119	104	105	129	121	97	107	102	1807
2–3	3–4	2–4	3–4	2–3	1–4	1	1–2	29–40
2.2–3.2	2.7–3.6	2.2–4.4	3.3–4.4	1.5–2.2	0.8–3.3	0.6	0.7–1.5	1.5–2.1
93	110	90	90	134	123	160	137	1894
10–12	4–6	2–9	10–13	7–11	7–14	5–12	8–13	112–163
4.7–5.7	1.9–2.8	1.0–4.6	4.6–5.9	2.7–4.3	3.2–6.4	1.9–4.5	3.3–5.4	3.0–4.4
212	214	195	219	255	220	267	239	3701

high or higher, and those in 1981 and 1983 are difficult to interpret. Part of the difficulty in recent years is that a larger proportion of studies of *peripheral blood* of infants is sent by mail to cytogenetic laboratories which often are not aware of the results of parental studies that may be done subsequently. Thus, in 1983 while there are four mutants in those under 30, there are seven cases that must be classified as of unknown origin. Independent knowledge leads to the estimate that about three-fourths of those of unknown status result from true mutation, but for purposes of surveillance we must consider the results in terms of ranges rather than making maximum likelihood estimates of mutation rates.

There are several additional points to consider in evaluating this episode. First, the introduction of prenatal diagnosis has had little effect upon detection of *mutant* translocations, especially in those under 35, so selective abortion of affected cases has not affected the rates of these events in live births.[3] [There is no marked association of Robertsonian translocations with maternal age (Hook 1984).] The estimated rate of Down syndrome live births with a mutant Robertsonian translocation is about 1 in 20,000 to 1 in 25,000. Adjusting for spontaneous fetal death between the time of amniocentesis and time of birth results in a predicted rate of about 1/14,000 to 1/17,500 fetuses at 16 to 20 weeks. The observed rate of mutants in Registry data on fetuses diagnosed prenatally between 1977 and 1984, was between 1/15,000 and 1/37,500. Thus, it appears unlikely that the use of prenatal diagnosis will have a marked effect in live births on the absolute numbers of those affected nor upon the proportion of cases, at least in those born to younger mothers.

3 There were two to five new mutants in 78,000 fetuses diagnosed prenatally in 8 years, but an estimated 60 mutant cases occurred in the approximately 1.5 million live births in these years.

Secondly, it should be emphasized that fluctuations in Robertsonian translocations diagnosed in live births could reflect differential changes in fetal survival rather than in the gametic mutation rate, although it is difficult to imagine factors which would affect fetal mortality of one genotypic group of Down syndrome cases but not another.

Lastly, this method of surveillance is (or at least was) very "efficient" for surveillance of Robertsonian translocations compared to the use of amniocentesis data in the following sense. The data on rates of abnormalities at amniocentesis are on a total of about 78,000 fetuses, whereas the admittedly less precise estimated mutation rates for Robertsonian translocations in the 1972–1977 period are on roughly 1.5 million live births. If the growing problem of lack of information on the parental karyotype of live-born cases can be solved, then this would enable us to continue to derive useful information for surveillance from this approach.

8 Cytogenetic Abnormalities in Fetuses of Parents Studied Because of Mutagen Exposure

Of the 78,567 fetuses on which data are available from 1977 to 1984, there were 447 (0.6%) studied because, among other reasons, a parent was exposed to radiation (232) or a putative chemical mutagen (215). These results were excluded from the surbeillance analysis, but even despite the small numbers involved, the results are of some interest in themselves.

There was only one numerical abnormality 47,+21 found. This was in a 41-year-old woman who had been studied, among other reasons, because of exposure to radiation. From the age distribution of cases, one to two numerical abnormalities would be expected. There were also one to two mutant structurally abnormal chromosomes among the 447. The observed rate of 2.2 to 4.4 per 1,000 is slightly higher than the expected rate of about 2.0 per 1,000, but of course not statistically different from it. Moreover, in one of the cases, the known de novo event, it was not certain whether radiation had occurred before or after conception.

Surprisingly, the greatest excess was among those with inherited, structurally abnormal rearrangements in whom environmental effects would be least expected. There were four such cases, a rate of 9 per 1,000 fetuses. The expected rate of inherited, structural abnormalities is about 2.9 per 1,000, so roughly 1.3 cases would be expected compared to the four observed in the 447 studied. The difference is not nominally significant at the 5% level, ($P \sim 0.10$) and may well be due to statistical fluctuation. Nevertheless, it is conceivable that environmental agents may influence *segregation* of cytogenetic abnormalities in carriers of balanced rearrangements, resulting in the observed (nonsignificant) trend. Clearly, further data on this point will be of great interest.

9 Specific Locus Mutations: Nature and Frequency in Live Births

With regard to specific locus mutations, some aspects of their population dynamics are pertinent to surveillance considerations. At least some individuals with whom I have discussed this issue have concluded that because mutation rates for Mendelian disorders are very low and most specific locus mutations detected with biochemical methods have little adverse effect, that society does not have to be concerned about the impact of specific locus mutation on our species, at least in the short run. For this reason it is important to review what quantitative data are avilable on this point.

If one assumes an average gene length of perhaps 1,000 base pairs (consistent with about 330 amino acid residues) then from the amount of DNA in the nucleus of human cells, one would predict 1.8 to 3.5 *million* gene length segments in the human genome in which specific locus mutations could occur (Vogel 1984). But a great deal of this DNA, perhaps as much as 95% to 99%, is thought to be apparently "inert" and/or redundant so that mutation or recombination in such regions has (presumably) little effect on the phenotype (Neel et al. 1986b). The precise number of gene segments actually functioning in the sense of producing or affecting formation of proteins pertinent to human development is still moot but a large number of different considerations, including extrapolations from lower organisms, suggest a range of 50,000 to 100,000 (McKusick 1983). It is mutation in these segments that alter regions which specify functional proteins and may have phenotypic significance.

Recent studies of electrophoretic variants in human populations are consistent with an "average" mutation rate of 0.3×10^{-5} per gamete per locus resulting in "electromorphs" (Neel et al. 1986a). Of course this is an underestimate of specific locus mutation at loci coding for functional proteins for several reasons. First, silent base substitutions resulting in proteins without qualitative electrophoretic variation are missed. Indeed, even a mutation with a severe effect, e.g., some base substitution mutations resulting in a "stop" codon, will not be detected because they will only result in the *absence* of a product, i.e., a null allele, giving 50% diminishment in activity (as one of the two alleles are lost) which will not produce qualitative alteration. Secondly, more extensive gene alterations, deletions, or other rearrangements may also produce a "null" allele. It is not clear what proportion the electromorphs are of the total spectrum of possible specific locus mutations at loci evaluated. On theoretical grounds it may be estimated that they constitute about one-third of the base substitution mutations (Neel et al. 1936a,b) and probably are no more than one-fourth of all specific locus mutations.

Thus, as manifest in live births, the average specific locus mutation rate resulting from base substitutions is roughly 10^{-5} per locus per gamete, and probably at least 1.3×10^{-5} if other types of specific locus mutation are considered. Lethal mutants resulting in embryonic or fetal death are not detected in these systems so that the rates would be even higher if all zygotes could be studied, not just those surviving live birth.

It may be noted incidentally, that most Mendelian *diseases* whose mutation rates had been studied individually have mutation rates of between 10^{-5} and 10^{-6} per gamete (Vogel and Rothenberg 1975). Some suspected that these rates are atypically high and that there has been ascertainment bias in the study of phenotypic traits with a high mutation rate because their greater frequency led to selection for study. On these grounds I believed the "true" average mutation rate to be clsoer to 10^{-6} to 10^{-7} per locus per gamete. Therefore, the estimates of about 10^{-5} derived from electrophoretic studies, which lack this bias, appeared to me initially to be surprisingly high. I suspect now that the apparent differences may be reconciled by the realization that the study of at least some phenotypic autosomal dominant "traits", e.g., Marfan's syndrome, with an average mutation rate of about 5×10^{-6} per gamete, does not reveal *all* mutations occurring at the locus coding for the gene, but only of course those which result in the known phenotype. Other mutations at the same locus may produce either no recognizable phenotypic alteration or a variant phenotype not recognized as allelic to the mutation resulting in Marfan's syndrome. Thus, a rate of 5×10^{-6} per gamete is highly likely to be an underestimate of all mutations occurring at the locus at which the Marfan's syndrome allele is located. The protein studies, on the other hand, reveal a much larger proportion of mutational variation at the loci investigated, and there is a rational theoretical mechanism for estimating and adjusting for the proportion of undetected mutants.[4]

It may be noted parenthetically that the mutations about which there is most concern because of their effect in the first generation, i.e., those which are dominant *and* deleterious, are *relatively* more likely to involve nonenzymatic structural proteins than enzymes. There are not only theroetical grounds for this based upon biochemical functional considereations, but also at least some indirect evidence from the following considerations.

A number of human Mendelian *disorders* have been characterized biochemically. Up to now there has been selective ascertainment of disorders with enzymatic abnormalities rather than with structural proteins, if only because more enzymes than nonenzymatic proteins were characterized and disorders may be detected relatively easily because of their enzyme activity (or lack of it) (Mohrenweiser 1983a,b). But there is no reason to expect any bias of the *ratio* of dominants to recessives *within* each biochemical group. I calculated from McKusick's tabulation of known Mendelian (i.e., specific locus) disorders with defined biochemical abnormality 17 were dominant and 172 were recessive, a ratio of 1:10. For autosomal disorders involving a nonenzymatic structural abnormality there are 23 dominant and 43 recessive conditions, a ratio of 1:1.5 and about a sevenfold relative difference, which is at least consistent with the theoretical expectation cited above (see Table 7).

In any event, from a (minimum) mutation rate of about 10^{-5} per gamete per locus, the presumed number of significant loci (5 to 10×10^{-4}), and recalling that (for autosomal traits) we must consider two gametes per individual, one derives a minimum estimate of one to two fresh, specific locus mutations per live birth! [Neel et al. (1986a,b) estimated that if one considers all new mutations including those in apparently nonfunctional segments, then there are 20 per individual.] A question of major concern is how many of these new mutant alleles are deleterious? Here, one must distin-

---

[4] Of course, if the disease is heterogeneous genetically, i.e., results from mutant alleles at two or more different loci, then the measured mutation rate for what is thought to be a single condition will be an overestimate of what occurs at a single locus. This problem arose with "hemophilia" before it was subdivided into hemophilia A (Factor VIII deficiency) and hemophilia B (Christmas disease or Factor IX deficiency), separate but very similar conditions resulting from two different genes that both happen to be X-linked. Admittedly, the mutation rate for hemophilia A (about 5×10^{-5}) is almost 20 times that for hemophilia B (about 3×10^{-6}) so the mutation rate for hemophilia A is still close to the mutation rate calculated on both conditions together. Nevertheless, this example at least illustrates the possible nature of the problem for Mendelian disorders not yet characterized biochemically.

Table 7. Type of inheritance of Mendelian biochemical disorders in humans by the nature of the identified gene product[a]

Biochemical disorder	Autosomal dominant	Autosomal recessive	X-linked	Total
Enzymatic	17 (8%)	172 (83%)	18 (9%) 207	207 (100%)
Nonenzymatic (structural protein)	23 (32%)	43 (59%)	7 (10%) 73	73 (100%)

[a] This table was calculated from data summerized by McKusick (1982). It should be noted that McKusick classified loci according to the genetic characteristics of the allele at the locus so that there is ambiguity in his categories particularly if there is more than one "disease" allele at a locus. For purposes of this analysis, however, this introduces little distortion. Also, it should be noted that because it has up to now been easier on the average to detect enzyme abnormalities than structural abnormalities, the ratio of these two categories should not be interpreted as likely to be representative of all human Mendelian biochemical abnormalities. There is, however, no reason to expect the distribution of mode of inheritance *within* each type of biochemical abnormality to be biased with regard to the comparison of the enzymatic and structural abnormalities.

guish between alleles whose effects are autosomal dominant (or X-linked), i.e., expressed in the first generation and the remainder, those that are true autosomal recessive with no deleterious effect unless present in double dose. The latter, while adding to the "genetic load", are of course unlikely to have effects immediately evident in the first generation, and their consequences may only appear in the distant future unless there is inbreeding. If only 1% of the mutant protein variants have significant deleterious effect as dominants – an estimate subject to considerable uncertainty but which appears conservatively low – then the proportion of live births with a fresh, deleterious dominant mutation is somewhere between 1% and 2% or 1/50 to 1/100 live births. This estimate, which ignores any impact of the (pure) recessives, either autosomal or X-linked, still appears extraordinarily high to me; only future investigations will reveal if the apparent, plausible numerical assumptions from which these calculations are derived are accurate. In any event they do provide quantitative grounds if any are needed further, for concern about the impact of germinal specific locus mutation in our species.[5]

[5] This calculation, moreover, does not include the deleterious X-linked recessive mutations *manifest in the first generation*. One might suspect initially, that these would be negligible compared to the (autosomal) dominants, no more than about 2.5% of the latter, because about 5% of loci in our genome are on the X-chromosome and of course for X-linked recessives, effects would appear only in the males in the first generation. But it should be recalled that since males are X-hemizygous, an average mutation on the X-chromosome is much more likely to be deleterious (in the male) than the average mutation in an autosome since there is no normal allele to "buffer" its effect. Moreover, the discussion above, suggesting that mutations in loci in structural proteins would occur relatively more frequently than mutations in loci for enzymatic proteins among deleterious dominants, appears unlikely to apply to deleterious, X-linked recessive conditions. In this regard it is of interest that the ratio of X-linked to autosomal dominant Mendelian disorders in McKusick's enumeration is as I calculated, 18 to 17 for those with a known enzymatic disorder, but 7 to 23 for those with a disorder of nonenzymatic, structural proteins. Thus, the structural mutations are represented *relatively* less frequently among the X-linked disorders than among the autosomal dominant diseases.

10 Surveillance of Specific Locus Mutations

There are at least three possible approaches to surveillance of specific locus mutations. One involves the systematic evaluation of cord blood or other specimens for the search for evidence of proteins present in the fetus but not in the blood of either parent. This has been described in great detail by Neel and colleagues (e.g. Neel 1981a,b; Neel et al. 1983) and at present this appears scientifically optimal. (Conceivably DNA probes for the detection of mutant alterations in genes directly may be possible in the near future and reveal even more variation.) Elsewhere I have characterized this as the "brute force" approach (Hook 1979). The main issue is the expense about which there has been some discussion, since the average cost of detection was, at least in 1981, about $1,000,000 per mutant. [See "Discussion" in Hook and Porter (1981), for comments on the issue of cost and efficiency.] Certainly Neel may be correct, however, in his suggestion that even at that price it may ultimately be more expensive for society *not* to have the information derived from such studies. Moreover, advances in technology may both diminish the cost and improve efficiency.

The second approach is an "opportunistic" variant of the latter approach, which involves the use of filter-paper blood spots from heel-stick specimens routinely collected from newborn infants for testing for phenylketonurea (Hook 1979; Vogel and Altland 1982). The advantage of this method is that it can be superimposed upon an existing system for specimen collection in many jurisdictions. The difficulty is that relatively few loci can be examined because of the quantitative limitations of the specimen. When I analyzed this possibility a few years ago, considering hemoglobin variants detected in sickle cell disease screening in New York State, it was not at all clear then even in a state as large as New York (roughly 250,000 live births a year) that one could detect a sufficient background number of hemoglobin mutants for useful surveillance, at least in the system then in use (Hook 1979). (One problem with hemoglobin variants is that a significant number of those observed will involve hemoglobin F which is not expressed in the adult, so that DNA probes would be necessary to detect mutation.) Again with increasing technological advances and funding this approach may become more efficient, for instance, by using isoelectric focusing methods solely to detect hemoglobin variants and ascertaining sickle cell disease separately through cellulose acetate electrophoresis (K.A. Pass, pers. commun.).

There is the additional expense and effort involved in characterization of observed variants and follow-up investigation of parents to exclude the inheritance of variants that would not normally be undertaken as part of routine surveillance. In an attempt to characterize hemoglobin variants from newborns in New York State, Pass found that he could only obtain the cooperation of 168 of 373 families in which such a variant was found (K.A. Pass, pers. commun.). Because the observed variant is not in itself an indication of infant disease, parents are much less likely to cooperate with such investigation than if, e.g., sickle cell disease or phenylketonurea is suspected. One could of course adjust mutation estimates for such cases wich cannot be investigated completely. But if there is such a large uncooperative proportion, then there will be considerable uncertainty in the precise mutation estimate, which thus introduces problems for surveillance similar to those discussed above for Robertsonian chromosome translocations.

A third possible approach is the use of "sentinel phenotypes" to monitor specific locus mutation (Sutton 1971). This is the scoring of morphological conditions usually detectable at birth that constitute examples of known single gene disorders that have resulted from a fresh mutation in affected individuals. Only autosomal dominant disorders (or X-linked disorders in which the maternal carrier state may be easily excluded) are useful for this approach. This method has been discussed for many years. A select review committee about 20 years ago suggested three conditions as prime candidates for such surveillance: achondroplasia, Apert's syndrome (acrocephalosyndactyly type 1), and aniridia (Sutton 1971). The collective frequency of infants with new mutations in the general population for these conditions might be somewhere between 1 and 25,000 and 1 in 50,000. The committee, nevertheless, citing problems with illegitimate births and proper diagnosis, was relatively pessimistic about the usefulness of this approach, although I believe the grounds for this were too stringent. (I do not believe that there are likely to be many fertile male achondroplastic dwarfs let alone fertile males with Apert's syndrome who are spawning cryptically any significant numbers of affected illegitimate children! While there are rare instances of nonpenetrance in parents, sporadic instances of these events are almost certainly the result of a fresh mutation.)

There are of course a number of additional disorders that might be considered for sentinel phenotypic surveillance.[6] One might include entities diagnosed after birth, in which case, as discussed above, some malignancies, e.g., bilateral retinoblastoma, could be included. The difficulty is that every conditions evaluated, each of which is quite rare, has its particular problems in ascertainment and nosology. Expressivity, penetrance, and nosologic definition vary considerably among candidate conditions.

If a knowledgeable dysmorphologist examined each affected child and separated each syndrome meticulously, one would have much greater reliance upon the accuracy of the reported condition. Holmes and colleagues (1981) as part of a systematic, consistent physical evaluation of 24,418 infants undertook an analysis of Mendelian disorders associated with morphologic abnormality and reported six presumed mutants. Holmes et al. (1981), however, described the screening experience as "masochistic". Neel (1981b) correctly emphasized the uncertainty in classifying the reported outcome in this study as examples of de novo dominants and thus new mutants so that it is not even certain if the observed rate of about 2.4 per 10,000 is valid in the population screened.

Despite the apprarent economies of this approach compared to biochemical screening (Hook 1981d) it is much less precise. Because of the meticulous diagnostic skills needed in sorting out "phenocopies" and in nosologic separation of specific mutant genetic syndromes, routine reporting systems of congenital malformations are not likely to provide sufficient sensitivity or specificity to use for surveillance of *mutations* without additional ad hoc efforts to characterize reported conditions. (As one example,

[6] One such list appears in Mulvihill and Czeizel (1983) which includes a good deal of discussion of the concept of "sentinel phenotypes". I disagree strongly, however, with the inclusion by these authors of some entities proposed for candidate conditions, and the suggestion that X-linked traits such as muscular dystrophy be included. This overlooks the serious problem of excluding inheritance from female carriers. (Admittedly, since that paper was written chemical probes have become available which diminish this problem for these conditions.)

because of superficial similarities, cases of thanatophoric dwarfism are often reported as instances of achondroplastic dwarfism.) The key point is not so much that 100% accuracy be achieved, as that there be *consistency* in diagnosis over time so that temporal variations in rate do not reflect simply variation in the knowledge and ability of diagnostic examiners. There is the additional problem that because of advances in the dissection of syndromes, the identification of entities as examples of autosomal dominants is highly dynamic. For example, the total number of entities identified by McKusick as confirmed autosomal dominants evolved from 269 in 1966 to 934 in 1982 (McKusick 1983). Even if all newborns were to be evaluated by experienced dysmorphologists, changes over time in identified dominant conditions would present problems.

This is not to dismiss the method of sentinel phenotypes. It is possible that by restriction to a small group of phenotypes which are well characterized, whose genetic "significance" is unquestioned, and which are easily and consistently identified shortly after birth, one might be able to generate consistent estimates. But there is still no agreement on which particular phenotypes to include. Nevertheless, it would appear worthwhile to begin to explore these possibilities in available data sets.

Acknowledgments. It thank Philip K. Cross for deriving some of the analyses presented in this paper and for other assistance with the Registry over many years; Janet Birch and Robert Dorn for expert clerical assistance; and Veronica Mott for faithful attention to detail in typing the manuscript. I also thank the many laboratory directors and their assistants who have contributed data to the Registry, some as far back as 1969, and numerous individuals within the New York State Department of Health who have been helpful both in the establishment and maintenance of the Registry. I thank Harvey Mohrenweiser for directing my attention to some of the significant differences in the detection between enzymatic and structural proteins. Over the years the interest and activity in mutation surveillance of James V. Neel have been a continuing spur to my own continuation in this activity. Lastly, my interest in chronic disease surveillance, in general, and surveillance of mutations, in particular, was initiated by a fascinating presentation by James Crow on this topic at a symposium at a meeting of the American Society of Human Genetics in Austin, Texas in October, 1968.

References

Albertini RJ (1985) Somatic gene mutations in vivo as indicated by the 6-thioguanine resistant T-lymphocytes in human blood. Mutat Res 150:411–422

Albertini RJ, Allen EF, Quinn AS, Albertini MR (1981) Human somatic cell mutation: In vivo variant lymphocyte frequencies as determined by 6-Thioguanine resistance. In: Hook EB, Porter IH (eds) Population and biological aspects of human mutation. Academic Press, London New York, pp 235–264

Albright SG, Hook EB (1980) Estimates of the likelihood that a Down's syndrome child of unknown genotype is a consequence of an inherited translocation. J Med Genet 17:273–275

Brewen JG, Preston RJ, Gengozian N (1975) Analysis of X-ray induced chromosome translocations in human and marmoset spermatogonial cells. Nature (London) 253:468–470

CDC (Center for Disease Control) (1985) Congenital malformation surveillance Rep. January 1981–December 1983

Edmunds LD, Layde JM, James LM, Flynt JW, Erickson JD, Oakley GP (1981) Congenital malformations surveillance: Two American systems. Int J Epidemiol 10:247–252

Ferguson-Smith MA, Yates JRW (1984) Maternal age specific rates for chromosome aberrations and factors influencing them: report of a collaborative European study on 52,965 amniocenteses. Prenat Diagn 4:5–44

Ford CE, Evans EP, Searle AG (1977) Failure of irradiation to induce Robertsonian translocations in germ cells of mice. In: Conference on mutations – their origin, nature, and potential relevance. Deutsche Forschung, Bonn, pp 102–108

Fleiss JL (1973) Statistical methods for rates and proportions. John Wiley & Sons, New York

Holmes LB, Vincent SE, Cook C, Cote KP (1981) Surveillance of newborn infants for malformations due to spontaneous germinal mutations. In: Hook EB, Porter IH (eds) Population and biological aspects of human mutation. Academic Press, London New York, pp 351–360

Hook EB (1978a) The ratio of de novo unbalanced translocations to 47, trisomy 21 Down syndrome: A new method for human mutation surveillance and an apparent recent change in mutation rate resulting in human interchange trisomies in one jurisdiction. Mutat Res 52: 427–439

Hook EB (1978b) Monitoring human mutations and consideration of dilemma posed by an apparent increase in one type of mutation rate. In: Morton NE, Chung CS (eds) Genetic epidemiology. Academic Press, London New York, pp 483–528

Hook EB (1979) Human germinal mutations: Monitoring for environmental effects. Soc Biol 26: 104–116

Hook EB (1981a) Human teratogenic and mutagenic markers in monitoring about point sources of pollution. Environ Res 25:178–203

Hook EB (1981b) Unbalanced Robertsonian translocations associated with Down syndrome or Patau syndrome: Chromosome subtype, proportion inherited, mutation rates and sex ratio. Human Genet 59:235–239

Hook EB (1981c) Interchange trisomic Down's syndrome and Patau's syndrome: General approaches to estimating mutation rates and epidemiological advantages for monitoring. In: Hook EB, Porter IH (eds) Population and biological aspects of human mutation. Academic Press, London New York, pp 167–190

Hook EB (1981d) Discussion. In: Hook EB, Porter IH (eds) Population and biological aspects of human mutation. Academic Press, London New York, pp 376–378

Hook EB (1982a) The epidemiology of Down syndrome. In: Pueschel SM (ed) Down syndrome: Advances in biomedicine and the behavioral sciences. Ware, Cambridge, pp 11–88

Hook EB (1982b) The value and limitations of clinical observations in assessing chemically-induced genetic damage in humans. In: Weinstein IB, Butterworth BE, Bridges BA (eds) Banbury Rep 13: Indicators of genotoxic exposure. Banbury Conf. Cold Spring Harbor Laboratory Press, New York, pp 19–30

Hook EB (1983) Perspectives in mutation epidemiology: 2. Contribution of chromosome abnormalities to human morbidity and mortality and some comments upon surveillance of chromosome mutation rates. Mutat Res 114:389–423

Hook EB (1984) Parental age and unbalanced Robertsonian translocations associated with Down syndrome and Patau syndrome. Comparison with maternal and paternal age effects for 47,+21 and 47,+13. Ann Human Genet 48:313–325 (See also 49:153, 1985)

Hook EB, Albright SG (1981) Mutation rates for unbalanced Robertsonian translocations associated with Down syndrome in New York State livebirths 1968–1977, and evidence for a temporal change. Am J Human Genet 33:443–454

Hook EB, Cross PK (1981) Temporal increase in the rate of Down syndrome livebirths to older mothers in New York State. J Med Genet 18:29–30

Hook EB, Cross PK (1982) Surveillance of human populations for germinal cytogenetic mutations. In: Sugimara T, Kondo S, Takebe H (eds) Environmental mutagens and carcinogens. Univ Press, Tokyo, and Alan Liss, Tokyo and New York, pp 613–621

Hook EB, Cross PK (1987) Rates of mutant and inherited structural cytogenetic abnormalities detected at amniocenteses: Results on about 63,000 fetuses. Ann Human Genet 51:27–55

Hook EB, Liss SM (1982) Age of ascertainment – a relative index of severity of cytogenetic disorders. J Chron Dis 35:207–209

Hook EB, Porter IH (eds) (1981) Population and biological aspects of human mutations. Academic Press, London New York, pp 376–378

Hook EB, Woodbury DF, Albright SG (1979) Rates of trisomy 18 in livebirths, stillbirths, and at amniocenteses. In: Epstein CJ, Crry CJ, Packman S, Hall B (eds) Genetic counseling: Risk communication and decision making. Birth Defect Original Article Series, vol 15, No 5C, pp 81–93

Hook EB, Cross PK, Schreinemachers D (1981) The evolution of the New York State Chromosome Registry. In: Hook EB, Porter IH (eds) Population and biological spects of human mutation. Academic Press, London New York, pp 389–428
Hook EB, Cross PK, Schreinemachers DM (1983) Chromosome abnormality rates at amniocentesis and in liveborn infants. JAMA 249:2034–2038
Hook EB, Cross PK, Regal RR (1984) The frequency of 47,+21, 47,+18, and 47,+13 at the uppermost extremes of maternal ages: Results on 56,094 fetuses studied prenatally and comparisons with data on livebirths. Human Genet 68:211–220
IPCS (International Programme on Chemical Safety) (1985) Guidelines for the study of genetic effects in human populations, (environmental health criteria 46). WHO, Geneva, 126 pp
Johnson KM, Huether CA, Hook EB, Crowe CA, Reeder BA, Sommer A, McCorquodale MM, Cross PK (1985) False positive and negative reporting of Down's syndrome on Ohio and New York birth certificates. Genet Epidemiol 2:123–131
Knudson AG (1971) Mutation and cancer: Statistical study of retinoblastoma. Proc Natl Acad Sci USA 68:820–823
Knudson AG, Strong LC (1972) Mutation and cancer: A model for Wilm's tumor of the kidney. J Natl Cancer Inst 48:313–324
Lamson SH, Hook EB (1981) Comparison of mathematical models for the maternal age dependence of Down's syndrome rates. Human Genet 59:232–234
Martin RH, Hildebrand K, Yamamoto J, Rademaker A, Barnes M, Douglas G, Arthur K, Ringrose T, Brown IS (1986) An increased frequency of human sperm chromosomal abnormalities after radiotherapy. Mutat Res 174:219–225
McKusick VA (1983) Foreword. In: Mendelian inheritance in man. Catalog of autosomal dominant, autosomal recessive, and X-linked phenotypes, 6th edn. The Johns Hopkins University Press, Baltimore, pp ix–lxvi
Mehes K (1973) Clinically recognisable autosomal anomalies in neonates. Lancet 1:1262–1263
Miller JR, Clemmesen J, Czeizel A, Evans HJ, Hook EB, Jansen JD, LeChat MF, Matsunaga E, Mulvihill JJ, Oftedal P (1983) Mutation epidemiology: Review and recommendations. Mutat Res 123:1–11
Mohrenweiser HW (1983a) Biochemical approaches to monitoring human populations for germinal mutation rates: II. Enzyme deficiency variants as a component of the estimated genetic risk. In: deSerres FJ, Sheridan W (eds) Utilization of mammalian specific locus studies in hazard evaluation and estimation of genetic risk. Plenum, New York, pp 55–69
Mohrenweiser HW (1983b) Enzyme-deficiency variants: Frequency and potential significance in human populations. Isozymes: Current topics in biological and medical research. Genet Evol 10:51–68
Mulvihill JJ, Czeizel A (1983) Perspectives in mutation epidemiology, 6: A 1983 view of sentinel phenotypes. Mutat Res 123:345–361
NICHD National Registry for Amniocentesis Study Group (1976) Midtrimester amniocentesis for prenatal diagnosis: safety and accuracy. JAMA 236:1471–1476
Neel JV (1981a) In quest of better ways to study human mutation rates. In: Hook EB, Porter IH (eds) Population and biological aspects of human mutation. Academic Press, London New York, pp 361–375
Neel JV (1981b) Discussion. In: Hook EB, Porter IH (eds) Population and biological aspects of human mutation. Academic Press, London New York, pp 376–378
Neel JV, Mohrenweiser H, Hanash S et al. (1983) Biochemical approaches to monitoring human populations for germinal mutation rates: I. Electrophoresis. In: deSerres FJ, Sheridan W (eds) Utilization of mammalian specific locus studies inhazard evaluation and estimation of genetic risk. Plenum, New York, pp 71–93
Neel JV, Mohrenweiser HW, Rothman ED, Naidu JM (1986a) A revised indirect estimate of mutation rates in Amerindians. Am J Human Genet 38:646–666
Neel JV, Satoh C, Gouki K, Fujita M, Takahaski N, Asakawa J-I, Hazama R (1986b) The rate with which spontaneous mutation rates alters the electrophoetic mobility of polypeptides. Proc Natl Acad Sci USA 83:389–393
Palmer CG (1981) Are there "hot spots" in the human genome? Evidence from breakpoint analysis in a collaborative study of germinal chromosome rearrangement. In: Hook EB, Porter IH (eds)

Population and biological aspects of human mutation. Academic Press, London New York, pp 147–166
Polan AK, Monetti C, Pohl R, Vallet HL, Vianna N (1985) The New York State congenital malformations, Pres Annu Meet Am Public Health Assoc, Washington DC
Sutton HE (1971) Report of the committee for the study of monitoring of human mutagenesis. Teratology 4:103–108
Vogel F (1984) Gene or point mutations. In: Obe G (ed) Mutations in man. Springer, Berlin Heidelberg New York, pp 101–127
Vogel F, Altland K (1982) Utilization of material from PKU-screening programs for mutation screening. Chemical mutagenesis, human population monitoring, and genetic risk assessment. In: Bora KC, Douglas GR, Nestmann ER (eds) Progr Mutat Res 3:143–158
Vogel F, Rothenberg R (1975) Spontaneous mutation in man. Adv Hum Genet 5:223–318

Addendum (July 1987): Since this paper was written, one additional limitation upon surveillance use of cytogenetic data from second trimester amniocentesis has become evident. Cytogenetic abnormalities detected after amniocentesis carried out because of suspected fetal pathology have increased because two clinical "screening" methods of pregnant women have expanded rapidly. Fetal ultrasound is now often done routinely to estimate fetal size and gestational age. Incidental observation of morphological abnormalities or other fetal pathology may lead to amniocentesis and eventually detection of associated abnormalities. Also, routine determination of maternal serum alpha fetoprotein (a.f.p.) is now widespread. This was introduced some years ago because of the association of high a.f.p. levels with neural tube defects. It was however, recently discovered that 47,+21, and perhaps other trisomies are associated with a *low* serum (a.f.p.) and such a finding may lead to amniocentesis. Studies carried out because of these or other grounds for suspect fetal pathology have been excluded in our analyses of data through 1984. (These 2196 include among others 307 with an ultrasound-detected abnormality, 69 with low a.f.p., 1796 because of concern regarding possible elevated amniotic a.f.p. Most of the latter include those which were reported simply as an "alpha fetoprotein study", and probably the vast bulk of these were studied only because of an affected relative, not because of an elevated serum a.f.p. value.) This should be a very small source of bias overall in the 1977 to 1984 experience, but these 2196 were excluded in any event, to be conservative. However, as the screening procedures extend to a very large proportion of the population, and amniocentesis is carried out on an increasing number because of suspect results, such exclusion will result in an increasing *underestimate* of "true" (maternal age-specific) rates of cytogenetic abnormalities, perhaps a serious underestimation. On the other hand, inclusion of such cases will probably result in an even more serious *overestimation* of rates. (The analyes for 1985 and 1986 are still not complete, so quantitative estimates cannot be given now of the differences involved.) A great deal of additional investigation is necessary to determine precisely how to adjust for this source of bias for the bulk of unbalanced cytogenetic abnormalities. This bias is not a serious factor for balanced rearrangements or for 47,XXX, 47,XXY, or 47,XYY karyotypes because there appears to be no associated fetal morphological abnormalities or altered serum alpha fetoprotein levels likely to lead to selective diagnosis of these conditions. For this reason it is likely that surveillance of cytogenetic abnormalities will be carried out, at least straightforwardly, in the future using registry data from amniocentesis only for 47,XXX, 47,XXY, 47,XYY karyotypes and balanced structural rearrangements.

An entirely separate issue is raised by the increasing use of first trimester diagnosis using chorionic villus studies (c.v.s.). Fetuses with abnormalities detected in c.v.s. studies which result in pregnancy termination will, of course, not be detected at amniocentesis. Nevertheless, data from c.v.s. may be readily integrated with that from amniocentesis in a registry and used for surveillance purposes. As there is a difference between gestational age at c.v.s. and at amniocentesis (roughly 8 weeks or 0.16 years), an adjustment for this is necessary in analysis of maternal-age-associated anomalies. For all unbalanced abnormalities, adjustment is necessary for the spontaneous loss that would be expected between the time of chorionic villus biopsy and amniocentesis.

8 Cytogenetic Study of the Offspring of Atomic Bomb Survivors, Hiroshima and Nagasaki

A. A. Awa[1], T. Honda[2], S. Neriishi[3], T. Sufuni[1,4], H. Shimba[1], K. Ohtaki[1], M. Nakano[1], Y. Kodama[1], M. Itoh[2], and H. B. Hamilton[5]

1 Introduction

A cytogenetic study of the children born to atomic bomb survivors in Hiroshima and Nagasaki and children born to unexposed parents was initiated in 1967 (Awa 1975; Awa et al. 1968). The study was expanded in 1976 as a part of the Genetic Platform Research Program at RERF (Radiation Effects Research Foundation), and has been continued to the present time in conjunction with the ongoing mortality study and biochemical genetics survey on the F_1 progeny (RERF Research Protocol 1975).

The main objective of the present study is to evaluate the radiation sensitivity of human germ-cell chromosomes by measuring the frequency of children with chromosome changes in structure or number induced by radiation in the germ cells of exposed parents. It is expected that stable chromosome aberrations, if induced in the germ cells, would be most likely transmitted to the offspring. Although there is no evidence of chromosome aneuploidy being induced by radiation exposure in humans, it is difficult to exclude the possibility that abnormalities, such as XYY and XXX, would be induced in the offspring.

The present chapter describes the results of somatic chromosome analysis of 8,322 children born to A-bomb survivors in Hiroshima and Nagasaki and 7,976 children born to parents who had received less than 1 rad (distally exposed) or were not in the cities (NIC) at the time of the bomb (ATB). Chromosome analyses were based mostly on nonbanded preparations throughout the study.

Because of the recent, extensive reassessment of A-bomb dosimetry by a US-Japan team of experts (1st and 2nd US-Japan Joint Workshop 1983, 1984), the present study samples have been divided into exposed and control groups based on the T65DR system that has been routinely used until recently at RERF (Milton and Shohoji 1968). The data base for the new DS86 dose system has been entered into the RERF

1 Radiation Effects Research Foundation, 5-2 Hijiyama Park, Minami-ward Hiroshima 732, Japan
2 Department of Radiobiology, Nagasaki, Japan
3 Department of Clinical Studies, Nagasaki, Japan
4 Biological Safety Research Center, National Institute of Hygienic Sciences, Tokyo, Japan
5 RERF Consultant

Cytogenetics. Ed. by G. Obe and A. Basler

computer; however, calculations of the individual dose estimates for each survivor are now in progress, but are not available at this time. For this reason, no attempt has been made to analyze the present data in terms of parental radiation doses.

2 Materials and Methods

The sample subjects of the present survey were selected primarily from the RERF F_1 mortality study cohort (Kato et al. 1966). This cohort includes children born between 1 May 1946 and 31 December 1958 to parents, one or both of whom were the residents of Hiroshima and Nagasaki ATB. The samples were later expanded to include children who were born after 1959 through the end of 1972. The contribution of the extended samples was about 10% of the total.

In this analysis, the exposed group consists of children born to parents, one or both of whom were located within 2000 m from the hypocenter and who had T65DR dose estimates of more than 1 rad. The control group consisted of children born to parents (1) one or both of whom were exposed distally (2500 m or more from the hypocenter) with estimated doses of less than 1 rad, or (2) were not present in the city ATB.

About 40% of the total individuals in the original samples were not included in this study, because they had died (5%), or migrated outside the contactable areas of both cities (35%). Of the remaining individuals, approximately 74% agreed to participate in this study. Thus, the participation rate of this survey was 45% of the total original sample.

After obtaining consent from the F_1 participants and, if necessary, their parents, they were invited to visit the RERF clinic. At that time they were interviewed by nurses to obtain medically-related information and a blood sample was drawn. If requested, a physical examination was performed.

Heparinized blood specimens, 1 to 2 ml per person, collected from each participant, were cultured for 2 days, and then harvested for chromosome preparations using the conventional Giemsa staining methods, the details of which have been described elsewhere (Awa et al. 1978).

In each case, ten well-spread metaphases were examined directly under the microscope, and three of them were photographed for detailed karyotype analysis. Cases with less than ten scorable metaphases were regarded as culture failure. The rate of failure was about 0.1% of the total blood samples in the two cities. When an abnormality was suspected, 100 or more cells were examined.

In addition to the conventional stain, both Q- and C-band preparations (Caspersson et al. 1971; Sumner 1972) were used as a routine procedure. The G-banding method of Seabright (1971) was also applied to cases suspected of having abnormality. When family studies were performed on probands with structural rearrangements, high resolution banding techniques (Yunis et al. 1978; Pai and Thomas 1980) were employed for the precise identification of breakpoints of the chromosomes involved in the aberrations. A description of the type of chromosome abnormalities was made following the standardized nomenclature of ISCN (1978, 1981, 1985).

Table 1. Number of F_1 children examined

			Number of cases			Number of parental couples
			Males	Females	Total	
Hiroshima	Control		2,477	2,635	5,112	4,242
	Exposed:	Father	638	636	1,274	968
		Mother	1,348	1,490	2,838	2,018
		Both	274	330	604	467
		Total	2,260	2,456	4,716	3,453
Nagasaki	Control		1,205	1,659	2,864	2,231
	Exposed:	Father	543	623	1,166	802
		Mother	912	1,123	2,035	1,305
		Both	199	206	405	263
		Total	1,654	1,952	3,606	2,370
Total	Control		3,682	4,294	7,976	6,473
	Exposed:	Father	1,181	1,259	2,440	1,770
		Mother	2,260	2,613	4,873	3,323
		Both	473	536	1,009	730
		Total	3,914	4,408	8,322	5,823

3 Results

3.1 Characteristics of the Study Sample

As shown in Table 1, the total number of children examined was 16,298 in the two cities; 9,828 in Hiroshima (4,716 exposed and 5,112 controls), and 6,470 in Nagasaki (3,606 exposed and 2,864 controls). The number of females predominated over males in both cities as well as in both the exposed and control groups. The exposed group was further divided into three categories by parental exposure status; i.e., children born to parents in which only the father was exposed, only the mother was exposed, or both parents were exposed.

Since efforts were made to collect as many cases as possible in the exposed group, the number of children per parental couple is considerably higher in the exposed group than in the controls (Table 1); 1.4 children per couple (or 8,322 in 5,823) in the exposed, and 1.2 per couple (or 7,976 in 6,473) in the controls, when the two cities were combined.

The mean age of the participants at examination was 24 in Hiroshima, and 23 in Nagasaki with a range between 12 and 38.

Table 2. Overall frequency of F_1 children with chromosome abnormalities (per 1,000)

			Sex chromosomes	Rearrangements Balanced	Rearrangements Unbalanced	Trisomy	Total	No. of cases examined
Hiroshima	Control		17 (3.33)	16 (3.13)	0	0	33 (6.46)	5,112
	Exposed:	Father	4 (3.14)	5 (3.92)	2 (1.57)	1 (0.78)	12 (9.42)	1,274
		Mother	5 (1.76)	7 (2.47)	1 (0.35)	0	13 (4.58)	2,838
		Both	3 (4.97)	2 (3.31)	0	0	5 (8.28)	604
		Total	12 (2.54)	14 (2.97)	3 (0.64)	1 (0.21)	30 (6.36)	4,716
Nagasaki	Control		7 (2.44)	9 (3.14)	2 (0.70)	0	18 (6.28)	2,864
	Exposed:	Father	3 (2.57)	3 (2.57)	1 (0.86)	0	7 (6.00)	1,166
		Mother	4 (1.97)	0	1 (0.49)	0	5 (2.46)	2,035
		Both	0	1 (2.47)	0	0	1 (2.47)	405
		Total	7 (1.94)	4 (1.11)	2 (0.55)	0	13 (3.61)	3,606
Total	Control		24 (3.01)	25 (3.13)	2 (0.25)	0	51 (6.39)	7,976
	Exposed:	Father	7 (2.87)	8 (3.28)	3 (1.23)	1 (0.41)	19 (7.79)	2,440
		Mother	9 (1.85)	7 (1.44)	2 (0.41)	0	18 (3.69)	4,873
		Both	3 (2.97)	3 (2.97)	0	0	6 (5.95)	1,009
		Total	19 (2.28)	18 (2.16)	5 (0.60)	1 (0.12)	43 (5.17)	8,322
Newborn infants[a]			127 (2.23)	110 (1.93)	34 (0.60)	82 (1.44)	353 (6.20)	56,952

[a] Cited from Hook and Hamerton (1977).

3.2 Types and Frequencies of Chromosome Abnormalities

The results of the cytogenetic observations are shown in Table 2, and every abnormal case is fully described in the Appendix Table (see pp. 181–183).

Chromosome abnormalities are classified into the following three groups; (1) sex chromosome abnormalities, (2) autosomal structural rearrangements, and (3) autosomal trisomics.

3.2.1 Sex Chromosome Abnormalities

As shown in Table 3, there were 43 cases with sex chromosome anomalies; 19 of 8,322 in the exposed group (2.28 per 1,000) and 24 of 7,976 in the controls (3.01 per 1,000). No increased frequency of sex chromosome abnormalities, ascribable to parental radiation exposure, was observed.

The majority of the abnormalities was due to sex chromosome aneuploidy, mostly XYY and XXY in males and XXX in females, which constituted 75% of the total sex

Table 3. Frequency of F_1 children with sex chromosome abnormalities (per 1,000)[a]

			Males					Females				
			XYY	XXY	Mosaic	Other	Total	X	XXX	Mosaic	Other	Total
Hiroshima	Control		5 (2.02)	4 (1.61)	0	2 (0.81)	11 (4.44)	0	3 (1.14)	3 (1.14)	0	6 (2.28)
	Exposed:	Father	0	1 (1.57)	1 (1.57)	0	2 (3.13)	0	2 (3.14)	0	0	2 (3.14)
		Mother	0	4 (2.97)	0	0	4 (2.97)	0	1 (0.67)	0	0	1 (0.67)
		Both	0	1 (3.65)	0	0	1 (3.65)	0	1 (3.03)	1 (3.03)	0	2 (6.06)
		Total	0	6 (2.65)	1 (0.44)	0	7 (3.10)	0	4 (1.63)	1 (0.41)	0	5 (2.04)
Nagasaki	Control		0	5 (4.15)	0	0	5 (4.15)	0	1 (0.60)	0	1 (0.60)	2 (1.21)
	Exposed:	Father	2 (3.68)	0	0	0	2 (3.68)	0	1 (1.61)	0	0	1 (1.61)
		Mother	1 (1.10)	1 (1.10)	0	1 (1.10)	3 (3.29)	0	0	1 (0.89)	0	1 (0.89)
		Both	0	0	0	0	0	0	0	0	0	0
		Total	3 (1.81)	1 (0.60)	0	1 (0.60)	5 (3.02)	0	1 (0.51)	1 (0.51)	0	2 (1.02)
Total	Control		5 (1.36)	9 (2.44)	0	2 (0.54)	16 (4.35)	0	4 (0.93)	3 (0.70)	1 (0.23)	8 (1.86)
	Exposed:	Father	2 (1.69)	1 (0.85)	1 (0.85)	0	4 (3.39)	0	3 (2.38)	0	0	3 (2.38)
		Mother	1 (0.44)	5 (2.21)	0	1 (0.44)	7 (3.10)	0	1 (0.38)	1 (0.38)	0	2 (0.77)
		Both	0	1 (2.11)	0	0	1 (2.11)	0	1 (1.87)	1 (1.87)	0	2 (3.73)
		Total	3 (0.77)	7 (1.79)	1 (0.26)	1 (0.26)	12 (3.07)	0	5 (1.13)	2 (0.45)	0	7 (1.59)
Newborn infants[b]			35 (0.93)	35 (0.93)	14 (0.37)	14 (0.37)	98 (2.59)	2 (0.10)	20 (1.04)	7 (0.37)	0	29 (1.51)
No. of cases							37,779					19,173

[a] The rates for the sex chromosome abnormalities apply only to the affected sex.
[b] Cited from Hook and Hamerton (1977).

anomaly cases. Except for mosaic situations, no female with 45,X was detected in any of the groups studied.

When the data from the two cities were combined, there were no significant differences between the frequencies of sex chromosome abnormalities in the exposed group when compared to the controls.

The frequency of mosaic cases was higher in females (2 in 4,408 exposed and 3 in 4,294 controls) than in males (1 in 3,914 exposed). Among the mosaics, there was a woman showing a mosaic of 45,X/46,X,r(X) in the Hiroshima controls. A family study revealed that the abnormality was identified as a de novo mutant since both parents showed the normal karyotype.

There were three cases with structural rearrangements involving sex chromosomes, all of which belonged to the control group; two unrelated males, each with a pericentric inversion of the Y-chromosome in Hiroshima, and a female with a distally deleted long arm of one of the X-chromosomes in Nagasaki.

One unusual observation was made in the Nagasaki series. A male child, born to an exposed mother, was found to have 46 chromosomes with two X-chromosomes (46,XX).

3.2.2 *Autosomal Structural Rearrangements*

3.2.2.1 Balanced. The majority of structural rearrangements involving autosomes were translocations of the Robertsonian or reciprocal types and pericentric inversions. They would, therefore, result in no loss or gain of chromosome material, and would be genetically balanced without any phenotypic effect. In the present study sample, there were occasionally two or more sibs in a family showing the identical karyotypic abnormality.

Among these cases, there were ten children (seven D/D and three D/G) from eight families with Robertsonian translocations in the exposed group, and six children (all D/D) from three families with similar rearrangements in the controls. By the same token, there were seven children from five families in the exposed group, and 13 children from 12 families in the controls with reciprocal translocations. Only one child in the exposed, and six children from five families in the controls had inversions.

An examination of Table 4 shows that an unusually high incidence of pericentric inversions was observed in the Hiroshima controls, and in Nagasaki a strikingly high rate of reciprocal translocations was observed in the controls, while no such cases were found in the exposed group.

As shown in the Appendix Table, chromosomes 5 and 8 were found to be involved more frequently than the others in the formation of reciprocal translocations. Furthermore, there was a frequent involvement of a chromosome 2 in pericentric inversions in Hiroshima.

In Table 5, pooled data on the frequencies of balanced structural rearrangements are tabulated in terms of the number of families in which they are observed as a function of total families included in the study. Although there is no significant difference in the frequencies of families with stable autosomal rearrangements, a relatively low frequency of rearrangements was observed in the exposed group in Nagasaki. The data

Table 4. Frequency of F_1 children with structural rearrangements (per 1,000)

		Balanced[a]					Unbalanced[a]					
		rob(D/D)	rob (D/G)	rcp	inv	Total	rob	rcp	del	supern	other	Total
Hiroshima	Control	4 (0.78)	0	6 (1.17)	6 (1.17)	16 (3.13)	0	0	0	0	0	0
	Exposed:											
	Father	2 (1.57)	0	3 (2.35)	0	5 (3.92)	0	0	0	1 (0.78)	1 (0.78)	2 (1.57)
	Mother	3 (1.06)	1 (0.35)	2 (0.70)	1 (0.35)	7 (2.47)	0	0	0	0	1 (0.35)	1 (0.35)
	Both	0	0	2 (3.31)	0	2 (3.31)	0	0	0	0	0	0
	Total	5 (1.06)	1 (0.21)	7 (1.48)	1 (0.21)	14 (2.97)	0	0	0	1 (0.21)	2 (0.42)	3 (0.64)
Nagasaki	Control	2 (0.70)	0	7 (2.44)	0	9 (3.14)	0	0	0	0	2 (0.70)	2 (0.70)
	Exposed:											
	Father	2 (1.72)	1 (0.86)	0	0	3 (2.57)	0	0	0	1 (0.86)	0	1 (0.86)
	Mother	0	0	0	0	0	0	0	0	0	1 (0.49)	1 (0.49)
	Both	0	1 (2.47)	0	0	1 (2.47)	0	0	0	0	0	0
	Total	2 (0.55)	2 (0.55)	0	0	4 (1.11)	0	0	0	1 (0.28)	1 (0.28)	2 (0.55)
Total	Control	6 (0.75)	0	13 (1.63)	6 (0.75)	25 (3.13)	0	0	0	0	2 (0.25)	2 (0.25)
	Exposed:											
	Father	4 (1.64)	1 (0.41)	3 (1.23)	0	8 (3.28)	0	0	0	2 (0.82)	1 (0.41)	3 (1.23)
	Mother	3 (0.62)	1 (0.21)	2 (0.41)	1 (0.21)	7 (1.44)	0	0	0	0	2 (0.41)	2 (0.41)
	Both	0	1 (0.99)	2 (1.98)	0	3(2.97)	0	0	0	0	0	0
	Total	7 (0.84)	3 (0.36)	7 (0.84)	1 (0.12)	18 (2.16)	0	0	0	2 (0.24)	3 (0.36)	5 (0.60)
Newborn infants[b] (56,952)		40 (0.70)	11 (0.19)	51 (0.90)	8 (0.14)	110 (1.93)	4 (0.07)	7 (0.12)	5 (0.09)	10 (0.18)	8 (0.14)	34 (0.60)

[a] Abbreviations: *rob* Robertsonian translocation; *rcp* reciprocal translocation; *inv* inversion; *del* deletion; *supern* supernumerary chromosome.
[b] Cited from Hook and Hamerton (1977).

Table 5. Frequency of autosomal structural rearrangements by parental couples

City	Group[a]	No. of parental couples	No. of rearrange-ments	Rate ($\times 10^{-3}$)
Hiroshima	E	3453	11	3.19
	C	4242	13	3.06
	T	7695	24	3.12
Nagasaki	E	2370	3	1.27
	C	2231	7	3.14
	T	4601	10	2.17
Combined	E	5823	14	2.40
	C	6473	20	3.09
	T	12296	34	2.77
Neonates[b]		59542	113	1.90

[a] Abbreviations: *E* exposed; *C* control; *T* total.
[b] Cited from Jacobs (1981).

were further compared with the frequencies of the same types of abnormalities derived from cytogenetic surveys on 59,542 consecutive newborn infants (Jacobs 1981). The results obtained in this study of both exposed and control groups were approximately 50% higher than that reported for neonatal surveys.

Family studies of abnormal cases were undertaken to determine whether the observed structural rearrangements arose de novo or were inherited from one or the other parent (Table 6). Family studies in all cases were not possible because of death of parents, or because parents did not wish to cooperate in this study. When two or more sibs in a family were found to carry the identical rearrangement, however, their abnormalities were judged to be inherited, even though a family study may not have been done.

The ample evidence indicates that the majority of cases with rearrangements are heritable. There were only two de novo mutants identified. Both mutants (one in the exposed and one in the control group) were seen in the Hiroshima group. None have been detected so far in Nagasaki. The gametic mutation rates on structural rearrangements derived from the combined data were estimated as 1.72×10^{-4} in the exposed group, and 1.40×10^{-4} in the controls, respectively. These values did not deviate significantly from the value of 1.88×10^{-4} per gamete per generation in the liveborn infant survey (Jacobs 1981) (see Addendum p. 178).

3.2.2.2 Unbalanced. There were seven abnormal cases which were placed in this category; five in the exposed and two in the controls. All of them were characterized by the presence of an extra-small metacentric element (or elements), termed as "mar" according to the standardized nomenclature system (ISCN 1978, 1981, 1985). The element (or elements) was present either as a "supernumerary" piece of chromosomal

Table 6. Parental origin of balanced structural rearrangements[a]

	Hiroshima[b]		Nagasaki[b]		H + N[b]		Neonates[c]
	E	C	E	C	E	C	
de novo	*1*	*1*	*0*	*0*	*1*	*1*	*18*
Inherited	*5*	*5*	*1*	*5*	*6*	*10*	*73*
Father	2	3	1	3	3	6	37
Mother	0	1	0	1	0	2	36
Undetermined	3	1	0	1	3	2	0
Total	*6*	*6*	*1*	*5*	*7*	*11*	*91*
Not studied	5	7	2	2	7	9	22
Mutation rate ($\times 10^{-4}$)	2.65	2.55	–	–	1.72	1.40	1.88

[a] Including reciprocal and Robertsonian translocations and pericentric inversions.
[b] Abbreviations: *E* exposed; *C* control.
[c] Cited from Jacobs (1981).

material, in addition to the normal chromosome complement in all cells, or was in the form of a mosaic (Table 4, Appendix Table). Five of the cases were mosaics (three exposed and two controls).

In one male observed in the Nagasaki controls, three extra minute elements, each of which differed in size and shape, existed as cell lines with different combinations of markers (Itoh et al. 1984). Family studies revealed that most of these abnormal cases of unbalanced type were found to arise as de novo mutants (see Appendix Table).

3.2.3 Autosomal Trisomics

There was only one male Downs syndrome case (47,XY,+21) born in 1966 to an exposed father in Hiroshima, whose age at cytogenetic examination was 15. No other trisomic cases, such as D- and E-trisomy, have been observed in the F_1 population.

A summary of the overall frequencies of cases with abnormalities by parental exposure and by city is shown in Table 7. It can be seen that for sex chromosome abnormalities and structural rearrangements, the frequencies are consistently higher in the controls than in the exposed, and also higher in Hiroshima than in Nagasaki. Yet none of these differences are significantly different.

The data from the two cities were combined, regardless of parental radiation exposure, and each of the frequencies was compared with that of corresponding abnormalities in the cytogenetic surveys on 56,952 consecutive liveborn infants (Hook and

Table 7. Frequency of cases with chromosome abnormalities in the F_1 population (per 1,000)

	Exposed	Control	Hiroshima	Nagasaki	Combined	Neonates[a]
No. of cases	8,322	7,976	9,828	6,470	16,298	56,952
Abnormalities:						
Sex chromosomes	2.28	3.01	2.95	2.16	2.64	2.23
Autosomal rearrangement						
Balanced	2.16	3.13	3.05	2.01	2.64	1.93
Unbalanced	0.60	0.25	0.31	0.62	0.43	0.60
Autosomal trisomy	0.12	0	0.10	0	0.06	1.44
Total	5.17	6.39	6.41	4.79	5.77	6.20
	(43)	(51)	(63)	(31)	(94)	(353)

[a] Cited from Hook and Hamerton (1977).

Hamerton 1977). Here again, the frequencies of both sex abnormalities and rearrangements were somewhat higher in the children included in this study as compared to the neonates. In contrast, the incidence of autosomal trisomics was strikingly decreased in our samples.

3.2.3.1 Heteromorphic Variants. In the course of the present screening, there was a variety of heterochromatic variants as detected with the conventional staining method. Some of these cases were reanalyzed by the application of C-, Q-, and other banding techniques, and the results have been already reported elsewhere (Sofuni et al. 1978, 1980; Sofuni and Awa 1982). Of the heteromorphic variants observed, cases with inv(9p+q-), 1qh+, 9qh+, 16qh+, (C-band variants at the constitutive heterochromatic regions), Dp+, Gp+, Dps+, Gps+, DP-, Gp- (Q-band variants either deleted or enlarged satellites and/or short arms of acrocentric chromosomes) were identified with high frequency.

4 Discussion

As mentioned previously, the main objective of this study was to demonstrate whether there is any measurable increase in the frequency of children with chromosome abnormalities that might be associated with A-bomb radiation exposure of parental germ-cell chromosomes. This cytogenetic evaluation was conducted by comparing the frequencies of children born to A-bomb survivors with chromosome abnormalities, especially structural rearrangements, with children born to nonexposed parents.

The present results show that there are no statistically significant increases in the frequency of chromosome abnormalities among the children of the exposed. In fact, the frequency of both sex chromosome aneuploidy and structural rearrangements was higher in the controls than in the exposed.

This does not imply that genetic effects ascribable to parental A-bomb exposure have not occurred. It simply indicates that such effects have not been detectable by the current cytogenetic methods used. The reason for the absence of cytogenetic effects of A-bomb radiation on the children of the survivors remains unresolved. One of the possible interpretations is that germ cells with chromosome damage could have been eliminated either in the course of gametogenesis or in the very early period of gestation as "unrecognized" spontaneous abortions.

Extensive cytogenetic surveys on consecutive liveborn infants, undertaken as international collaborative studies in several European and North American countries, have provided very useful and relevant information on the natural incidence of various types of constitutional chromosome abnormalities in the human population (Sergovich et al. 1969; Lubs and Ruddle 1970; Friedrich and Nielsen 1973; Jacobs et al. 1974; Hamerton et al. 1975; Nielsen and Sillesen 1975; Walzer and Gerald 1977; also refer to Hook and Hamerton 1977 for review).

Cytogenetic data based on these surveys were compared with our present findings (Tables 2 to 4, and 7). It is apparent that the frequencies of both sex chromosome abnormalities and balanced rearrangements are slightly higher in our study than in the neonatal surveys. The situation is reversed when one compares unbalanced rearrangements and autosomal trisomics. It is known that most, but not all, situations where autosomal trisomy as well as unbalanced rearrangements are associated with physical malformations, are incompatible with life. Since the mean age of our study subjects was 24 years at examination, it is conceivable that most conceptions with these abnormalities do not survive for 2 or 3 decades even if they survived to term.

Recently, attempts have been made to reevaluate cytogenetic surveys of neonates using banding techniques in order to obtain more precise information on the frequency of abnormalities associated with structural chromosomal changes in the general human population (Lin et al. 1976; Buckton et al. 1980; Hansteen et al. 1982; Nielsen et al. 1982). The results obtained in such studies are discordant. Some studies indicated an increase in the frequency of reciprocal translocations (Hansteen et al. 1982), and inversions (Hansteen et al. 1982; Nielsen et al. 1982), or an increase in the Q-band variants (Lin et al. 1976). In contrast, the frequency of all types of chromosome abnormalities detected when G-banding techniques were used have been found to be similar to that observed previously when conventional stain preparations were used (Buckton et al. 1980). In our experience, there seemed to be no difference in the detectability of structural rearrangements between conventional stain and G- and Q-banding methods, although the latter were found to be more efficient in detecting heteromorphic variants than the conventional method.

Maeda et al. (1978) reported the results of a cytogenetic survey on 2,626 consecutive liveborn infants in Japan. They found nine cases with sex chromosome abnormalities (0.30%) and ten autosomal abnormalities, including five with Robertsonian translocation of the balanced type (0.19%), one with 13-trisomy associated with Robertsonian translocation (0.04%), one with a supernumerary marker (0.04%), and three with 21-trisomy (0.11%). Although the sample size of this survey was rather small, the results agreed to a certain extent with our findings with respect to the frequency of sex chromosome abnormalities and types and patterns of autosomal abnormalities.

Table 8. Frequency of autosomal structural rearrangements

City	Group[a]	No. of children	No. of rearrangements	Rate ($\times 10^{-3}$)
Hiroshima	E	4716	14	2.97
	C	5112	16	3.13
	T	9828	30	3.05
Nagasaki	E	3606	4	1.11
	C	2864	9	3.14
	T	6470	13	2.01
Combined	E	8322	18	2.16
	C	7976	25	3.13
	T	16298	43	2.64
Neonates[b]		59542	113	1.90

[a] Abbreviations: *E* exposed; *C* control; *T* total.
[b] Cited from Jacobs (1981).

The origin and cytogenetic features of an additional marker chromosome, often designated as a supernumerary chromosome, have been studied extensively (refer to Buckton et al. 1985). This marker chromosome is chracterized by a small metacentric element. Often it has a satellite (or satellites) at one or both distal ends. With regard to the presence of cases in the F_1 carrying such a marker, it was interesting to observe that (1) the supernumerary was seen in all cells or existed in the form of a mosaic, and (2) the majority of the supernumeraries were do novo mutants without any phenotypic change. This finding led to the presumption that the origin of the markers in our cases might arise from a product of a Robertsonian translocation between the short arms of the D- and G-chromosomes.

There has been discrepancy concerning a possible association between maternal radiation and trisomy-21 (Uchida et al. 1961; Schull and Neel 1962; Uchida 1977). In the A-bomb exposed population, Schull and Neel (1962) reported that the frequency of children with Down syndrome born to exposed mothers (fathers not exposed) was 0.54 per 1,000 (3 in 5,579), and 1.27 per 1,000 (12 in 9,440) in children born to nonexposed mothers. These results suggest that no association exists between Down syndrome and maternal radiation. The present results do not indicate an increase in children with 21-trisomy or aneuploidy involving any other chromosome.

Finally, when the new A-bomb radiation dosimetry system (DS86) becomes available for estimating the dose for individual survivors, the present data will be reanalyzed to determine the relationship between the frequency of chromosome abnormalities observed in children as a function of parental gonadal dose.

Table 9. Parental origin of balanced structural rearrangements[a]

	Hiroshima		Nagasaki		H + N		Neonates[b]
	E	C	E	C	E	C	
de novo	*1*	*1*	*0*	*0*	*1*	*1*	*18*
Inherited	*8*	*8*	*2*	*7*	*10*	*15*	*73*
Father	2	4	2	4	4	8	37
Mother	0	1	0	1	0	2	36
Undetermined	6	3	0	2	6	5	0
Total	*9*	*9*	*2*	*7*	*11*	*16*	*91*
(Not studied)	(5)	(7)	(2)	(2)	(7)	(9)	(22)
Gametic mutation rate ($\times 10^{-4}$)	1.65	1.74	–	–	0.98	0.98	1.88

[a] Including reciprocal and Robertsonian translocations plus pericentric inversions.
[b] Cited from Jacobs (1981).
Abbreviations: *E* exposed; *C* control; *H* Hiroshima; *N* Nagasaki.

Addendum. In the test, we computed the gametic mutation rates of structural rearrangements of balanced type in our population in terms of the frequency of families with abnormality as a function of total families studied. However, it seems more appropriate to estimate the chromosomal mutation rate using the following formula:

$$\text{Mutation rate} = \frac{\text{Number of individuals with abnormality} \times \text{Proportion who are likely to have de novo mutations}}{\text{Total number examined}}$$

For the gametic mutation rate, the value derived from the above formula is further divided by 2 (Jacobs 1981). The gametic mutation rates on structural rearrangements derived from the combined data in Tables 8 and 9 were estimated as 0.98×10^{-4} in the exposed, and 0.98×10^{-4} in the controls, respectively. These values were much lower than the value of 1.88×10^{-4} per gamete per generation in the liveborn infant survey (Jacobs 1981).

Acknowledgments. We are grateful to the Hiroshima and Nagasaki citizens for their willingness to voluntarily participate in this survey. Without their cooperation this study would not have been possible. Our thanks are due also to the following RERF staff members for their continued support throughout this survey: Drs. M. Otake and S. Fujita, Department of Statistics, Dr. Y. Shimizu, Department of Epidemiology, for sample selection and statistical advices; Drs. T. Amano, M. Soda, R. Hazama, and their associaes in the Nursing Section, Department of Clinical Studies, for their clinical assistance; public health nurses and clinical contactors in the Department of Research Support who obtained consent of the participants and arrangements for their visit to RERF. We are very much indebted to Messrs. S. Iida, K. Tanabe, Mrs. Y. Urakawa, and their colleagues in the Cytogenetics Laboratories in Hiroshima and Nagasaki for their continued technical support including blood cultures, chromosome preparations, microscope and karyotype analysis, and photographic works. We wish to thank Dr. C.W. Edington, Vice-Chairman of RERF, for his encouragement and for going through the manuscript.

References

Awa AA (1975) Review of thirty years study of Hiroshima and Nagasaki atomic bomb survivors. II Biological effect. B Genetic effects. 2 Cytogenetic study. J Radiat Res 16 (Suppl):75–81

Awa AA, Bloom AD, Yoshida MC, Neriishi S, Archer PG (1968) Cytogenetic study of the offspring of atom bomb survivors. Nature (London) 218:367–368

Awa AA, Sofuni T, Honda T, Itoh M, Neriishi S, Otake M (1978) Relationship between the radiation dose and chromosome aberrations in atomic bomb survivors of Hiroshima and Nagasaki. J Radiat Res 19:126–140

Buckton KE, O'Riordan ML, Ratcliffe S, Slight J, Mitchell M, McBeath S, Keay AJ, Barr D, Short M (1980) A G-band study of chromosome in liveborn infants. Ann Human Genet 43:227–239

Buckton KE, Spowart G, Newton MS, Evans HJ (1985) Forty four probands with an additional "marker" chromosome. Human Genet 69:353–370

Caspersson T, Lomakka G, Zech L (1971) The 24 fluorescence pattern of the human metaphase chromosomes – distinguishing characters and variability. Hereditas 67:89–102

Friedrich U, Nielsen J (1973) Chromosome studies in 5,049 consecutive newborn children. Clin Genet 4:333–343

Hamerton JL, Canning N, Ray M, Smith S (1975) A cytogenetic survey of 14069 newborn infants. I Incidence of chromosome abnormalities. Clin Genet 8:223–243

Handsteen I-L, Varslot K, Steen-Johnsen, Langard S (1982) Cytogenetic screening of a newborn population. Clin Genet 21:309–314

Hook EB, Hamerton JL (1977) The frequency of chromosome abnormalities detected in consecutive newborn studies – differences between studies – results by sex and by severity of phenotypic involvement. In: Hook EB, Porter IH (eds) Population cytogenetics – Studies in humans. Academic Press, London New York, pp 63–79

ISCN (1978) An international system for human cytogenetic nomenclature (1978) In: Lindsten JE, Klinger HP, Hamerton JL (eds) Birth defects: original article series, vol 14 (8). Natl Found, New York; also in: Cytogenet Cell Genet 21:309–404

ISCN (1981) An international system for human cytogenetic nomenclature (1981) High resolution banding. In: Harnden DG, Lindsten JE, Buckton KE, Klinger HP (eds) Birth defects: original article series, vol 17 (5). March of Dimes, Birth Defects Found, New York; also in: Cytogenet Cell Genet 31:1–28

ISCN (1985) An international system for human cytogenetic nomenclature (1985) In: Harnden DG, Klinger HP (eds) puglished in collaboration with Cytogenet Cell Genet (Karger, Basel 1985); also in: Birth defects: original article series, vol 21 (1). March of Dimes Birth Defects Found, New York

Itoh M, Soda M, Honda T (1984) Unstable behavior and mosaicism of extra minute chromosome in man. Nagasaki Med J 59:441–445

Jacobs PA (1981) Mutation rates of structural chromosome rearrangements in man. Am J Human Genet 33:44–54

Jacobs PA, Melville M, Ratcliffe S, Keay AJ, Syme J (1974) A cytogenetic survey of 11680 newborn infants. Ann Human Genet 37:359–376

Kato H, Schull WJ, Neel JV (1966) A cohort-type study of survival in the children of parents exposed to atomic bombings. Am J Human Genet 18:339–373

Lin CC, Gedeon MM, Griffith P, Smink WK, Newton DR, Wilkie L, Sewell LM (1976) Chromosome analysis on 930 consecutive newborn children using quinacrine fluorescent banding technique. Human Genet 31:315–328

Lubs HA, Ruddle FH (1970) Chromosomal abnormalities in the human population: Estimation of rates based on New Haven newborn study. Science 169:495–497

Maeda T, Ono M, Takada M, Kato Y, Nishida M, Jobo T, Adachi H, Taguchi A (1978) A cytogenetic survey of consecutive liveborn infants – incidence and type of chromosome abnormalities. Jpn J Hum Genet 23:217–224

Milton RC, Shohoji T (1968) Tentative 1965 radiation dose (T65D) estimation for atomic bomb survivors; Hiroshima and Nagasaki. ABCC Teech Rep 1–68

Nielsen J, Sillesen I (1975) Incidence of chromosome aberrations among 11148 newborn children. Humangenetik 30:1–12

Nielsen J, Wohlert M, Faaborg-Andersen J, Hansen KB, Hvidman L, Klag-Olsen B, Moulvad I, Videbech P (1982) Incidence of chromosome abnormalities in newborn children. Comparison between incidences in 1969–1974 and 1980–1982 in the same area. Human Genet 61:98–101

Pai GS, Thomas GH (1980) A new R-banding technique in clinical cytogenetics. Human Genet 54: 41–45

RERF Research Protocol (1975) Research plan for RERF studies of the potential genetic effects of atomic radiation; Hiroshima and Nagasaki. RERF RP 4–75

Schull WJ, Neel JV (1962) Maternal radiation and mongolism. Lancet 1:537–538

Seabright M (1971) A rapid banding technique for human chromosomes. Lancet 2:971–972

Sergovich FR, Valentine GH, Chen ATL, Kinch RAH, Smout MS (1969) Chromosome aberrations in 2159 consecutive newborn babies. New Engl J Med 280:851–855

Sofuni T, Awa AA (1982) Chromosome heteromorphisms in the Japanese. III Frequency of C-band variants. RERF Tech Rep 4–82

Sofuni T, Tanabe K, Awa AA (1978) Chromosome heteromorphisms in the Japanese. I Banding patterns of Dp+ and Gp+ by Q- and C-staining methods. RERF Tech Rep 8–78

Sofuni T, Tanabe K, Awa AA (1980) Chromosome heteromorphisms in the Japanese. II Nucleolus organizer regions of variant chromosomes in D and G groups. Human Genet 55:265–270

Sumner AT (1972) A simple technique for demonstrating centromeric heterochromatin. Exp Cell Res 75:304–306

Uchida IA (1977) Maternal radiation and trisomy 21. In: Hook EB, Porter IH (eds) Population cytogenetics – Studies in humans. Academic Press, London New York, pp 285–299

1st US-Japan joint Worksh Reassessment of atomic bomb radiation dosimetry in Hiroshima and Nagasaki. Radiation Effects Research Foundation, Hiroshima, 1983

2nd US-Japan joint Worksh Reassessment of atomic bomb radiation dosimetry in Hiroshima and Nagasaki. Radiation Effects Research Foundation, Hiroshima, 1984

Walzer S, Gerald PS (1977) A chromosome survey of 13751 male newborns. In: Hook EB, Porter IH (eds) Population cytogenetics – studies in humans. Academic Press, London New York, pp 45–61

Yunis JJ, Sawyer JR, Ball DW (1978) The characterization of high resolution G banded chromosomes of man. Chromosoma 67:293–307

Appendix Table. Family studies of abnormal cases of the unbalanced type

Case No.	Sex	Age ATE	Year of birth	Age at birth Mo	Age at birth Fa	Exposure status	Type of abnormality	Remarks
			Hiroshima					
FH3321	M	28	1948	32	38	C	A1. 47,XYY	
FH4305	M	22	1956	30	38	C	" "	
FH4998	M	29	1949	25	29	C	" "	
FH7308	M	17	1964	28	32	C	" "	
FH9913	M	15	1969	25	27	C	" "	
FH0757	M	17	1953	26	34	Mo	A2. 47,XXY	
FH0801	M	22	1948	23	27	Fa	" "	
FH0815	M	13	1957	31	48	B	" "	
FH4223	M	22	1956	34	31	Mo	" "	
FH4727	M	29	1949	33	38	Mo	" "	
FH5953	M	26	1953	21	24	C	" "	
FH6415	M	30	1950	32	40	C	" "	
FH6662	M	28	1952	27	31	C	" "	
FH8632	M	24	1958	24	28	Mo	" "	
FH8718	M	33	1949	35	41	C	" "	
FH8454	M	17	1965	25	30	Fa	A3. 46,XY/47,XYY	46:48 cells, 47:152 cells
FH6505	M	24	1956	32	38	C	A4. 46,X,inv(Y)(p11.2q11.2)	
FH7417	M	23	1958	33	31	C	" 46,X,inv(Y)(p11.2q11.23)pat	
FH0492	F	21	1948	25	29	Fa	A6. 47,XXX	
FH1870	F	15	1957	34	33	B	" "	
FH3886	F	30	1947	25	27	C	" "	
FH5852	F	21	1958	30	36	Mo	" "	
FH8291	F	24	1958	23	27	Fa	" "	F_2: one (liveborn)
FH8912	F	16	1967	37	39	C	" "	Sib (dizygotic co-twin)—normal
FH9339	F	15	1967	28	30	C	" "	
FH0092	F	18	1949	33	36	B	A7. 45,X/47,XXX	45:95 cells, 47:4 cells
FH3033	F	22	1954	33	38	C	" "	45:82 cells, 47:15 cells
FH8020	F	24	1957	24	28	C	" "	45:44 cells, 47:49 cells. F_2: one (liveborn), also pregnant at examination
FH8590	F	33	1948	25	38	C	" 45,X/46,X,r(x) §	§ *de novo* mutant. 45:141 cells, 46:59 cells

Appendix Table (continued)

Case No.	Sex	Age ATE	Year of birth	Age at birth Mo	Age at birth Fa	Exposure status	Type of abnormality	Remarks
FH0381	M	18	1951	21	30	Fa	B1. 45,XY,t(DqDq)	
FH1988	F	16	1956	27	32	Mo	" 45,XX,t(DqDq)	Sibs
FH2443	F	21	1952	23	28	Mo	" 45,XX, "	
FH4989	M	28	1951	29	39	Mo	" 45,XY,t(13q14q)	
FH6456	M	30	1949	24	27	C	" 45,XY,t(13q14q)	
FH6664	F	29	1951	26	29	C	" 45,XX, "	Sibs
FH6757	F	26	1954	29	32	C	" 45,XX, "	
FH6538	F	30	1950	20	23	C	" 45,XX,t(14q15q)	
FH8096	F	16	1966	30	35	Fa	" 45,XX,t(13q14q)	
FH3496	F	22	1954	26	34	Mo	B2. 45,XX,t(14q21q)	
FH0405	F	18	1951	29	43	Fa	B3. 46,XX,t(5;11)(q13;p15)	Sibs
FH8076	M	33	1949	27	41	Fa	" 46,XY, "	
FH2209	F	26	1947	34	36	B	" 46,XX,t(5;17)(p13;q25) §	§ *de novo* mutant
FH2210	F	26	1947	23	29	B	" 46,XX,t(5;8)(q22;p11.2)pat	
FH2690	M	17	1957	23	27	Fa	" 46,XY,t(5;8)(q22;p11.2)pat	
FH4493	M	29	1949	32	34	C	" 46,XY,t(6;12)(q15;q22)	
FH4595	M	23	1955	27	30	C	" 46,XY,t(1;6)(q21;p21)	
FH6611	F	29	1951	27	32	Mo	" 46,XX,t(3;8)(q21;q24.1)	Sibs
FH6771	M	26	1954	30	35	Mo	" 46,XY, "	
FH7342	F	32	1949	27	36	C	" 46,XX,t(12;13)(q21.32;p12.3) §	§ *de novo* mutant
FH9219	M	25	1957	25	31	C	B3. 46,XY,t(5;7)(q15;p21)pat	
FH9876	F	13	1970	29	30	C	" 46,XX,t(5;8)(q22;p11.2)pat	
FH9922	M	13	1971	25	30	C	" 46,XY,t(6;8)(q13;q22)	
FH1705	F	25	1946	34	35	C	B4. 46,XX,inv(2p+q−)	
FH5672	M	32	1947	27	32	C	" 46,XY,inv(18)(p11.32q11.2)mat	
FH6099	M	27	1953	27	38	Mo	" 46,XY,inv(2)(p11.2q13)	
FH7331	F	16	1964	27	33	C	" 46,XX,inv(2)(p11.2q13)pat	Sibs
FH7333	F	14	1966	29	34	C	" 46,XX, "	
FH9118	M	35	1948	39	45	C	" 46,XY,inv(2)(p11.2q14.2)	
FH9254	M	34	1949	27	34	C	" 46,XY,inv(5)(p15.3q13)	
FH7361	F	24	1957	24	25	Fa	B8. 47,XX,+mar* § (*minute)	§ *de novo* mutant. F_2:one (liveborn)
FH6169	F	23	1956	28	28	Mo	B9. 46,XX/47,XX,+mar* (*minute)	46:9 cells, 47:91 cells
FH8980	M	14	1969	25	28	Fa	" 46,XY,del(18)(p11.1)/46,XY,i(18q)	del(18p):181 cells, i(18q):7 cells
FH7594	M	15	1966	25	25	Fa	C1. 47,XY,+21 §	§ *de novo* mutant

Nagasaki

FN0217	M	13	1956	37	33	Mo	A1. 47,XYY	
FN0486	M	21	1948	30	35	Fa	" "	
FN0970	M	19	1951	26	28	Fa	" "	
FN1717	M	26	1948	31	39	C	A2. 47,XXY	
FN2696	M	18	1957	27	27	C	" "	
FN4025	M	30	1948	27	31	Mo	" "	
FN4171	M	25	1953	22	19	C	" "	
FN4317	M	28	1951	40	45	C	" "	
FN6865	M	14	1970	25	29	C	" "	
FN3338	M	21	1956	25	30	Mo	A4. 46,XX	[XX male]
FN3252	F	22	1955	24	27	Fa	A6. 47,XXX	
FN4237	F	31	1947	20	24	C	" "	
FN2120	F	18	1956	26	24	Mo	A7. 46,XX/47,XXX	46:3 cells, 47:96 cells
FN3202	F	21	1956	33	37	C	A8. 46,XXq−	
FN4935	F	32	1948	32	46	C	B1. 45,XX,t(13q14q)pat	Sibs
FN4936	F	28	1952	36	50	C	" 45,XX, "	
FN5837	M	13	1969	32	34	Fa	" 45,XY,t(13q14q)pat	Sibs
FN5859	F	17	1965	29	31	Fa	" 45,XX, "	
FN0870	M	18	1952	23	31	B	B2. 45,XY,t(DqGq)	
FN6530	F	14	1969	29	36	Fa	" 45,XX,t(14q21q)	
FN1638	M	24	1950	21	22	C	B3. 46,XY,t(Cq−;17q+)pat	
FN3068	F	28	1948	40	44	C	" 46,XX,t(18q+;20q−)	
FN3948	M	21	1957	33	41	C	" 46,XY,t(1p−;12q+)mat	
FN4419	F	32	1947	22	28	C	" 46,XX,t(2p−;8p+)	Sibs
FN4891	F	31	1949	24	30	C	" 46,XX, "	
FN4724	M	29	1951	24	29	C	" 46,XY,t(Bq−;Cq+)pat	
FN5049	F	29	1951	28	35	C	" 46,XX,t(2q−;13q+)	
FN2357	F	26	1949	23	46	Fa	B8. 47,XX,+mar* mat (*minute)	
FN3990	M	22	1957	40	37	C	B9. 46,XY/47,XY,+mar* § (*minute)	§*de novo* mutant. 46:18 cells, 47:31 cells
FN4121	M	30	1948	20	29	Mo	" 46,XY/47,XY,+mar* § (*minute)	§*de novo* mutant. 46:21 cells, 47:9 cells
FN6023	M	31	1951	23	24	C	" 46,XY/47,XY,+mar/48,XY,+mar x2 §	§*de novo* mutant. Complex mosaic consisting of cell lines with 2 or 3 different minute markers of varying combinations.

[Abbreviations] ATE: At the time of examination. Age at birth: Parental age. Mo: Mother. Fa: Father. Exposure status: B=Both parents exposed. Fa=Father exposed. Mo=Mother exposed. C=Control.

9 Chromosome Studies of Human in Vitro Fertilization (IVF) Failures

H. SCHMIADY, H. KENTENICH, and M. STAUBER[1]

1 Introduction

Nowadays, in vitro fertilization (IVF) and embryo transfer (ET) is an established method in human medicine for the treatment of certain forms of infertility (e.g. for review Beier and Lindner 1983; Trounson and Wood 1984). Nevertheless, the success rate in IVF programs is still relatively low. While the rate of fertilization of human oocytes is relatively high (∿70%), the rate of clinical pregnancies per transfer is approximately 14%, with an abortion rate of approximately 28% (current status of IVF in Germany, Switzerland, and Austria; for details see Szalay et al. 1986). Therefore, factors which limit the rate of ongoing pregnancies appear to be related to a high incidence of embryonic failure to implant or to develop beyond the preimplantation stage. One factor in the failure of implantation or in early developmental arrest might be the presence of lethal chromosome abnormalities, which are manifested in over 50% of spontaneous abortions (Boué et al. 1975; Hassold et al. 1980). To determine the frequency of chromosome abnormalities in man, IVF provides the opportunity to perform chromosome analysis of human oocytes and preimplantation embryos, and preliminary results on cytogenetic research have been recently reported (Edwards 1977, 1983; Wramsby et al. 1982; Angell et al. 1983, 1986a,b; Rudak et al. 1984; Michelmann 1985; Vogel et al. 1985; Zenzes et al. 1985). However, because of ethical and legal restrictions in many countries, normal embryos should not be analyzed. Therefore, in our IVF program we used only unfertilized oocytes, oocytes with developmental disorders, and polyploid oocytes for chromosome analysis. Polyploid oocytes are not retransferred to the uterus of the patient because of the possibility of producing a potentially abortive conceptus.

The results of such an analysis are reported in this chapter.

1 Frauenklinik Charlottenburg der Freien Universität Berlin, Pulsstraße 4, 1000 Berlin (West) 19

Cytogenetics. Ed. by G. Obe and A. Basler

2 Procedures

2.1 Patients

Oocytes were obtained from women (up to an age of 40 years) undergoing laparoscopy in our IVF program. In the period of this study from July 1985 to July 1986, 153 laparoscopies were performed on 132 women. The majority of patients (> 90%) suffered from tubal infertility.

2.2 Induction of Ovulation

Stimulation of multiple follicular development in patients who participated in our IVF program was induced either by clomiphene citrate (100 mg/day, days 2 to 6 or days 5 to 9) and additional hMG (human menopausal gonadotropin) according to the individual response or by hMG alone, starting at day 3. Ovulation was induced after injection of 5,000–10,000 IU of hCG (human chorionic gonadotropin) when the leading follicle had attained a diameter of 19–20 mm by ultrasonic scanning and when the level of 17β-estradiol (E2) in serum had reached more than 300 pg/ml/follicle > 15 mm.

2.3 Oocyte Culture and Fertilization

After 34–36 h hCG administration, follicles were aspirated by laparoscopy. The degree of maturity of preovulatory oocytes (mature – intermediate – immature) was determined by the size and dispersion of the cumulus cell mass surrounding the oocyte and the condition of the corona radiata and granulosa cells as described by Veeck et al. (1983). Oocytes were preincubated in Ham's F-10 medium supplemented with 10% umbilical-cord serum for 4–24 h (depending on the maturity stage of the oocytes) at 37°C in a CO_2 incubator. Each oocyte was inseminated with 200,000 to 250,000 motile sperms. A standard semen analysis was performed after liquefaction of the ejaculate. The semen sample was washed twice with 2 ml Ham's F-10 medium and centrifuged at 200 × g for 10 min. After the second centrifugation and removal of seminal plasma, the pellet was resuspended in 1 ml culture medium and incubated at 37°C for at least 30 min. The supernatant containing the motile sperms was then removed and used for insemination.

For further incubation the cells were transferred to a gas atmosphere of 5% O_2, 5% CO_2, and 90% N_2 at 37°C, and 14–20 h after insemination the oocytes were examined for the presence of polar bodies and pronuclei. In most cases adhering granulosa cells have to be gently removed either by gentle pipetting or dissection to determine the number of polar bodies and pronuclei. Oocytes observed to have two or more pronuclei were defined as being fertilized. In some cases oocytes were reinseminated if they exhibited a polar body but no pronuclei.

Embryos at the 2- to 6-cell stage that had developed regularly from the diploid eggs were retransferred to the patient's uterus 40–48 h after insemination (in some cases, oocytes in the pronuclei stage were also retransferred).

Table 1. Efficiency of fertilization and transfer procedures

No. of laparoscopies	153
No. of patients	132
Total No. of oocytes	845
Oocytes per laparoscopy	5.5
No. of inseminated oocytes	716
No. of fertilized oocytes	365
No. of multipronuclear oocytes	27 (3 pn[a])
	3 (4–5 pn)
Embryo transfer	
No. of embryos at 2- to 6-cell stage	302
No. of oocytes transferred in pronuclei stage	33
No. of embryo transfers	119
Transfer per laparoscopy	78%
No. of clinical pregnancies	26
Clinical pregnancy per laparoscopy	17%
Clinical pregnancy per transfer	22%

[a] *pn* pronuclei.

2.4 Preparation of Oocytes for Chromosome Analysis

Unfertilized oocytes and oocytes with two pronuclei showing no cleavage were fixed without Colcemid according to the method of Tarkowski (1966). Triploid oocytes were treated with Colcemid (0.4 μg/ml for 5–20 h). The oocytes were hypotonically treated with 1% sodium citrate solution at room temperature for 8–10 min, transferred to precleaned slides and fixed by one or two drops of cold fixative (methanol/acetic acid 3:1). The preparations were allowed to dry on a hot plate at 37°C to enhance chromosome spreading. Fixed preparations were stained with Giemsa.

3 Efficiency of Fertilization and Transfer Procedures

The data presenting the status of our IVF program are summarized in Table 1. The fertilization rate of the inseminated oocytes was 51% with a rate of polyploidy of approximately 8%. In 119 cases embryos could be retransferred resulting in a clinical pregnancy rate of 22%.

4 Chromosome Analysis of Inseminated Oocytes Showing Neither Formation of Pronuclei Nor Cell Cleavage

A random sample of 177 oocytes of the 351 oocytes, which were unfertilized or failed to cleave, were prepared for chromosome analysis. Degenerated oocytes (characterized by dark ooplasm) were excluded. Thirty metaphases could not be analyzed exactly because of inferior quality (chromosome clumping or lack of chromosomes). The

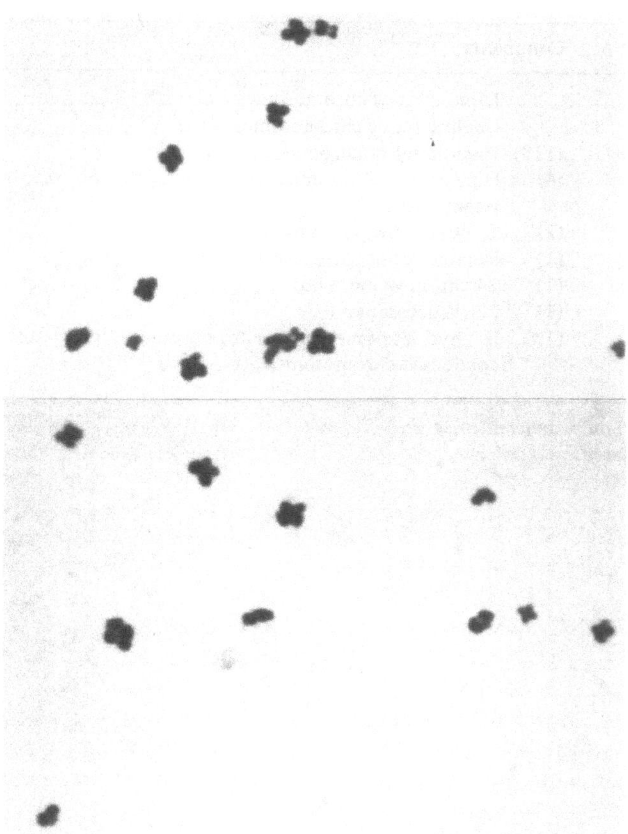

Fig. 1. A mature unfertilized oocyte exhibiting both chromosomes of the polar body (*pb*) and of the metaphase II (*mII chr*); *gc* nucleus from a granulosa cell still adhering at the zona pellucida

Table 2. Chromosome analysis of oocytes exhibiting no signs of fertilization or cleavage[a]

Observations	No.	Comments	
−pb −pn (metaphase I)	3		Diploid set of chromosomes
+pb −pn (prophase II)	3		Haploid set of chromosomes
(metaphase II)	141	(119)	Haploid set of chromosomes
(? pb)		(4)	Diploid set of chromosomes (nondisjunction; acentric fragments)
		(2)[b]	Triploid set of chromosomes
		(1)	Haploid set of chromosomes + acentric fragment
		(1)	24 chromosomes (+G) Aneuploidy
		(1)	22 chromosomes (-C)
		(13)	Haploid set of metaphase chromosomes + "prematurely condensed chromosomes" (G_1-PCC)

[a] Abbreviations: *pb* polar body; *pn* pronuclei.
[b] These oocytes were reinseminated.

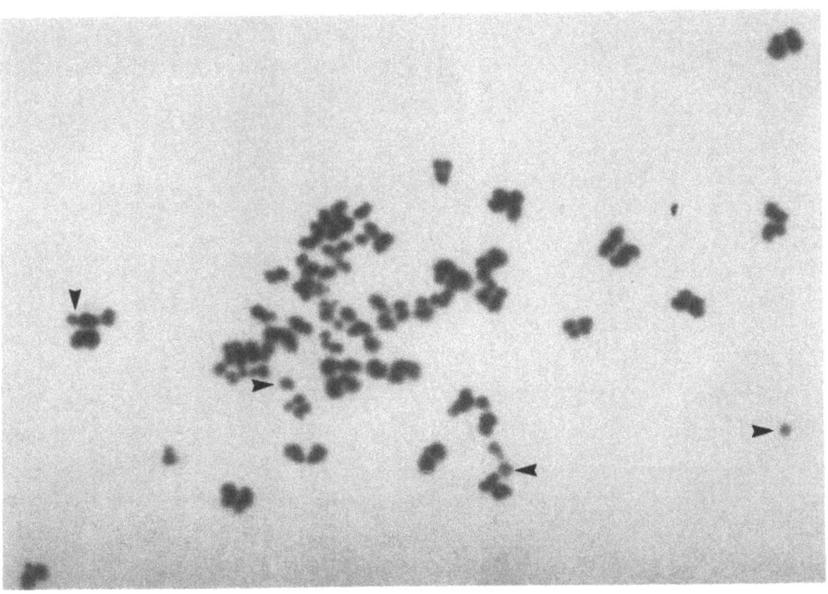

Fig. 2. Representative part of a diploid metaphase plate from a human oocyte as a result of nondisjuction showing acentric fragments (*arrows*)

results of the analysis of 147 oocytes are summarized in Table 2. Three oocytes (2.0%) showed no extrusion of the polar body, and consequently they had not completed meiosis I. Twenty-two (15%) of those oocytes which had obviously extruded the polar body but failed in the observation of forming pronuclei (Fig. 1) were characterized by some chromosome disorders: four oocytes with distinct polar bodies were diploid as a result of nondisjunction in the first meiotic division with one of these

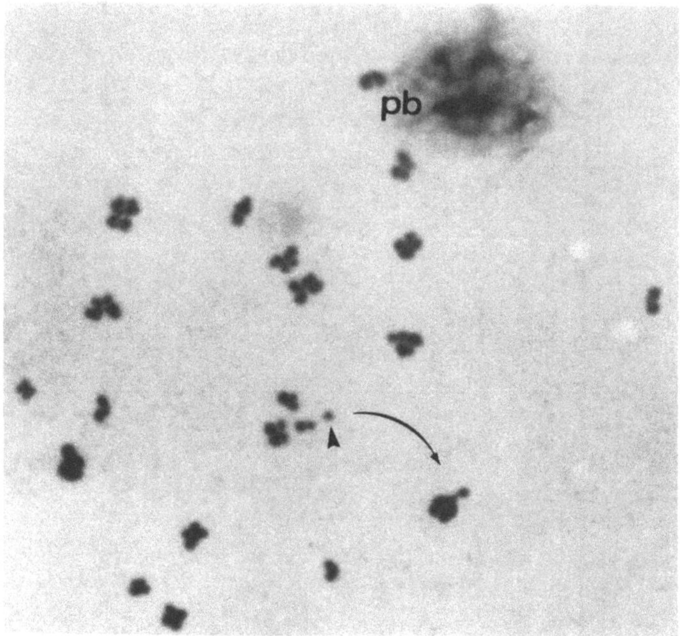

Fig. 3. Metaphase II chromosomes from an oocyte showing an acentric fragment (*arrowhead*) which might belong to the marked chromosome (*arrow*); *pb* polar body

oocytes showing acentric chromosome fragments (Fig. 2). Two other oocytes exhibited a triploid chromosome complement. In the latter case the oocytes were reinseminated because the first attempt was unsuccessful. Obviously, fertilization of the oocytes still occurred, but they failed to cleave and became arrested in the metaphase of the first mitotic division. Another oocyte was characterized by a haploid set of metaphase chromosomes showing an acentric fragment (Fig. 3).

Recently Angell et al. (1986a) reported on a metaphase from a blastomere of a 6-cell embryo showing acentric fragments. The authors concluded that the error arose as a chromatid break in the oocyte after fertilization. From our results it becomes evident that such fragments arise also during the maturation of the oocyte.

Aneuploidy was observed in two cases: one set had an extra chromosome of the G-group; the other set missed a chromosome of the C-group.

Thirteen oocytes exhibited also a paternal set of so-called prematurely condensed chromosomes of the G_1-phase (G_1-PCC) besides the haploid set of metaphase II chromosomes and the diffuse first polar body (Fig. 4). The phenomenon of premature chromosome condensation (for review see Rao et al. 1982) was observed by Johnson and Rao (1970) in experimentally fused cells which were at different stages of the cell cycle (mitosis and interphase). Under the influence of mitotic factors from the cytoplasm the premature chromosome condensation is induced in the interphase cells. In the present cases the oocytes did not become activated after sperm penetration but

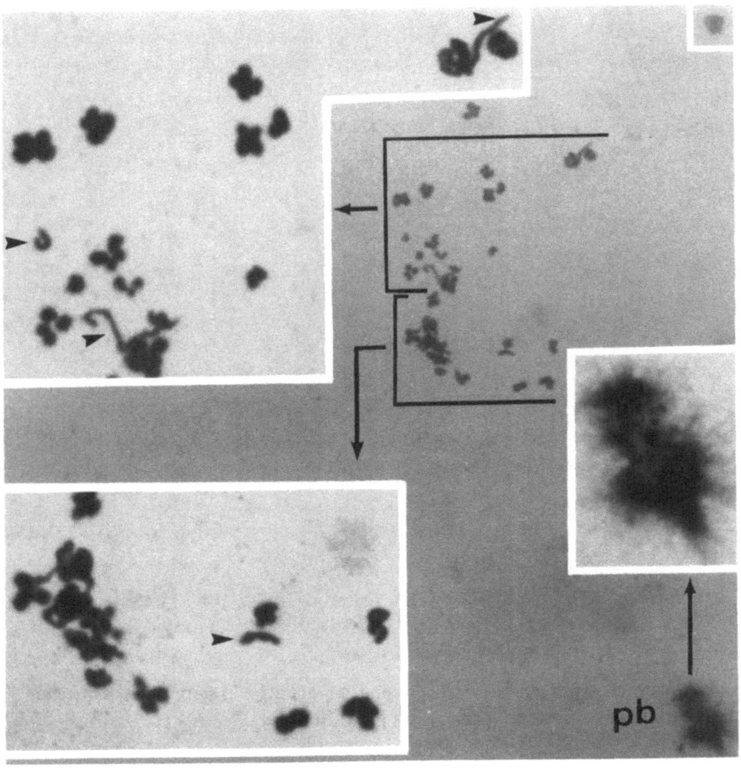

Fig. 4. Prematurely condensed sperm chromosomes of the G_1-phase (G_1-PCC) from an in vitro fertilized oocyte (*arrowheads* point to some prematurely condensed chromosomes) besides the maternal metaphase II chromosomes and the polar body (*pb*)

remained arrested at metaphase II. The chromosome condensing factors were still present, preventing the sperm nuclei from transforming into pronuclei but inducing the condensation of the sperm nuclei into single-stranded chromosomes (G_1-PCC). Recently, in one case a Y-chromosome had been identified after Q-banding, confirming the paternal origin of these elements (Schmiady et al. 1986). No correlation with sperm parameters could be found (75% of these oocytes were inseminated with sperm of normal quality) and all the oocytes demonstrating this phenomenon were in a mature or intermediate stage at laparoscopy.

The causes for the failure of fertilization are not unequivocally known. Besides the immaturity of some oocytes, insufficient sperm quality might be the main cause because about half of the oocytes were inseminated with sperm of reduced quality (oligozoospermy, asthenozoospermy, teratozoospermy, or mixed forms).

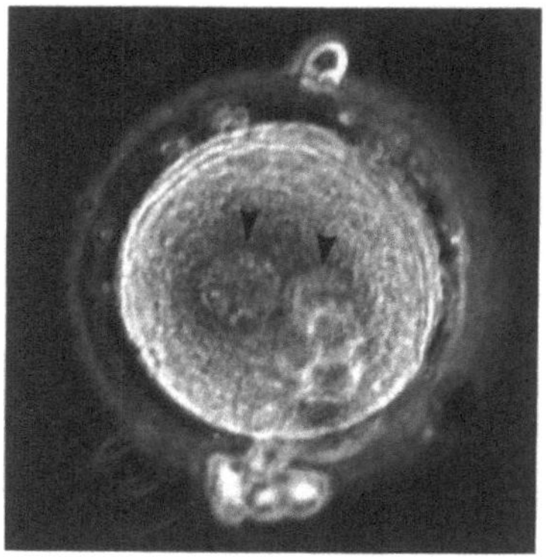

Fig. 5. An in vitro fertilized oocyte with two pronuclei (*arrowheads*) 16 h after insemination (× 240)

5 Chromosome Analysis of Fertilized Oocytes Exhibiting Developmental Arrest

Developmental arrest of 15 (4.1%) fertilized eggs which had formed two pronuclei (Fig. 5) was observed, with about 40% of them demonstrating still visible pronuclei 40 h after insemination (normally the pronuclei will have "disappeared" about 20 h after insemination). About half of the oocytes were fertilized with sperm of normal quality. They were defined as mature or intermediate at laparoscopy and were not different from the retransferred eggs of the same patients.

Cytologic analysis of ten eggs revealed that the arrest of development occurred at different stages of the cell cycle: from interphase to metaphase of the first mitotic division (Table 3). Asynchronous development of the pronuclei (Fig. 6) was observed in five eggs. One egg was characterized by a tetraploid chromosome set (Fig. 7). Obviously, this phenomenon resulted from nondisjunction in both the oocyte and the penetrating sperm.

A marked asynchrony in pronuclear morphology was also reported by Zenzes et al. (1985); one pronucleus was in early prophase, whereas the second pronucleus presented metaphasic chromosomes. The authors found asynchrony in pronuclear morphogenesis manifested among nonmature occytes and concluded that this might be

Table 3. Stages of developmental arrest of ten oocytes characterized by polar bodies and two pronuclei

	Developmental arrest in		
	Interphase–transition to prophase	Prophase	Metaphase of first mitotic division
No. of oocytes with (without) visible pronuclei 40 h after insemination	2^{asyn}[a] (1^{asyn})	1 (2; 2^{asyn})	1 (1)

[a] asyn: indicates asynchronous development of the pronuclei.

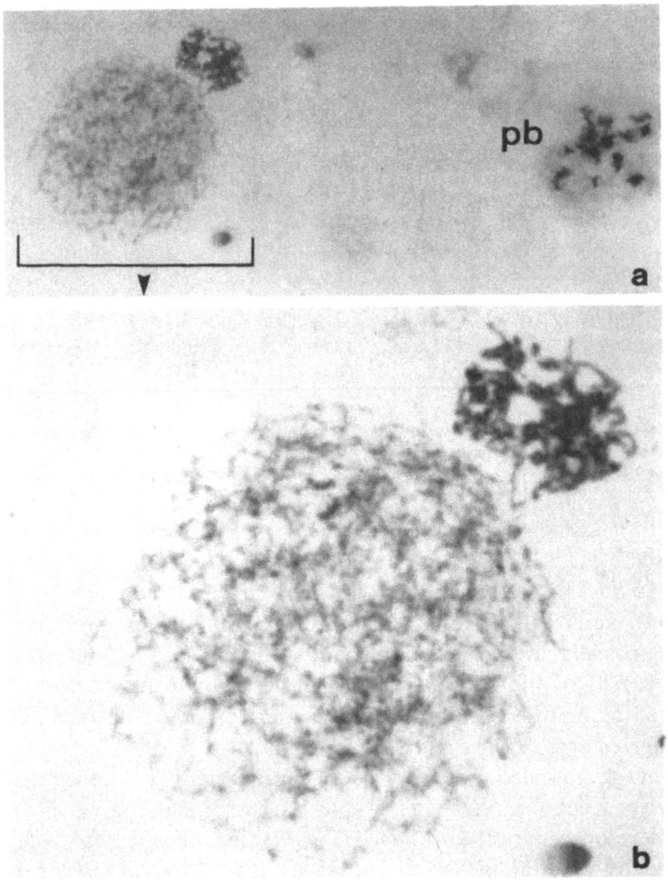

Fig. 6a,b. Developmental arrest of an in vitro fertilized oocyte. Note the remarkable asynchrony in the morphology cf the two pronuclei **a** which are shown at higher magnification in **b**. The diffuse structure of the polar bodies (*pb*) and some spermatozoa are also visible (**a**)

the main cause of developmental arrest. We have observed that developmental arrest can also occur in mature or intermediate oocytes.

6 Polyploidy

Multiple pronuclei as a consequence of polyspermy (Edwards et al. 1981) have been often reported in IVF oocytes. Different causes such as the method of ovarian stimulation, maturity stage of the oocytes, and sperm concentration are discussed: some authors believe that the method of stimulation of follicular development correlates with the frequency of polyploidy (Wentz et al. 1983), whereas others have found no correlation (Al-Hasani et al. 1984; Diamond et al. 1985). Trounson et al. (1982) and Al-Hasani et al. (1984) attributed the relatively high rates (about 30%) of polyspermy

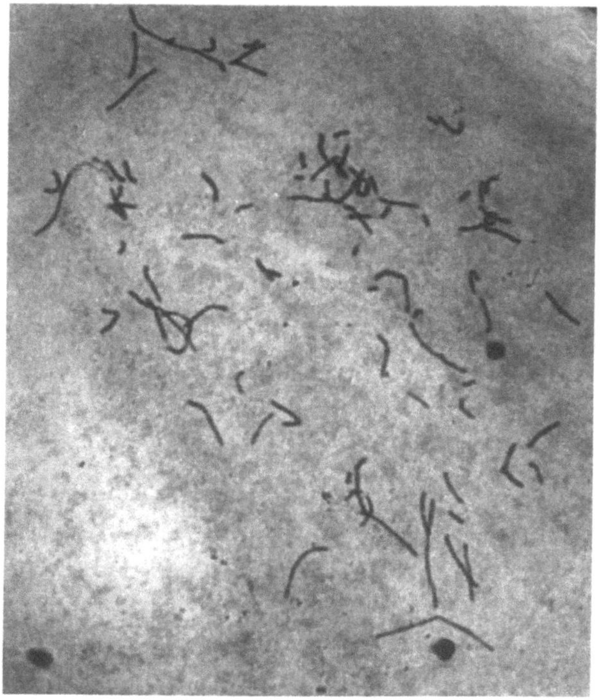

Fig. 7. Tetraploid chromosome set of an in vitro fertilized oocyte exhibiting two pronuclei followed by developmental arrest

to oocyte immaturity, while Diamond et al. (1985) found no difference in the rate of polyspermy in mature and immature oocytes. The latter authors also found no difference in the rate of polyspermy when inseminating the oocytes with different concentrations of sperm (up to 0.5×10^6 motile sperms), whereas Al-Hasani et al. (1984) observed an increase in polyspermy with higher concentrations of sperm. However, these authors used about threefold concentrations. We have used routinely 0.2×10^6 to 0.25×10^6 motile sperms per oocyte.

Recently, it has been assumed that a genetic disposition could also be a cause for polyploidy (Michelmann 1985). We have observed triploidy in two stimulation cycles from four patients, which possibly agrees with such an idea.

The frequency of polyploidy (mainly triploidy) in our IVF program was about 8%, which is in the range (6–10%) reported by other IVF teams. Polyploid oocytes can show a normal cleavage rate and a normal morphology, indicating the necessity for the determination of pronuclei prior to the first cleavage division. Polyploidy in mammals is generally considered to be inconsistent with normal preimplantation development. Therefore, such oocytes will not be retransferred.

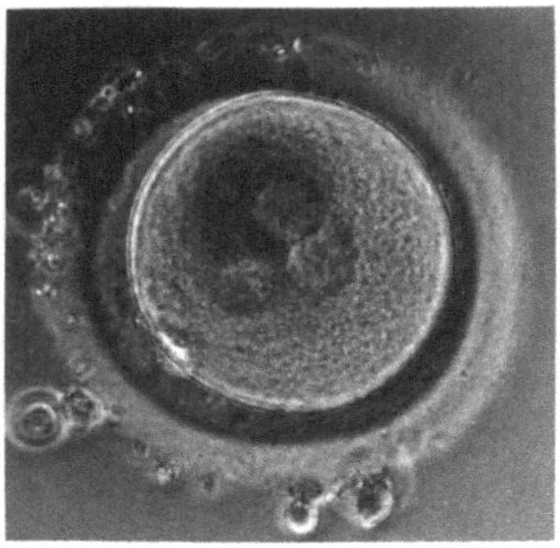

Fig. 8. An in vitro fertilized oocyte with three pronuclei (triploidy) 16 h after insemination (× 240)

6.1 Chromosome Analysis of Polyploid Oocytes

In a preliminary analysis we prepared 10 polyploid oocytes (9 × triploid, 1 × tetraploid) characterized by the number of their pronuclei (Fig. 8). All oocytes were in a mature or intermediate stage at laparoscopy. Two oocytes were characterized by asynchronous development of the pronuclei resulting in one case in the induction of "premature chromosome condensation" (S-PCC) in obviously two male pronuclei. These results have been discussed in detail elsewhere (Schmiady et al., in prep.). In five cases the chromosome analysis confirmed the triploid state, whereby in one case a haploid and a diploid chromosome set could be discerned with the latter having originated obviously from the two penetrated sperm nuclei (Fig. 9). Two eggs which were prepared in the 2- and 3-cell stage exhibited only interphase or early prophase structures. In one tetraploid oocyte four distinct haploid chromosome sets were found.

Rudak et al. (1984) have shown that if triploid oocytes were fixed before the first cleavage division, it is possible to analyze the chromosome constitution of the gametes resulting in some findings of chromosome aberrations. Grossly abnormal chromosome findings in triploid human embryos have been recently reported by Angell et al. (1986b).

7 Conclusions

Chromosome analysis of unfertilized human oocytes within an IVF program showed that chromosome abnormalities (structural alterations, aneuploidy, and diploidy as consequence of nondisjunction) occurred during maturation of the oocytes. After in vitro fertilization the phenomenon of premature chromosome condensation (PCC), based on the marked asynchrony between male and female nuclei, was observed. Asynchrony in the development of the pronuclei, which was also noticed in about 50% of those eggs which had formed two distinct pronuclei but failed to cleave, might, therefore, be a cause for early developmental arrest, as was also concluded by other authors (Zenzes et al. 1985). Although the influence of the IVF-culturing techniques

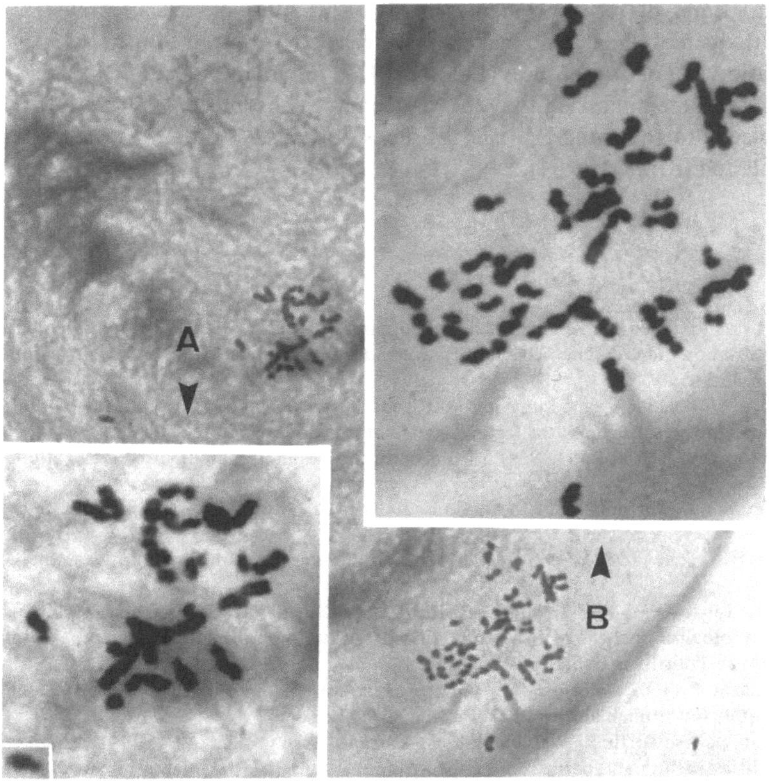

Fig. 9. Chromosome set of a triploid oocyte. Note the two regions with a haploid (*"A"*) and a diploid chromosome set (*"B"*). The latter might originate from the two sperm nuclei

on the rate of chromosome abnormalities is unknown, it seems unlikely that it has an effect, because most of the stimulated oocytes had reached metaphase II. However, hyperstimulation might be deleterious to the oocytes or preimplantation embryos. In the mouse IVF system no significant increase of aneuploidy was found but, rather, more triploid metaphases when the gonadotropin dose was increased (Sato and Marrs 1984). Belkien et al. (1986) concluded from hormonal and cytological analyses that low LH (luteinizing hormone) concentrations in follicular fluid indicate abnormal development.

The frequency of chromosome abnormalities (including PCC and the two reinseminated triploid oocytes) of those oocytes characterized by polar body, but no formation of pronuclei, was about 15%. This frequency should be still higher in fertilized oocytes because chromosome abnormalities of the sperm (Martin et al. 1982; Brandriff et al. 1984) have to be calculated, too. Therefore, some chromosome abnormalities of the kind we have found are present in human embryos at a relatively high rate (see Angell et al. 1983, 1986a) and might result in early embryonic loss and probably contribute to the low success rate after embryo transfer.

If the aberrant types of fertilization presented in this study also occur in vivo at about the same frequency, they could make a significant contribution to the relatively high preimplantation loss of conceptuses (Leridon 1977; Edmonds et al. 1982).

Finally, it should be emphasized that the frequency of chromosome anomalies in over a 1000 babies born following IVF and ET does not appear to be higher than in the normal population (Seppälä 1984).

References

Al-Hasani S, Ven H van der, Diedrich K, Hamerich U, Lehmann F, Krebs D (1984) Polyploidien bei der in-vitro-Fertilisation menschlicher Eizellen: Häufigkeit und mögliche Ursachen. Geburtshilfe Frauenheilkd 44:395–399

Angell RR, Aitken RJ, Look PFA van, Lumsden MA, Templeton AA (1983) Chromosome abnormalities in human embryos after in vitro fertilization. Nature (London) 303:336–338

Angell RR, Templeton AA, Aitken RJ (1986a) Chromosome studies in human in vitro fertilization. Human Genet 72:333–339

Angell RR, Templeton AA, Messinis IE (1986b) Consequences of polyspermy in man. Cytogenet Cell Genet 42:1–7

Beier HM, Lindner HR (eds) (1983) Fertilization of the human egg. Springer, Berlin Heidelberg New York

Belkien L, Bordt J, Zenzes TM, Kan I, Michel E, Hölzle C, Schneider HPG, Nieschlag E (1986) Hormonal and cytological analysis of in vitro fertilized eggs with abnormal development. J In Vitro Fertil Embryo Transfer 3 (Abstr 46):135

Boué J, Boué A, Lazar P (1975) Retrospective and prospective epidemiological studies of 1500 karyotyped spontaneous human abortions. Teratology 12:11–26

Brandriff B, Gordon L, Ashworth L, Watchmaker G, Carrano A, Wyrobek AJ (1984) Chromosomal abnormalities in human sperm: comparisons among four healthy men. Human Genet 66:192–201

Diamond MP, Rogers BJ, Webster BW, Vaughn WK, Wentz AC (1985) Polyspermy: effect of varying stimulation protocols and inseminating sperm concentrations. Fertil Steril 43:777–780

Edmonds DK, Lindsay KS, Miller JF, Williamson E, Wood PJ (1982) Early embryonic mortality in women. Fertil Steril 38:447–453

Edwards RG (1977) Early human development. From the oocyte to implantation. In: Philipp EE, Barnes J, Newton M (eds) Scientific foundations of obstetrics and gynaecology. Heinemann, London, pp 175–252

Edwards RG (1983) Chromosome abnormalities in human embryos. Nature (London) 303:283

Edwards RG, Purdy JM, Steptoe PC, Walters DE (1981) The growth of human preimplantation embryos in vitro. Am J Obstet Gynecol 141:408

Hassold T, Chen N, Funkhouser J, Joss T, Manual B, Matsuura J, Matsuyama A, Wilson C, Yamane JA, Jacobs PA (1980) A cytogenetic study of 1000 spontaneous abortions. Ann Human Genet 44:151–176

Johnson RT, Rao PN (1970) Mammalian cell fusion: induction of premature chromosome condensation in interphase nuclei. Nature (London) 226:717–722

Leridon H (1977) Human fertility. Univ Press, Chicago

Martin RH, Lin CC, Balkan W, Burns K (1982) Direct chromosomal analysis of human spermatozoa: preliminary results from 18 normal men. Am J Human Genet 34:459–468

Michelmann HW (1985) Chromosomenuntersuchungen an frühen menschlichen Embryonalstadien. Fortschr Med 103:265–267

Rao PN, Johnson RT, Sperling K (eds) (1982) Premature chromosome condensation. Application in basic, clinical, and mutation research. Academic Press, London New York

Rudak E, Dor J, Mashiach S, Nebel L, Goldman B (1984) Chromosome analysis of multipronuclear human oocytes fertilized in vitro. Fertil Steril 41:538–545

Sato F, Marrs RP (1984) Effect of ovarian hyperstimulation on chromosome aberration and sister chromatid exchange in in vitro fertilization. Fertil Steril 41 (Abstr 56)
Schmiady H, Sperling K, Kentenich H, Stauber M (1986) Prematurely condensed human sperm chromosomes after in vitro fertilization (IVF) Human Genet 74:441–443
Schmiady H, Kentenich H, Dincer C, Maaßen V, Stauber M (in prep.) Premature chromosome condensation in triploid human oocytes after in vitro fertilization
Seppälä M (1984) The world collaborative report on in vitro fertilization and embryo replacement: current state of the art in January 1984. Ann NY Acad Sci 442:558–563
Szalay S, Spernol R, Nachtigall (1986) Der gegenwärtige Stand der in-vitro-Fertilisation in der Bundesrepublik Deutschland, der Schweiz und Österreich. Fertilität 2:18–23
Tarkowski AK (1966) An air-drying method for chromosome preparations from mouse eggs. Cytogenetics 5:394–400
Trounson AO, Wood C (eds) (1984) In vitro fertilization and embryo transfer. Churchill Livingstone, Edinburgh London Melbourne New York
Trounson AO, Mohr LR, Wood C, Leeton JF (1982) Effect of delayed insemination on in vitro fertilization, culture and transfer of human embryos. J Reprod Fertil 64:285–294
Veeck LL, Wortham JWE, Witmyer J, Sandow BA, Acosta AA, Garcia JE, Jones GS, Jones HW (1983) Maturation and fertilization of morphologically immature human oocytes in a program of in vitro fertilization. Fertil Steril 39:594–602
Vogel R, Krüger C, Stauber M, Maaßsen V, Dincer C, Spielmann H (1985) In-vitro–Fertilisierung: Chromosomenanalyse unbefruchteter menschlicher Eizellen. Geburtshilfe Frauenheilkd 45: 382–385
Wentz AC, Repp JE, Maxson WS, Pittaway DE, Torbit CA (1983) The problem of polyspermy in in vitro fertilization. Fertil Steril 40:748–754
Wramsby H, Hansson A, Leidholm P (1982) Chromosomal preprations from in vitro inseminated human oocytes. In: Hafez ESE, Semm K (eds) In vitro fertilization and embryo transfe. MTP, Lancaster, pp 263–271
Zenzes MT, Belkien L, Bordt J, Kan I, Schneider HPG, Nieschlag E (1985) Cytologic investigation of human in vitro fertilization failures. Fertil Steril 43:883–891

10 Chromosome Analyses Using Human Spermatozoa *

M. T. ZENZES

1 Introduction

In the last 2 decades the development and refinement of cytogenetic techniques have contributed important information on the chromosome basis of human reproduction. The introduction of chromosome banding methods (Caspersson et al. 1971) permits the precise detection of chromosome rearrangements and their parental contribution at conception.

Natural and man-made environmental hazards, to which human populations are increasingly exposed, may produce genetic damage (in Sorsa and Norppa 1986; see Chap. 17). The recognized sources of man-made hazards have been steadily extended from the testing of nuclear weapons, to nuclear energy, medical uses of X-rays both for diagnostic and therapeutic purposes and to many groups of chemically-active substances used in the chemical industry (alkylating agents) in medicine (antibiotics and chemotherapeutic agents) and in agriculture (herbicides and pesticides). Obviously, it becomes of utmost importance to assess the levels of genetic damage in the human female and male gametes.

At the chromosome level, damage is detectable by changes in the number and structure of chromosomes. These changes, which may occur during gametogenesis or fertilization, are the most important cause of human reproductive failure, and are manifested in sterility, low fertility and a very high rate of mortality among human conceptuses (Carr 1971; Jacobs 1972; Boué et al. 1975; Carr and Gedeon 1977; Bond and Chandley 1983).

Chromosome analysis of human oocytes and spermatocytes will provide an estimate of first meiotic non-disjunction. To assess the total levels of non-disjunction arising at both meiotic divisions, an analysis must be made in a post-meiotic germ cell stage (i.e. ejaculated spermatozoa) or in the first cleavage. In vitro fertilization (IVF/ET) programs offer a unique opportunity for conducting chromosome analyses on unfertilized oocytes after attempted inseminations and on non-transferable embryos with morphological abnormalities or developmental arrest (Zenzes et al. 1985). This important information is the subject of another chapter in this book (see Chap. 9).

---

* This chapter is dedicated to the memory of a friend and cytogeneticist, Dr. Barbara Meer

1 Max-Planck-Gesellschaft, Klinische Forschungsgruppe für Reproduktionsmedizin, Steinfurter Str. 107, D–4400 Münster, FRG

Cytogenetics. Ed. by G. Obe and A. Basler

Chromosomes from human spermatozoa can only be visualized after fertilization. The environment provided by the ovum cytoplasm activates the spermatozoon to a point where its chromosomes can be visualized directly. The lack of species specificity of oocyte cytoplasmic factors among mammalian and amphibian species (Yanagimachi 1981; Ohsumi et al. 1986) makes them surrogates of the human ovum.

Ova from the golden hamster *Mesocricetus aureatus*, when free of their zona pellucida, permit entry of capacitated human spermatozoa (Yanagimachi et al. 1976). If the penetrated ova are maintained under controlled culture conditions, pronuclear chromosomes in metaphase from both the hamster and the human can be directly visualized in the ovum cytoplasm (Rudak et al. 1978). This exciting experimental approach permits, for the first time, carrying out chromosome analysis on human spermatozoa, and has opened a new era of knowledge on the contribution of the human male gamete to pregnancy loss.

Once the method is well established, it can be applied to the assessment of the effects of natural and man-made hazards to the genome of the male gamete; for genetic counselling of couples in which the male carries balanced translocations; in the analysis of meiotic segregation in the spermatozoa of male carriers; to determine some chromosome causes of male infertility.

In the following sections a review of this new area of research is presented.

2 Properties of the Mammalian Gametes at the Time of Fertilization

At the time of fertilization the female and male gametes must have completed their processes of maturation which will render them capable of interaction.

2.1 Oocyte Maturation

Meiotic maturation is of fundamental importance since it is only after reaching metaphase II that the female gamete is competent to ensure fertilization, cleavage and cellular differentiation (Thibault 1977). Maturation involves not only the meiotic progression from the dictyotene stage to metaphase II but also, subsequent to breakdown of the germinal vesicle, the appearance of cytoplasmic factors and macromolecules upon which the progress of early embryogenesis depends (Wassarmann 1983).

The ability of the ovum cytoplasm to dissociate the nuclear sperm membrane and totally decondense the penetrated spermatozoon appears to be acquired gradually between germinal vesicle breakdown and metaphase II (Yanagimachi and Usui 1972). The transformation of decondensing sperm into a male pronucleus depends upon the activity of a "male pronucleus growth factor" in the ovum cytoplasm, which appears at the time of germinal vesicle breakdown (Yanagimachi and Usui 1972). Its effect disappears or becomes inactive at later stages of development (Usui and Yanagimachi 1976; Schubeus et al. 1986). In polyspermic zona-free hamster ova, multiple decondensing sperm heads fail to develop into male pronuclei, whereas the ovum meiotic chromosomes form a female pronucleus (Yanagimachi 1981), suggesting that the factor promoting the female pronucleus development is different from the male pronucleus growth factor.

2.2 Sperm Maturation

Freshly ejaculated mammalian spermatozoa do not have fertilizing ability. This is acquired by a process of maturation during passage of spermatozoa in the epididymis (Austin 1985) and during their sojourn in the female tract. The latter process, referred to as capacitation, involves biochemical changes in the plasma membrane (Yanagimachi 1981) and prepares the spermatozoon for the acrosome reaction. This involves multiple-point fusions between the plasmalemma and the outer acrosomal membrane (Bedford and Cooper 1978), and renders the inner acrosomal membrane capable of fusing with the ovum plasma membrane.

2.3 Activation of the Gametes

At the time of fertilization the female and male gametes are genetically inactive. Fertilization provides the trigger for the onset of genetic acitivity.

Fusion of the spermatozoon with the ovum activates processes essential for zygote development as well as the mechanisms that prevent polyspermy. Upon penetration gross alterations within the ovum cytoplasm occur. The main events include resumption of the ovum meiosis, activation of cytoplasmic factors responsible for decondensation of the spermatozoon, transformation of the paternally derived chromatin, formation of the pronuclei, synthesis of DNA, recondensation of pronuclear chromatin and cleavage (Longo 1985).

3 Analysis of the Methodology for Human Sperm Chromosomes

The hamster ovum supports development of male pronuclei from a variety of mammalian species including human spermatozoa (Yanagimachi et al. 1976). This method was modified by Rudak et al. (1978). By use of controlled culture conditions, penetrated ova develop until the pronuclear metaphase stage. Up to three discrete haploid chromosome sets from human spermatozoa can be visualized in the ovum cytoplasm. Experience has shown, however, that the establishment of this technique on a routine basis is difficult and only a few groups of investigators have so far succeeded. The method requires skill, experience and a large number of donors willing to be available for repeated testing of their semen samples.

In the experimental procedure the following conditions are essential: (1) the population of ova has to be large in number (about 300 ova per test) and homogeneous in quality (ova with equal penetrability); (2) capacitation in vitro of human sperm must be adequate so as to provide large numbers of spermatozoa capable of penetrating; (3) a high penetration rate, (ideally 100% penetrated ova) should be achieved; and (4) the conditions for monospermic fertilization, or close to it, must be selected for each semen sample. A brief analysis of these methodological requirements is presented below. A summary of the procedure is shown in Fig. 1.

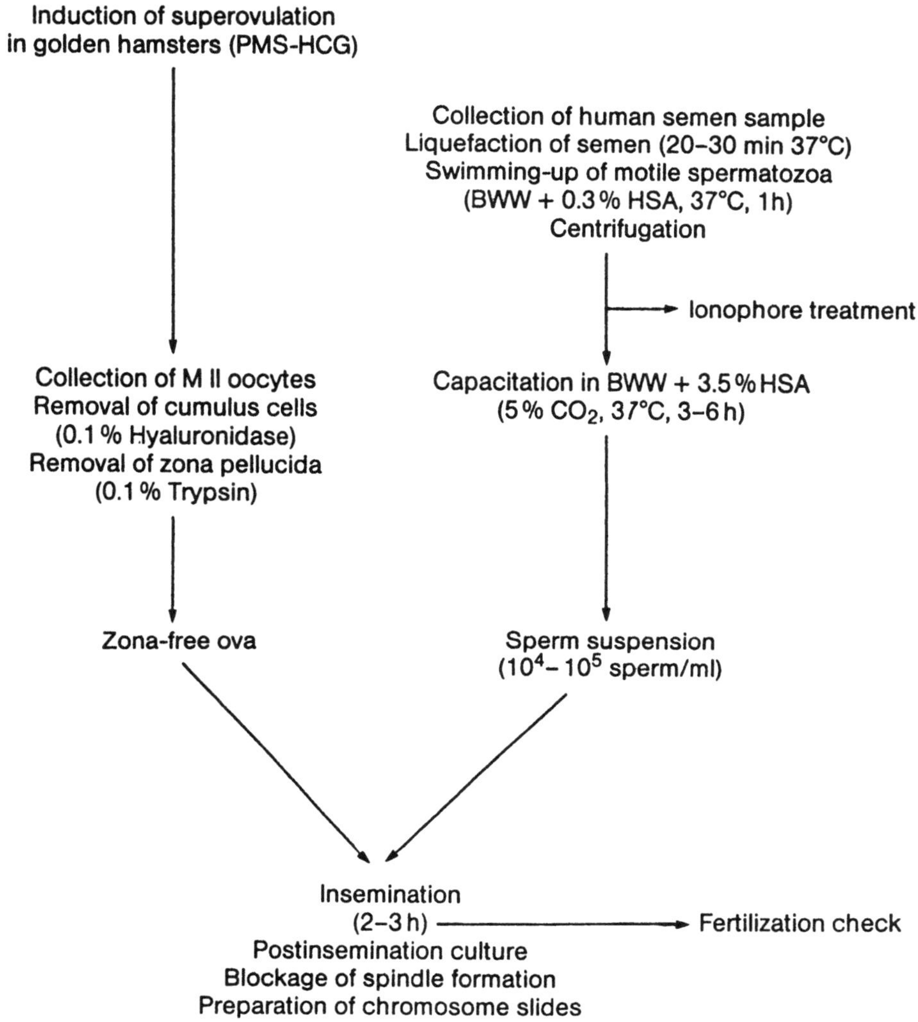

Fig. 1. Summary of the procedure for preparing human sperm chromosomes (in time sequence from top to bottom)

3.1 Preparation of Hamster Ova

Since the environment provided by the cytoplasm of the mature ovum is essential for the reactivation of the sperm nucleus, the preparation of a large, homogeneous population of healthy and mature ova, is required. When young female hamsters are maintained under controlled conditions of temperature, relative humidity, a diet supplemented with lettuce or carrots together with a light cycle of 14 h on, 10 h off, the yield of ova after superovulation increases. If the superovulation protocols are kept constant, namely, animals are similar in age, with an estrous cycle of 4 days, pregnant mares' serum (PMS) administration (25–30 IU, i.p. injection) is given on day 1 of estruation followed, after 62 h, by human chorionic gonadotropin (HCG) (25–30 IU, i.p. injection), the collection of ova from the oviduct 15 to 17 h later yields 40–60 ova per hamster (Martin 1983).

The preparation of ova must be done under controlled conditions of temperature (37°C) and light (use of a red filter). Two practitioners, each using three animals, should be able to prepare about 300 ova, the minimal number required for each experiment. The period between animal dissection and co-incubation of zona-free ova with sperm should last no longer than 30 min. Minimization of the time of ova handling is important to reduce degenerative processes and spontaneous activation.

3.2 Capacitation in Vitro of Human Spermatozoa

A successful rate of fertilization depends on how well spermatozoa can be capacitated in vitro. As removal of seminal plasma components is part of capacitation (Oliphant et al. 1985), preincubation of fresh, ejaculated semen samples during prolonged periods of time in simple, chemically-defined media, e.g. BWW (Biggers et al. 1971), containing glucose, pyruvate, lactate and human serum albumin (HSA), renders a proportion of the spermatozoa capable of penetrating. In these culture conditions 10 to 20% of the spermatozoa undergo acrosome reaction (Talbot and Chacon 1982). A modification of the original technique of Rudak et al. (1978) uses the swim-up method described by Overstreet et al. (1980). This method separates highly motile spermatozoa, thereby increasing the penetration efficiency (Chaudhuri and Yanagimachi 1984; Wramsby and Hansson 1984; Benet et al. 1986; Kamiguchi and Mikamo 1986).

Multiple testing of several batches of HSA is preferable since the capacitation rate of spermatozoa can be affected by a sort of affinity between HSA and individual semen samples (Kamiguchi and Mikamo 1986). A strategy to eliminate dependency to HSA is to incubate spermatozoa in a buffer containing fresh hen's yolk at 4°C 24 to 72 h (Brandriff et al. 1985b). This method allows the use of a semen sample on two occasions, thus increasing the number of chromosome spreads.

The time course of capacitation varies considerably from donor to donor, with times for maximal penetration indices (average number of sperm per ovum) ranging from 2 to 22 h (Perrault and Rogers 1982). For this reason the time required for maximal fertilizing ability has to be determined individually. Another alternative is a brief treatment of spermatozoa, before incubation in BWW, with the divalent metal cation ionophore A23187, which appears to accelerate sperm capacitation, consequently minimizing the inter-individual variability in capacitation time (Kamiguchi and Mikamo 1986; Wramsby and Yanagimachi 1986). To secure adequate capacitation it is preferable to conduct multiple testing using different concentrations of ionophore, because of inter-donor variation in the susceptibility of spermatozoa to ionophore.

3.3 Co-Incubation of the Gametes

Removal of the zona pellucida disturbs the mechanisms that prevent polyspermy, so that several spermatozoa can penetrate a zona-free ovum. The lack of zona pellucida has the advantage that it permits visualization of several sperm chromosome sets. But heavy polyspermy can be a limiting factor, as eggs containing more than five decondensing sperm heads may result in poor development of sperm pronuclei (Hirao and Yanagimachi 1979).

The time of co-incubation of zona-free ova with capacitated spermatozoa usually lasts 2–3 h (Rudak et al. 1978; Martin et al. 1983; Brandriff et al. 1984; Chaudhuri and Yanagimachi 1984; Tomkins et al. 1984; Wramsby and Hansson 1984). After 0.5 to 1.5 h, some ova are examined to verify fertilization. Penetrated ova contain at least one swollen sperm head or one pronucleus with an accompanying tail: activated (parthenogenetic) ova show one or two pronuclei. Since the fusion speed may vary from donor to donor, the co-incubation period must be assessed individually in relation to the penetration rate. For example: (1) if after 1.5 h at least half of the ova contain decondensing sperm heads with tails, they are removed from the sperm suspension, if less than half of the ova contain swollen sperm heads, the co-incubation time is prolonged for up to 3 h (Brandriff et al. 1984). (2) If inseminated ova have an average of seven attached spermatozoa, their subsequent transfer to the culture medium will usually guarantee visualization of sperm chromosomes (Wramsby and Yanagimachi 1986).

Theoretically, semen samples which produce a high penetration rate with low numbers of penetrating spermatozoa per ovum are best suited for obtaining sperm chromosomes. As the penetration rate increases linearly with increasing concentrations of spermatozoa (Martin and Taylor 1983), manipulations with sperm concentrations can be effective. However, when the penetration rate reaches 100% the average number of spermatozoa/ovum may increase accordingly, as there is a positive correlation between the two parameters (Zenzes et al. 1986). With concentrations of spermatozoa below 10^4 motile sperm/ml the penetration rate is drastically reduced, sometimes to zero.

3.4 Success Rate in Karyotyping

The ultimate goal of this technique is to obtain a large number of male pronuclear chromosome spreads suitable for analysis. This requires adequate condensation of chromosomes with good spreading and complete blockage of karyogamy to prevent overlapping of human and hamster pronuclear chromosomes. Zygotes having two to three analyzable sperm chromosome complements will contribute to a higher rate of successful karyotyping. As shown in Table 1, the mean number of karyotyped complements per semen sample varies greatly among the different groups of investigators. The means range from 8 to 121 (Rudak et al. 1978; Martin et al. 1982, 1983; Brandriff et al. 1985a; Sèle and Pellestor 1985, Sèle et al. 1985; Benet et al. 1986; Kamiguchi and Mikamo 1986; Jenderny and Röhrborn 1986). The differences reside, as noted earlier, in inter-individual variations in penetration rate, which yields variable numbers

Table 1. Rates of successful karyotyping among several groups

Study	No. of donors	No. of semen samples	No. of karyotypes per sample[a]	Total
Rudak et al. (1978)	1	–	10	60
Martin et al. (1982)	18	–	13.3[b]	240
Martin et al. 1983)	33	–	31	1000
Brandriff et al. (1984)	4	30	30.3[b]	909
Brandriff et al. (1985a)	9	34	45.9[b]	1559
Sèle and Pellestor (1985)	8	–	12.5[b]	100
Sèle et al. (1985)	7	–	10	70
Benet et al. (1986)	1	–	16.6	76
Kamiguchi and Mikamo (1986)	4	9	121.2	1091
Jenderny and Röhrborn (1986)	6	10	8.3	129

[a] Mean number.
[b] Calculated from their data.

of chromosome spreads. In addition, the different methods used to capacitate spermatozoa may result in different penetration rates. The most efficient appears to be that of Kamiguchi and Mikamo (1986) with the largest number of karyotypes per semen sample.

Banding techniques are required for precise chromosome analysis. These are especially important for detecting chromosome structural rearrangements; they also permit the determination of the type of segregation that is actually taking place. To secure good banding the treatment has to be adapted to the amount of cytoplasm observed prior to staining, and of course, it is better applied to high quality spreads with prometaphasic chromosomes. Karyotypes with Q-, C-, R- and G-bands have been successfully obtained (Rudak et al. 1978; Rudak 1981; Martin et al. 1982; Brandriff et al. 1985a; Sèle et al. 1985; Benet et al. 1986; Jenderny and Röhrborn 1986). Usually individual chromosomes can be identified by their banding patterns, but in a proportion of spreads, one or more chromosomes can be identified only according to the Denver group classification (Denver Conference 1960).

3.5 Other Experimental Approaches

Monospermic fertilization, or a situation very similar to it, is secured by injecting spermatozoa microsurgically. Male pronuclei are visualized after microinjection of hamster or human spermatozoa into the cytoplasm of hamster ova (Uehara and Yanagimachi 1976; Perrault 1985). Another alternative is to inject capacitated human spermatozoa into the perivitelline space of hamster zona-intact ova, while a portion of each batch of ova is freed of the zona pellucida and is co-incubated with capacitated spermatozoa. Results from our experiments showed pronuclear development in the two groups of ova. Sperm pronuclear chromosomes were observed with the incubation technique but not with the injection technique, as a large number of ova were destroyed in the manipulation (M.T. Zenzes, T.P. Haromy, M. Metka, W. Kocak, I. Freund, unpubl.). Even in the hands of an experienced practitioner this technical difficulty makes the injection technique unsuitable for routine work.

The cytoplasm from mature ova of the clawed frog *Xenopus laevis* also supports pronuclear development of human spermatozoa (Ohsumi et al. 1986). This animal model has several advantages over the hamster: (1) many more ova can be obtained after superovulation; (2) ova are much larger in size; and (3) there is no need to sacrifice the animals. Injection of human spermatozoa into ova of *Xenopus* results in pronuclear formation and, occasionally, recondensing chromatin threads or chromosomelike structures are seen (Ohsumi et al. 1986). Cell-free extracts prepared from many ova of *Xenopus* produce sperm decondensation, pronuclear formation and sometimes visualization of chromosome threads using either demembrenated or capacitated membrane-intact human spermatozoa (Gordon et al. 1985; Goddard and Zenzes 1986). This experimental approach appears promising as it would allow the analysis of large numbers of sperm chromosome complements and of spermatozoa that lack fertilizing ability. The method still requires development but can become more efficient and simple. A simple technique will speed the advancement of knowledge in this area of research.

4 Time Course of Human Sperm Transformation in the Hamster Ovum

The time when zona-free hamster ova are placed into the drops of medium containing capacitated sperm is considered as zero-hour post-insemination (0-h p.i.). Shortly after mixing, spermatozoa associate with the ovum plasma membrane. Since many thousands of them are used with relatively few ova, by the end of the insemination time (3-h p.i.) ova look like pin cushions. Only acrosome-reacted spermatozoa fuse with the ovum plasma membrane (Koehler et al. 1983) and incorporate in the ovum cytoplasm. Once inside, the nuclear envelope of the sperm disappears and the sperm nucleus rapidly decondenses. There is considerable inter-individual variation in the maximal time of sperm decondensation, ranging among four donors from 2 to 7 h (Schubeus et al. 1986).

At sperm entry ova resume meiosis and form the female pronucleus. Decondensing sperm at later stages of chromatin dispersion form a nuclear envelope which encompasses the paternally-derived chromatin. Sperm pronuclei are visualized within 2–4 h p.i. (Brandriff et al. 1982; Schubeus et al. 1986). Polyspermic ova may show up to 12 decondensing sperm heads (Brandriff et al. 1982; Zenzes et al. 1986), but no more than six (sometimes none) achieve a pronuclear stage of development.

The chromatin of spermatozoa at the time of fertilization is in G_1 phase. Chromosome replication takes place during the S-phase intervening between fertilization and cleavage. Utilizing an R-banding procedure in which 5-bromodeoxyuridine is incorporated into DNA, Balkan and Martin (1982) found that the male pronuclear chromatin is in mid-to-late S-phase 5.0–6.5 h p.i. After this time the sperm chromatin recondenses with the concomitant breakdown of the pronuclear envelope. Chromosomes begin to be visible by about 12 h p.i. Usually, ova contain discrete haploid sets of both hamster and human chromosomes with up to three sperm complements. In these polyspermic ova asynchrony in the stage of chromatin decondensation may occur, but

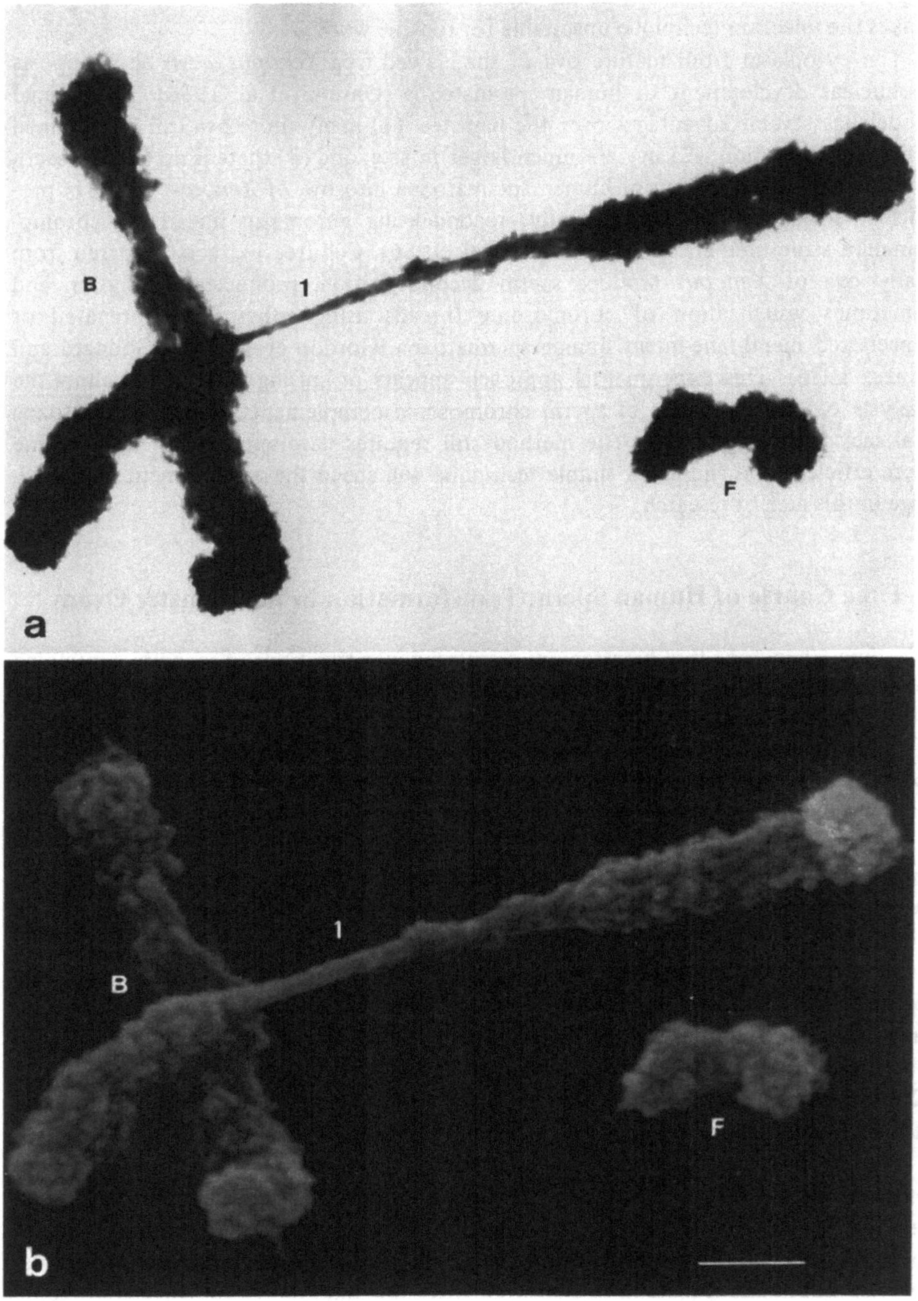

Fig. 2a,b. Electron micrographs of three human sperm chromosomes: *1* group B and group F. Note highly decondensed centromeric heterochromatin of chromosome 1. **a** Transmission electron microscopy (TEM). **b** Scanning electron microscopy (SEM). Bar = 1 μm. (Photos courtesy of Dr. J. Navarro, Universidad Autônoma de Barcelona, Spain)

by syngamy, synchronization is achieved (Brandriff et al. 1982). First cleavage occurs about 16-h p.i. (Yanagimachi 1984), 13 h being approximately the earliest time of visualization of the two-cell stage (Brandriff et al. 1982). Addition of Colcemide or Vinblastine will prevent karyogamy and first cleavage spindle formation, and consequently, mingling of the male and female chromosome sets.

4.1 Morphological Patterns of Sperm Pronuclear Chromosomes

The morphology of human sperm pronuclear chromosomes in metaphase shows frequently, but inconsistently, a gap in the centromeric region of one or more chromosomes when using C-, R- and Q-bands or solid Giemsa (Rudak et al. 1978; Martin et al. 1983; Brandriff et al. 1984; Sèle et al. 1985). In some preparations the heterochromatic regions of chromosomes 1, 9, 16 and the Y-chromosome seem to be differentially less condensed than the rest of the chromatin. This is particularly evident in C-banded preparations. Satellite associations involving the nucleolus organizer loci of acrocentric chromosomes appear also evident (Rudak et al. 1978).

Scanning (SEM) and transmission (TEM) electron microscopy techniques have permitted the analysis of gaps on preparations of pronuclear sperm chromosomes (Chernos et al. 1986; Navarro et al. 1986), and revealed that the chromosomes are continuous, linked by fibers of different diameter and length. The gaps are assumed to be regions of undermethylated DNA. Figure 2 shows a continuous chromosome No. 1 prepared with both SEM and TEM.

4.2 Frequency of Spermatozoa with Normal Chromosome Complements

In Table 2 data on 5234 spermatozoa, from 89 normal men, show that the overall proportion of spermatozoa with normal karyotypes is 89.7 ± 0.4% (weighted $\overline{X}$ ± standard error, SE), with a range from 81.0% to 93.3%. Is this proportion, which is based only on spermatozoa which have successfully capacitated, acrosome-reacted and penetrated, an unbiased estimate of the true proportion in the total sperm population? If selection takes place, which spermatozoa will be represented by chromosome complements? Furthermore, do morphological abnormalities in spermatozoa reflect chromosome imbalance? The answers to these questions are unknown. It seems likely that selection based on the above-mentioned sperm-physiological functions may take place. The swim-up technique yields highly motile spermatozoa and may bias the results. In this case, the question whether the swim-up technique selects genetically imbalanced sperm was addressed by analyzing both the swim-up fraction and the rest of the ejaculate (Brandriff et al. 1986c). An improvement in penetration rate was observed in highly motile spermatozoa from the swim-up fraction, but there was no difference between the two populations of spermatozoa regarding their chromosome complements. Up to 40% of spermatozoa in the ejaculate of fertile men may exhibit distinguishable morphological abnormalities with light microscopy (WHO 1980). There is no evidence that morphologically abnormal spermatozoa carry abnormal chromosome complements. It is likely that such penetrating spermatozoa are present

Table 2. Frequency of sperm chromosome abnormalities

Author	No. of donors	Sperm analyzed (No.)	Abnormal complements: Aneuploid (%)	Hyperhaploid (%)	Hypohaploid (%)	Structural (%)	Sex chrom. X (%)	Sex chrom. Y (%)
Rudak et al. (1978)	1	60	3 (1)[a] (5.0)	2 (3.3)	1 (1)[a] (1.7)	1 (1.7)	34 (56.7)	26 (43.3)
Martin et al. (1982)	18	240	18 (7.5)	13 (5.4)	5 (2.1)	4 (1.7)	143 (59.6)	97 (40.4)
Martin et al. (1983)	33	1000	52 (7)[a] (5.2)	24 (2.4)	27 (1)[b] (2.7)	33 (3.3)	527[c] (53.9)	450[c] (46.1)
Brandriff et al. (1985a)	11	2468	41 (4) (1.7)	18 (0.7)	23 (0.9)	190 (7.7)	1256[d] (50.1)	1251[d] (49.9)
Sèle et al. (1985)	7	70	9 (12.9)	4 (5.7)	5 (7.1)	1 (1.4)	36 (51.4)	34 (48.6)
Sèle and Pellestor (1985)	8	100	17 (17.0)	5 (5.0)	12 (12.0)	2 (2.0)	55 (55.0)	45 (45.0)
Benet et al. (1986)	1	76	5 (6.6)	1 (1.3)	4 (5.3)	2 (3.9)	42[e] (54.7)	34[e] (45.3)
Jenderny and Röhrborn (1986)	6	129	2 (1.6)	1 (0.78)	1 (0.78)	8 (6.2)	67 (52.0)	62 (48.0)
Kamiguchi and Mikamo (1986)	4	1091	10 (2)[a] (0.9)	5 (2)[a] (0.5)	5 (0.5)	142 (13.0)	571[f] (53.2)	502[f] (46.8)

[a] Aneuploids with structural anomalies.
[b] Double aneuploid
[c] From 977 complements.
[d] From 2507 complements.
[e] Calculated from their data in %.
[f] From 1973 complements.

in a metaphase spread, unless the abnormality involves a phenotypic dysfunction affecting their fertilizing ability. In this respect the question whether there is effective genetic selection can be investigated using techniques which do not require sperm penetration.

4.3 X- and Y-Bearing Sperm Ratios

There has long been an interest in the proportions of X- and Y-bearing spermatozoa. Using the fluorescent technique, X- or Y-sperm have been recognized by the absence or presence respectively, of a fluorescent body (Barlow and Vosa 1970). Spermatozoa with a fluorescent body (Y-bearing) have a smaller DNA content than those lacking this body (Sumner and Robinson 1976). Using this method, deviations from the 1:1 theoretical expectation, with excess of X-sperm, were observed (Schwinger et al. 1976; Klasen and Schmid 1981; Joseph et al. 1984). From these data the average percentage of Y-sperm in normal men is 43.3 ± 2.2 (weighted $\bar{X}$ ± SE; my calculation). This proportion is significantly less than the expected 50%. Using ejaculates from seven males, decondensing spermatozoa from both the swim-up fraction and the remaining ejaculate were evaluated 4 h after incubation with ova, for presence or absence of a fluorescent body. A total of 115 decondensing spermatozoa were assessed. Both fractions showed similar proportions of decondensing spermatozoa with (F) and without (0F) a fluorescent body; both fractions had an excess ($p < 0.01$) of 0F- (X-bearing) spermatozoa (M.T. Zenzes, B. Schumann, unpubl.).

The analysis of human sperm chromosome complements provides direct evidence on the proportions of X- and Y-bearing spermatozoa. Data from nine studies are shown in Table 2. Chromosome complements of 5232 human spermatozoa occur with an overall frequency for X-bearing sperm of 52.2% ± 0.7 and for Y-bearing sperm of 47.8% ± 0.7 (weighted mean proportions). The overall frequencies of X- and Y-sperm differ significantly ($p < 0.005$). The ratio of X-/Y-bearing sperm is 1.0919.

It is interesting and significant that the results using two different techniques are comparable: both show an excess of X-bearing sperm. Data on sperm chromosome complements are, of course, more reliable as these allow direct analysis (Martin 1985) and scoring is not subjective. Those studies in Table 2, with a low number of spermatozoa (60 to 100), do not show a significant difference in the proportion of X- and Y-sperm, but the difference is significant in the large-scale studies with 240 to 1091 spermatozoa (Martin et al. 1983, 1984; Kamiguchi and Mikamo 1986). In the studies of Brandriff et al. (1984, 1985a), which include 2507 spermatozoa, an excess of X-bearing sperm was observed in some but not in all individuals. Semen samples from six men processed for Y-enrichment with an albumin density gradient procedure showed, in both the processed and the unprocessed fraction, a significant excess of X-sperm (Brandriff et al. 1986d).

The reasons for these consistent deviations from the theoretical 1:1 expectation are not known, but do not seem to be due to technical biases.

5 Spontaneous Chromosome Abnormalities of Human Spermatozoa

Based on estimates of spontaneous meiotic non-disjunction frequencies in differentially stained chromosome regions, it was argued by Pawlowiski and Pearson (1972) that when all chromosomes are taken into account, up to 40% of spermatozoa in the ejacultes of normal men can be aneuploid. This is an extremely high value and shows the need for techniques for direct and precise analyses of human sperm genomes. The method introduced by Rudak et al. (1978) now permits the analysis of sperm chromosomes with about the same precision as from human somatic chromosomes.

5.1 Frequency and Type of Chromosome Abnormalities in Spermatozoa of Normal Men

Nine studies contributed data on the frequency of numerical and structural abnormalities in spermatozoa of normal men (Rudak et al. 1978; Martin et al. 1983, 1984; Brandriff et al. 1984, 1985a; Sèle and Pellestor 1985; Sèle et al. 1985; Benet et al. 1986; Jenderny and Röhrborn 1986; Kamiguchi and Mikamo 1986). Volunteers for sperm chromosome analyses were young healthy men selected for having no history of infertility, radiotherapy or chemotherapy and who were occasional alcohol drinkers or cigarette smokers.

The results from these studies, presented in Table 2, provide data for chromosome complements of 5234 spermatozoa from 89 normal men. The chromosome complements range from 60 to 2468 per study. The overall frequency of spermatozoa with abnormal chromosome complements is 10.3% ± 0.4 (weighted $\bar{X}$ ± SE). If there is an equal contribution of chromosome abnormalities from the female and the male gametes, this value matches well with the level of chromosome abnormality at conception of around 20% estimated by Ford (1975).

The overall frequency of numerical abnormalities in these studies is 6.5% ± 1.8 (unweighted $\bar{X}$ ± SE), ranging from 0.9% to 17.0%, and that of structural abnormalities is 4.5% ± 1.3 ranging from 1.4% to 13.0%. The different cytological techniques and the evaluation of the resulting karyotypes preclude the use of weighted, overall frequencies. Instead, unweighted mean frequencies are used to give equal importance to each study.

The studies differ in the proportion of numerical versus structural chromosome abnormalities. An elevated frequency of structural abnormalities, relative to aneuploidies, was observed in three studies (Brandriff et al. 1985a; Jenderny and Röhrborn 1986; Kamiguchi and Mikamo 1986), while the opposite was found in six studies (Rudak et al. 1978; Martin et al. 1982, 1983; Sèle and Pellestor 1985; Sèle et al. 1985; Benet et al. 1986). Some of the differences in the small studies are clearly not statistically significant. The discrepancy between studies is also due to great individual variations, but differences in technique and the small sample size may also contribute.

Among structural abnormalities, chromosome breaks are the most frequent. These may result from (1) chromatid breaks that replicate during S-phase after fertilization or (2) chromosome breaks prior to S-phase, which may occur after meiosis. Chromatid breaks, acentric fragments, deletions, translocations and exchanges involving complex formations are also found. Repeated breaks in one chromosome occur more frequently than expected by chance. The possibility exists that these breaks might be induced in the hamster cytoplasm. As there is a great individual variation in their incidence, this may well be due to individual differences in susceptibility. As more data become available the question of variability in chromosome structural abnormalities will be clarified.

Numerical abnormalities were found in a total of 157 spermatozoa, as shown in Table 2. Analyses of individual chromosomes from 118 spermatozoa were carried out in five studies (Rudak et al. 1978; Martin et al. 1982, 1983; Brandriff et al. 1985a; Sèle et al. 1985). In a proportion of these aneuploid complements identification could only be conducted in the Denver A-G groups. From the 118 aneuploid complements,

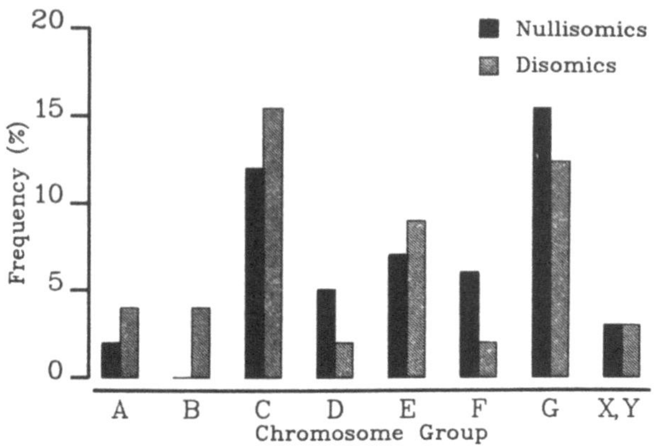

Fig. 3. Distribution of numerical abnormalities among chromosome groups. (Data of Rudak et al. 1978; Martin et al. 1982, 1983; Brandriff et al. 1985a; Sèle et al. 1985)

98 have one chromosome extra or missing (identified by number or group), 14 complements have two and 6 complements have three extra or missing chromosomes. In addition, some complements with numerical abnormalities have also chromosome breaks. The number of hypohaploid complements is slightly higher than hyperhaploid (62 vs 56); these proportions do not deviate significantly from the 1:1 ratio theoretically expected for non-disjunction. All chromosome groups from A to G are represented among the 118 aneuploid complements. The distribution of aneuploidies among the A- to G-groups is, however, unequal, probably due to group differences in the rate of non-disjunction. Group G is represented most frequently (33/118 = 28.0%), followed by group C (31/118 = 26.3%) and group E (21/118 = 17.8%). Aneuploidies of the sex chromosomes including 24,XXXX; 24,XY; 22.0-X or -Y are less frequent (6/118 = 5.1%). These findings confirm the prediction of increased initial frequency of non-disjunction for chromosomes 16, 21 and 22 from studies on spontaneous abortions (Jacobs 1975).

Figure 3 shows the frequency of disomic and nullisomic complements and their distribution in the A-G groups. Data are provided from the 98 complements with one chromosome extra or missing, as mentioned above. Complements nullisomic for a single chromosome of group B were not found (but do occur with double or triple nullisomies). The frequencies of disomic and nullisomic complements do not differ signficantly within chromosome groups. This is in marked contrast to studies of spontaneous abortions and newborns where monosomy of an autosome is rarely seen. Complements disomic or nullicomic for groups C and G are the most frequent, followed by complements disomic or nullisomic for group E. The higher frequency of aneuploidies in chromosomes with heterochromatic, polymorphic, centromeric regions may have a causal relationship. As spermatozoa with the aneuploid complements 24,Y,+21 and 24,X,+21 were detected in several studies, this is the most conclusive evidence for the paternal contribution to trisomy 21. With the use of polymorphic chromosome markers it has been estimated that paternal non-disjunction for chromosome 21 may be responsible for as many as 20–30% cases of trisomy 21 (Stene et al.

Table 3. Distribution by chromosome group of double and triple aneuploidies among 20 complements. (Data of Martin et al. 1982, 1983; Brandriff et al. 1984, 1985a; Sèle and Pellestor 1985)

Group	Double			Triple		
	Hypo-haploid	Hyper-haploid	Total	Hypo-haploid	Hyper-haploid	Total
A	2/14	1/14	3	1/6	–	1
B	2/14	–	2	1/6	4/6	5
C	3/14	–	3	2/6	2/6	4
D	4/14	2/14	6	4/6	1/6	5
E	4/14	–	4	1/6	–	1
F	3/14	–	3	1/6	–	1
G	5/14	–	5	1/6	–	1
X or Y	2/14	–	2	–	–	–
Total			28/14			18/6

1981). Further studies on sperm chromosomes will provide data to confirm and determine these levels.

Double and triple aneuploidies were found respectively, in 14 and 6 chromosome complements. Their distribution in the A- to G-groups is shown in Table 3. As is the case for single aneuploidies, double aneuploidies appear to be more frequent in some groups such as G, D and E. Although the sample size is low (20 spreads), the results are comparable with studies on abortuses in which double trisomies for groups G, E and C are more frequent (Boué et al. 1975). From the few data available it would appear that the events leading to double and triple aneuploidy are independent. These findings are important and show, for the first time, that chromosome abnormalities may result from different non-disjunctions in the same gamete or from a non-disjunction in each of the gametes that participate in the formation of a zygote.

As spermatozoa nullisomic for each chromosome group occur in the ejaculate, this is further evidence of lack of selection on the genetic content at the level of fertilization. To my knowledge, there are no data available on chromosome abnormalities in pre- or peri-implantation stages; such data are needed to clarify the contribution of nullisomic spermatozoa to problems of implantation and subsequent loss. There is some direct evidence from in vitro fertilized human eggs in first and second cleavage, showing that monosomy occurs in about half of the aneuploid embryos (Angell et al. 1986). As autosomal monosomy is rarely present in spontaneous abortions (Hassold et al. 1980) and is virtually non-existent in human populations (Hassold and Jacobs 1984), monosomic human embryos must be lost very early in development, as they are in mice (Gropp 1973).

It is encouraging that this method will provide reliable evidence on the contribution of the human male gamete to abnormalities of the zygote. An important question is whether this contribution is less than or equal to that of the female gamete. In the Chinese hamster 0.7% of male pronuclear chromosomes are abnormal versus 2.1% in the female (Mikamo and Kamiguchi 1983).

Table 4. Frequency of chromosome abnormalities in males with proven or unproven fertility

Study	Proven		Unproven	
	No. of men	Mean % (number)	No. of men	Mean % (number)
Martin et al. (1982)	5	6.9 (10/144)	5	12.5 (12/96)
Martin et al. (1983)	21	7.0 (42/602)	12	10.8 (43/398)[a]
Brandriff et al. (1985a)	6	9.7 (163/1676)	4	8.2 (65/792)
Kamiguchi and Mikamo (1986)	2	13.2 (45/341)	2	14.3 (107/750)

[a] $p < 0.05$.

5.2 Comparison Between Individuals with Proven or Unproven Fertility

Fifty-seven males, 34 with proven and 23 with unproven fertility, were compared for their incidence of sperm chromosome abnormalities. Data from the four studies in Table 4 show no significant difference between the two groups. There are occasionally individual differences in the incidence of abnormalities which presumably are not related to fertility status. Only in one study (Martin et al. 1983) was the difference in frequency of abnormalities significantly higher in males with unproven fertility. This is because two males of unproven fertility show a significantly higher frequency of complements with chromosome breaks. If these two are not considered, the difference between the two groups of individuals is no longer significant. Since most of the normal males with unproven fertility are, in fact, fertile we do not expect differences between these two groups.

5.3 Effect of Age on the Frequency of Sperm Chromosome Abnormalities

Circumstantial evidence of an increased frequency of trisomy 21 with advanced paternal age, for a given maternal age, comes from refined statistical analyses (Stene and Stene 1977; Stene et al. 1977; Matsunaga et al. 1978). Data from prenatal diagnosis, due to advanced maternal age, have shown that fathers above 39 years of age have an elevated risk of having trisomy 21 offspring with a strong effect from the age of 41 (Wagenbichler et al. 1976; Mattei et al. 1979).

Chromosome analyses of spermatozoa from aged men will provide the evidence for an age effect and its contribution to pregnancy loss or birth defects. Such a contribution is now fairly well recognized for the human female. A recent study has addressed this question (Martin 1986; Martin and Rademaker 1986). It includes 30 men with proven fertility, stratified by age in six categories (20–24, 25–29, 30–34, 35–39,

40–44 and 45+). A total of 1582 sperm chromosome complements were analyzed blindly with a minimum of 30 complements per donor. The results showed no correlation between paternal age and aneuploid sperm, but there was a correlation with structural abnormalities ($r = 0.6467$, $p = 0.0001$). Brandriff et al. (1985a) reported on a higher incidence of structural abnormalities in their oldest donor (49-years-old). The interpretation of these results is that there is a risk of increased frequency of structural abnormalities with advanced paternal age.

6 Sperm Chromosome Abnormalities in Infertile Men

Chromosome abnormalities make a significant contribution to male infertility either through spermatogenic impairment or by increasing levels of abortion through the production of genetically imbalanced spermatozoa.

Chromosome abnormalities responsible for reproductive failure can occur de novo during gametogenesis or as heritable chromosome imbalance. The former are mainly aneuploidies of the sex chromosomes and may have severe consequences for gonadal development. These are mainly represented by the 47,XXY karyotype followed by the 47,XYY karyotype. Inherited chromosome imbalance renders males subfertile or infertile by two different mechanisms. Firstly, by segregation into unbalanced genomes from a structural rearrangement leading to subfertility by pregnancy loss. Secondly, by spermatogenic impairment or arrest so that production of spermatozoa is affected, lowered or eliminated.

6.1 Sperm Chromosome Complements in XYY Men

Men with an XYY sex chromosome constitution may have gonosomal XY/XYY mosaicism in their gonadal tissue. Loss of the extra Y-chromosome is assumed to occur in a primitive germ cell, followed by selective proliferation of the XY cell line and failure of XYY spermatocytes to reach diakinesis (Evans et al. 1970). However, XYY spermatocytes from three XYY men have been reported (Tettenborn et al. 1970; Hultèn and Pearson 1971; Luciani et al. 1973). Analyses of sperm chromosomes of XYY men are important to determine the meiotic genomes, if produced, by spermatogenesis of the XYY line.

Recently, a study has analyzed sperm chromosome complements from a 27-year-old XYY man (Martin and Benet 1986). He was oligozoospermic with a large number of immature cells. His wife (46,XX) had two first trimester spontaneous abortions. Seventy-five sperm chromosome complements were analyzed and all of these contained one sex chromosome. Fourty (53%) were 23,X and 35 (47%) were 23,Y. Structural and numerical anomalies were within the normal range. This study agreed with the hypothesis that an extra Y-chromosome is eliminated in the majority of the germ-line cells. Since the XYY condition has a wide spectrum of gonadal development, (Skakkebaek et al. 1973), ranging from fertile men with normal sperm count to azoospermy, population studies on XYY men are needed to adequately test the validity of this hypothesis.

6.2 Structural Chromosome Rearrangements

Carriers of balanced chromosome rearrangements have an increased risk of recurrent fetal wastage by production of gametes with unbalanced structural rearrangements. Analyses of their sperm chromosomes will show the way the rearranged chromosomes pair and subsequently segregate at meiosis. The direct assessments of the frequency and type of imbalanced sperm complements and their segregation have practical applications for the genetic counselling of parents carrying the abnormalities. The most desirable goal in genetic counselling is to be able to specify precisely the risk for each specific translocation.

Analyses of sperm chromosomes have been performed in one male heterozygous for an inversion (Balkan et al. 1983) and in eight males heterozygous for different translocations (Balkan and Martin 1983a,b; Burns and Chaganti 1984; Martin 1984; Brandriff et al. 1986a; Templado et al. 1986).

A pericentric inversion of chromosome No. 3, inv (3), was detected in the analysis of sperm chromosomes of a control donor (Balkan et al. 1983). One hundred eleven complements were analyzed. From these, 64 had the normal chromosome 3, while 47 had the inverted one. This segregation was not significantly different from the theoretically expected 1:1 ratio. Spermatozoa with chromosome imbalance caused by a crossover within the inversion were not seen. This agrees with the finding of chromosomally normal progeny born to male carriers of this inversion. Since no recombinant chromosomes were found in this study, the question arises whether this arrangement represents a true inversion or is a polymorphism.

Heterozygosity for a reciprocal translocation may result in different meiotic products according to the type of segregation. A 2:2 segregation occurs in three ways: in alternate segregation 50% of the gametes produced have a normal chromosome complement, and the other half contains the translocated chromosomes in a balanced karyotype. In adjacent I segregation homologous centromeres pass to opposite poles, whereas in adjacent II, they pass to the same pole. These three types of segregation do not occur with equal frequency, adjacent II occurring less often. Chromosomally abnormal offspring of reciprocal translocation may also result from 3:1 or 4:0 segregations, the latter occurring less frequently. In humans the exact frequencies of gametes produced with unbalanced genomes are not known as many of the conceptuses are lost. Therefore, the analysis of sperm chromosomes can have a prognostic value.

Data on the meiotic segregation in spermatozoa from seven males heterozygous for different translocations are shown in Table 5. Different chromosomes are involved in these translocations. These include: males 1: t(5;18), 2: t(6;14) and 3: t(14;21) studied by Balkan and Martin 1983a,b); male 4: t(11;22) by Martin (1984); males 5: t(8;15) and 6: t(3;16) by Brandriff et al. (1986a); and male 7: t(2;5) by Templado et al. (1986). The ge of these males ranged from 25 to 39 years.

In male 1 the 5;18 translocation was ascertained during the investigation of his two daughters who had congenital abnormalities. Male 2 had a 6;14 translocation inherited from his father. In both males the proportions of normal to balanced complements did not differ significantly from the 1:1 ratio. The frequencies of unbalanced sperm, 19.4% and 31.6%, respectively, are higher than those reported from prenatal diagnosis studies. No adjacent II segregation was seen which may indicate that alternate segrega-

Table 5. Meiotic segregation in six males heterozygous for different autosome translocations. (Data of Balkan and Martin 1983a,b; Martin 1984; Brandriff et al. 1986a; Templado et al. 1986)

Male No.	Karyotype	Spreads analyzed	Complements Normal (%)	Balanced (%)	Unbalanced (%)	Products Nr.	Segragation
1	46,XY,-5,-18, +t(5;18) (p15;q21)	31	14 (45.2)	11 (35.5)	6 (19.4)	5 1	Adjacent I 3:1
2	46,XY,-6,-14, +t(6;14) (p24;q22)	19	8 (42.1)	5 (26.3)	6 (31.6)	6	Adjacent I
3	45,XY,-14,-21, +t(14;21)	24	17 (70.8)	4 (16.7)	3 (12.5)	3	Adjacent I
4	46,XY,-11,-22, +t(11;22) (q23;q11)	13	1 (7.7)	2 (15.4)	10 (76.9)	5 3 2	Adjacent I Adjacent II 3:1
5	46,XY,t (8;15) (p22;q21)	226	43 (19.0)	41 (18.1)	142 (62.8)	86 48 8	Adjacent I Adjacent II 3:1
6	46,XY,t (3;15) (p23;q24)	201	40 (19.9)	35 (17.4)	126 (62.7)	83 33 10	Adjacent I Adjacent II 3:1
7	46,XY,t (2;5) (p112;q15)	57[a]	5 (18.5)	4 (14.8)	18 (66.7)	2 2 14	Adjacent I Adjacent II 3:1

[a] 30 spreads from 2:2 segregation could not be identified.

tion is more frequent than adjacent. The two translocations produced equal proportions of normal and balanced complements and in both alternate segregation was more frequent than adjacent (Balkan and Martin 1983a).

Male 3, heterozygous for a 14;21 Robertsonian translocation, was ascertained after his brother had had a child with Down syndrome due to an unbalanced segregation of a D;G translocation. In this male there was an appreciable deviation between normal and balanced complements. The proportion of spermatozoa with balanced or unbalanced complements was low, probably owing to reduced viability (Balkan and Martin 1983b).

Male 4, heterozygous for t(11;22), was ascertained during a family study after his first cousin had a child with multiple congenital abnormalities and an unbalanced form of the translocation. He had two children who also carried the balanced translocation. All types of segregations were present. The frequency of unbalanced sperm is extremely high and may reflect the lethality of the unbalanced chromosome segregations (Martin 1984).

Male 5, heterozygous for t(8;15), was ascertained during the investigation following the birth of a daughter with multiple congenital abnormalities who died at 4 days of age and carried an unbalanced form of the translocation. This birth was preceded by two spontaneous abortions, followed by the birth of a son who carried the balanced form of the translocation and a daughter with a normal karyotype. Male 6 was heterozygous for t(3;16). His mother was a balanced carrier of this translocations. In both males normal and balanced complements occurred in equal proportions but the frequencies of the two types of adjacent segregations differed considerably. In both translocations 63% of the chromosome complements were unbalanced and 3:1 segregation was low (Brandriff et al. 1986a).

Male 7 was heterozygous for a reciprocal translocation t(2;5). He had a son with ambiguous genitalia carrying the translocation and a 45,X,t(2;5)/46,XX,t(2;5) karyotype. A significantly high number of this male's sperm chromosomes were unbalanced and resulted from adjacent I, adjacent II and 3:1 segregations. This was the first case where the 3:1 segregation was high (Templado et al. 1986).

Male 8, who is not shown in Table 5 because of missing data, was a double translocation heterozygote with the chromosome complement 46,XY,t(5;11)(p13;q23); t(7;14)(q11;q24). His two living progeny inherited the t(7;14) translocation. Among the sperm karyotyped rearranged products of t(7;14) were more frequent than those of t(5;11). Products of alternate, adjacent I, 3:1 and 4:0 segregation of each quadrivalent were recognized. Each translocation segregated independently of the other; t(5:11) showed alternate more often, while t(7;14) exhibited adjacent I segregation more commonly (Burns and Chaganti 1984).

Among these males each chromosome rearrangement behaves differently with respect to the frequency and type of chromosomally abnormal sperm. In general, the frequencies of abnormalities seem to be higher in spermatozoa than in abortuses or liveborn offspring. This is an important finding as there is an apparent discrepancy between the frequency of unbalanced zygotes, which are expected from the theoretical segregation of gametes of a balanced carrier, and the relatively small number of unbalanced fetuses and newborns observed (Zuffardi and Tiepolo 1982). The collection of more data will permit the evaluation of the recurrence risk of the different types of rearrangements.

7 Induced Chromosome Aberrations

Chromosome damage can be induced by various natural and synthetic clastogens. The sperm chromosome technique can be applied to monitor the effect of these hazards on the male gamete genome and on male fertility. It will permit one to assess the extent of damage sustained, and the DNA repair capacity of germ cells during meiosis.

A few preliminary studies have started to determine the type and frequency of chromosome abnormalities after exposure to various agents, such as radiotherapy and chemotherapy, used in cancer treatment (Burns and Chaganti 1985; Genescà et al. 1986; Jenderny and Röhrborn 1986; Martin et al. 1986), ionizing radiation in vitro (Brandriff et al. 1986b) and cryopreservation (Chernos and Martin 1986).

7.1 Effects of Radiotherapy and Chemotherapy on Sperm Chromosomes

Sperm chromosome analyses after radiotherapy or chemotherapy provide relevant information on the type and incidence of chromosome abnormalities after sperm restitution following radiotherapy. The analyses on sperm chromosomes may detect a relationship between radiation dose, the time passed after radiation and resulting sperm chromosome constitution.

The effect of radiotherapy on semen parameters and sperm function, as assessed by their penetration ability, was analyzed by Martin et al. (1985) and Freund et al. (1987). After recovery of spermatogenesis, penetration ability was still poor as manifested by low penetration rates (Martin et al. 1985) and penetration indices (Freund et al. 1987). In both studies restitution of testicular function was found to be time-dependent.

Martin et al. (1986) studied 13 patients with cancer before radiotherapy (RT) and at regular intervals after RT. The men were 19 to 47 years old and received testicular radiation doses of 0.4 to 5.0 gy. Their penetration rates before and after RT were low and this limited the number of sperm chromosome spreads for analysis. The frequency of abnormal chromosome complements before RT was 0% (0/9). After RT the majority of men were azoospermic for 24 months but complements from four men could be analyzed. In the first 12 months the frequency of abnormalities was 13% (1/8) and at 24 months it was 13% (7/55). By 36 months after RT most men had recovered sperm production and the frequency of abnormalities in eight men was 21% (18/86). This frequency is significantly higher than that in control donors (8.5%). For individual men the range was 6% to 67%. There was a significant correlation between testicular radiation dose and the frequency of sperm chromosome abnormalities. These abnormalities were both numerical and structural but with an excess of hypohaploid complements. The data base is small and must be confirmed by other studies.

Burns and Chaganti (1985) studied two patients with Hodgkin's disease who had been treated by chemotherapy. The patients were oligozoospermic (total sperm count $7–41 \times 10^6$) and had penetration rates of 50–90% prior to treatment. Analyses of over 130 metaphase spreads revealed only hamster genomes, suggesting failure of their sperm to decondense. Another man underwent a barium X-ray series 1 month prior to sperm analysis. His sperm chromosomes exhibited extensive damage (gaps, breaks, chromatid rearrangements and pulverization) in most cells which persisted over 9 months. In each of the fertilized oocytes with damaged human sperm chromosomes, the hamster chromosomes were undamaged.

Genescà et al. (1986) performed analyses on sperm chromosomes in a man with Wilms' tumor treated with chemotherapy (actinomycin-D) and radiotherapy (6500 rad at skin level). Fifty-one sperm chromosome complements were analyzed and 13.7% of these had structural abnormalities. This proportion is significantly higher than the control values. Four of the seven structurally abnormal metaphases had the same aberration, an isochromatid breakage in the heterochromatid region of chromosome 1. Studies on lymphocyte chromosomes using chemical agents have shown that breaks are often concentrated in heterochromatic regions of chromosomes (Shaw and Cohen 1965).

Jenderny and Röhrborn (1986) showed preliminary results in men treated with chemotherapy with a considerable increase in sperm chromosome abnormalities compared to control values.

These different studies are the first evidence that radiation and chemical agents do increase the frequency of chromosome abnormalities in human spermatozoa and that total elimination of meiotic and post-meiotic products does not take place. As these data show, men exposed to mutagenic agents may be at increased risk of reproductive failure and, among couples with a family still to be completed, should be advised accordingly.

7.2 Effects of Irradiation in Vitro

Among fertile donors, the incidence of chromosome abnormalities in a given man is significantly higher in his sperm than in lymphocytes. Brandriff et al. (1986b) reported on sperm chromosome structural abnormalities ranging from 2% to 14.5%; the lymphocyte chromosomes of ten of these same men ranged from 0.0 to 2.5%. Only in the individual with the lowest frequency of sperm chromosome abnormalities did the frequencies in the two cell types correspond. These authors also compared the effects of in vitro irradiation in sperm and lymphocyte chromosomes from one donor. Using ^{137}Cs both type of cells were irradiated with 100, 200 and 400 rad respectively, and analyzed for structural aberrations. Frequencies increased in a dose-dependent manner and to the same levels in both cell types but the spectra of aberrations were different. Lymphocytes showed more translocations and dicentrics than open breaks. In sperm the opposite was observed with more breaks than exchanges occurring.

7.3 Effect of Cryopreservation

Chernos and Martin (1986) analyzed the effect of cryopreservation on the frequency and type of chromosome abnormalities. The preliminary data consist of 187 karyotypes from fresh semen and 233 karyotypes from frozen semen from six normal men. Structural abnormality rates were not significantly different pre- and post-freeze (10.2% vs 9.9%). Aneuploidy rates were 12.3% pre-freeze and 5.6% post-freeze. However, most of these numerical abnormalities were hypoploidies. Based on a doubling of hyperploidy, the aneuploidy rate was not considered significantly different pre- and post-freeze. The sex ratio for normal spreads was unaffected by freezing, being skewed towards more X-bearing sperm both pre- and post-freeze. The results suggest that freezing does not alter the frequency of chromosome abnormality nor does it influence the sex ratio. These observations have important application in artificial insemination and for in vitro fertilization using cryopreserved semen.

8 Concluding Remarks

As shown in this brief review of the literature of sperm chromosomes, a new era has begun in which direct assessment of the contribution of the male gamete to pregnancy loss is now possible. This technique has a wide range of applications. It will bring a wealth of new knowledge useful for controlling and reducing the effects of some of the causes of reproductive failure and should significantly reduce human suffering.

Acknowledgments. I thank Prof. T.E. Reed, University of Toronto, for reading the manuscript and for important comments and suggestions. My sincere gratitude to Ms. Bettina Schuhmann for her help during a difficult time and for excellent technical, photographic and typing assistance.

References

Angell RA, Templeton AA, Aitken RJ (1986) Chromosome studies in human in vitro fertilization. Human Genet 72:333–339

Austin CR (1985) Sperm maturation in the male and female genital tracts. In: Metz CB, Monroy A (eds) Biology of fertilization, vol 2: Biology of sperm. Academic Press, London New York, pp 121–147

Balkan W, Martin RH (1982) Timing of human sperm chromosome replication following fertilization of hamster eggs in vitro. Gamete Res 6:115–119

Balkan W, Martin RH (1983a) Chromosome segregation into the spermatozoa of two men heterozygous for different reciprocal translocations. Human Genet 63:345–348

Balkan W, Martin RH (1983b) Segregation of chromosomes into the spermatozoa of a man heterozygous for a 14;21 Robertsonian translocation. Am J Med Genet 16:169–172

Balkan W, Burns K, Martin RH (1983) Sperm chromosome analysis of a man heterozygous for a pericentric inversion of chromosome 3. Cytogenet Cell Genet 35:295–297

Barlow P, Vosa CG (1970) The Y chromosome in human spermatozoa. Nature (London) 226: 961–962

Bedford JM, Cooper GW (1978) Membrane fusion events in the fertilization of vertebrate eggs. Cell Surf Rev 5:65–125

Benet J, Genescà A, Navarro J, Egoscue J, Templado C (1986) G-banding of human sperm chromosomes. Human Genet 73:181–182

Biggers JD, Whitten WK, Whittingham DG (1971) The culture of mouse embryos in vitro. In: Daniel JC (ed) Methods in mammalian embryology. Freeman, San Francisco, p 86

Bond DJ, Chandley AC (1983) Aneuploidy. In: Fraser Roberts JA, Carter CO, Motulsky AG (eds) Oxford monographs on medical genetics, No 11. Univ Press, Oxford

Bouè J, Bouè A, Lazar P (1975) Retrospective and prospective epidemiological studies of 1500 karyotyped spontaneous human abortions. Teratology 12:11–26

Brandriff B, Gordon L, Watchmaker G, Summers L, Wyrobek A (1982) Gamete interaction and developmental progression to the 2-cell stage in the human-sperm/hamster-egg interspecific cross. J Cell Biol 95:154a

Brandriff B, Gordon L, Ashworth L, Watchmaker G, Carrano A, Warobek A (1984) Chromosomal abnomralities in human sperm: Comparisons among four healthy men. Human Genet 66: 193–201

Brandriff B, Gordon L, Ashworth L, Watchmaker G, Moore II D, Wyrobek AJ, Carrano AV (1985a) Chromosomes of human sperm: Variability among normal individuals. Human Genet 70:18–24

Brandriff B, Gordon L, Watchmaker G (1985b) Human sperm chromosomes obtained from hamster eggs after sperm capacitation in TEST-yolk buffer. Gamete Res 11:253–259

Brandriff B, Gordon L, Ashworth LK, Littman V, Watchmaker G, Carrano AV (1986a) Cytogenetics of human sperm: Meiotic segregation in two translocation carriers. Am J Human Genet 38:197–208

Brandriff BF, Gordon LA, Carrano AV (1986b) Frequencies of chromosome aberrations in human lymphocytes and sperm. 7th Int Congr Human Genetics, Berlin, Sept 22–26, 1986, 5385L
Brandriff BF, Gordon LA, Haendel S, Ashworth LK, Carrano AV (1986c) The chromosomal constitution of human sperm selected for motility. Fertil Steril 46:686–690
Brandriff BF, Gordon LA, Haendel S, Singer S, Moore II DH, Gledhill BL (1986d) Sex chromosome ratio determined by karyotypic analysis in albumin-isolated human sperm. Fertil Steril 46:678–685
Burns JP, Chaganti RSK (1984) Meiotic segregation in a double translocation heterozygote. Am J Human Genet (Suppl 36):87S
Burns JP, Chaganti RSK (1985) Evaluation of the effects of chemotherapy and radiation on sperm chromosomes using the humster system. Am J Human Genet (Suppl 37):127A
Carr DH (1971) Chromosomes and abortion. Adv Human Genet 2:201–258
Carr DH, Gedeon M (1977) Population genetics of human abortuses. In: Hook EB, Porter IH (eds) Population cytogenetics: studies in human. Academic Press, London New York
Caspersson T, Lomakka G, Zech L (1971) 24 fluorescence patterns of human metaphase chromosomes – distinguishing characters and variability. Hereditas 67:89–102
Chaudhuri JP, Yanagimachi R (1984) An improved method to visualize human sperm chromosomes using zona-free hamster eggs. Gamete Res 10:233–239
Chernos JE, Martin RH (1986) The effects of cryopreservation on the frequency and type of human sperm chromosome abnormalities. Am J Human Genet (Suppl 39):A 109
Chernos JE, Rattner JB, Martin RH (1986) An investigation of human sperm pronuclear chromosome "gaps" using scanning electron microscopy. Cytogenet Cell Genet 42:57–61
Denver Conference (1960) A proposed standard system of nomenclature of human mitotic chromosomes. Am J Human Genet 12:384–388
Evans EP, Ford CE, Chaganti RSK, Blank CE, Hunter H (1970) XY spermatocytes in an XYY male. Lancet 1:719–720
Ford CE (1975) The time in development at which gross genome imbalance is expressed. In: Balls M, Wild AE (eds) The early development of mammals. Br Soc Dev Biol, Symp 2. Univ Press, Cambridge, pp 285–304
Freund I, Zenzes MT, Müller RP, Pötter R, Knuth UA, Nieschlag E (1987) Testicular function in eight seminoma patients after unilateral orchiectomy and radiatherapy. Int J Androl 10: 447–455
Genescà A, Benet J, Caballín MR, Miró R, Navarro J, Templado C, Egozcue J (1986) Sperm chromosome studies in a patient treated for Wilms' tumor. 7th Int Congr Human Genetics, Berlin, Sept 22–26, 1986, p 167
Goddard RJ, Zenzes MT (1986) The decondensation of human sperm with extract of *Xenopus laevis* eggs. Human Reprod 226 a
Gordon K, Brown DB, Ruddle FH (1985) In vitro activation of human sperm induced by amphibian egg extract. Exp Cell Res 157:409–418
Gropp A (1973) Fetal mortality due to aneuploidy and irregular meiotic segregation in the mouse. In: Boué A, Thibault C (eds) Proc Symp Institut National de la Santé et de la Récherche Médicale, Paris, pp 255–268
Hassold TJ, Jacobs PA (1984) Trisomy in man. Annu Rev Genet 18:69–97
Hassold TJ, Chen N, Funkhouser J (1980) A cytogenetic study of 1000 spontaneous abortions. Ann Human Genet 44:151–178
Hirao Y, Yanagimachi R (1979) Development of pronuclei in polyspermic eggs of the golden hamster: Is there any limit to the number of sperm heads that are capable of developing into male pronuclei? Zool Mag (Tokyo) 88:24–33
Hultèn M, Pearson PL (1971) Fluorescent evidence for spermatocytes with two Y chromosomes in an XYY male. Ann Human Genet 34:273–276
Jacobs PA (1972) Chromosome mutations: Frequency at birth in humans. Human Genet 16: 137–140
Jacobs PA (1975) The load due to chromosome abnormalities in man. In: Salzano FM (ed) The role of natural selection in human evolution. Elsevier/North-Holland Biomedical Press, Amsterdam, pp 337–352

Jenderny J, Röhrborn G (1986) Preliminary results: Cytogenetic analysis of G-banded human sperm chromosomes. 7th Int Congr Human Genetics, Berlin, Sept 22–26, 1986, p 147

Joseph AM, Gosden JR, Chandley AC (1984) Estimation of aneuploidy levels in human spermatozoa using chromosome specific probes and in situ hybridisation. Human Genet 66:234–238

Kamiguchi Y, Mikamo K (1986) An improved, efficient method for analyzing human sperm chromosomes using zona-free hamster ova. Am J Human Genet 38:724–740

Klasen M, Schmid M (1981) An improved method for Y-body identification and confirmation of a high incidence of YY sperm nuclei. Human Genet 58:156–161

Hoehler JK, Smith WD, Stenchever MA (1983) Attachment of acrosome intact sperm to the plasma membrane of zona-free hamster eggs. Fertil Steril 39:413–414

Longo FJ (1985) Pronuclear events during fertilization. In: Metz CB, Monroy A (eds) Biology of fertilization, vol 3: The fertilization response of the egg. Academic Press, London New York, pp 252–288

Luciani JM, Vagner-Capodano AM, Devictor-Vuillet M, Aubert L, Stahl A (1973) Presumptive fluorescent evidence for a spermatocyte with X+Y+Y diakinetic univalents in an XYY male. Clin Genet 4:415–416

Martin RH (1983) A detailed methodology for obtaining preparations of human sperm chromosmes. Cytogenet Cell Genet 35:459–468

Martin RH (1984) Analysis of human sperm chromosome complements from a male heterozygous for a reciprocal translocation t(11;22) (q23;q11). Clin Genet 25:357–361

Martin RH (1985) Recognition of the Y chromosome in human spermatozoa. In: Sandberg AA (ed) The Y chromosome. Liss, New York, pp 403–417

Martin RH (1986) Sperm chromosomal abnormalities in normal men: Studies on variation, the effect of age, sperm morphology and hamster egg penetration rates. 7th Int Congr Human Genetics, Berlin, Sept 22–26, 1986

Martin RH, Benet J (1986) Analysis of sperm chromosome complements in a 47, XYY male. 7th Int Congr Human Genetics, Berlin, Sept 22–26, 1986, p 149

Martin RH, Rademaker A (1986) The effect of age on the frequency of sperm chromosome abnormalities in normal men. Am J Human Genet (Suppl 39)

Martin RH, Taylor PJ (1983) Effect of sperm concentration in the zona-free hamster ova penetration assay. Fertil Steril 39:379–381

Martin RH, Lin CC, Balkan W, Burns K (1982) Direct chromosomal analysis of human spermatozoa: Preliminary results from 18 normal men. Am J Human Genet 34:459–468

Martin RH, Balkan W, Burns K, Rademaker AW, Lin CC, Rudd NL (1983) The chromosome constitution of 1000 human spermatozoa. Human Genet 63:305–309

Martin RH, Rademaker A, Barnes M, Arthur K, Ringrose T, Douglas G (1985) A prospective serial study of the effects of radiotherapy on semen parameters and hamster egg penetration rates. Clin Invest Med 8 239–243

Martin RH, Hildebrand K, Yamamoto J, Rademaker A, Barnes M, Douglas G, Arthur K, Ringrose T, Brown IS (1986) An increased frequency of human sperm chromosomal abnormalities after radiotherapy. Mutat Res 174:219–225

Matsunaga E, Tonomura A, Oishi H, Kikuchi Y (1978) Reexamination of paternal age effect in Down's syndrome. Human Genet 40:259–268

Mattei JF, Mattei MG, Ayme S, Giraud F (1979) Origin of the extra chromosome in trisomy 21. Human Genet 46:107–110

Mikamo K, Kamiguchi Y (1983) Primary incidences of spontaneous chromosomal anomalies and their origins and causal mechanisms in the Chinese hamster. Mutat Res 108:265–278

Navarro J, Templado C, Benet J, Genescà A, Eogzcue J (1986) A technique to study human sperm metaphases by transmission electron microscopy. Abstr 9th Int Chromosome Conf, Marseilles, June 18–21, 1986, p 133

Ohsumi K, Katagari C, Yanagimachi R (1986) Development of pronuclei from human spermatozoa injected microsurgically into frog (Xenopus) eggs. J Exp Zool 237:319–325

Oliphant G, Reynolds AB, Thomas TS (1985) Sperm surface components involved in the control of the acrosome reaction. Am J Anat 174:269–283

Overstreet JW, Yanagimachi R, Katz DF, Hayashi K, Hanson FW (1980) Penetration of human spermatozoa into the human zona pellucida and the zona-free hamster egg: A study of fertile donors and infertile patients. Fertil Steril 33:534–542

Pawlowiski IH, Pearson PL (1972) Chromosomal aneuploidy in human spermatozoa. Human Genet 16:119–122

Perreault SD (1985) Formation of chromosomes by heterologous sperm microinjected into golden hamster oocytes. J Androl 6:32 P

Perreault SD, Rogers BJ (1982) Capacitation pattern of human spermatozoa. Fertil Steril 38: 258–260

Rudak E (1981) Interspecific fertilization. In: Jagiello G, Nogel HJ (eds) Bioregulators of reproduction. PS Biomedical Sciences Symposia Series. Academic Press, London New York, pp 167–186

Rudak E, Jacobs PA, Yanagimachi R (1978) Direct analysis of the chromosome constitution of human spermatozoa. Nature (London) 274:911–913

Schubeus P, Zenzes MT, Schuhmann B, Freund I (1986) Time course of human sperm transformation in the cytoplasm of zona-free ova. Human Reprod 8:529–532

Schwinger E, Ites J, Korte B (1976) Studies on frequency of Y chromatin in human sperm. Human Genet 34:265–270

Sèle B, Pellestor F (1985) La cytogénétique des gamètes humains: Application à l'étude des anomalies de la fécondation. J Gynecol Obstet Biol Reprod 14:997–1003

Sèle B, Pellestor F, Jalbert P, Estrade C, Ostorero C, Gelas M (1985) Analyse cytogénétique des pronucleus à partir du modèle de fécondation interspécifique homme-hamster. Ann Genet 28: 81–85

Shaw MW, Cohen MM (1965) Chromosome exchanges in human leukocytes induced by mitomycin C. Genetics 51:181–190

Skakkebaek NE, Hultèn M, Jacobsen P, Mikkelsen M (1973) Quantification of human seminiferous epithelium II. Histological studies in eight 47,XYY men. J Reprod Fertil 32:391–401

Sorsa M, Norppa H (eds) (1986) Monitoring of occupational genotoxicans. Symp Progress in Clinical and Biological Research, vol 207, Helsinki, June 1985. Liss, New York

Stene J, Stene E (1977) Statistical methods for detecting a moderate paternal age effect on incidence of disorder when a maternal one is present. Ann Human Genet 40:343–353

Stene J, Fischer G, Stene E, Mikkelsen M, Petersen E (1977) Paternal age effect in Down's syndrome. Ann Human Genet 40:299–306

Stene J, Stene E, Stengel-Rutkowski S, Murken JD (1981) Paternal age and incidence of chromosomal aberrations: Prenatal diagnosis data. 6th Int Congr Human Genetics, Jerusalem, pp 2–12

Sumner AT, Robinson JA, (1976) A difference in dry mass between the heads of X- and Y-bearing human spermatozoa. J Reprod Fertil 48:9–15

Talbot P, Chacon RS (1982) Ultrastructural observations on binding and membrane fusion between human sperm and zona pellucida-free hamster oocytes. Fertil Steril 37:240–248

Templado C, Navarro J, Benet J, Perez MM, Genescà A, Egozcue J (1986) Human sperm chromosome studies in a reciprocal t(2;5). 7th Int Congr Human Genetics, Berlin, Sept 22–26, 1986, p 147

Tettenborn U, Gropp A, Murken JD, Tinnefeld W, Fuhrmann W, Schwinger E (1970) Meiosis and testicular histology in XYY males. Lancet 2:267–268

Thibault C (1977) Are follicular maturation and oocyte maturation independent processes? J Reprod Fertil 51:1–15

Tomkins PT, Carrol C, Houghton JA (1984) Analysis of human sperm chromosomes and factors involved in their formation in a heterospecific in vitro environment. J Embryol Exp Morphol (Suppl) 82:202

Uehara T, Yangimachi R (1976) Microsurgical injection of spermatozoa into hamster eggs with subsequent transformation of sperm nuclei into male pronuclei. Biol Reprod 15:467–470

Usui N, Yanagimachi R (1976) Behaviour of hamster sperm nuclei incorporated into eggs at various stages of maturation, fertilization and early development: The appearance and disappearance of factors involved in sperm chromatin decondensation in egg cytoplasm. J Ultrastruct Res 57: 276–288

Wagenbichler P, Killian W, Rett A, Schnedl W (1976) Origin of the extra chromosome number 21 in Down's syndrome. Human Genet 32:13–16

Wassarman PM (1983) Oogenesis: Synthetic events in the developing mammalian egg. In: Hartmann JF (ed) Mechanism and control of animal fertilization. Academic Press, London New York, Chap 1

WHO – World Health Organization (1980) In: Belsey MA, Eliasson R, Gallegos AJ, Moghissi KS, Paulsen CA, Prasad MRN (eds) Laboratory manual for the examination of humen semen and semen-cervical mucus interaction. Press concern, Singapore

Wramsby H, Hansson A (1984) Technique to examine human sperm chromosomes using zona-free hamster eggs. Arch Androl 12:79–83

Wramsby H, Yanagimachi R (1986) Human sperm chromosomes in zona-free hamster eggs: Pretreatment of spermatozoa with calcium ionophore facilitates later examination of chromosomes. Gamete Res 15:295–303

Yanagimachi R (1981) Mechanisms of fertilization in mammals. In: Mastroianni L Jr, Biggers JD (eds) Fertilization and embryonic development in vitro. Plenum, New York, Chap 5

Yanagimachi R (1984) Zona-free hamster eggs: Their use in assessing fertilizing capacity and examining chromosomes of human spermatozoa. Gamete Res 10:187–232

Yanagimachi R, Usui N (1972) The appearance and disappearance of factors involved in sperm chromatin decondensation in the hamster egg. J Cell Biol 55:293 a

Yanagimachi R, Yanagimachi H, Rogers BJ (1976) The use of zona-free animal ova as a test system for the assessment of the fertilizing capacity of human spermatozoa. Biol Reprod 15:471–476

Zenzes MT, Belkien L, Bordt J, Kan I, Schneider HPG, Nieschlag E (1985) Cytologic investigation of human in vitro fertilization failures. Fertil Steril 43:883–891

Zenzes MT, Kan I, Hano R, Schubeus P, Nieschlag E (1986) Role of the hamster ovum penetration test in infertility diagnosis. Int J Androl (Suppl) 6:113–122

Zuffardi D, Tiepolo L (1982) Frequencies and types of chromosome abnormalities associated with human male infertility. In: Crosignani PG, Rubin RL, Fraccaro M (eds) Genetic control of gamete production and function. Proc Serono clinical colloquia in reproduction 3. Academic Press, Grune & Stratton, London New York, pp 260–272

11 Class II Restriction Endonucleases *

C. KESSLER [1]

1 Introduction

The availability of a large variety of class II restriction endonucleases with different site specificities is the basis of recombinant DNA technology and numerous analytical applications. The importance of this enzyme class is demonstrated by recent findings in the detailed analysis of gene structure and the rapid succession of spectacular advances in genetic engineering (Malcolm 1981; O'Connor et al. 1984).

The existence of restriction endonucleases was postulated in 1953 by Bertani and Weigle as well as Luria and Human on the basis of genetic experiments (Luria and Human 1952; Bertani and Weigle 1953). The data were obtained by the investigation of host-controlled restriction and modification of bacteriophage lambda in the closely related *E. coli* strains C and K12. Bertani and Weigle observed that after infection of *E. coli* K12 cells with lambda phages isolated from *E. coli* C, most of the invaded DNA became degraded. However, a small portion of the DNA was still intact and thus survived. Surprisingly, these few phages can reinfect *E. coli* K12 with high efficiency. Some years later Linn and Arber demonstrated that the breakdown of the largest portion of the invading phage DNA is mainly due to cell-specific nucleases, the restriction endonucleases (Linn and Arber 1968). These DNA-digesting enzymes cut the double-stranded DNA after binding to enzyme-specific recognition sequences. In contrast, protection of the small portion of DNA is achieved by cell-specific DNA-methyltransferases, the restriction methylases. These DNA-modifying enzymes protect also the chromosomal host DNA against degradation by the nucleases. Protection is achieved by the transfer of a methyl group of S-adenosyl-methionine (SAM) to certain adenine or cytosine residues of the enzyme-specific recognition sequence. The N^6-methyladenine, 5-methyl-cytosine or 4-methyl-cytosine nucleotides formed by the DNA-methyltransferase inhibit the activity of the corresponding nucleases because both enzymes are of identical sequence specificity (Fig. 1) (Janulaitis et al. 1982; McClelland and Nelson 1985).

Restriction endonucleases and methylases are thus organism-specific enzyme pairs. The specific interaction between the two enzymes is the molecular basis of the restriction modification phenomenon. The antagonistic action of the restriction endonuclease

* Dedicated to Prof. Dr. G.R. Hartmann

1 Boehringer Mannheim GmbH, Biochemical Research Center Penzberg, Department of Molecular Biology, D–8122 Penzberg, FRG

Cytogenetics. Ed. by G. Obe and A. Basler

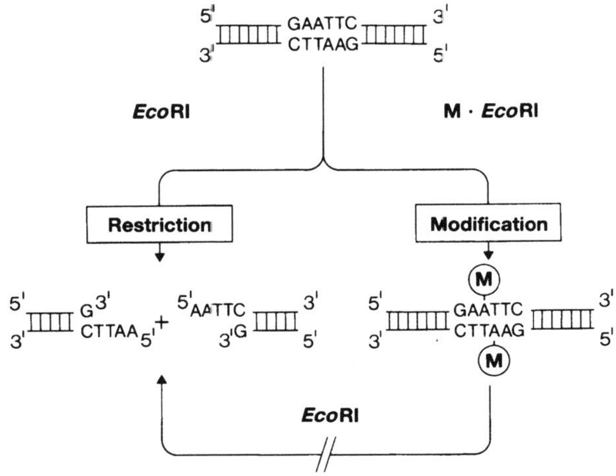

Fig. 1. Antagonistic action of restriction endonuclease *Eco*RI and restriction methylase M·*Eco*RI as constituents of a restriction-modification system. The nuclease *Eco*RI cleaves both DNA strands specifically within the common recognition sequence GAATTC between the 5′-flanking guanine and adenine residues. The corresponding methylase M·*Eco*RI modifies specifically the internal adenine residue of both DNA strands resulting in the formation of N^6-methyladenine. The activity of the nuclease is completely inhibited by N^6-methyladenine formation

and methylase of identical sequence specificity thus represents a simple bacterial immune system.

2 Classification of Restriction Modification Systems

When searching for the enzymes that are the basis of the various restriction modification systems, Linn and Arber as well as Meselson and Yuan discovered in 1968 one of the class I enzymes (Linn and Arber 1968; Meselson and Yuan 1968). With *Hin*dII, the first example of a class II restriction endonuclease was found in 1970 by Smith and Wilcox in *Haemophilus influenzae*, serotype d (Smith and Wilcox 1970; Kelly and Smith 1970). In later years, also enzymes of class III were found (Kauc and Piekarowicz 1978), which have properties between those of class I and class II (Table 1).

Class I enzymes are coded by the three gene segments *hsd*S, *hsd*R and *hsd*M expressing the functions for sequence specificity, restriction and modification methylation. The corresponding endonucleases and DNA-methyltransferases form enzyme complexes of high molecular weights. These enzyme complexes require magnesium ions, ATP and SAM as essential cofactors. Class I enzymes bind to and methylate specific sequences, but cut unspecifically after translocation into the 3′-direction (Modrich 1979; Endlich and Linn 1981; Yuan 1981; Bickle 1982). An example for a class I system is the *Eco*B restriction modification system. Whereas the early steps of SAM-binding, enzyme activation and formation of initial enzyme DNA complexes, are common to both restriction and modification, both reaction pathways diverge after

Table 1. Classification of restriction endonucleases in class I, II and III on the basis of the specificity of binding and cleavage; the different specificities of both reactions are exemplified for the enzymes *Eco*B, *Eco*RI and *Eco*PI

	Class I	Class II	Class III
Complementation groups	3	2	2
Genes	*hsd*S, *hsd*R, *hsd*M	*nuc/met*	*res, mod*
Enzymes encoded by	Chromosome	Chromosome, Plasmid/ Chromosome, Plasmid	Plasmid, Phage
Molecular weight (kD)	~ 450	$31^1/38^2$	$(106 + 73)^3$
Subunit composition	$(\alpha_2\beta_4\lambda_2)^2$	$(\alpha_2)^1/\alpha^2$	$(\alpha\beta)^3$
Cofactors for restriction	Mg^{2+}, ATP, SAM	Mg^{2+}	Mg^{2+}, ATP (SAM)
Cofactors for methylation	SAM (Mg^{2+}, ATP)	SAM	SAM (Mg^{2+}, ATP)
Restriction/methylation reaction	mutually exclusive	separate reactions	simultaneous
ATPase activity	+	–/–	–
Enzymatic turnover	–	+/+	–
DNA translocation	+	–/–	–
Specificity of recognition	specific	specific	specific
Specificity of cleavage	unspecific	specific	rather specific
Specificity of methylation	specific	specific	specific
Site of cleavage	~ 100–1000 bp downstream of recognition site	within or very near of recognition site	25–27 bp downstream of recognition site
Site of methylation	within the recognition site	within the recognition site	within the recognition site
Example	*EcoB*	*EcoRI/M·EcoRI*	*EcoP1*
Site of Recognition	TGA(N)$_8$ TGCT(N)$_{100-1000}$	G↓AATTC	AGACC(N)$_{25-27}$
Site of methylation	M TGA(N)$_8$ TGCT	M GAATTC	M AGACC

[1] for *EcoRI* [2] for M · *EcoRI* [3] for *EcoP15* [4] for *EcoB* [5] only in one strand

stable complex formation. ATP leads to the release of the enzyme from modified DNA, whereas in the case of unmodified recognition complexes, ATP stimulates translocation. This reaction is coupled to ATP hydrolysis. Sequential cleavage of both DNA strands occurs in a limited sequence stretch some 100 to 1000 base pairs away from the binding site.

Class II enzymes are expressed as two single proteins from two different genes acting as separate enzymes. Class II endonucleases mostly act as dimers of identical subunits and hydrolyze both DNA strands at fixed phosphodiester bonds within or very near the recognition sequence. Their activity depends only on magnesium ions. The corresponding class II DNA-methyltransferases act as monomeric proteins and methylate both DNA strands independently of the presence of the respective nuclease in an SAM-dependent reaction. An example for a class II system is the *Eco*RI restriction and modification system. Class II restriction endonucleases are by far the most important tools for the synthesis of recombinant DNA because of their absolute sequence specificity for both binding and cleavage reactions (Nathans and Smith 1975; Smith 1979; Nath and Azzolina 1981; Wells et al. 1982; Modrich and Roberts 1982; Fuchs and Blakesley 1983).

Class III enzymes are also expressed from two genes. However, as in the case of Class I enzymes, the corresponding endonucleases and DNA-methyltransferases form enzyme complexes in which both activities act simultaneously. Class III enzymes recognize specific sequences, but cleave rather specifically only 25 to 27 base pairs downstream of this region. The endonuclease activity is dependent on magnesium ions and ATP, whereas methylation occurs in the presence of SAM (Kauc and Piekarowicz 1978; Modrich 1979; Bickle 1982; Hadi et al. 1983; Iida et al. 1983).

After the discovery of the first class II enzyme by Smith and Wilcox there was a rapid increase in the number of different enzymes found in other organisms. Class II restriction endonucleases and methylases are ubiquitous in prokaryotic cells (Kessler et al. 1985; McClelland and Nelson 1985; Roberts 1985; Kessler and Höltke 1986). In accordance with the proposals made by Smith and Nathans, the various enzymes are abbreviated by taking the first letters of the bacterial classifications (Smith and Nathans 1973). The various endonucleases and DNA-methyltransferases occur in almost all genera of the two kingdoms of eu- and archaebacteria. In particular, type II enzymes are known both in Gram-positive and Gram-negative eubacteria, in cyanobacteria as well as in thermophilic, halophilic, methanogenic and sulphur-dependent archaebacteria. In contrast, eukaryotic restriction modification systems have not been described as yet. Specifically cleaving eukaryotic endonucleases like the HO nuclease in yeast or the retroviral endonucleases have other functions and are characterized by more complex recognition sequences (Duyk et al. 1983; Kostriken et al. 1985).

Besides class II restriction modification systems of single specificity a whole series of multiple component systems were also discovered containing several restriction modification systems of different specificity. Particularly complex systems, which contain up to seven different specificities, are found in certain species of *Nostoc* (*Nsp*7524I-V) and in *Neisseria gonorrhoeae* (*Ngo*I-VII) (Reaston et al. 1982; Korch et al. 1983). Methods for separating the various enzyme species include size fractionation, column chromatography with anion and cation exchangers as well as hydrophobic or affinity materials applying DNA or DNA-like heparin as ligands (Greene et al. 1978; Pirrotta and Bickle 1980).

In 1986 634 different class II restriction endonucleases with at least 111 different specificities are known which have been isolated from 564 different bacterial strains (Appendix 1, see pp. 251–266). The corresponding restriction methylases have been characterized only for the most important systems. As complete description of the currently known enzymes can be found elsewhere (Roberts 1985; Kessler and Höltke 1986).

The active enzymes of the class II restriction endonucleases consist of protein dimers of identical subunits, whereas the corresponding DNA-methyltransferases act as monomeric proteins. The molecular weights under denaturing conditions are in the range between 20 kD and 70 kD for both the class II endonucleases and the DNA-methyltransferases. The only protein of larger molecular weight known so far is *Bst*I with about 400 kD (Modrich 1979; Modrich and Roberts 1982).

As an example, the *Eco*RI endonuclease monomer is a 277 amino acid protein with a molecular weight of 31,063 daltons which is post-translationally processed by the removal of the terminal N-formylmethionine. The corresponding methylase protein consists of 326 amino acids resulting in a molecular weight of 38,048 daltons. This protein is also processed by the removal of the N-terminal dipeptide fMet-Ala (Greene et al. 1981; Newman et al. 1981).

Examples of cloned class II restriction modification systems are *Bsu*RI from *Bacillus subtilis* (Trautner et al. 1980; Kiss et al. 1985), *Eco*RI (Greene et al. 1981; Newman et al. 1981; Cheng et al. 1984; Botterman and Zabeau 1985; Haqqi et al. 1985; Luke and Halford 1985), *Eco*RII (Kosykh et al. 1980) and *Eco*RV (Bougueleret et al. 1984; 1985; Kraev et al. 1985) from *Escherichia coli, Hha*II from *Haemophilus parainfluenzae* (Schoner et al. 1983; Kelly et al. 1985), *Pae*R7 from *Pseudomonas aeruginosa* (Gingeras and Brooks 1983; Theriault et al. 1985), *Pst*I from *Providentia stuartii* (Walder et al. 1981, 1984; Lee and Rho 1985) and *Pvu*II from *Proteus vulgaris* (Blumenthal et al. 1985). Gene mapping in the various systems demonstrates that proteins for both endonuclease and DNA-methyltransferase may be either plasmid or chromosomally coded. Further analysis of the two genes shows that they are arranged either in tandem to form a bicistronic transcription unit or expressed as independent genes (Trautner et al. 1980; Greene et al. 1981; Newman et al. 1981; Kiss et al. 1985). Comparison of the known protein and gene sequences of corresponding endonucleases and DNA-methyltransferases reveals that there is no sequence homology which is statistically significant (Newman et al. 1981; Kiss et al. 1985). Circular dichroism studies indicate also extensive differences at the secondary structure level (Newman et al. 1981). These observations led to the conclusion that the two proteins have different evolutionary origins and interact with DNA in a different way.

3 Activity of Class II Restriction Endonucleases

An enzyme unit is normally defined as the amount of protein that completely digests 1 μg of the DNA from the bacteriophage λ within 1 h at 37°C. To analyze the completeness of the reaction, the fragment mixture is separated in 0.5–2% agarose gels by horizontal or vertical electrophoresis (Fig. 2). The use of λ DNA as the standard test substrate has two advantages. As this DNA is 48,502 base pairs in length, it contains cleavage sites for almost all known restriction endonucleases which allows a comparable activity determination (Sanger et al. 1982; Daniels et al. 1985a,b; Kessler and Höltke 1986). Because the complete sequence of λ DNA is known any cleavage pattern can be calculated in advance by the computer.

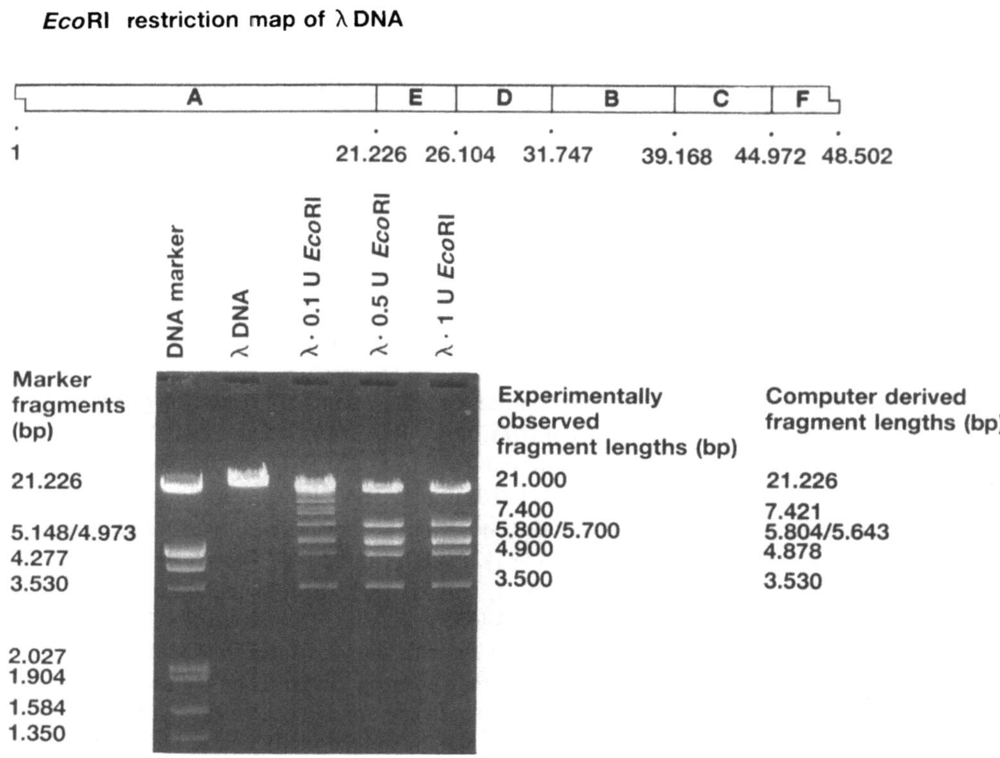

Fig. 2. Determination of the volume activity of *Eco*RI. For precise activity determinations 1 μg λ DNA (lane 2) is incubated at 37°C with increasing amounts of *Eco*RI for 1 h. The amount of enzyme activity which completely cleaves 1 μg λ DNA under the assay conditions is defined as one unit (lane 5). One unit *Eco*RI cleaves λ DNA into six fragments *A–F*, 31,226, 7,421, 5,804, 5,643, 4,878 and 3,530 base pairs in length. Incubation with less than one unit *Eco*RI results in partial digestion products which migrate as intermediate bands during electrophoresis (lanes 3 and 4). Separation of the fragment mixture is obtained by electrophoresis in 1% agarose gels. Subsequently, the separated fragments are stained with the intercalating fluorescent dye ethidium bromide. The dye can be visualized by irradiation with shortwave UV-light. The sizes of the individual bands are determined by comparison with DNA marker fragments of known molecular weight (lane 1)

3.1 Reaction Parameters

The activity of *Eco*RI as well as other class II restriction endonucleases is not only dependent on the presence of specific recognition sites and essential divalent cations, but also very strongly on reaction parameters such as temperature (T), ionic strength (I) and pH value (pH) (Wells et al. 1981). The activity of class II restriction endonucleases is also frequently stimulated by the presence of reducing agents like β-mercaptoethanol or dithiothreitol.

On pBR322 *Eco*RI has a temperature optimum of 40°C. The apparent activation energy for the overall enzymatic reaction can be calculated from the temperature dependence to be 5.1 KJ mol^{-1}. The observed inactivation of the enzyme at temperatures above 45–50°C was found to be irreversible (Pohl et al. 1982). The pH optimum for a specific *Eco*RI cleavage is 7.5, the optimal NaCl concentration 150 mM (Woodhead et al. 1981).

Often the optimal temperature for the enzymatic activity corresponds to the optimal growth temperature for the organism from which the enzyme is isolated (Wells et al. 1981). Accordingly, enzymes like *Bcl*I, *Bst*EII, *Mae*I–III or *Taq*I with high temperature optima are isolated from thermophilic organisms (Sato et al. 1977; Bingham et al. 1978; Lautenberger et al. 1980; Schmid et al. 1984). Enzymes are also known which require for maximal activity higher ionic strength such as *Mae*III with 350 mM NaCl or higher pH values such as *Bgl*I with pH 9.5 (Bickle and Ineichen 1980; Schmid et al. 1984).

3.2 Additional Structural Requirements Influencing Activity

Besides the incubation conditions, structural properties of the substrate also strongly influence the activity of class II restriction endonucleases at individual cleavage sites. In this context important parameters have been demonstrated to be:

1. G:C content and base distribution in natural DNA;
2. size of the DNA and cleavage frequencies;
3. length and base composition of the flanking sequences;
4. position of the cleavage sites with respect to each other;
5. DNA tertiary structure;
6. modification of DNA and protein attachment.

In natural DNA the G:C content varies broadly. Whereas DNA from *Micrococcus lysodeikticus* contains 72% G:C residues, DNA from *Euglena gracialis* chloroplasts is composed of only 25% G:C residues (Vanyushin et al. 1968; Gray and Hallick 1977). Although this DNA consists of about 130 kb there is not a single cleavage site for *Sma*I recognizing only G:C base pairs [CCC/GGG]. Furthermore, in natural DNA the base distribution is non-statistical. In eukaryotic SV40 DNA, the dinucleotide CpG occurs only extremely seldom (Nussinov 1980). As a result thereof the cleavage frequency for class II restriction endonucleases like *Mae*II, *Cfo*I, *Hpa*I, *Taq*I or *Fnu*DII, which contain the CpG dinucleotide in their tetranucleotide recognition sequences, is extremely low (Buchmann et al. 1980; Kessler and Höltke 1986). In contrast, the number of cleavage sites on the similar sized prokaryotic pBR322 DNA are about tenfold higher and show values that are also obtained with enzymes such as *Alu*I or *Hae*III lacking any CpG dinucleotides in their tetranucleotide recognition sequences (Kessler and Höltke 1986) (Table 2).

The influence of the DNA length can be demonstrated by the *Sal*I cleavage of pBR322 DNA (Arrand et al. 1978). The length of this plasmid is only about 10% of the length of the bacteriophage λ DNA (Sutcliffe 1979; Sanger et al. 1982; Peden 1983). Whereas λ only has two cleavage sites, the much smaller pBR322 DNA still has one cleavage site. As the cleavage frequency is fivefold increased with pBR322 DNA, for digestion of the same amount of the two DNAs with *Sal*I about a fivefold greater quantity of the enzyme must be used for the plasmid DNA as compared with the phage DNA.

Table 2. Cleavage frequencies of class II restriction endonucleases on eukaryotic SV40 DNA and prokaryotic pBR322 DNA; described are those enzymes recognizing CG- or CG-containing tetranucleotides. No enzyme is yet known for TGCA

Enzyme	Recognition sequence	Cleavage frequency on SV40 DNA	Cleavage frequency on pBR322 DNA
*Mae*II	ACGT	0	10
*Cfo*I	GCGC	2	31
*Hpa*II	CCGG	1	26
*Taq*I	TCGA	1	7
*Fnu*DII	CGCG	0	23
*Alu*I	AGCT	34	16
*Hae*III	GGCC	18	22

The influence of the flanking sequences is demonstrated by the observation that short oligonucleotides consisting, for example, only of the hexanucleotide *Eco*RI recognition sequence GAATTC are not cleaved. Even if several hexanucleotides were ligated to longer DNA molecules to stabilize the double strand – which would be completely denatured in the short hexanucleotide at 37°C – they could not be digested with *Eco*RI. Only if the recognition sequence is extended on both ends by one or more flanking nucleotides, can the enzyme exert its activity. However, also in these cases higher quantities of enzyme and longer reaction times are necessary. Normal reaction rates can only be achieved by increasing the length of the flanking sequences.

As an additional parameter the base composition within the flanking sequences has a distinct influence on the actual cleavage rates at the individual recognition sites. This effect was first observed with *Eco*RI which digests the various sites on λ or Ad2 DNA with tenfold different reaction rates (Forsblom et al. 1976; Nath and Azzolina 1981). A similar preferential digestion has also been observed with *Hin*dIII on λ, *Hga*I on ϕX174 and *Pst*I on plasmid pSM1 DNA (Brown and Smith 1977; Armstrong and Bauer 1982). In the latter example significant resistance to cleavage was conferred to G:C-rich flanking regions.

More significant differences by individual cleavage rates were observed on pBR322 DNA with *Nae*I recognizing the G:C-rich sequence GCC/GGC (Comb and Wilson, in Kessler and Höltke 1986). *Nae*I cuts pBR322 DNA at four different sites two of which are digested rapidly and the third five to ten times more slowly. The complete cleavage at the fourth site, however, occurs only at a 50 times slower rate. Identification of the resistant cleavage site is possible on a linear pBR322·*Pvu*II fragment (Fig. 3). After incubation of this fragment with a fivefold excess of *Nae*I, complete cleavage occurs at positions 401, 769 and 949. The uncomplete digestion at position 1,283 results in a 1137 base pair partial *Nae*I/*Pvu*II fragment representing pBR322 DNA between position 929 and 2066. For a complete digestion of this remaining fragment at position 1,283, resulting in the final 783 and 354 base pair fragments, a 50-fold excess of enzyme has to be applied (Grosskopf and Kessler, unpubl. results). Similar slow cleavage

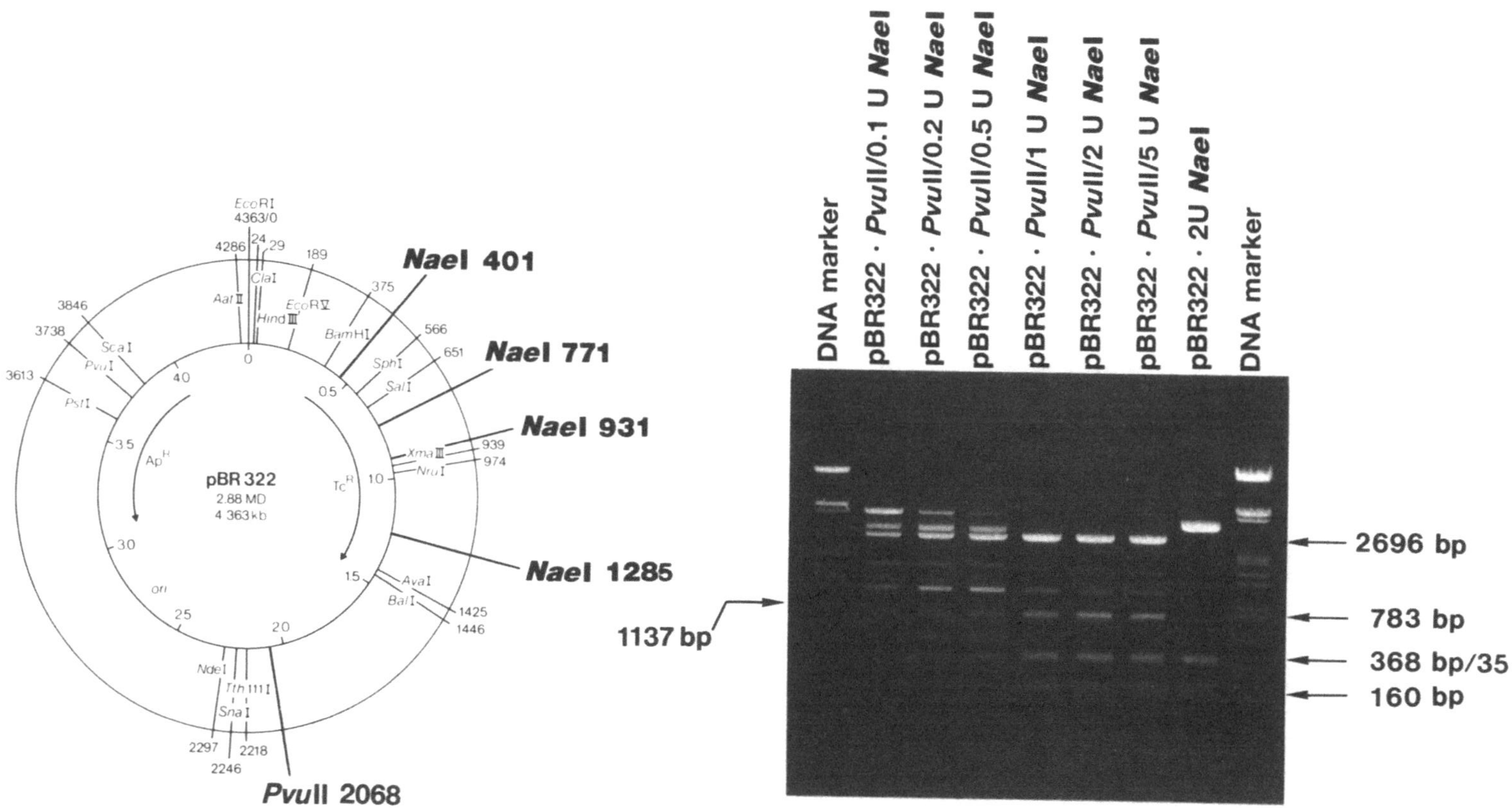

Fig. 3. Activity of *Nae*I on pBR322 DNA. *Nae*I digests pBR322 DNA at positions 403, 771, 932 and 1,285. The cleavage position 1,285 represents the "slow" cleavage site

sites were observed also for the enzymes *Nar*I [GG/CGCC], *Sac*II [CCG/CGG] and *Xma*III [C/GGCCG/] recognizing sequences with alternating guanosine and cytidine residues (Kunkel et al. 1979; Ehrlich and Wang 1981; Kessler and Höltke 1986). In all these examples the slower reaction rates at particular sites are attributed to the sequences within the flanking regions. It is, however, not yet known if a particular sequence element or an altered DNA conformation is responsible for the strong reduction of the cleavage reaction.

A similar decrease in the cleavage rate is observed if two or more recognition sites that alone would be readily digested are only a few nucleotides apart from each other. After digestion at one of the two sites, the essential flanking sequences are missing for an efficient cleavage at the neighbouring position. Therefore, after digestion at the first cleavage site the DNA can be hydrolyzed at a second site only at an extremely slow rate. This is true for the multiple cloning sites of the M13mp derivatives that are frequently used for sequencing studies. In these polylinker regions the recognition sequences for several restriction endonucleases are located directly next to each other (Yanish-Perron et al. 1985).

4 Specificity of Class II Restriction Endonucleases

The specificity of the presently known class II restriction endonucleases are defined by three criteria:

1. recognition sequence;
2. position of cleavage site;
3. influence of methylation.

However, only for a limited number of enzymes are all three characteristics of enzyme specificity already known. For many class II restriction endonucleases only the recognition sequences have so far been determined (Roberts 1985; Kessler and Höltke 1986).

4.1 Palindromic Recognition Sequences

For the majority of class II enzymes the recognition sequence is characterized by a common structural property, a twofold axis of rotational symmetry. Such symmetric recognition sequences are called palindromes. In a palindromic sequence a horizontal complementary arrangement exists in addition to the normal base pairing between A:T and G:C residues; for example, in the *Eco*RI recognition sequence G/AATTC the 5′- and the 3′-flanking nucleotides G and C correspond to each other in both strands.

Most palindromic recognition sequences are tetra-, penta- and hexanucleotides. However, with *Not*I an enzyme has recently been found, which recognizes a well-defined octanucleotide (Schildkraut et al., in Kessler and Höltke 1986). In addition, palindromic recognition sequences of extended length exist, for example, for *Bst*XI, in which only the flanking nucleotides are specifically recognized. In most of these cases

Table 3. Assignment of all unambiguously defined palindromic tetra- and hexanucleotides to known class II restriction endonculeases[a]

	AT	GC	CG	TA	
A		*Alu*I	*Mae*II		T
G	*Sau*3A	*Hae*III	*Cfo*I	*Rsa*I	C
C	*Nla*III	*Fnu*DII	*Hpa*II	*Mae*I	G
T			*Taq*I		A
AA		*Hind*III		*Ssp*I	TT
GA	*Eco*RI	*Sac*I	*Aat*II	*Eco*RV	TC
CA		*Pvu*II	*Pma*CI	*Nde*I	TG
TA			*Sna*BI		TA
AG	*Bgl*II	*Stu*I	*Eco*47III	*Sca*I	CT
GG	*Bam*HI	*Apa*I	*Nar*I	*Asp*718	CC
CG	*Pvu*I	*Xma*III		*Spl*I	CG
TG	*Bcl*I	*Bal*I	*Mst*I		CA
AC		*Mlu*I		*Spe*I	GT
GC	*Sph*I	*Bss*HII	*Nae*I	*Nhe*I	GC
CC	*Nco*I	*Sac*II	*Sma*I	*Avr*II	GG
TC	*Bsp*HI	*Nru*I	*Bsp*MII	*Xba*I	GA
AT		*Nsi*I	*Cla*I		AT
GT	*Sna*I	*Apa*LI	*Sal*I	*Hpa*I	AC
CT		*Pst*I	*Xho*I	*Afl*II	AG
TT			*Mla*I	*Dra*I	AA
	AT	GC	CG	TA	

[a] The various sequences are generated by the stepwise expansion of the dinucleotides A, G, C or T on the left-hand side as well as by the complementary nucleotides on the right-hand side. The repetition of the permutation steps results in the possible hexanucleotides. If more than one isoschizomeric enzyme is known, only the most frequently used nuclease is listed. Blank fields show those sequences that are not yet covered by a known enzyme.

the flanking trinucleotides are unambiguously defined. However, one case of non-equivocally defined flanking trinucleotides is known with *Dra*II [(^A_G)G/GNCC(^T_C)] (de Wit et al. 1985; Grosskopf et al. 1985). *Sfi*I [GGCCNNNN/NGGCC] is an example for flanking tetranucleotides (Qiang and Schildkraut 1984).

Palindromic recognition sequences can be arranged according to homologies within their recognition sequences (Table 3). In this approach the sequences are arranged by starting from the shortest central palindromes AT, GC, CG and TA. Tetranucleotide palindromes are created by extending each of the dinucleotide sequences with the addition of one of the four nucleotides in the A ⇒ G ⇒ C ⇒ T order on the left side and the addition of the complementary nucleotide on the right side. Repeated application of this cycle generates hexa- and octanucleotide palindromes. Of the 16 possible tetranucleotides, 11 sequences are currently covered by corresponding enzymes. For the 64 possible hexanucleotides 49 enzyme specificities are known (Kessler and Höltke 1986).

The positions of cleavage sites for class II restriction endonucleases are within or directly next to the recognition sequence (Table 4). As with the recognition itself, the positions in the two complementary strands are arranged in rotational symmetry. Either blunt double-stranded ends or single-stranded 5′- or 3′-protruding ends are formed. The lengths of the terminal single-stranded regions depend on the position of the cleavage sites within the recognition sequence. Independent of the cleavage positions all class II restriction endonucleases form fragments which have 5′-phosphates and 3′-hydroxyls at the fragment ends. The only exception so far known is *Nci*I, which produces 3′-phosphorylated fragment termini (Hu and Marschel, in Kessler and Höltke 1986).

Some class II restriction endonucleases are able to digest DNA/RNA hybrids. It was shown for *Alu*I, *Hae*III, *Hha*I and *Taq*I that these enzymes cut the DNA strand of the hybrid and the correct recognition sequences (Molloy and Symons 1980).

Sequence-specific digestion of single-stranded DNA has also been shown for a number of enzymes. However, as with double-stranded DNA base-paired recognition sequences are necessary for cleavage of single-stranded DNA. Therefore, refolding effects between different recognition sites resulting in transiently formed secondary canonical structures play a decisive role. The cleavage rates depend mainly on the stability of these double-stranded structures (Yoo and Agarwal 1980; Hofer et al. 1982; Nishigaki et al. 1985).

4.2 Non-Palindromic Recognition Sequences

In addition to class II enzymes acting on palindromic sequences class IIS enzymes like *Fok*I, *Hga*I, *Mbo*II or *Mnl*I have been found recognizing non-symmetric sequences. These enzymes digest the double-stranded DNA at precise distances downstream from their recognition sites (Table 2) (Sugisaki 1978; Brown et al. 1980a; Sugisaki and Kanazawa 1981; Zabeau et al., in Kessler and Höltke 1986). The spatial separation between the recognition and cleavage reaction was utilized to digest any predetermined sequence in single-stranded DNAs like those of the various M13 mp derivatives (Pohajska and Szybalski 1985; Szybalski 1985). Universal cleavage is achieved by applying a specifically designed adapter. This synthetic DNA element is constructed from a constant double-stranded domain harboring the recognition sequence of the class IIS enzymes

Table 4. Possible structures of recognition sequences and different positions of the cleavage sites[a]

Enzyme	Recognition sequence	End produced
*Hae*III	GG/CC CC/GG	blunt ended
*Eco*RI	G/AATT-C C-TTAA/G	5'-protruding
*Pst*I	C-TGCA/G G/ACGT-C	3'-protruding
*Not*I	GC/GGCC-GC CG-CCGG/CG	5'-protruding
*Eco*RII	/CC ($^{A}_{T}$) GG GG ($^{A}_{T}$) CC/	5'-protruding
*Sau*I	CC/TNA-GG GG-ANT/CC	5'-protruding
*Xmn*I	GAANN/NNTTC CTTNN/NNAAG	blunt ended
*Sfi*I	GGCCN-NNN/NGGCC CCGGN/NNN-NCCGG	3'-protruding
*Bst*XI	CCAN-NNNN/NTGG GGTN/NNNN-NACC	3'-protruding
*Mnl*I	CCTCNNNNNNN/N GGAGNNNNNNN/N	blunt ended
*Hga*I	GACGCNNNNN/NNNNN-N CTGCGNNNNN-NNNNN/N	5'-protruding
*Mbo*II	GAAGANNNNNNN-N/N CTTCTNNNNNNN/N-N	3'-protruding
*Asp*718	G/GTAC-C C-CATG/G	5'-protruding
*Kpn*I	G-GTAC/C C/CATG-G	3'-protruding

[a] In the group of unambiguously defined palindromes examples of tetra-, hexa- and octanucleotides are known. Penta- to dodecanucleotides have been described for the second group with only flanking palindromic sequences. The third group describes class II as well as class IIS enzymes with nonsymmetric recognition sequences. The class II enzymes of the first and second group cut within or directly beside the recognition sequence producing blunt, 5'- or 3'-protruding ends. Class IIS enzymes always cut downstream of their recognition sequence.

and a variable single-stranded domain complementary to the sequence to be cleaved. The linearized single-stranded target DNA can be coverted to double-stranded DNA by a Klenow-catalyzed polymerization reaction using the synthetic deoxy-oligonucleotide as primer. By this approach those double-stranded DNA molecules are created which are precisely cleaved at any position. Thus, the new method permits the design of novel enzyme specificities by combining class IIS restriction endonucleases with bivalent DNA adapter oligonucleotides mediating the novel sequence specificities.

4.3 Isoschizomers

Class II restriction endonucleases which are isolated from various organisms but recognize identical sequences are known as isoschizomers. Isoschizomeric enzymes, however, may have various cleavage sites within the recognition sequence or may show different sensitivity to particular modified bases. With 36 different enzymes all recognizing the sequence GATC the greatest variability of isoschizomers is known (Kessler and Höltke 1986). In this collection of isoschizomers well-known examples for enzymes with different methylation sensitivities are *Dpn*I, *Mbo*I and *Sau*3AI (Gelinas et al. 1977a; Lacks and Greenberg 1977; Dreiseikelmann et al. 1979). Whereas *Sau*3AI is insensitive to methylation at the internal adenine residue, *Mbo*I is completely inhibited by this methylation. In contrast, *Dpn*I is fully dependent on this modification as this enzyme does not cleave the unmethylated GATC sequence. pBR322 DNA, isolated from wild-type *E. coli* cells, is highly methylated at its adenosine residues by *E. coli dam*-methylase (Hattman et al. 1978a; Urieli-Shoval et al. 1983). Consequently, *Sau*3AI and *Dpn*I will cleave this DNA, whereas *Mbo*I is not able to digest this substrate.

Another important pair of isoschizomers, *Hpa*II and *Msp*I, shows a different sensitivity with respect to cytosine methylation. Modification of the central C-residue within their common tetranucleotide recognition sequence CCGG renders the cleavage site resistant against digestion with *Hpa*II but sensitive to cleavage with *Msp*I (Mann and Smith 1977; Jentsch et al. 1981). This ability of discrimination between methylated and unmethylated cytosine residues is used to detect methylated CpG-dinucleotides which are thought to be involved in eukaryotic gene regulation (Razin and Riggs 1980; Gruenbaum et al. 1981b; Wigler 1981).

An example for an isoschizomeric enzyme pair with different cleavage sites is *Asp*718 and *Kpn*I. Both enzymes recognize the sequence GGTACC but cleave either between 5′-terminal G-residues or the 3′-terminal C-residues (Tomassini et al. 1978; Bolton et al. 1985). The 5′-terminal protruding ends of *Asp*718 fragments may be efficiently labelled with T4 polynucleotide kinase, whereas the 3′-ends are suitable substrates for 3′-end-labelling reactions with Klenow enzyme. Labelling of the 3′-protruding ends of *Kpn*I fragments can be obtained via the tailing reaction with terminal transferase.

In the most commonly used cloning vehicles there are often no restriction sites for particular enzymes. In such cases cloning can be achieved by the use of enzymes from an enzyme family. Members of an enzyme family all produce identical single-stranded fragment ends, whilst the complete recognition sequences of individual enzymes are

Table 5. Enzyme families of GATC- and CTAG-specificity

GATC Family		CTAG Family	
*Sau*3A	/GATC	*Mae*I	C/TAG
*Bgl*II	A/GATCT	*Spe*I	A/CTAGT
*Bam*HI	G/GATCC	*Nhe*I	G/CTAGC
*Bcl*I	T/GATCA	*Avr*II	C/CTAGG
*Xho*II	(A/G)/GATC(T/C)	*Xba*I	T/CTAGA

different in the flanking nucleotides. The best known enzyme family is the GATC-family, whose members are *Sau*3AI, *Bgl*II, *Bam*HI, *Bcl*I and *Xho*II (Table 5) (Kessler and Höltke 1986). All these enzymes produce single-stranded ends of the sequence GATC. Whereas the respective hexanucleotide recognition sequences are lost after recombination of fragments produced by different enzymes of this family, the common internal tetranucleotide remains in all possible combinations. In this way, for example, *Bgl*II fragments which are cloned into the *Bam*HI site of pBR322 DNA can be recovered from the vector by cutting with *Sau*3AI. Another enzyme family exists for the sequence CTAG with the enzymes *Mae*I, *Spe*I, *Nhe*I, *Avr*II and *Xba*I (Kessler and Höltke 1986).

5 Changes in Sequence Specificity

For many enzymes such as *Eco*RI, a relaxation of the specificity occurs under altered solvent conditions (Appendix 2, see p. 267) (Hsu and Berg 1978; Tikchonenko et al. 1978; Malyguine et al. 1980; Woodhead et al. 1981; George and Chirikjian 1982; Shinomiya et al. 1982; Fuchs and Blakesley 1983). This ability of *Eco*RI is known as *Eco*RI* activity. Relaxed specificity results in a marked increase in the number of small fragments, whereas the large fragments disappear (Fig. 4). From sequencing data it was deduced that the subcanonical sequences differ from the canonical recognition sites by at least one base pair (Gardner 1982). Relaxation of both flanking positions yield as the predominant *Eco*RI* specificity the tetranucleotide AATT. However, relaxation of the internal positions has also been observed.

The experimentally measured hydrolysis rates at all possible subcanonical sequences have been categorized by the hierarchical arrangement of base pairs GC $\gg$ AT $\sim$ TA $\gg$ CG at the flanking position and AT $\gg$ GC,CG $\gg$ TA at the two internal positions of the half-side sequence GAA being in contact with each *Eco*RI monomer (Rosenberg and Greene 1982). This observed hierarchy is explained by the loss of one or two hydrogen bonds predominantly between particular amino acid side chains of the *Eco*RI monomer and the 6-O and N^7 position of the flanking guanine or the 6-NH_2 and N^7 position of the internal adenine residues.

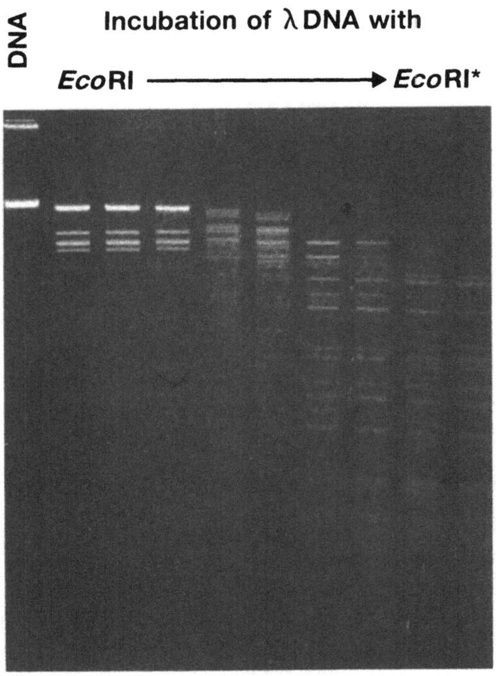

EcoRI incubation conditions

100 mM Tris-Hcl, pH 7.5
50 mM NaCl
10 mM $MgCl_2$
0 % glycerol
1 unit EcoRI

Reaction time 1 hour

EcoRI* incubation conditions

25 mM Tris-Hcl, pH 8.5
2 mM $MgCl_2$
12 % glycerol
100 units EcoRI

Reaction time 16 hours

Fig. 4. Relaxation of *Eco*RI specificity GAATTC both by decreased ionic strength and by increased pH value or increased concentration of the enzyme as well as glycerol. Under normal conditions the GAATTC-specific *Eco*RI fragment pattern of λ DNA is generated. By relaxation of the specificity additional cleavage sites with shorter recognition sequences (e.g. AATT) are recognized and cleaved by *Eco*RI. As a result smaller fragments are formed under *Eco*RI* conditions

Relaxation of the specificity of class II restriction endonucleases has generally been attributed to the effects of an altered reaction environment on the nature of complexes formed between the DNA helix and the *Eco*RI dimer. As the *Eco*RI* is an inherent enzyme property, it cannot be eliminated by more extensive purification. However, relaxation of the specificity can be suppressed by using suitable buffer conditions which counteract the relaxation effect (Tikchonenko et al. 1978; Malyguine et al. 1980; Woodhead et al. 1981). The reaction conditions that prevent relaxation are characterized by

1. high ionic strength;
2. decreased pH value;
3. use of Mg^{2+} ions rather than Mn^{2+}, Co^{2+} or Zn^{2+} ions;
4. low enzyme concentrations;
5. low concentrations of glycerol or other organic solvents such DMSO, which destabilize the double-stranded DNA helix;
6. short incubation periods.

In addition, the incubation temperature is also an important parameter for preventing or favouring relaxed specificity. As can be shown for *Pst*I, when increasing the incubation temperature from 37° to 42°C the relaxed specificity can be suppressed nearly totally even with higher concentrations and longer incubation periods. In contrast, lowering the incubation temperature to 30°C strongly favours *Pst*I* activity. After further lowering of the incubation temperature to 25° or 20°C the relaxed activity becomes predominant even under normal incubation conditions. Analogous observations have been made for the enzymes *Bam*HI, *Bst*EII, *Eco*RI, *Sph*I, *Sal*I and *Taq*I (Rexer, unpubl. results).

Additional changes in the activity and specificity of class II restriction endonucleases are attributed to DNA modifications like DNA methylation (McClelland and Nelson 1985; Kessler and Höltke 1986). Certain enzymes are inhibited not only by the corresponding methylase counterpart, but also by DNA methylation mediated by site-specific *E. coli dam* [GATC] or *dcm*I methylase [$CC(\frac{A}{T})GG$] (May and Hattman 1975; Urieli-Shoval et al. 1983). These two methylases are not part of restriction modification systems but have other biological functions like the regulatory role of *dam* methylase in mismatch repair and DNA replication (Meijer et al. 1979; Sugimoto et al. 1979). In addition, *dcm*II, *dcm*III and *dcm*IV are also known in *E. coli* as minor species (Hattman 1981).

Total overlapping of the recognition sequence of *dam* or *dcm*I methylase and particular class II restriction endonucleases can result in complete inhibition of activity on DNA isolated from *E. coli* (Appendix 3, see pp. 268–269). Well-known examples of different *dam* influences at N^6-methylated adenine residues at GATC sequences are the already mentioned isoschizomers *Dpn*I, *Mbo*I and *Sau*3AI (Gelinas et al. 1977a; Lacks and Greenberg 1977; Dreiseikelmann et al. 1979). Similar considerations apply for the *dcm*I methylase which inhibits *Eco*RII but activates *Apy*I by 5C-methylation of the internal cytosine residue (Boyer et al. 1973; Gruenbaum et al. 1981a). However, *Bst*NI is not affected by this particular methylation (Gruenbaum et al. 1981a; Petrusyte and Janulaitis 1982). *Dam* or *dcm*I inhibitory effects can be eliminated by isolation of the DNA from *E. coli* cells in which both methylase genes have been inactivated by mutation (May and Hattman 1975; Hattman et al. 1978a; Hattman 1981).

Partial overlapping with the *dam* recognition sequence which is observed for a variety of enzymes, also frequently causes inhibition of activity (McClelland et al. 1984; Nelson et al. 1984). As an example, *Cla*I digestion will be inhibited after *dam* methylation specifically at the heptanucleotide sequences GATCGAT or ATCGATC (Mayer et al. 1981). These overlapping sequences occur on λ DNA at positions 15,584, 31,991 and 32,964. All the other *Cla*I sites are not affected. Therefore, by sequential action of isolated *dam* methylase and *Cla*I endonuclease a new octanucleotide sequence specificity is created:

$$\begin{array}{c}(A)\\(G)\\(T)\end{array}ATCGAT\begin{array}{c}(T)\\(G)\\(A)\end{array}$$

In contrast, the above mentioned heptanucleotides will be specifically cleaved with endonuclease *Dpn*I after methylation of DNA with *Cla*I methylase (McClelland 1981). Using this approach the other *Dpn*I cleavage sites remain stable. In this way, both enzyme pairs complement each other in creating new sequence specificities.

Another approach for creating novel cleavage sites applies the permutation of generated fragment ends with 5′-protruding termini (Panayotatos 1984). Sites such as *Eco*RI, *Bam*HI or *Sal*I can be rendered blunt-ended by filling-in the protruding ends with deoxynucleoside triphosphates and Klenow enzyme. Ligation creates a new symmetry center which corresponds to the ligation point of the two filled-in ends. If, for example, a *Sal*I site (G/TCGAC) is subjected to treatment, a 10-base-pair palindromic sequence GTCGATCGAC with a novel *Pvu*I site (CGAT/CG) has been created. After *Pvu*I cleavage a third restriction site may be generated by treatment with nuclease S1. After ligation a novel cleavage site is created again because the resulting octanucleotide GTCGCGAC contains a novel *Nru*I cleavage site (TCG/CGA). Analogous conversions of restriction sites hold true for all the other enzymes generating fragments with 5′-protruding ends. Interconversion of restriction sites could generate unique novel cloning sites without the need of synthetic linkers. This should improve the flexibility of genetic engeneering experiments.

Bacteriophage-induced DNA modifications are also known to influence strongly the activity of class II restriction endonucleases. The respective phage genomes contain all or most of the cytosine or thymine residues substituted at the 5-C or 5-T positions (Huang et al. 1982). Substitution of cytosine by 5-methyl-cytosine (*Xanthomonas oryzae* phage XP12), glycosylated 5-hydroxymethyl-cytosine (*Escherichia coli* phage T4) or thymine by phosphoglucuronated and glycosylated 5-(4′,5′-dihydroxy-pentyl)-uracil (*Bacillus subtilus* phage SP15) renders the DNA resistant to almost all of the analyzed class II restriction endonucleases. The only exception is *Taq*I, which fragmentates the modified DNAs extensively. Substitution of thymine by 5-hydroxy-uracil or uracil (*Bacillus subtilis* phages SPO1 or PBS1) results only in a reduced cleavage rate but not in a complete inhibition of the various enzyme activities.

Other inhibitors of class II restriction endonucleases are DNA-binding agents like ethidium-bromide, actinomycin D, proflavine, distamycin A or neotrypsin (Goppelt et al. 1981; Nilsson et al. 1982; Österlund et al. 1982). Spermine or spermidine also have an inhibitory effect in high concentrations, whilst low concentrations of these substances have a stimulatory effect. This stimulation is similar to that observed after addition of proteins like *E. coli* RNA polymerase or T4 gene 32 protein which bind tightly to single-stranded termini generated by many class II restriction endonucleases (Pingoud et al. 1984; Dombroski and Morgan 1985; Kuosmanen and Pösö 1985; Pingoud 1985).

6 Novel Class II Restriction Endonucleases

New sequence specificities are also obtained by screening for novel class II restriction endonucleases.

The first method for finding new enzymes consists of examining those bacterial strains which are resistant to certain bacteriophages. As an alternative procedure, representatives of previously unexamined families from the wide range of eu- and archaebacteria are screened systematically for the presence of new class II restriction endonucleases. As an example, in a total of 348 microorganisms screened, 105 novel enzymes were found predominantly in Gram-negative bacteria (Kessler et al. 1986). To detect the novel specificities, aliquots of bacterial extracts are incubated with a DNA of known sequence like λ DNA for various periods under standard incubation conditions. The fragment mixtures are analyzed on agarose gels.

Experimentally determined (E) and computer derived (C) *Dra*II fragment lengths (bp)

λ DNA E	λ DNA C	SV40 DNA E	SV40 DNA C
26.000	25.982	3.000	3.033
20.000	19.676	1.700	1.671
2.800	2.816	550	539
28*	28		

pBR322 DNA E	pBR322 DNA C	pBR 328 DNA E	pBR 328 DNA C
2.850	2.863	2.700	2.731
900	915	2.150	2.176
550	543		
42*	42		

a

Cleavage pattern of *Dra*I, *Dra*II and *Dra*III

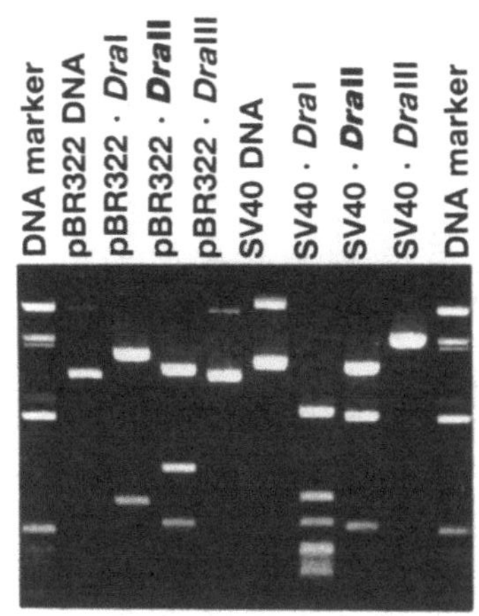

b

***Dra*II cleavage position**

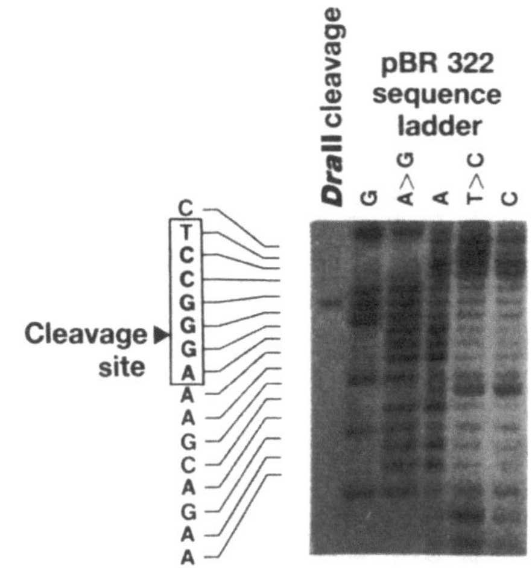

c

* Length determination in sequencing gels

Fig. 5a–c. Determination of the recognition and cleavage specificity of *Dra*II. **a** Comparison of the calculated and experimentally observed *Dra*II fragment lengths of λ, pBR322 and SV40 DNA. **b** Gel electrophoresis separation of the fragment mixture after digestion of pBR322 and SV40 DNA with *Dra*I, *Dra*II or *Dra*III. The lengths of the individual fragments can be determined by comparison with DNA marker fragments of known molecular weight. **c** Determination of the *Dra*II cleavage position within one of the four recognition sequences on pBR322 DNA. The *Dra*II cleavage position is obtained by comparison of the *Dra*II-specific band with the bands of the pBR322 sequence ladder representing the recognition sequence AGGGCCT. (Data from Grosskopf et al. 1985)

The determination of a new sequence specificity is performed with computer assistance by correlating experimentally observed and predicted fragment patterns (Kessler, unpubl. results). In a following step the corresponding fragments obtained with other DNAs, for example, pBR322 or SV40 DNA, are analyzed. In a third step, the precise cleavage positions with the recognition sequence are fixed in sequencing gels by accurate fragment length determination of both strands of a particular DNA fragment created with the novel enzyme.

The three steps of characterization of the specificity and cleavage positions of a new enzyme are exemplified by means of the recently discovered class II restriction endonuclease *Mae*I (Schmid et al. 1984). This new enzyme can be isolated from the archaebacterium *Mathanococcus aeolicus* PL15/H along with *Mae*II and *Mae*III found to be also novel specificities. *Mae*I recognizes the tetranucleotide sequence C/TAG and cleaves both DNA strands specifically between the 5′-flanking cytosine and thymine nucleotides. *Mae*II recognizes also a tetranucleotide sequence (A/CGT), whereas *Mae*III acts on a pentanucleotide sequence on the pentanucleotide sequence GTNAC.

Novel class II restriction endonucleases with more complex recognition sequences are *Dra*II and *Dra*III which were recently isolated in addition to the already known main activity *Dra*I as minor species from *Deinococcus radiophilus*. *Dra*II recognizes a novel type of a heptanucleotide $[(^{G}_{A})G/GNCC(^{C}_{T})]$ with ambiguities in the flanking trinucleotides (Fig. 5a–c). *Dra*III is characterized by a nonanucleotide [CACNNN/GTG] with three internal undefined nucleotides (Grosskopf et al. 1985; Wit et al. 1985).

These examples illustrate that enzymes with novel sequence specificities may be found by screening still unexplored parts of the microbial world. The search for new enzymes complement the various described methods for the generation of novel sequence specificities which are essential for the highest flexibility in the construction of recombinant DNA.

7 Mechanisms of Action

The mechanism of action of class II restriction endonucleases was analyzed in greatest detail for the cleavage of DNA with *Eco*RI. In these studies either a plasmid DNA or a short oligonucleotide of varying length is used in most cases as model substrate.

7.1 Molecular Reaction Course

By analysis of the kinetic data of the *Eco*RI cleavage reaction under various incubation conditions the following reaction mechanism was deduced (Modrich and Zabel 1976; Rubin and Modrich 1978; Hinsch and Kula 1981).

$$\begin{array}{ccccccccc} E + S & \underset{k_{-1}}{\overset{k_1}{\rightleftharpoons}} & E - S & \underset{k_{-2}}{\overset{k_2}{\rightleftharpoons}} & E - I & \underset{k_{-3}}{\overset{k_3}{\rightleftharpoons}} & E - P & \underset{k_{-4}}{\overset{k_4}{\rightleftharpoons}} & E + P \\ & & & & k_{-5} \Big\uparrow\Big\downarrow k_5 & & & & \\ & & & & E + I & & & & \end{array}$$

E = enzyme; S = DNA substrate; I = open-nicked intermediate; P = DNA product.

The kinetic data indicate that specific cleavage of the two phosphodiester bonds consists of four different reaction steps:

1. Binding of the enzyme E to specific recognition sites of the DNA substrate S by facilitated diffusion or an intersegment transfer mechanism after unspecific binding to random sequences.
2. Cleavage of the first phosphodiester bond resulting in an open-nicked intermediate I.
3. Cleavage of the second phosphodiester bond resulting in the completely digested DNA product P.
4. Dissociation of the enzyme from the final DNA product P.

As a parallel reaction to product formation dissociation of the enzyme from the intermediate also occurs as shown with pBR322 DNA as substrate.

Inspection of the reaction products after *Eco*RI hydrolysis of a synthetic octamer with the chiral phosphothionate group at the cleavage position within the *Eco*RI hexanucleotide sequence GAATTC indicates that the hydrolysis reaction proceeds with an inversion of the configuration of phosphorus. This result is compatible with a direct enzyme-catalyzed nucleophilic attack of H_2O at phosphorus without involvement of a covalent enzyme intermediate (Connolly et al. 1984).

By analysis of the reaction rates at 37°C under steady-state conditions the Michaelis-Menten constant K_M was calculated to be between $0.5-10 \times 10^{-9}$ varying with the nature of the substrate and ionic strength. For the turnover number values between 1 and 8 min^{-1} were observed per enzyme dimer under analogous conditions (Modrich and Zabel 1976; Berkner and Folk 1977; Rubin and Modrich 1978; Hinsch and Kula 1981; Langowski et al. 1981; Woodhead et al. 1981).

Measurement of the *Eco*RI cleavage rates on pBR322 DNA in the initial phase of the reaction shows that the rate-limiting step of the overall reaction is strongly dependent on the enzyme concentration (Langowski et al. 1981). Separate determination of the catalytic constants for the cleavage of the first and second phosphodiester bond was possible by fast quench-flow experiments. At enzyme concentrations higher than the concentration of the specific *Eco*RI sites, the dissociation reaction of the enzyme from the final DNA product becomes rate-limiting. Under these conditions the reaction constants of both cleavage reactions reaches its maximum. However, at enzyme concentrations comparable or lower than substrate the first cleavage reaction becomes rate-limiting. The difference is explained by the observation that at low enzyme concentrations a significant fraction of the enzyme is trapped by non-specific binding at random sequences and thus become inactivated (Langowski et al. 1980, 1981).

Specific and non-specific binding of *Eco*RI occurs in the presence as well as absence of the equivalent Mg^{2+} ions. In the presence of Mg^{2+} ions specific binding is favoured by at least a factor of 50 (Jack et al. 1980; Rosenberg et al. 1980; Woodhead and Malcolm 1980; Gardner et al. 1982; Jen-Jacobsen et al. 1983; Terry et al. 1983). Even in the absence of Mg^{2+} ions *Eco*RI is able to form site-specific complexes with DNA which are significantly reduced by *Eco*RI methylation at the central adenine residue of its recognition site (Berkner and Folk 1977; Jack et al. 1980; Rosenberg et al. 1980; Woodbury et al. 1980; Terry et al. 1983). These specific complexes found in the absence of equivalent cations are rather stable, since they dissociate at 37°C with a linear molecule of 6,200 base pairs with a first-order rate constant K_{-1} of 7×10^{-4} s^{-1}

corresponding to a half-life of $t_{1/2}$ = 16 min (Jack et al. 1980, 1982). A further stabilization of the enzyme-DNA complexes is achieved by a marked reduction of the flanking sequences. A 34-base-pair oligomer containing the *Eco*RI site is nearly tenfold more stable ($K_{-1} = 8 \times 10^{-5}\ s^{-1}$; $t_{1/2}$ = 140 min) (Jack et al. 1982). For non-specific binding of *Eco*RI to random sequences in the absence of Mg^{2+} ions dissociation constants between 10^{-3} and 10^{-6} M have been measured (Langowski et al. 1981; Woodhead and Malcolm 1980).

The nature of unspecific interactions is not yet understood. However, unspecific binding is thought to be the initial event of *Eco*RI-DNA interaction, whereas in the following step the *Eco*RI-specific hexanucleotide sequence GAATTC is recognized by one-dimensional facilitated diffusion or intersegment DNA transfer (Langowski et al. 1983; Terry et al. 1985). For *Eco*RI the mean diffusion length was determined to be about 1,000 base pairs at 1 mM $MgCl_2$ (Ehbrecht et al. 1985). This suggests a hybride mechanism in which transfers between distal segments by dissociation and reassociation processes are coupled with sliding of the enzyme over short distances (Fried and Crothers 1984).

7.2 Enzyme-DNA Interactions

The majority of class II restriction endonucleases exists in solution as oligomers of identical subunits and often several aggregation states like dimers and tetramers have frequently been observed (Modrich and Zabel 1976; Rubin and Modrich 1980; Woodhead and Malcolm 1981). The only exceptions so far examined are *Bgl*I and *Bst*I where only monomeric species have been found (Kleid et al. 1976; Lee and Chirikjian 1979). In the case of *Eco*RI an equilibrium constant of about 10^{-7} M was determined for the tetramer/dimer transition. At the analytical concentration of 10^{-10} M the *Eco*RI dimer is the stable confirmation of the protein in solution. Gel-filtration analysis of *Eco*RI-DNA complexes also indicates that the dimer is the active species for a site-specific cleavage of DNA (Modrich and Zabel 1976; Rubin and Modrich 1978, 1980; Jen-Jacobsen et al. 1983).

Analysis of the marked effects of salt concentration on the equilibrium binding constant indicates that in the specific attachment of the *Eco*RI dimer both electrostatic and non-electrostatic interactions between certain amino acids of the two subunits and the DNA helix contribute to specific binding (Lu et al. 1981; Jen-Jacobsen et al. 1983; Poulsen et al. 1985):

1. Formation of coordinated ion pairs between basic amino acid residues and negatively charged phosphate groups of the DNA backbone.
2. Formation of hydrogen bonds between amino acid residues with hydrogen donor and acceptor capabilities and particular bases of the *Eco*RI-specific recognition sequences GAATTC.

Alkylation interference and protection methods have been utilized to deduce both electrostatic and non-electrostatic DNA contacts involved in specific complex formation between *Eco*RI and its hexanucleotide recognition sequence (Lu et al. 1981). The analysis of ion-pair formation by ethylation interference suggests four major

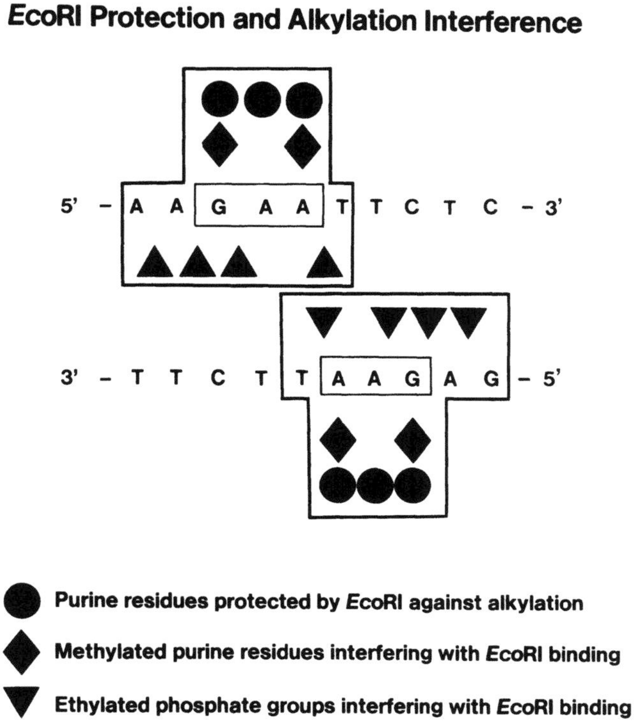

Fig. 6. Protection of the base residues and phosphate groups after specific complex formation between *Eco*RI and the recognition sequence GAATTC. The nucleotide triplets GAA specifically interact with *Eco*RI. After *Eco*RI binding only bases within this triplet are protected from methylation with dimethyl sulphate (●). In addition, specific binding of *Eco*RI to the recognition sequence is inhibited by methylation with dimethylsulphate (♦). Specifically bound *Eco*RI protects certain phosphate groups of the DNA backbone from ethylation with ethyl-nitroso-urea also outside of the recognition sequence (▾). (Data from Lu et al. 1981)

phosphate contacts between each *Eco*RI subunit and each DNA strand (Fig. 6). Thus, a total of eight ion pairs within a stretch of ten base pairs is formed with each *Eco*RI dimer. Two of the four phosphate groups protected by each *Eco*RI subunit are located outside of the enzyme-specific recognition sequence. Quantitative analysis of the protective actions shows that the flanking phosphate groups are the ones most strongly protected by the bound enzyme. This protein-DNA interaction outside of the recognition sequence strengthens the marked influence of the flanking sequences which had already been observed in the different cleavage frequencies at individual cleavage sites.

Mutation of *Eco*RI at the arginine 187 position abolishes six of the eight ion-pair formations of the protein dimer. The maintenance of the non-electrostatic component of the binding as well as the normal catalytic activity shows that the affected ionic interactions are highly localized and act in a coordinated manner during the interaction with DNA (Greene et al. 1981; Jen-Jacobsen et al. 1983).

Information on the nature of sequence-specific interaction between the *Eco*RI dimer and particular bases of the *Eco*RI recognition sequence was obtained by methylation protection and interference studies (Lu et al. 1981). Analysis of the protection of particular bases against methylation with dimethyl sulphate clearly demonstrates that each of the two *Eco*RI subunits protects only the GAA triplet of the recognition sequence (Fig. 6). Bases flanking this triplet are not protected. Analogous results are obtained in interference experiments. This suggests that only the bases of the GAA triplet are responsible for sequence-specific complex formation. Quantitative analysis shows that N^7-methylation of the outer guanine residue projecting into the major groove of the DNA helix causes greatest interference and protection. The N^3-position of both internal adenine residues also contributes to interference and protection. However, the 2-NH_2-group of the flanking guanine residue exposed in the minor groove and the 5-CH_3-groups of the four thymine residues are non-essential for sequence recognition (Modrich and Rubin 1977; Halford and Johnson 1980).

A representation of the spatial distribution of the contact points along the helix leads to the observation that most implicated determinants are accessible only from one side of the helix. Implicated major groove determinants are also present on this surface, and potential minor groove contacts are located just behind the sugar-phosphate backbone (Siebenlist et al. 1980). The physical proximity of the contact points of the *Eco*RI dimer and particular base residues predominantly within the major groove of the DNA helix indicates that only a small number of amino acid side chains is involved in forming the necessary hydrogen bonds. This is consistent with the observation, that two α-helices of each *Eco*RI monomer are inserted laterally into the major groove and that the recognition contacts result from a limited number of side chains of these α-helices (Greene et al. 1981; Newman et al. 1981; Manavalan et al. 1984; Poulsen et al. 1985). The binding of the two *Eco*RI subunits within the major groove also explains the inhibitory effects of the 6-NH_2-methylation at the central adenine residue by *Eco*RI DNA-methyltransferase. This methyl group is also exposed into the major groove so that the affinity of the restriction endonuclease to the modified recognition sequence is considerably reduced (Modrich 1979; Jack et al. 1980; Rosenberg et al. 1980). A further indication that the *Eco*RI dimer binds to the major groove of the DNA helix is deduced from the resistance of modified bacteriophage T4 or XP12 DNA to *Eco*RI cleavage (Kaplan and Nierlich 1975; Berkner and Folk 1977; Huang et al. 1982). This resistance is caused by 5C-modified cytosine residues which are located in the major groove. Binding of *Eco*RI exclusively to the major groove was also deduced by the examination of the hydrogen bonds which are lost on the relaxation of sequence specificity (Gardner et al. 1982; Rosenberg and Greene 1982).

7.3 Crystallographic Analysis

The binding model has been confirmed by X-ray structure analysis (Rosenberg et al. 1978; Young et al. 1981; Frederick et al. 1984; Grable et al. 1984). Crystals have been obtained for both purified *Eco*RI and complexes between the enzyme and short oligonucleotides. With GAATTC as oligonucleotide the unit cell contains four endo-

Structure of EcoRI·TCGCGAATTCGCG Complex

Fig. 7. Electron density map of the crystalline complex between *Eco*RI dimers and the synthetic, double-stranded tridecamer

5′-TCGC*GAATTC*GCG-3′
3′-AGCG*CTTAAG*CGC-5′.

The two *Eco*RI subunits are located exactly within the major groove of the DNA helix. Interaction between the two subunits and the bases occurs only in the region of the two GAA-triplets. In contrast, the DNA backbone is covered by the two subunits also in the flanking regions. Partial unwinding of the DNA helix resulting in kinks is a prerequisite for specific binding of the *Eco*RI dimer resulting in a stable enzyme-DNA complex. (Data from Frederick et al. 1984; courtesy of J.M. Rosenberg)

nuclease monomers and two duplex DNA fragments in an assymetric unit. This stoichiometry of one *Eco*RI dimer per duplex recognition sequence reflects the mode of specific action of the enzyme with DNA in solution (Young et al. 1981). Analogous results have been deduced from X-ray analysis of the co-crystalline recognition complex between *Eco*RI and the synthetic tridecamer

5′-TCGC*GAATTC*GCG-3′
3′-AGCG*CTTAAG*CGC-5′.

*Eco*RI and the oligonucleotide co-crystallize in the space group P321 as already observed with the DNA hexamer with four monomer subunits and two DNA duplexes in an assymetric unit (Frederick et al. 1984). The 3 Å electron density map shows that *Eco*RI interacts with the DNA in two different ways (Fig. 7):

1. Direct sequence-specific interaction with the bases of the recognition sequence. This interaction is mediated by hydrogen bond formation between the enzyme and DNA.
2. Indirect sequence-selective interaction with the flanking region of different base compositions. The actual binding energy and thus the particular reaction rate at the various cleavage sites is modulated by this interaction via ion-pair formation and sequence-dependent differences in the DNA conformation.

A more precise analysis of the structure confirms that *Eco*RI binds in form of a dimer with rotational symmetry to the symmetric recognition sequence of the DNA helix. Contacts between each subunit and the DNA helix are found exclusively within the major groove. The segments of the backbone additionally covered by both enzyme subunits are located in deep semi-cylindrical grooves of the protein. These enzyme grooves, however, cover not only those phosphate groups flanking the hydrolyzed phosphodiester bonds, but also those outside the recognition sequence. The X-ray analysis also shows that by binding of the *Eco*RI dimer the compact B-form of the DNA helix is markedly disturbed. This helix unwinding results in permanent torsional kinks. These DNA kinks are caused by bending the DNA helix left and right to the GAA triplet by 12° or 23° respectively. Owing to these enzyme-induced torsional kinks the two *Eco*RI subunits fit precisely into the expanded major groove. The observation of kinks in *Eco*RI-DNA co-crystals is in accordance with the definitive distortion of the DNA in the specific *Eco*RI-DNA recognition sequences in solution. No measurable distortion is measured in the non-specific interaction of *Eco*RI and DNA (Kim et al. 1984).

The predominant structural elements within the two protein subunits involved in protein-DNA interaction are two exposed α-helices. Within each subunit the two α-helices are arranged with rotational symmetry to each other. Certain amino acid side chains of this helix superstructure are in close contact to both DNA strands. Two exposed amino acid side chains again arranged with rotational symmetry appeared to be aspecially important. The exact nature of the involved amino acid side chains is, however, not yet known.

In summary, X-ray analysis of the *Eco*rRI-DNA complex shows that at least for this enzyme and probably also for other class II restriction endonucleases, specific recognition of palindromic DNA elements has its origin in the molecular structure of the protein subunits. The structural motive of a helical superstructure has been also described to be the decisive structural element for other sequence-specific binding proteins like the λ-repressor, cro-protein and CAP (catabolite gene activator protein) (Anderson et al. 1981; McKay and Steitz 1981; McKay et al. 1982; Pabo and Lewis 1982). In these cases, too, the formation of hydrogen bonds between side chains of two exposed α-helices and particular bases of partially palindromic recognition sequences are responsible for the site-specific binding of the regulatory proteins to the DNA helix (Ohlendorf et al. 1983; Takeda et al. 1983).

The elucidation of further crystal structures of complexes between DNA and class II restriction endonucleases of other sequence specificity will show whether the formation of symmetrically arranged enzyme dimers with two exposed α-helices is a more general structural principle for sequence-specific binding of proteins to DNA.

Acknowledgements. The author is grateful to R. Becker-Hirth for helpful discussions, H. Effgen, H. Greb-Emrich and S. Starl for proving the references and S. Metzger-Bergel for typing the manuscript.

Appendix 1. Recognition sequences of important class II restriction endonucleases and methylases

Position No.	Recognition sequence[a]	Enzyme[b]	Number of recognition sites[c] λ	Ad2	SV40	φX174	M13mp7	pBR322	pBR328	Microorganism[d]	Reference(s)[e]	Commercial[f] availability
A. Class II enzymes with palindromic recognition sequences												
A.1 Tetra-, hexa- or octanucleotide recognition sequences												
A.1.1 Internal AT-palindromes												
1	AATT	–	188	0	39	25	65	8	13	–	–	–
2	AAATTT	–	16	13	4	5	3	0	0	–	–	–
3	o+ + G/AATTC	*Eco*RI	5	5	1	0	2	1	1	*Escherichia coli* RY13	Jack et al. (1980), Kessler and Höltke (1986)	A, B, C, D, E, F
	M GAATTC	M·*Eco*RI	5	5	1	0	2	1	1	*Escherichia coli* RY13	Jack et al. (1980), Grandoni and Schildkraut in Kessler and Höltke (1986)	D
4	CAATTG	–	8	4	4	1	0	0	0	–	–	–
5	TAATTA	–	8	4	2	1	3	0	0	–	–	–
6	M GA/TC	*Dpn*I	116	87	8	0	8	22	27	*Diplococcus pneumoniae*	Lacks and Greenberg (1975, 1977)	B, C, D, E
	+ o /GATC	*Mbo*I	116	87	8	0	8	22	27	*Moraxella bovis*	Gelinas et al. (1977a), Dreiseikelmann et al. (1979)	C, D, F

a–f Footnotes see p. 253.

Appendix 1 (continued)

Position No.	Recognition sequence[a]	Enzyme[b]	Number of recognition sites[c] λ	Ad2	SV40	φX174	M13mp7	pBR322	pBR328	Microorganism[d]	Reference(s)[e]	Commercial[f] availability
	+ GATC ·····	*Nde*II	116	87	8	0	8	22	27	*Neisseria denitrificans*	Watson et al. (1982), Visentin et al. in Kessler and Höltke (1986)	B, C
	o + /GATC ·····	*Sau*3AI	116	87	8	0	8	22	27	*Staphylococcus aureus*	Sussenbach et al. (1976), Dreiseikelmann et al. (1979)	A, B, C, D, E
	M GATC ·····	M·*Eco dam*	116	87	8	0	8	22	27	*Escherichia coli* HB101	Hattman et al. (1978), Brooks et al. (1983), Smith and Kelly (1984)	D
7	o + A/GATCT ·····	*Bgl*II	6	11	0	0	1	0	0	*Bacillus globigii*	Pirrotta (1976), Duncan et al. (1978)	A, B, C, D, E, F
8	o + o ······· G/GATCC	*Bam*HI	5	3	1	0	2	1	1	*Bacillus amyloliquefaciens* H	Wilson and Young (1975), Roberts et al. (1977)	A, B, C, D, E, F
	M ··· GGATCC ·····	M·*Bam*HI	5	3	1	0	2	1	1	*Bacillus amyloliquefaciens* RUB500	Hattman et al. (1978b), Nardone et al. (1984)	D
	o + ··· (A)/GATC(T) (G) (C) ·····	*Xho*II	21	22	3	0	4	8	7	*Xanthomonas holcicola*	Olson in Kessler and Höltke (1986)	B, D
9	o + CGAT/CG ······	*Pvu*I	3	7	0	0	1	1	1	*Proteus vulgaris*	Gingeras et al. (1981)	A, B, C, D, F

10	T/GATCA (+ over T, o over C; GATC dotted underlined)	*Bcl*I	8	5	1	0	0	0	1	*Bacillus caldolyticus*	Bingham et al. (1978)	B, C, D, F
11	CATG/	*Nla*III	181	183	16	22	15	26	27	*Neisseria lactamica*	Qiang et al. in Kessler and Höltke (1986)	D
12	ACATGT	–	2	9	0	0	3	1	0	–	–	–
13	GCATG/C (o over C)	*Sph*I	6	8	2	0	0	1	1	*Streptomyces phaeochromogenes*	Fuchs et al. (1980), Kessler et al. (1981)	B, C, D, E, F

[a] Only important class II restriction endonucleases and methylases are listed. Recognition sequences with a dyad symmetry are specified only for the (5'→3') strand. The cleavage site of a restriction endonuclease is indicated by the *slash symbol*. For class IIS restriction endonucleases recognizing sequences without a dyad symmetry and cleaving outside of their recognition sequences, the sequences for both strands are listed. In these cases the sites of cleavage are identified for both strands by *small numbers* as *subscripts* after the recognition sequence. For methylases the position of methylation is marked by the symbol *M*.

The symbols $\overset{+}{A}$ or $\overset{+}{C}$ indicate the inhibition of a restriction endonuclease by a N^6-methyladenine or 5-methylcytosine residue within the recognition sequence. The symbols $\overset{m}{A}$ or $\overset{m}{C}$ indicate that N^6-methyladenine or 5-methylcytosine residues within the recognition sequence are a prerequisite for the enzymatic activity of the restriction endonuclease. The symbols $\overset{o}{A}$ or $\overset{o}{C}$ show that digestion of the DNA with the restriction endonuclease is not at all influenced by a N^6-methyladenine or 5-methylcytosine residue within the recognition sequence. For 4-methylcytosine residues see Janulaitis et al. (1982).

Dotted underlining indicates the recognition sequence of the *E. coli dam*-coded methylase M·*Eco dam* (5'-G$\overset{M}{A}$TC-3') or a part of this sequence. *Dotted overlining* marks the recognition sequence of the *E. coli dcm*I-coded methylase M·*Eco dcm*I (5'-C$\overset{M}{C}$$\binom{A}{T}$GG-3'), often also designated M·*Eco mec*, or a part of this sequence.

[b] The nomenclature of the enzymes refers to the original publications. Methylases are specified by the symbol *M*· in front of the enzymes' names. Only the most important isoschizomer(s) are listed which are in most cases commercially available.

[c] The numbers of recognition sites for each restriction endonuclease and methylase on the various DNAs have been obtained by computer analysis of the published DNA sequences of the bacteriophages λ, φX174 and M13mp7, the viruses Ad2 and SV40, strain 776 and the plasmids pBR322 and pBR328 (for references see Kessler and Höltke 1986).

[d] The strain refers to the microorganism used as the source for the isolation of the restriction endonuclease or methylase.

[e] The reference(s) refer to the isolation and sequence specificity of the restriction endonuclease or methylase.

[f] The availability of restriction endonucleases and methylases by one of the listed main commercial suppliers is specified within *square brackets: A*, Amersham Buchler, Buckinghamshire (UK); *B*, Boehringer Mannheim GmbH, Mannheim (FRG); *C*, BRL, Bethesda Research Laboratories Inc., Gaithersburg, Maryland (USA); *D*, NEBL, New England Biolabs, Beverly, Massachusetts (USA); *E*, PB, Promega Biotec, Madison, Wisconsin (USA); *F*, Pharmacia P-L Biochemicals Inc., Milwaukee, Wisconsin (USA).

Appendix 1 (continued)

Position No.	Recognition sequence[a]	Enzyme[b]	Number of recognition sites[c] λ	Ad2	SV40	φX174	M13mp7	pBR322	pBR328	Microorganism[d]	Reference(s)[e]	Commercial[f] availability
14	+ C/CATGG	*Nco*I	4	20	3	0	0	0	1	*Nocardia corallina*	Langdale et al. in Kessler and Höltke (1986)	B, C, D
	C/C(A)(T)/(T)(A)GG	*Sty*I	10	44	8	0	0	1	2	*Escherichia coli* KM201[pST27 hsd, S-a]	Mise and Nakajima (1985)	B, D
15	TCATGA (dots under GA)	–	8	3	2	3	1	4	4	–	–	–
16	TATA	–	113	65	17	11	16	8	13	–	–	–
17	ATATAT	–	11	3	0	0	4	1	3	–	–	–
18	GTATAC	*Sna*I	3	3	0	0	0	1	0	*Sphaerotilus natans* C	Pope et al. in Kessler and Höltke (1986)	–
19	CTATAG	–	0	4	2	1	0	0	0	–	–	–
20	TTATAA	–	12	4	3	1	2	0	0	–	–	–
A.1.2 Internal GC-palidromes												
21	+ AG/CT	*Alu*I	143	158	34	24	24	16	14	*Arthrobacter luteus*	Roberts et al. (1976a)	A, B, C, D, E, F
	M AGCT	M·*Alu*I	143	158	34	24	24	16	14	*Arthrobacter lutues*	Kramarov and Smolyaninov (1981)	D
22	+ + A/AGCTT	*Hind*III	6	12	6	0	0	1	1	*Haemophilus influenzae* R_d	Old et al. (1975)	A, B, C, D, E, F

23	o + GAGCT/C	*Sac*I	2	16	0	0	0	0	0	*Streptomyces achromogenes*	Arrand et al. in Kessler and Höltke (1986)	A, B, C, D, E, F
24	+ CAG/CTG	*Pvu*II	15	24	3	0	3	1	1	*Proteus vulgaris*	Gingeras et al. (1981)	A, B, C, D, E, F
25	TAGCTA	–	2	4	2	0	3	0	0	–	–	–
26	+o GG/CC	*Hae*III	149	216	18	11	15	22	30	*Haemophilus aegyptius*	Middleton et al. (1972), Bron and Murray (1975)	A, B, C, D, E, F
	M GGCC	M·*Hae*III	149	216	18	11	15	22	30	*Haemophilus aegyptius*	Mann and Smith (1977)	D
27	o AGG/CCT	*Stu*I	6	11	7	1	0	0	0	*Streptomyces tubercidicus*	Shimotsu et al. (1980)	A, B, C, D
28	+ GGGCC/C	*Apa*I	1	12	1	0	0	0	0	*Acetobacter pasteurianus* sub. pasteurianus	Seurinck et al. (1983)	B, D, E, F
	+ G(A)(G)GC(T)(C)/C	*Ban*II	7	57	2	0	1	2	2	*Bacillus aneurinolyticus*	Sugisaki et al. (1982)	B, D, F
29	+ C/GGCCG	*Xma*III	2	19	0	0	0	1	2	*Xanthomonas malvacearum*	Kunkel et al. (1979)	C, D
	+ o GC/GGCCGC	*Not*I	0	7	0	0	0	0	0	*Nocardia otitidiscaviarum*	Schildkraut et al. in Kessler and Höltke (1986)	B, D
	+ (T)(C)/GGCC(A)(T)	*Eae*I	39	70	0	2	3	6	7	*Enterobacter aerogenes*	Whitehead and Brown (1983), Jacobs and Brown, pers.commun.)	B

Appendix 1 (continued)

Position No.	Recognition sequence[a]	Enzyme[b]	Number of recognition sites[c] λ	Ad2	SV40	φX174	M13mp7	pBR322	pBR328	Microorganism[d]	Reference(s)[e]	Commercial[f] availability
30	TGG/CCA (+ above G; dots above CCA)	*Bal*I	18	17	0	0	1	1	1	*Brevibacterium albidum*	Gelinas et al. (1977b)	A, C, D
31	CG/CG	*Fnu*DII	157	303	0	14	18	23	24	*Fusobacterium nucleatum* D	Lui et al. (1979)	D
32	A/CGCGT (o above A)	*Mlu*I	7	5	0	2	0	0	0	*Micrococcus luteus*	Sugisaki and Kanazawa (1981)	A, B, C, D, F
	A/C(A/G)(T/C)GT	*Afl*III	20	25	0	2	3	1	0	*Anabaena flos-aquae*	Whitehead and Brown (1982)	B
33	G/CGCGC	*Bss*HII	6	52	0	1	0	0	0	*Bacillus stearothermophilus* H3	Langdale et al. in Kessler and Höltke (1986)	D
34	CCGC/GG (+ above first C)	*Sac*II	4	33	0	1	0	0	0	*Streptomyces achromogenes*	Arrand et al. in Kessler and Höltke (1986)	B, D, E
35	TCG/CGA (+ above A; dots below GA)	*Nru*I	5	5	0	2	0	1	2	*Nocardia rubra*	Comb and Schildkraut in Kessler and Höltke (1986)	B, C, D
36	TGCA	–	272	206	36	18	15	21	19	–	–	–
37	ATGCA/T	*Nsi*I	14	9	3	0	0	0	0	*Neisseria sicca*	Comb et al. in Kessler and Höltke (1986)	B, C, D
38	G/TGCAC	*Apa*LI	4	7	0	1	1	3	2	*Acetobacter pasteurianus*	Yamada and Murakami (1985)	–

	G(T)GC(A)/C	*Hgi*AI	28	38	0	3	2	8	7	*Herpetosiphon giganteus* HP1023	Brown et al. (1980b)	D
	+ . . . . (A) (T) G(G)GC(C)/C (T) (A)	*Bsp*1286	38	105	4	3	4	10	10	*Bacillus sphaericus*	Myers and Roberts in Kessler and Höltke (1986)	D
39	+ + CTGCA/G	*Pst*I	28	30	2	1	1	1	1	*Providencia stuartii* 164	Smith et al. (1976), Brown and Smith (1976)	A, B, C, D, E, F
	CTGCAG	M·*Pst*I	28	30	2	1	1	1	1	*Providencia stuartii* 164	Walder et al. (1981)	D
40	TTGCAA	–	13	23	3	2	2	1	1	–	–	
A.1.3 Internal CG-palidnromes												
41	+ A/CGT	*Mae*II	143	83	0	19	22	10	12	*Methanococcus aeolicus* PL-15/H	Schmid et al. (1984)	B
42	AACGTT	–	7	3	0	3	2	4	5	–	–	–
43	GACGT/C	*Aat*II	10	3	0	1	0	1	1	*Acetobacter aceti*	Sugisaki et al. (1982)	B, D, F
44	CACGTG	–	3	10	0	0	0	0	0	–	–	–
45	TAC/GTA	*Sna*BI	1	0	0	0	1	0	0	*Sphaerotilus natans*	Borsetti et al. in Kessler and Höltke (1986)	B, D
46	+ + GCG/C	*Cfo*I	215	375	2	18	25	31	25	*Clostridium formicoaceticum*	Makula and Meagher (1980)	B, C, E
	+ + GCG/C	*Hha*I	215	375	2	18	25	31	25	*Haemophilus haemolyticus*	Roberts et al. (1976b)	A, C, D, F

Appendix 1 (continued)

Position No.	Recognition sequence[a]	Enzyme[b]	Number of recognition sites[c] λ	Ad2	SV40	φX174	M13mp7	pBR322	pBR328	Microorganism[d]	Reference(s)[e]	Commercial[f] availability
	M GCGC	M·*Hha*I	215	375	2	18	25	31	25	*Haemophilus haemolyticus*	Mann and Smith in Kessler and Höltke (1986)	D
47	AGC/GCT	*Eco*47III	2	13	1	0	2	4	3	*Escherichia coli* RFL47	Janulaitis et al. (1983a)	–
48	o GG/CGCC (dots over CCC)	*Nar*I	1	20	0	2	1	4	5	*Nocardia argentiensis*	Comb et al. in Kessler and Höltke (1986)	B, C, D
	+ $^{(A)}_{(G)}$GCGC/$^{(T)}_{(C)}$ (dots over GC/(T))	*Hae*II	48	76	1	8	6	11	9	*Haemophilus aegyptius*	Roberts et al. (1975), Tu et al. (1976)	A, B, C, D, F
	+ + G$^{(A)}_{(G)}$/CG$^{(T)}_{(C)}$C (dots over G(T)C)	*Aha*II	40	44	0	7	1	6	7	*Aphanotheke halophytica*	Whitehead and Brown (1982)	D
	G/G$^{(T)(A)}_{(C)(G)}$CC (dots over (A)CC)	*Ban*I	25	57	1	3	6	9	12	*Bacillus aneurinolyticus*	Sugisaki et al. (1982)	B, D, F
49	CGCGCG	–	1	48	0	0	1	1	0	–	–	–
50	TGCGCA	*Fsp*I	15	17	0	1	0	4	3	*Fischerella* species	Szekeres (pers. commun.)	D
51	+ C/CGG	*Hpa*II	328	171	1	5	19	26	33	*Haemophilus parainfluenzae*	Sharp et al. (1973)	B, C, D, E, F

	M CCGG	M·*Hpa*II	328	171	1	5	19	26	33	*Haemophilus parainfluenzae*	Mann and Smith (1977)	B, C
	+ o C/CGG	*Msp*I	328	171	1	5	19	26	33	*Moraxella* species	Van Montagu et al. in Kessler and Höltke (1986)	B, C, D, F
	M CCGG	M·*Msp*I	328	171	1	5	19	26	33	*Moraxella* species	Jentsch et al. (1981)	D
52	ACCGGT	–	13	5	0	0	0	0	1	–	–	–
53	+ + GCC/GGC	*Nae*I	1	13	1	0	1	4	6	*Nocardia aerocolonigenes*	Comb and Wilson in Kessler and Höltke (1986)	B, D
	(A) (T) CCGG (G) (C)	*Cfr*10I	61	40	1	0	1	7	10	*Citrobacter freundii*	Janulaitis et al. (1983b)	–
54	+ CCC/GGG	*Sma*I	3	12	0	0	0	0	0	*Serratia marcescens* S_b	Endow and Roberts (1977)	A, B, C, D, E, F
	+o C/CCGGG	*Xma*I	3	12	0	0	0	0	0	*Xanthomonas malvacearum*	Endow and Roberts (1977)	B, D, F
	o + + (T) (A) C/ CG G (C) (G)	*Ava*I	8	40	0	1	1	1	1	*Anabaena variabilis*	Murray et al. (1976), Hughes and Murray (1980)	A, B, C, D, F
55	T/CCGGA ··	*Bsp*MII	24	8	0	0	0	1	1	*Bacillus* species M	Morgan (pers. commun.)	D
56	o + T/CGA	*Taq*I	121	50	1	10	14	7	13	*Thermus aquaticus* YTI	Sato et al. (1977)	B, C, D, E, F
	M TCGA ··	M·*Taq*I	121	50	1	10	14	7	13	*Thermus aquaticus* YTI	Sato et al. (1980), McClelland (1981)	D

Appendix 1 (continued)

Position No.	Recognition sequence[a]	Enzyme[b]	Number of recognition sites[c] λ	Ad2	SV40	φX174	M13mp7	pBR322	pBR328	Microorganism[d]	Reference(s)[e]	Commercial[f] availability
57	AT/CGAT (+ above C; dots under CGAT)	*Cla*I	15	2	0	0	2	1	1	*Caryophanon latum* L	Mayer et al. (1981)	B, C, D
	ATCGAT (M above C; dots under CGAT)	M·*Cla*I	15	2	0	0	2	1	1	*Caryophanon latum* L	McClelland (1981)	D
58	GTCGAC (+ +)	*Sal*I	2	3	0	0	2	1	1	*Streptomyces albus* G	Arrand et al. (1978)	A, B, C, D, E, F
	GT(T)(C)/(A)(G)AC (+)	*Hinc*II	35	25	7	13	2	2	2	*Haemophilus influenzae* R_c	Olson et al. in Kessler and Höltke (1986)	A, C, D, E, F
	GT(T)(C)/(A)(G)AC (+o)	*Hind*II	35	25	7	13	2	2	2	*Haemophilus influenzae* R_d	Smith and Wilcox (1970), Kelly and Smith (1970)	B
	GT/(A)(C)(T)(G)AC (+)	*Acc*I	9	17	1	2	2	2	1	*Acinetobacter calcoaceticus*	Zabeau and Roberts in Kessler and Höltke (1986)	A, B, C, D, F
59	C/TCGAG (+ +)	*Xho*I	1	6	0	1	0	0	0	*Xanthomonas holcicola*	Gingeras et al. (1978b)	A, B, C, D, E, F
60	TT/CGAA	*Fsp*II	7	1	0	0	0	0	2	*Fischerella* species	Szekeres (pers. commun.)	–

A.1.4 Internal TA-palindromes

Position No.	Recognition sequence	Enzyme	λ	Ad2	SV40	φX174	M13mp7	pBR322	pBR328	Microorganism	Reference(s)	Commercial availability
61	ATAT	–	230	81	21	20	42	9	19	–	–	–
62	AAT/ATT	*Ssp*I	20	5	6	1	6	1	2	*Sphaerotilus* species	Grandoni and Schildkraut in Kessler and Höltke (1986)	B, D

63	GAT/ATC (+ over T)	*Eco*RV	21	9	1	0	0	1	1	*Escherichia coli* J62[pLG74]	Kholmina et al. (1980), Schildkraut et al. (1984)	A, B, C, D, E
64	CA/TATG	*Nde*I	7	2	2	0	3	1	0	*Neisseria dentrificans*	Watson et al. (1982)	B, C, D
65	TATATA	–	4	5	0	0	1	1	1	–	–	–
66	GT/AC	*Rsa*I	113	83	12	11	18	3	4	*Rhodopseudomonas sphaeroides*	Lynn et al. (1980)	B, C, D
67	AGT/ACT	*Sca*I	5	5	0	0	0	1	2	*Streptomyces caespitosus*	Kita et al. (1985), Takahashi et al. (1985)	A, B, D, E
68	G/GTACC (+ and dots over ACC)	*Asp*718	2	8	1	0	0	0	0	*Achromobacter* species 718	Bolton et al. (1985), Kessler et al. (1986)	B
	GGTAC/C (o o and dots over TAC)	*Kpn*I	2	8	1	0	0	0	0	*Klebsiella pneumoniae* OK8	Smith et al. (1976), Tomassini et al. (1978)	A, B, C, D, E, F
69	CGTACG	–	1	4	0	2	0	0	0	–	–	–
70	TGTACA	–	5	5	2	0	1	0	0	–	–	–
71	C/TAG	*Mae*I	13	54	12	3	4	5	4	*Methanococcus aeolicus* PL-15/H	Schmid et al. (1984)	B
72	A/CTAGT	*Spe*I	0	3	0	0	0	0	0	*Sphaerotilus* species	Comb and Schildkraut in Kessler and Höltke (1986)	B, D
73	G/CTAGC	*Nhe*I	1	4	0	0	0	1	1	*Neisseria mucosa*	Comb and Schildkraut in Kessler and Höltke (1986)	B, D

Appendix 1 (continued)

Position No.	Recognition sequence[a]	Enzyme[b]	Number of recognition sites[c] λ	Ad2	SV40	φX174	M13mp7	pBR322	pBR328	Microorganism[d]	Reference(s)[e]	Commercial[f] availability
74	C/CTAGG	*Avr*II	2	2	2	0	0	0	0	*Anabaena variabilis* uw	Rosenvold in Kessler and Höltke (1986)	D
75	T/CTAGA (+ over C and final A; dots under GA)	*Xba*I	1	5	8	0	0	0	0	*Xanthomonas badrii*	Zain and Roberts (1977)	A, B, C, D, E, F
76	TTAA	–	195	115	47	35	61	15	17	–	–	–
77	ATTAAT	–	17	3	3	2	6	1	1	–	–	–
78	GTT/AAC (+o over AA)	*Hpa*I	14	6	4	3	0	0	0	*Haemophilus parainfluenzae*	Sharp et al. (1973) Garfin and Goodman (1974)	A, B, C, D, E, F
79	C/TTAAG	*Afl*II	3	4	1	2	0	0	0	*Anabaena flos-aquae*	Whitehead and Brown (1983)	A, B
80	TTT/AAA	*Dra*I	13	12	12	2	5	3	5	*Deinococcus radiophilus*	Purvis and Moseley (1983)	A, B, C, D, E, F
A.2 Class II enzymes with penta- or heptanucleotide recognition sequences												
81	G/G(A/T)CC (+ and dots over the final C)	*Ava*II	35	73	6	1	1	8	7	*Anabaena variabilis*	Murray et al. (1976) Hughes and Murray (1980)	A, B, C, D, F
	(A/G)G/G(A/T)CC(T/C) (dots over CC(T))	*Ppu*MI	3	23	1	0	0	2	0	*Pseudomonas putida* M	Morgan (pers. commun.)	D

	CG/G(A)(T)CCG	*Rsr*II	5	2	0	0	0	0	0	*Rhodopseudomonas sphaeroides* G30	O'Connor et al. (1984)	D
82	+m .... CC/(A)(T)GG	*Apy*I	71	136	17	2	7	6	10	*Arthrobacter pyridinolis*	Gruenbaum et al. (1981a)	B
	oo .... CC/(A)(T)GG	*Bst*NI	71	136	17	2	7	6	10	*Bacillus stearothermophilus*	Schildkraut and Comb in Kessler and Höltke (1986)	D
	o+ .... /CC(A)(T)GG	*Eco*RII	71	136	17	2	7	6	10	*Escherichia coli* R245	Boyer et al. (1973), Bogdarina et al. (1976)	C
83	+ CC/(G)(C)GG	*Nci*I	114	97	0	1	4	10	10	*Neisseria cinerea*	Watson et al. (1980)	B, C, D
84	+ + G/ANTC	*Hin*fI	148	72	10	21	26	10	10	*Haemophilus influenzae* R_f	Middleton et al. in Kessler and Höltke (1986)	A, B, C, D, E, F
85	++ ... G/GNCC	*Cfr*13I	74	164	11	2	4	15	16	*Citrobacter freundii* RFL13	Janulaitis et al. (1983b)	A
	++ ... G/GNCC	*Sau*96I	74	164	11	2	4	15	16	*Staphylococcus aureus* PS96	Sussenbach et al. (1978)	B, C, D
	+ ..... (A)(G)G/GNCC(T)(C)	*Dra*II	3	44	3	0	0	4	2	*Deinococcus radiophilus*	Grosskopf et al. (1985), Wit et al. (1985)	B
86	+ GC/NGC	*Fnu*4HI	380	411	24	31	15	42	37	*Fusobacterium nucleatum* 4H	Leung et al. (1979)	D

Appendix 1 (continued)

Position No.	Recognition sequence[a]	Enzyme[b]	Number of recognition sites[c] λ	Ad2	SV40	ϕX174	M13mp7	pBR322	pBR328	Microorganism[d]	Reference(s)[e]	Commercial[f] availability
87	o+ CCNGG (dotted)	*Scr*FI	185	233	17	3	11	16	20	*Streptococcus cremoris* F	Fitzgerlad et al. (1982)	D
88	/GTNAC	*Mae*III	156	118	14	17	24	17	18	*Methanococcus aeolicus* PL-15/H	Schmid et al. (1984)	B
	G/GTNACC	*Bst*EII	13	10	0	0	0	0	0	*Bacillus stearothermophilus* ET	Lautenberger et al. (1980)	B, C, D, F
89	+ C/TNAG	*Dde*I	104	97	20	14	29	8	9	*Desulfovibrio desulfuricans* Norway	Makula and Meagher (1980)	B, C, D, E, F
	CC/TNAGG	*Sau*I	2	7	0	0	1	0	1	*Streptomyces aureofaciens* IKA18/4	Timko et al. (1981)	B
A.3 Class II enzymes with recognition sequences containing internal $(N)_x$ sequences												
90	GGN/NCC (dotted)	*Nla*IV	82	178	16	6	6	24	26	*Neisseria lactamica*	Qiang et al. in Kessler and Höltke (1986)	D
91	CACNNN/GTG	*Dra*III	10	10	1	1	1	0	0	*Deinococcus radiophilus*	Grosskopf et al. (1985), Wit et al. (1985)	B
92	GAANN/NNTTC	*Asp*700	24	5	0	3	2	2	1	*Achromobacter* species 700	Bolton et al. (1985), Kessler et al. (1986)	B
93	+o GAANN/NNTTC	*Xmn*I	24	5	0	3	2	2	1	*Xanthomonas manihotis* 7AS1	Lin et al. (1980)	D

94	+o GCC$(N)_4$/NGGC	*Bgl*I	29	20	1	0	1	3	5	*Bacillus globigii*	Duncan et al. (1978), Bickle and Ineichen (1980), Van Heuverswyn and Fiers (1980)	B, C, D, F
95	+ + CCA$(N)_5$/NTGG	*Bst*XI	13	10	1	3	0	0	0	*Bacillus stearothermophilus* X1	Langdale et al. in Kessler and Höltke (1986)	D
96	o oo GGCC$(N)_4$/NGGCC	*Sfi*I	0	3	1	0	0	0	0	*Streptomyces fimbriatus*	Qiang and Schildkraut (1984)	D

B. Class II enzymes with non-palindromic recognition sequences
B.1 Class II enzymes

97	GAATG-CN/N CTTAC/GN-N	*Bsm*I	46	10	4	3	1	1	3	*Bacillus stearothermophilus* NUB36	Myers and Roberts in Kessler and Höltke (1986)	D
98	TGACN/N-NGTCN ACTGN-N/NCAGN	*Tth*111I	1	6	0	0	0	2	0	*Thermus thermophilus* 111	Shinomiya and Sato (1980)	A, D, F

B.2 Class IIS enzymes

99	+ GCAGC$(N)_8$ CGTCG$(N)_{12}$	*Bbv*I	199	179	22	14	8	21	11	*Bacillus brevis*	Gingeras et al. (1978a)	D
100	ACCTGC$(N)_4$ TGGACG$(N)_8$	*Bsp*MI	41	39	0	3	4	1	2	*Bacillus* species M	Morgan (pers. commun.)	D
101	(+) GGATG$(N)_9$ CCTAC$(N)_{13}$	*Fok*I	150	78	11	8	4	12	11	*Flavobacterium okeanokoites*	Sugisaki and Kanazawa (1981)	A, B, D

Appendix 1 (continued)

Position No.	Recognition sequence[a]	Enzyme[b]	Number of recognition sites[c] λ	Ad2	SV40	φX174	M13mp7	pBR322	pBR328	Microorganism[d]	Reference(s)[e]	Commercial[f] availability
102	+ GACGC(N)$_5$ CTGCG(N)$_{10}$	*Hga*I	102	87	0	14	7	11	12	*Haemophilus gallinarum*	Sugisaki (1978)	D
103	+ (+) GGTGA(N)$_8$ CCACT(N)$_7$ · ·	*Hph*I	168	99	4	9	18	12	18	*Haemophilus parahaemolyticus*	Kleid et al. (1976)	D
104	+ GAAGA(N)$_8$ CTTCT(N)$_7$ · ·	*Mbo*II	130	113	16	11	12	11	12	*Moraxella bovis*	Endow (1977), Gelinas et al. (1977a), Brown et al. (1980a)	C, D
105	CCTC(N)$_7$ GGAG(N)$_7$	*Mnl*I	262	397	51	34	61	26	30	*Moraxella nonliquefaciens*	Zabeau et al. in Kessler and Höltke (1986)	D
106	o GCATC(N)$_5$ CGTAG(N)$_9$	*Sfa*NI	169	84	6	12	7	22	17	*Streptococcus faecalis*	Sciaky and Roberts in Kessler and Höltke (1986)	D

Appendix 2. Important restriction endonucleases with known relaxed specificities[a]

Restriction endonuclease	Recognition sequence	Relaxed specificities
*Eco*RI	G/AATTC	N/AATTN or (A)(A) (T)(T) (G)(G)AT(C)(C)
*Sau*3AI[b]	/GATC	GAGC or CATC
*Bam*HI	G/GATCC	GGATCN or G(A)/(G)ATCC
*Sph*I	GCATG/C	nd
*Hin*dIII	A/AGCTT	nd
*Pvu*II	CAG/CTG	CCGCTG or CATCTG or CAGATG or CAGGTG or CAGCGG
*Hae*III	GG/CC	nd
*Pst*I	CTGCA/G	nd
*Hha*I	GCG/C	nd
*Ava*I	C/(T)/(C)CG(A)/(G)	nd
*Taq*I	T/CGA	nd
*Sal*I	GTCGAC	nd
*Eco*RV	GAT/ATC	nd
*Sca*I	AGT/ACT	nd
*Kpn*I	GGTAC/C	nd
*Xba*I	T/CTAGA	nd
*Hpa*I	GTT/AAC	nd
*Apy*I	CC/(A)/(T)GG	nd
*Dde*I	C/TNAG	nd
*Bst*EII	G/GTNACC	nd
*Tth*lllI	TGACN/N-NGTCN ACTGN-N/NCAGN	NACN/N-NGTCN NTGN-N/NCAGN or GACN/N-NNTCN CTGN-N/NNAGN or GACN/N-NGNCN CTGN-N/NCNGN

[a] Corresponding references are given in Kessler and Höltke (1986). For explanation of symbols see footnote [a] of Appendix 1.

[b] *Sau*3AI, when used in excess amounts, introduces nicks in double-stranded DNA at certain sequences similar to the *Sau*3AI site such as 5′-GAGC-3′ and 5′-CATC-3′ but not at 5′-GCTC-3′, 5′-GCTC-3′, 5′-GTTC-3′, 5′-GAAC-3′ or 5′-GATT-3′ sites.

Appendix 3. *Dam-* and *dcm*I-sensitivity of important restriction endonucleases[a]

Restriction endonuclease(s)	Recognition sequence	Restriction endonuclease(s)	Recognition sequence
1. *dam*-sensitivity			
*Mbo*I	+ GATC	*Nru*I	+ TCGCGA ...
*Bcl*I	+ TGATCA	*Taq*I	+ TCGA ...
*Cla*I	+ ATCGAT ...	*Xba*I	+ TCTAGA ...
*Mbo*I	+ GAAGA CTTCT ...		
2. *dam*-insensitivity			
*Sau*3AI	o GATC	*Bam*HI	o GGATCC .....
*Dpn*I	m GATC	*Xho*II	o (A) (T) GATC (G) (C) .....
*Bgl*II	o AGATCT	*Pvu*I	o CGATCG .....
3. *dcm*I-sensitivity			
*Aha*II	+ .... (A) (T) G CG C (G) (C)	*Cfr*13I	+ .. GGNCC
*Asp*718	+ .. GGTACC	*Dra*II	+ ..... (A) (T) GGNCC (G) (C)
*Eco*RII	+ .... (A) CC GG (T)	*Bgl*I	+ ...... GCCNNNNNGGC
4. *dcm*-insensitivity			
*Bam*HI	o .. GGATCC	*Apy*I	m .... (A) CC GG (T)

[a] Corresponding references are given in Kessler and Höltke (1986).

Appendix 3 (continued)

Restriction endonuclease(s)	Recognition sequence	Restriction endonuclease(s)	Recognition sequence
*Nar*I	o . . . GGCGCC	*Bst*EII	o . . . GGTNACC
*Kpn*I	o . . . GGTACC	*Bst*NI	. o . . . . CC(A/T)GG

References

Anderson WF, Ohlendorf DH, Takeda Y, Matthews BW (1981) Structure of the cro repressor from bacteriophage λ and its interaction with DNA. Nature (London) 290:754–758

Armstrong K, Bauer WR (1982) Preferential site-dependent cleavage by restriction endonuclease *Pst*I. Nucleic Acids Res 10:993–1007

Arrand JR, Myers PA, Roberts RJ (1978) A new restriction endonuclease from *Streptomyces albus* G. J Mol Biol 118:127–135

Arrand JR, Myers PA, Roberts RJ. Cited in Kessler and Höltke (1986)

Berkner KL, Folk WR (1977) *Eco*RI cleavage and methylation of DNAs containing modified pyrimidines in the recognition sequence. J Biol Chem 252:3185–3193

Bertani G, Weigle JJ (1953) Host controlled variation in bacterial viruses. J Bacteriol 65:113–121

Bickle TA (1982) The ATP-dependent restriction endonucleases. In: Linn SM, Roberts RJ (eds) Nucleases. Cold Spring Harbor Lab, New York, pp 85–108

Bickle TA, Ineichen K (1980) The DNA sequence recognized by *Bgl*I. Gene 9:205–212

Bickle TA, Pirrotta V, Imber R (1980) Purification and properties of the *Bgl*I and II endonucleases. In: Colowick SP, Kaplan NO (eds) Methods in enzymology, vol 65. Academic Press, London New York, pp 132–138

Bingham AHA, Atkinson T, Sciaky D, Roberts RJ (1978) A specific endonuclease from *Bacillus caldolyticus*. Nucleic Acids Res 5:3457–3467

Blumenthal RM, Gregory SA, Cooperider JS (1985) Cloning of a restriction modification system from *Proteus vulgaris* and its use in analyzing a methylase-sensitive phenotype in *Escherichia coli*. J Bacteriol 164:501–509

Bogdarina IG, Vagabova LM, Buryanov YaI (1976) DNA-cytosine methylation in *E. coli* MRE 600 cells. FEBS Lett 68:177–180

Bolton BJ, Comer MJ, Kessler C (unpublished results)

Bolton BJ, Nesch G, Comer MJ, Wolf W, Kessler C (1985) *Asp*718 from a non-pathogenic species of the genus *Achromobacter*: a *Kpn*I isoschizomer generating DNA-fragments with 5′-protruding ends. FEBS Lett 182:130–134

Borsetti R, Grandoni R, Schildkraut I. Cited in Kessler and Höltke (1986)

Botterman J, Zabeau M (1985) High-level production of the *Eco*RI endonuclease under the control of the P_L promoter of bacteriophage lambda. Gene 37:229–239

Bougueleret L, Schwarzstein M, Tsugita A, Zabeau M (1984) Characterization of the genes coding for the *Eco*RV restriction and modification system of *Escherichia coli*. Nucleic Acids Res 12: 3659–3676

Bougueleret L, Tenchini ML, Botterman J, Zabeau M (1985) Overproduction of the *Eco*RV endonuclease and methylase. Nucleic Acids Res 13:3823–3839

Boyer HW, Chow LT, Dugaiczyk A, Hedgpeth J, Goodman HM (1973) DNA substrate site for the *Eco*RII restriction endonuclease and modification methylase. Nature New Biol 244:40–43

Bron S, Murray K (1975) Restriction and modification in *B. subtilis*. Nucleotide sequence recognized by restriction endonuclease R·*Bsu*R from strain R. Mol Gen Genet 143:25–33

Brooks JE, Blumenthal RM, Gingeras TR (1983) The isolation and characterization of the *Escherichia coli* DNA adenine methylase (*dam*) gene. Nucleic Acids Res 11:837–851

Brown NL, Smith M (1976) The mapping and sequence determination of the single site in ϕX174*am*3 replicative form DNA cleaved by restriction endonuclease *Pst*I. FEBS Lett 65: 284–287

Brown NL, Smith M (1977) Cleavage specificity of the restriction endonuclease isolated from *Haemophilus gallinarum* (*Hga*I). Proc Natl Acad Sci USA 74:3213–3216

Brown NL, Hutchison III CA, Smith M (1980a) The specific non-symmetrical sequence recognized by restriction endonuclease *Mbo*II. J Mol Biol 140:143–148

Brown NL, McClelland M, Whitehead PR (1980b) *Hgi*AI: A restriction endonuclease from *Herpetosiphon giganteus* HP1023. Gene 9:49–68

Buchmann AR, Burnett L, Berg P (1980) The SV40 nucleotide sequence. In: Tooze J (ed) DNA tumor viruses. Cold Spring Harbor Lab, New York, pp 799–829

Cheng S-C, Kim R, King K, Kim S-H, Modrich P (1984) Isolation of gram quantities of *Eco*RI restriction and modification enzymes from an overproducing strain. J Biol Chem 259:11571–11575

Comb DG, Schildkraut I. Cited in Kessler and Höltke (1986)

Comb DG, Wilson G. Cited in Kessler and Höltke (1986)

Comb DG, Parker P, Grandoni R, Schildkraut I. Cited in Kessler and Höltke (1986)

Comb DG, Schildkraut I, Wilson G, Greenough L. Cited in Kessler and Höltke (1986)

Connolly BA, Eckstein F, Pingoud A (1984) The stereochemical course of the restriction endonuclease *Eco*RI-catalyzed reaction. J Biol Chem 259:10760–10763

Daniels DL, Schroeder JL, Szybalski W, Sanger F, Blattner FR (1985a) A molecular map of coliphage *Lambda*. In; Hendrix RW, Roberts JW, Stahl FW, Weisberg RA (eds) Lambda II. Cold Spring Harbor Lab, New York, pp 469–517

Daniels DL, Schroeder W, Szybalski W, Sanger F, Coulson AR, Hong GF, Hill DF, Petersen GB, Blattner FR (1985b) Complete annotated *Lambda* sequence. In: Hendrix RW, Roberts JW, Stahl FW, Weisberg RA (eds) Lambda II. Cold Spring Harbor Lab, New York, pp 519–676

Dombroski DF, Morgan AR (1985) Restriction nuclease digestions driven to completion by *Escherichia coli* RNA polymerase and T4 gene 32 protein. J Biol Chem 260:415–417

Dreiseikelmann B, Eichenlaub R, Wackernagel W (1979) The effect of differential methylation by *Escherichia coli* of plasmid DNA and phage T7 and λ DNA on the cleavage by restriction endonuclease *Mbo*I from *Moraxella bovis*. Biochim Biophys Acta 562:418–428

Dugaiczyk A, Hedgpeth J, Boyer HW, Goodman HM (1974) Physical identity of the SV40 deoxyribonucleic acid sequence recognized by the *Eco*RI restriction endonuclease and modification methylase. Biochemistry 13:503–512

Duncan CH, Wilson GA, Young FE (1978) Biochemical and genetic properties of site-specific restriction endonucleases in *Bacillus globigii*. J Bacteriol 134:338–344

Duyk G, Leis J, Longiaru M, Skalka AM (1983) Selective cleavage in the avian retroviral long terminal repeat sequence by the endonuclease associated with the $\alpha\beta$ form of avian reverse transcriptase. Proc Natl Acad Sci USA 80:6745–6749

Ehbrecht H-J, Pingoud A, Urbanke C, Maass G, Gualerzi C (1985) Linear diffusion of restriction endonucleases on DNA. J Biol Chem 260:6160–6166

Ehrlich M, Wang RY-H (1981) 5-Methylcytosine in eukaryotic DNA. Science 212:1350–1357

Endlich B, Linn S (1981) Type I restriction enzymes. In: Boyer PD (ed) The enzymes, vol 14. Academic Press, London New York, pp 137–156

Endow SA (1977) Analysis of *Drosophila melanogaster* satellite IV with restriction endonuclease *Mbo*II. J Mol Biol 114:441–449

Endow SA, Roberts RJ (1977) Two restriction-like enzymes from *Xanthomonas malvacearum*. J Mol Biol 112:521–529

Fitzgerlad GF, Daly C, Brown LR, Gingeras TR (1982) *Scr*FI: a new sequence-specific endonuclease from *Streptococcus cremoris*. Nucleic Acids Res 10:8171–8179

Forsblom S, Rigler R, Ehrenberg M, Pettersson U, Philipson L (1976) Kinetic studies on the cleavage of adenovirus DNA by restriction endonuclease *Eco*RI. Nucleic Acids Res 3:3255–3269

Frederick CA, Grable J, Melia M, Samudzi C, Jen-Jacobson L, Wang B-C, Greene P, Boyer HW, Rosenberg JM (1984) Kinked DNA in crystalline complex with *Eco*RI endonuclease. Nature (London) 309:327–331

Fried MG, Crothers DM (1984) Kinetics and mechanism in the reaction of gene regulatory proteins with DNA. J Mol Biol 172:263–282

Fuchs LY, Covarrubias L, Escalante L, Sanchez S, Bolivar F (1980) Characterization of a site-specific restriction endonuclease *Sph*I from *Streptomyces phaeochromogenes.* Gene 10:39–46

Fuchs R, Blakesley R (1983) Guide to the use of type II restriction endonucleases. In: Colowick SP, Kaplan NO (eds) Methods in enzymology, vol 100. Academic Press, London New York, pp 3–38

Gardner RC, Horwarth AJ, Messing J, Shepherd RJ (1982) Cloning and sequencing of restriction fragments generated by *Eco*RI. DNA 1:109–115

Garfin DE, Goodman HM (1974) Nucleotide sequences at the cleavage sites of two restriction endonucleases from *Haemophilus parainfluenzae.* Biochem Biophys Res Commun 59:108–116

Gelinas RE, Myers PA, Roberts RJ (1977a) Two sequence-specific endonucleases from *Moraxella bovis.* J Mol Biol 114:169–180

Gelinas RE, Myers PA, Weiss GH, Roberts RJ, Murray K (1977b) A specific endonuclease from *Brevibacterium albidum.* J Mol Biol 114:433–440

George J, Chirikjian JG (1982) Sequence-specific endonuclease *Bam*HI: Relaxation of sequence recognition. Proc Natl Acad Sci USA 79:2432–2436

Gingeras TR, Brooks JE (1983) Cloned restriction/modification system from *Pseudomonas aeruginosa.* Proc Natl Acad Sci USA 80:402–406

Gingeras TR, Milazzo JP, Roberts RJ (1978a) A computer assisted method for the determination of restriction enzyme recognition sites. Nucleic Acids Res 5:4105–4127

Gingeras TR, Meyers PA, Olson JA, Hanberg FA, Roberts RJ (1978b) A new specific endonuclease present in *Xanthomonas holcicola, Xanthomonas papavericola* and *Brevibacterium luteum.* J Mol Biol 118:113–122

Gingeras TR, Greenough L, Schildkraut I, Roberts RJ (1981) Two new restriction endonucleases from *Proteus vulgaris.* Nucleic Acids Res 9:4525–4536

Goppelt M, Pingoud A, Maass G, Mayer H, Köster H, Frank R (1980) The interaction of the *Eco*RI restriction endonuclease with its substrate. A physicochemical study employing natural and synthetic oligonucleotides and polynucleotides. Eur J Biochem 104:101–107

Goppelt M, Langowski J, Pingoud A, Haupt W, Urbanke C, Mayer H, Maass G (1981) The effect of several nucleic acid binding drugs on the cleavage of d(GGAATTCC) and pBR322 by the *Eco*RI restriction endonuclease. Nucleic Acids Res 9:6115–6127

Grable J, Frederick CA, Samudzi C, Jen-Jacobsen L, Lesser D, Greene P, Boyer HW, Itakura K, Rosenberg JM (1984) Two-fold symmetry of crystalline DNA-*Eco*RI endonuclease recognition complexes. J Biomol Struct Dyn 1:1149–1160

Grandoni RP, Schildkraut I. Cited in Kessler and Höltke (1986)

Gray PW, Hallick RB (1977) Restriction endonuclease map of *Euglena gracilis* chloroplast DNA. Biochemistry 16:1665–1671

Greene PJ, Betlach MC, Boyer HW, Goodman HM (1974) The *Eco*RI restriction endonuclease. Methods Mol Biol 7:87–111

Greene PJ, Heynecker HL, Bolivar F, Rodriguez RL, Betlach MC, Covarrubias AA, Backman K, Russel DJ, Tait R, Boyer HW (1978) A general method for the purification of restriction enzymes. Nucleic Acids Res 5:2373–2380

Greene PJ, Gupta M, Boyer HW, Brown WE, Rosenberg JM (1981) Sequence analysis of the DNA encoding the *Eco*RI endonuclease and methylase. J Biol Chem 256:2143–2153

Grosskopf R, Kessler C (unpublished results)

Grosskopf R, Wolf W, Kessler C (1985) Two new restriction endonucleases *Dra*II and *Dra*III from *Deinococcus radiophilus.* Nucleic Acids Res 13:1517–1528

Gruenbaum Y, Cedar H, Razin A (1981a) Restriction enzyme digestion of hemimethylated DNA. Nucleic Acids Res 9:2509–2515

Gruenbaum Y, Stein R, Cedar H, Razin A (1981b) Methylation of CpG sequences in eukaryotic DNA. FEBS Lett 124:67–71

Hadi SM, Bächi B, Iida S, Bickle TA (1983) DNA restriction-modification enzymes of phage P1 and plasmid p15B. Subunit functions and structural homologies. J Mol Biol 165:19–34

Halford SE, Johnson NP (1980) The *Eco*RI restriction endonuclease with bacteriophage λ DNA. Equilibrium binding studies. Biochem J 191:593–604

Haqqi TM, Ahmad S, Ahmad NS, Ahmad M, Hasnain A-U, Siddiqi M, Hadi SM (1985) Cloning and expression of *Eco*RI specific restriction modification system. Ind J Biochem Biophys 22:252–254

Hattman S (1981) DNA methylation. In: Boyer PD (ed) The enzymes, vol 14. Academic Press, London New York, pp 517–548

Hattman S, Brooks JE, Masurekar M (1978a) Sequence Specificity of the P1 modification methylase (M·*Eco*P1) and the DNA methylase (M·*Eco dam*) controlled by the *Escherichia coli dam* gene. J Mol Biol 126:367–380

Hattman S, Keister T, Gottehrer A (1978b) Sequence specificity of DNA methylases from *Bacillus amyloliquefaciens* and *Bacillus brevis*. J Mol Biol 124:701–711

Hedgpeth J, Goodman HM, Boyer HW (1972) DNA nucleotide sequence restricted by the RI endonuclease. Proc Natl Acad Sci USA 69:3448–3452

Hinsch B, Kula M-R (1981) Reaction kinetics of some important site-specific endonucleases. Nucleic Acids Res 9:3159–3174

Hofer B, Ruhe G, Koch A, Köster H (1982) Primary and secondary structure specificity of the cleavage of 'single-stranded' DNA by endonuclease *Hin*fI. Nucleic Acids Res 10:2763–2773

Hsu M-t, Berg P (1978) Altering the specificity of restriction endonuclease: Effect of replacing Mg^{2+} with Mn^{2+}. Biochemistry 17:131–138

Hu AW, Marschel AH. Cited in Kessler and Höltke (1986)

Huang L-H, Farnet CM, Ehrlich KC, Ehrlich M (1982) Digestion of highly modified bacteriophage DNA by restriction endonucleases. Nucleic Acids Res 10:1579–1591

Hughes SG, Murray K (1980) The nucleotide sequences recognized by endonucleases *Ava*I and *Ava*II from *Anabaena variabilis*. Biochem J 185:65–76

Iida S, Meyer J, Bächi B, Stalhammar-Carlemalm M, Schrickel S, Bickle TA, Arber W (1983) DNA restriction-modification genes of phage P1 and plasmid p15B. Structure and in vitro transcription. J Mol Biol 165:1–18

Jack WE, Rubin RA, Newman A, Modrich P (1980) Structures and mechanisms of *Eco*RI DNA restriction and modification enzymes. In: Chirikjian JG (ed) Gene amplification and analysis, vol 1: Restriction endonucleases. Elsevier/North Holland Biomedical Press, Amsterdam New York, pp 165–179

Jack WE, Terry BJ, Modrich P (1982) Involvement of outside DNA sequences in the major kinetic path by which *Eco*RI endonuclease locates and leaves its recognition sequence. Proc Natl Acad Sci USA 79:4010–4014

Jacobs D, Brown NL (personal communication)

Janulaitis A, Povilionis P, Sasnauskas K (1982) Cloning of the modification methylase gene of *Bacillus centrosporus* in *Escherichia coli*. Gene 20:197–204

Janulaitis A, Petrusyte M, Butkus V (1983a) Three sequence-specific endonucleases from *Escherichia coli* RFL47. FEBS Lett 161:213–216

Janulaitis AA, Stakenas PS, Bitinaite YuB, Jaskelyavichene BP (1983b) Distribution of specific endodeoxynucleases in various strains of *Citrobacter freundii*. Dokl Acad Nauk SSSR 271: 483–485

Jen-Jacobsen L, Kurpiewski L, Lesser D, Grable J, Boyer HW, Rosenberg JM, Greene PJ (1983) Coordinate ion pair formation between *Eco*RI endonuclease and DNA. J Biol Chem 258: 14638–14646

Jentsch S, Günthert U, Trautner TA (1981) DNA methyltransferases affecting the sequence 5'CCGG. Nucleic Acids Res 9:2753–2759

Kaplan DA, Nierlich DP (1975) Cleavage of nonglucosylated bacteriophage T4 deoxyribonucleic acid by restriction endonuclease *Eco*RI. J Biol Chem 250:2395–2397

Kauc L, Piekarowicz A (1978) Purification and properties of a new restriction endonuclease from *Haemophilus influenzae* Rf. Eur J Biochem 92:417–426

Kelly S, Kaddurah-Daoukh R, Smith HO (1985) Purification of the *Hha*II restriction endonuclease from overproducer *Escherichia coli* clone. J Biol Chem 260:15339–15344

Kelly TJ Jr, Smith HO (1970) A restriction enzyme from *Haemophilus influenzae*. II. Base sequence of the recognition site. J Mol Biol 51:393–409

Kessler C (unpublished results)

Kessler C, Nesch G, Brack R (1981) *Sph*I restriction map of bacteriophage lambda DNA. Gene 16: 321–323

Kessler C, Neumaier PS, Wolf W (1985) Recognition sequences of restriction endonucleases and methylases – a review. Gene 33:1–102

Kessler C, Höltke H-J (1986) Specificity of restriction endonucleases and methylases – a review (edition 2). Gene 47:1–153

Kessler C, Bolton BJ, Comer MJ (1986) Screening for novel type II restriction endonucleases. J Cell Biochem Supp 10D:100, Abstr 058

Kholmina GV, Rebentish BA, Skoblov YuS, Mironov AA, Yankovskii NK, Kozlov YuI, Glatman LI, Moroz AF, Debabov VG (1980) Isolation and characterization of a new site-specific endonuclease *Eco*RV. Dokl Acad Nauk SSSR 253:495–497

Kim R, Modrich P, Kim S-H (1984) 'Interactive' recognition in *Eco*RI restriction enzyme-DNA complex. Nucleic Acids Res 12:7285–7292

Kiss A, Posfai G, Keller CC, Venetianer P, Roberts RJ (1985) Nucleotide sequence of the *Bsu*RI restriction-modification system. Nucleic Acids Res 13:6403–6421

Kita K, Hiraoka N, Kimizuka F, Obayashi A, Kojima H, Takahashi H, Saito H (1985) Interaction of the restriction endonuclease *Sca*I with its substrates. Nucleic Acids Res 13:7015–7024

Kleid D, Humayun Z, Jeffrey A, Ptashne M (1976) Novel properties of a restriction endonuclease isolated from *Haemophilus parahaemolyticus*. Proc Natl Acad Sci USA 73:293–297

Koncz C, Kiss A, Venetianer P (1978) Biochemical characterization of the restriction-modification system of *Bacillus sphaericus*. Eur J Biochem 89:523–529

Korch C, Hagblom P, Normark S (1983) Sequence-specific DNA modification in *Neisseria gonorrhoeae*. J Bacteriol 155:1324–1332

Kostriken R, Zoller M, Heffron F (1985) HO nuclease: a site-specific double-strand endonuclease essential to mating type interconversion. In: Leive L (ed) Microbiology-1985. Cold Spring Harbor Lab, New York, pp 295–298

Kosykh VG, Buryanov YaI, Bayev AA (1980) Molecular cloning of *Eco*RII endonuclease and methylase genes. Mol Gen Genet 178:717–718

Kraev AS, Kravets AN, Chernov BK, Skryabin KG, Baev AA (1985) The *Eco*RV restriction-modiication system: Genes, enzymes, synthetic substrates. Mol Biol (Moscow) 19:278–284

Kramarov VM, Smolyaninov VV (1981) DNA methylase from *Arthrobacter luteus* screens DNA from the action of site-specific endonuclease *Alu*I. Biokhimiya 46:1526–1529

Kunkel LM, Silberklang M, McCarthy BJ (1979) A third restriction endonuclease from *Xanthomonas malvacearum*. J Mol Biol 132:133–139

Kuosmanen M, Pösö H (1985) Inhibition of the activity of restriction endonucleases by spermidine and spermine. FEBS Lett 179:17–20

Lacks S, Greenberg B (1975) A deoxyribonuclease of *Diplococcus pneumoniae* specific for methylated DNA. J Biol Chem 250:4060–4066

Lacks S, Greenberg B (1977) Complementary specificity of restriction endonucleases of *Diplococcus pneumoniae* with respect to DNA methylation. J Mol Biol 114:153–168

Langdale JA, Myers PA, Roberts RJ. Cited in Kessler and Höltke (1986)

Langowski J, Pingoud A, Goppelt M, Maass G (1980) Inhibition of *Eco*RI action by polynucleotides. A characterization of the non-specific binding of the enzyme to DNA. Nucleic Acids Res 8:4727–4736

Langowski J, Urbanke C, Pingoud A, Maass G (1981) Transient cleavage kinetics of the *Eco*RI restriction endonuclease measured in a pulsed quench-flow apparatus: enzyme concentration-dependent activity change. Nucleic Acids Res 9:3483–3490

Langowski J, Alves J, Pingoud A, Maass G (1983) Does the specific recognition of DNA by the restriction endonuclease *Eco*RI involve a linear diffusion step? Investigation of the processivity of the *Eco*RI endonuclease. Nucleic Acids Res 11:501–513

Lautenberger JA, Edgell MH, Hutchison III CA (1980) The nucleotide sequence recognized by the *Bst*EII restriction endonuclease. Gene 12:171–174

Lee SH, Rho HM (1985) Cloning of *Pst*I methylase gene. Korean J Genet 7:42–48

Lee YH, Chirikjian JG (1979) Sequence-specific endonuclease *Bgl*I. Modification of lysine and arginine residues of the homogeneous enzyme. J Biol Chem 254:6838–6841

Leung DW, Lui ACP, Merilees H, McBride BC, Smith M (1979) A restriction enzyme from *Fusobacterium manihotis* 4H which recognizes GCNGC. Nucleic Acids Res 6:17–25

Lin B-C, Chien M-C, Lou S-Y (1980) A sequence-specific endonuclease (*Xmn*I) from *Xanthomonas manihotis*. Nucleic Acids Res 8:6189–8198

Linn S, Arber W (1968) Host specificity of DNA produced by *Escherichia coli*. X. In vitro restriction of phage fd replicative form. Proc Natl Acad Sci USA 59:1300–1306

Lu A-L, Jack WE, Modrich P (1981) DNA determinants important in sequence recognition by *Eco*RI endonuclease. J Biol Chem 256:13200–13206

Lui ACP, McBride BC, Vovis GF, Smith M (1979) Site specific endonuclease from *Fusobacterium nucleatum*. Nucleic Acids Res 6:1–15

Luke PA, Halford SE (1985) Solubility of the *Eco*RI restriction endonuclease and its purification from an over-producing strain. Gene 37:241–246

Luria SE, Human ML (1952) A nonhereditary, host-induced variation of bacterial viruses. J Bacteriol 64:557–569

Lynn SP, Cohen LK, Kaplan S, Gardner JF (1980) *Rsa*I: A new sequence-specific endonuclease activity from *Rhodopseudomonas sphaeroides*. J Bacteriol 142:380–383

Makula RA, Meagher RB (1980) A new restriction endonuclease from the anaerobic bacterium, *Desulfovibrio desulfuricans*, Norway. Nucleic Acids Res 8:3125–3131

Malcolm ADB (1981) The use of restriction enzymes in genetic engineering. In: Williamson R (ed) Genetic engineering, vol 2. Academic Press, London New York, pp 129–173

Malyguine E, Vannier P, Yot P (1980) Alteration of the specificity of restriction endonucleases in the presence of organic solvents. Gene 8:163–177

Manvalan P, Johnson WC Jr, Modrich P (1984) Prediction of secondary structure for *Eco*RI endonuclease. J Biol Chem 259:11666–11667

Mann MB, Smith HO (1977) Specificity of *Hpa*II and *Hae*III DNA methylases. Nucleic Acids Res 4:4211–4221

Mann MB, Smith HO. Cited in Kessler and Höltke (1986)

May MS, Hattman S (1975) Deoxyribonucleic acid-cytosine methylation by host- and plasmid-controlled enzymes. J Bacteriol 122:129–138

Mayer H, Grosschedl R, Schütte H, Hobom G (1981) *Cla*I, a new restriction endonuclease from *Caryophanon latum* L. Nucleic Acids Res 9:4833–4845

McClelland M (1981) Purification and characterization of two new modification methylases; M*Cla*I from *Caryophanon latum* L and M*Taq*I from *Thermus aquaticus* YTI. Nucleic Acids Res 9:6795–6804

McClelland M, Nelson M (1985) The effect of site specific methylation on restriction endonuclease digestion. Nucleic Acids Res 13 (Suppl):r201–r207

McClelland M, Kessler LG, Bittner M (1984) Site-specific cleavage of DNA at 8- and 10-base-pair sequences. Proc Natl Acad Sci USA 81:983–987

McKay DB, Steitz TA (1981) Structure of catabolite gene activator protein at 2.9-Å resolution suggests binding to left-handed B-DNA. Nature (London) 290:744–749

McKay DB, Weber IT, Steitz TA (1982) Structure of catabolite gene activator protein at 2.9-Å resolution. Incorporation of amino acid sequence and interactions with cyclic AMP. J Biol Chem 257:9518–9524

Meijer M, Beck E, Hansen FG, Bergmans HEN, Messer W, Meyenburg K von, Schaller H (1979) Nucleotide sequence of the origin of replication of the *Escherichia coli* K-12 chromosome. Proc Natl Acad Sci USA 76:580–584

Meselson M, Yuan R (1968) DNA restriction enzyme from *E. coli*. Nature (London) 217:1110–1114

Middleton JH, Edgell MH, Hutchison III CA (1972) Specific fragments of ϕX174 deoxyribonucleic acid produced by a restriction enzyme from *Haemophilus aegyptius*, endonuclease Z. J Virol 10:42–50

Middleton JH, Stankus PV, Edgell MH, Hutchison CA III. Cited in Kessler and Höltke (1986)

Miller PS, Cheng DM, Dreon N, Jayaraman K, Kan L-S, Leutzinger EE, Pulford SM, Ts'o POP (1980) Preparation of a decadeoxyribonucleotide helix for studies by nuclear magnetic resonance. Biochemistry 19:4688–4698

Mise K, Nakajima K (1985) Purification of a new restriction endonuclease, *Sty*I, from *Escherichia coli* carrying the *hsd*$^+$ minimplasmid. Gene 33:357–361

Modrich P (1979) Structures and mechanisms of DNA restriction and modification enzymes. Q Rev Biophys 12:315–369

Modrich P, Roberts RJ (1982) Type-II restriction and modification enzymes. In: Linn SM, Roberts RJ (eds) Nucleases. Cold Spring Harbor Lab, New York, pp 109–154

Modrich P, Rubin RA (1977) Role of the 2-amino group of deoxyguanosine in sequence recognition by *Eco*RI restriction and modification enzymes. J Biol Chem 252:7273–7278

Modrich P, Zabel D (1976) *Eco*RI endonuclease. Physical and catalytic properties of the homogeneous enzyme. J Biol Chem 251:5866–5874

Molloy PL, Symons RH (1980) Cleavage of DNA.RNA hybrids by type II restriction enzymes. Nucleic Acids Res 8:2939–2946

Morgan R (personal communications by R.J. Roberts)

Murray K, Hughes SG, Brown JS, Bruce SA (1976) Isolation and characterization of two sequence-specific endonucleases from *Anabaena variabilis*. Biochem J 159:317–322

Myers PA, Roberts RJ. Cited in Kessler and Höltke (1986)

Nardone G, George J, Chirikjian JG (1984) Sequence-specific *Bam*HI methylase. Purification and characterization. J Biol Chem 259:10357–10362

Nath K, Azzolina BA (1981) Cleavage properties of site-specific restriction endonucleases. In: Chirikjian JG (ed) Gene amplification and analysis, vol 1: Restriction endonucleases. Elsevier North Holland Biomedical Press, Amsterdam New York, pp 113–130

Nathans D, Smith HO (1975) Restriction endonucleases in the analysis and restructuring of DNA molecules. Annu Rev Biochem 44:273–293

Nelson M, Christ C, Schildkraut I (1984) Alteration of apparent restriction endonuclease recognition specificities by DNA methylases. Nucleic Acids Res 12:5165–5173

Newman AK, Rubin RA, Kim S-H, Modrich P (1981) DNA sequences of structural genes for *Eco*RI DNA restriction and modification enzymes. J Biol Chem 256:2131–2139

Nilsson M-G, Skarped C, Magnusson G (1982) Structure at restriction endonuclease *Mbo*I cleavage sites protected by actinomycin D or distamycin A. FEBS Lett 145:360–364

Nishigaki K, Kaneko Y, Wakuda H, Husimi Y, Tanaka T (1985) Type II restriction endonucleases cleave single-stranded DNAs in general. Nucleic Acids Res 13:5747–5760

Nussinov R (1980) Some rules in the ordering of nucleotides in the DNA. Nucleic Acids Res 8: 4545–4562

O'Connor CD, Metcalf E, Wrighton CJ, Harris TJR, Saunders JR (1984) *Rsr*II – a novel restriction endonuclease with a heptanucleotide recognition site. Nucleic Acids Res 12:6701–6708

Österlund M, Luthman H, Nilsson SV, Magnusson G (1982) Ethidium-bromide-inhibited restriction endonuclease cleaves one strand of circular DNA. Gene 20:121–125

Ohlendorf DH, Anderson WF, Takeda Y, Matthews BW (1983) High resolution structural studies of cro repressor protein and implications for DNA recognition. J Biomol Struct Dyn 1:553–563

Old R, Murray K, Roizes G (1975) Recognition sequence of restriction endonuclease III from *Hemophilus influenzae*. J Mol Biol 92:331–339

Olson JA, Myers PA, Roberts RJ. Cited in Kessler and Höltke (1986)

Pabo CO, Lewis M (1982) The operator binding domain of λ repressor: structure and DNA recognition. Nature (London) 298:443–447

Panayotatos N (1984) Practical consequences of restriction site symmetry. Gene 31:291–294

Peden KWC (1983) Revised sequence of the tetracycline-resistance gene of pBR322. Gene 22: 277–280

Petrusyte M, Janulaitis A (1982) Isolation and some properties of the restriction endonuclease *Bcn*I from *Bacillus centrosporus*. Eur J Biochem 121:377–382

Pingoud A (1985) Spermidine increases the accuracy of type II restriction endonucleases: Suppression of cleavage at degenerate, non-symmetrical sites. Eur J Biochem 147:105–109

Pingoud A, Urbanke C, Alves J, Ehbrecht H-J, Zabeau M, Gualerzi C (1984) Effect of polyamines and basic proteins on cleavage of DNA by restriction endonucleases. Biochemistry 23:5697–5703

Pirrotta V (1976) Two restriction endonucleases from *Bacillus globigii.* Nucleic Acids Res 3:1747–1760

Pirrotta V, Bickle TA (1980) General purification schemes for restriction endonucleases. In: Colowick SP, Kaplan NO (eds) Methods in enzymology, vol 65. Academic Press, London New York, pp 89–95

Podhajska AJ, Szybalski W (1985) Conversion of the *Fok*I endonuclease to a universal restriction enzyme: cleavage of phage M13mp7 DNA at predetermined sites. Gene 40:175–182

Pohl FM, Thomae R, Karst A (1982) Temperature dependence of the activity of DNA-modifying enzymes: endonucleases and DNA ligase. Eur J Biochem 123:141–152

Pope A, Lynn SP, Gardner JF. Cited in Kessler and Höltke (1986)

Poulsen M, Johnson PH, Loew G (1985) Peptide-nucleic acid interactions: Possible recognition determinants of *Eco*RI endonuclease. In: Molecular basis of cancer, Pt B: Macromolecular recognition, chemotherapy, and immunology. Liss, New York, pp 77–90

Purvis IJ, Moseley BEB (1983) Isolation and characterization of *Dra*I, a type II restriction endonuclease recognizing a sequence containing only A:T basepairs, and inhibition of its activity by uv irradiation of substrate DNA. Nucleic Acids Res 11:5467–5474

Qiang B-Q, Schildkraut I (1984) A type II restriction endonuclease with an eight nucleotide specificity from *Streptomyces fimbriatus.* Nucleic Acids Res 12:4507–4516

Qiang B-Q, Schildkraut I, Visentin L. Cited in Kessler and Höltke (1986)

Razin A, Riggs AD (1980) DNA methylation and gene function. Science 210:604–610

Reaston J, Duyvesteyn MGC, Waard A de (1982) *Nostoc* PCC7524, a cyanobacterium which contains five sequence-specific deoxyribonucleases. Gene 20:103–110

Rexer B (unpublished results)

Roberts RJ (1976) Restriction endonucleases. CRC Crit Rev Biochem 4:123–164

Roberts RJ (1985) Restriction and modification enzymes and their recognition sequences. Nucleic Acids Res 13 (Suppl):r165–r200

Roberts RJ, Breitmeyer JB, Tabachnik NF, Myers PA (1975) A second specific endonuclease from *Haemophilus aegyptius.* J Mol Biol 91:121–123

Roberts RJ, Myers PA, Morrison A, Murray K (1976a) A specific endonuclease from *Arthrobacter luteus.* J Mol Biol 102:157–165

Roberts RJ, Myers PA, Morrison A, Murray K (1976b) A specific endonuclease from *Haemophilus haemolyticus.* J Mol Biol 103:199–208

Roberts RJ, Wilson GA, Young FE (1977) Recognition sequence of specific endonuclease *Bam*HI from *Bacillus amyloliquefaciens* H. Nature (London) 265:82–84

Rosenberg JM, Greene P (1982) *Eco*RI specificity and hydrogen bonding. DNA 1:117–124

Rosenberg JM, Dickerson RE, Greene PJ, Boyer HW (1978) Preliminary X-ray diffraction analysis of crystalline *Eco*RI endonuclease. J Mol Biol 122:241–245

Rosenberg JM, Boyer HW, Greene P (1980) The structure and function of *Eco*RI restriction endonuclease. In: Chirikjian JG (ed) Gene amplification and analysis, vol 1: Restriction endonucleases. Elsevier/North Holland Biomedical Press, Amsterdam New York, pp 131–164

Rosenvold EC. Cited in Kessler and Höltke (1986)

Rubin RA, Modrich P (1978) Substrate dependence of the mechanism of *Eco*RI endonuclease. Nucleic Acids Res 5:2991–2997

Rubin RA, Modrich P (1980) Purification and porperties of *Eco*RI endonuclease. In: Colowick SP, Kaplan NO (eds) Methods in enzymology, vol 65. Academic Press, London New York, pp 96–108

Sanger F, Coulson AR, Hong GF, Hill DF, Petersen GB (1982) Nucleotide sequence of bacteriophage *Lambda* DNA. J Mol Biol 162:729–773

Sato S, Hutchison III CA, Harris JI (1977) A thermostable sequence-specific endonuclease from *Thermus aquaticus.* Proc Natl Acad Sci USA 74:542–546

Sato S, Nakazawa K, Shinomiya T (1980) A DNA methylase from *Thermus thermophilus* HB8. J Biochem 88:737–747
Schildkraut I, Comb D. Cited in Kessler and Höltke (1986)
Schildkraut I, Banner CDB, Rhodes CS, Parekh S (1984) The cleavage site for the restriction endonuclease *Eco*RV is 5'-GAT/ATC-3'. Gene 27:327–329
Schildkraut I, Wise R, Borsetti R, Qiang B-Q. Cited in Kessler and Höltke (1986)
Schmid K, Thomm M, Laminet A, Laue FG, Kessler C, Stetter KO, Schmitt R (1984) Three new restriction endonucleases *Mae*I, *Mae*II and *Mae*III from *Methanococcus aeolicus*. Nucleic Acids Res 12:2619–2628
Schoner B, Kelly S, Smith HO (1983) The nucleotide sequence of *Hha*II restriction and modification genes from *Haemophilus haemolyticus*. Gene 24:227–236
Sciaky D, Roberts RJ. Cited in Kessler and Höltke (1986)
Seurinck J, Voorde A van de, Montagu M van (1983) A new restriction endonuclease from *Acetobacter pasteurianus*. Nucleic Acids Res 11:4409–4415
Sharp PA, Sudgen B, Sambrook J (1973) Detection of two restriction endonuclease activities in *Haemophilus parainfluenzae* using analytical agarose-ethidium bromide electrophoresis. Biochemistry 12:3055–3063
Shimotsu H, Takahashi H, Saito H (1980) A new site-specific endonuclease *Stu*I from *Streptomyces tubercidicus*. Gene 11:219–225
Shinomiya T, Sato S (1980) A site specific endonuclease from *Thermus thermophilus* 111, *Tth*111I. Nucleic Acids Res 8:43–56
Shinomiya T, Kobayashi M, Sato S, Uchida T (1982) A new aspect of a restriction endonuclease *Tth*111I. It has a degenerated specificity (*Tth*111I*). J Biochem 92:1823–1832
Siebenlist U, Simpson RB, Gilbert W (1980) *E. coli* RNA polymerase interacts homologously with two different promoters. Cell 20:269–281
Smith DI, Blattner FR, Davies J (1976) The isolation and partial characterization of a new restriction endonuclease from *Providencia stuartii*. Nucleic Acids Res 3:343–353
Smith HO (1979) Nucleotide sequence specificity of restriction endonucleases. Science 205:455–462
Smith HO, Kelly SV (1984) Methylases of the type II restriction-modification systems. In: Cedar H, Riggs AD, Razin A (eds) DNA methylation. Springer, Berlin Heidelberg New York Tokyo, pp 39–71
Smith HO, Nathans D (1973) A suggested nomenclature for bacterial host modification and restriction systems and their enzymes. J Mol Biol 81:419–423
Smith HO, Wilcox KW (1970) A restriction enzyme from *Haemophilus influenzae*. I. Purification and general properties. J Mol Biol 51:379–391
Sugimoto K, Oka A, Sugisaka H, Takanami M, Nishimura A, Yasuda Y, Hirota Y (1979) Nucleotide sequence of *Escherichia coli* K-12 replication origin. Proc Natl Acad Sci USA 76:575–579
Sugisaki H (1978) Recognition sequence of a restriction endonuclease from *Haemophilus gallinarum*. Gene 3:17–28
Sugisaki H, Kanazawa S (1981) New restriction endonucleases from *Flavobacterium okeanokoites* (*Fok*I) and *Micrococcus luteus* (*Mlu*I). Gene 16:73–78
Sugisaki H, Maekawa Y, Kanazawa S, Takanami M (1982) New restriction endonucleases from *Acetobacter aceti* and *Bacillus aneurinolyticus*. Nucleic Acids Res 10:5747–5752
Sussenbach JS, Monfoort CH, Schiphof R, Stobberingh EE (1976) A restriction endonuclease from *Staphylococcus aureus*. Nucleic Acids Res 3:3193–3202
Sussenbach JS, Steenbergh PH, Rost JA, Leeuwen WJ van, Embden JDA van (1978) A second site-specific restriction endonuclease from *Staphylococcus aureus*. Nucleic Acids Res 5:1153–1163
Sutcliffe JG (1979) Complete nucleotide sequence of the *Escherichia coli* plasmid pBR322. Cold Spring Harbor Symp Quant Biol 43:77–90
Szekeres M, pers commun
Szybalski W (1985) Universal restriction endonucleases: designing novel cleavage specificities by combining adapter oligodeoxynucleotide and enzyme moieties. Gene 40:169–173

Takahashi H, Kojima H, Saito H (1985) A new site-specific endonuclease, *Sca*I, from *Streptomyces caespitosus.* Biochem J 231:229–232
Takeda Y, Ohlendorf DH, Anderson WF, Mathews BW (1983) DNA-binding proteins. Science 221:1020–1026
Terry BJ, Jack WE, Rubin RA, Modrich P (1983) Thermodynamic parameters governing interaction of *Eco*RI endonuclease with specific and nonspecific DNA sequences. J Biol Chem 258: 9820–9825
Terry BJ, Jack WE, Modrich P (1985) Facilitated diffusion during catalysis by *Eco*RI endonuclease. Nonspecific interactions in *Eco*RI catalysis. J Biol Chem 260:13130–13137
Theriault G, Roy PH, Howard KA, Benner JS, Brooks JE, Waters AF, Gingeras TR (1985) Nucleotide sequence of the *Pae*R7 restriction/modification system and partial characterization of its protein products. Nucleic Acids Res 13:8441–8461
Thomas M, Davis RW (1975) Studies on the cleavage of bacteriophage *Lambda* DNA with *Eco*RI restriction endonuclease. J Mol Biol 91:315–328
Tikchonenko TI, Kramarov EV, Zavizion BA, Naroditsky BS (1978) *Eco*RI* activity: Enzyme modification or activation of accompanying endonuclease? Gene 4:195–212
Timko J, Horwitz AH, Zelinka J, Wilcox G (1981) Characterization of a site-specific restriction endonuclease from *Streptomyces aureofaciens.* J Bacteriol 145:873–877
Tomassini J, Roychoudhury R, Wu R, Roberts RJ (1978) Recognition sequence of restriction endonuclease *Kpn*I from *Klebsiella pneumoniae.* Nucleic Acids Res 5:4055–4064
Trautner TA, Pawlek B, Günthert U, Canosi U, Jentsch S, Freund M (1980) Restriction and modification in *Bacillus subtilis:* Identification of a gene in the temperate phage SPβ coding for a *Bsu*R specific modification methyltransferase. Mol Gen Genet 180:361–367
Tu C-PD, Roychoudhury R, Wu R (1976) Nucleotide recognition sequence at the cleavage site of *Haemophilus aegyptius* II (*Hae*II) restriction endonuclease. Biochem Biophys Res Commun 72:355–362
Urieli-Shoval S, Gruenbaum Y, Razin A (1983) Sequence and substrate specificity of isolated DNA methylases from *Escherichia coli* C. J Bacteriol 153:274–280
Van Heuverswyn H, Fiers W (1980) Recognition sequence for the restriction endonuclease *Bgl*I from *Bacillus globigii.* Gene 9:195–203
Van Montagu M, Sciaky D, Myers PA, Roberts RJ. Cited in Kessler and Höltke (1986)
Vanyushin BF, Belozersky AN, Kokurina NA, Kadirova DK (1968) 5-methylcytosine and 6-methylaminopurine in bacterial DNA. Nature (London) 218:1066–1067
Visentin LP, Watson RJ, Martin S, Zuker M. Cited in Kessler and Höltke (1986)
Wagner R Jr, Meselson M (1976) Repair tracts in mismatched DNA heteroduplexes. Proc Natl Acad Sci USA 73:4135–4139
Walder RY, Hartley JL, Donelson JE, Walder JA (1981) Cloning and expression of the *Pst*I restriction-modification system in *Escherichia coli.* Proc Natl Acad Sci USA 78:1503–1507
Walder RY, Walder JA, Donelson JE (1984) The organization and complete nucleotide sequence of the *Pst*I restriction-modification system. J Biol Chem 259:8015–8026
Watson R, Zuker M, Martin SM, Visentin LP (1980) A new site-specific endonuclease from *Neisseria cinerea.* FEBS Lett 118:47–50
Watson RJ, Schildkraut I, Qiang B-Q, Martin SM, Visentin LP (1982) *Nde*I: a restriction endonuclease from *Neisseria denitrificans* which cleaves DNA at 5'-CATATG-3' sequences. FEBS Lett 150:114–116
Wells RD, Klein RD, Singleton CK (1981) Type II restriction enzymes. In: Boyer PD (ed) The enzymes, vol 14. Academic Press, London New York, pp 157–191
Whitehead PR, Brown NL (1982) *Aha*III: A restriction endonuclease with a recognition sequence containing only A:T basepairs. FEBS Lett 143:296–300
Whitehead PR, Brown NL (1983) *Eae*I: a restriction endonuclease from *Enterobacter aerogenes.* FEBS Lett 155:97–101
Wigler MH (1981) The inheritance of methylation patterns in vertebrates. Cell 24:285–286
Wilson GA, Young FE (1975) Isolation of a sequence-specific endonuclease (*Bam*HI) from *Bacillus amyloliquefaciens* H. J Mol Biol 97:123–125

Wit CM de, Dekker BMM, Neele AC, Waard A de (1985) Purification and characterization of endonucleases *Dra*II and III from *Deinococcus radiophilus*. FEBS Lett 180:219–223

Woodbury CP Jr, Hagenbüchle O, Hippel PH von (1980) DNA site recognition and reduced specificity of the *Eco*RI endonuclease. J Biol Chem 225:11534–11546

Woodhead JL, Malcolm ADB (1980) Non-specific binding of restriction endonuclease *Eco*RI to DNA. Nucleic Acids Res 8:389–402

Woodhead JL, Malcolm ADB (1981) The essential carboxyl group in restriction endonuclease *Eco*RI. Eur J Biochem 120:125–128

Woodhead JL, Bhave N, Malcolm ADB (1981) Cation dependence of restriction endonuclease *Eco*RI activity. Eur J Biochem 115:293–296

Yamada Y, Murakami M (1985) A new restriction endonuclease from *Acetobacter pasteurianus* (*Apa*LI). Agric Biol Chem 49:3627–3629

Yanisch-Perron C, Vieira J, Messing J (1985) Improved M13 phage cloning vectors and host strains: nucleotide sequences of the M13mp18 and pUC19 vectors. Gene 33:103–119

Yoo OJ, Agarwal KL (1980) Cleavage of single strand oligonucleotides and bacteriophage ϕX174 DNA by *Msp*I endonuclease. J Biol Chem 255:10559–10562

Yuan R (1981) Structure and mechanism of multifunctional restriction endonucleases. Annu Rev Biochem 50:285–315

Young T-S, Kim S-H, Modrich P, Beth A, Jay E (1981) Preliminary X-ray diffraction studies of *Eco*RI restriction endonuclease-DNA complex. J Mol Biol 145:607–610

Zabeau M, Roberts RJ (1979) The role of restriction endonucleases in molecular genetics. In: Taylor JH (ed) Molecular genetics, Pt III: Chromosome structure. Academic Press, London New York, pp 1–63

Zabeau M, Roberts RJ. Cited in Kessler and Höltke (1986)

Zabeau M, Greene R, Myers PA, Roberts RJ. Cited in Kessler and Höltke (1986)

Zain BS, Roberts RJ (1977) A new specific endonuclease from *Xanthomonas badrii*. J Mol Biol 115:249–255

12 Analysis of the Eukaryotic Chromosome Organization with Restriction Endonucleases

N. O. Bianchi and M. S. Bianchi[1]

1 Introduction

In chromosomes of metaphase cells spread on slides, digestion by certain restriction enzymes produces a general decrease in the intensity of the stain in the chromosomes and the appearance of bands with a pattern characteristic for each enzyme (Lima-de-Faria et al. 1980; Miller et al. 1983).

It has been demonstrated that both conventional C-banding methods and chromosome digestion with micrococcal endonuclease induce chromosome banding by a selective extraction of DNA from nonbanded chromosome regions (Comings et al. 1973; Sahasrabuddhe et al. 1978). It has been proposed, therefore, that restriction enzymes induce chromosome banding by a similar mechanism. Miller et al. (1983) suggested that DNA regions rich in recognition sites for a given restriction enzyme are nicked and that the DNA fragments are then extracted from the chromosomes. The DNA segments extracted are thought to be 100–200 base pairs (bp) long, whereas fragments longer than 1,000 bp would remain in the fixed chromatin (Miller et al. 1983). According to this hypothesis the chromosome bands induced by restriction enzymes would represent chromosome regions having a very low frequency of restriction sites or none at all. Although this assumption lacks direct experimental evidence, it has received support from some reported data. Mouse chromosomes show C-banding after digestion with *AluI* (N.O. Bianchi, unpublished). Moreover, when synchronized mouse mitotic cells are suspended in saline solution and digested with *AluI,* 85% of the DNA is released; the fraction resistant to this enzyme digestion corresponds to centromeric regions highly enriched in satellite DNA, which is known to lack recognition sequences for *AluI* (Horz et al. 1974; Lica and Hamkalo 1983).

HaeIII endonuclease induces the appearance of C- and G-banding in human chromosomes (Miller et al. 1983; Bianchi et al. 1984). Since the restriction site for this enzyme comprises the base sequence GGCC, and since the chromosome regions between G-bands (interbands) are rich in guanine, it has been proposed that *HaeIII* preferentially digests the interband regions and thus enhances the appearance of G-bands (Miller et al. 1983). Recently, it has been suggested that *HindIII* and *EcoRI* also produce G-banding in human chromosomes (Mezzanotte et al. 1983). The restriction sites of these enzymes, however, are not rich in guanine (AAGCTT and GAATTC

1 IMBICE, C.C. 403, 1900 La Plata, Argentina

Cytogenetics. Ed. by G. Obe and A. Basler

respectively). Therefore, it was proposed that the structural organization of chromatin and not the frequency of restriction sites is the primary cause of the chromosome banding pattern resulting from treatments with restriction endonucleases.

Thus far, it has been reported that restriction enzymes may induce C-, G- or NOR-banding; moreover, some restriction enzymes can also induce the appearance of chromosome gaps. In this presentation we shall identify the restriction enzyme banding as Re-banding. The type of banding induced will be placed in parentheses after the symbol "Re"; the name of the enzyme-producing banding will appear in parentheses following the banding type connotation. Thus, for instance, the C-banding induced by *AluI* will be indicated as: Re(C)(*AluI*)-banding.

Table 1 lists the endonucleases and animal species studied for restriction enzyme chromosome banding. In this report we shall analyze the mechanisms by which restriction enzymes induce longitudinal chromosome differentiation. Moreover, we shall also describe the pattern of restriction enzyme banding in human chromosomes and we shall discuss the use of these enzymes in obtaining information on several aspects of the chromosome structure and organization.

2 Mechanism of Re-Banding

Most data on the mechanism of restriction endonuclease banding have been obtained by using the technique of sequential enzyme digestion (SED) of chromosome preparations (M.S. Bianchi et al. 1985; N.O. Bianchi et al. 1985). In this method the cells are labelled with (^{14}C)-thymidine for one to three cycles. After fixation in 3/1 methanol/acetic acid, cell spreads are prepared by air drying. Chromosome preparations are thereafter sequentially digested with: (1) a test restriction endonuclease for 2 h at 37°C (30 U in 20 μl assay buffer), (2) proteinase K (0.5 mg/ml in phosphate buffer), (3) DNase I (1000 U Kunitz/ml), (4) residual debris are recovered by scraping off the slides with a razor blade. Radioactivity in each fraction is measured by scintillation counting, expressed as percentage of the total radioactivity and assumed to represent the relative amount of DNA extracted during each of the sequential treatments.

Table 2 illustrates the DNA extracted by *AluI, HaeIII, EcoRI* and *AvaI*. The first two enzymes, which induce decreased chromosome staining and banding, extracted about 50–60% of the chromosome DNA. On the other hand, *EcoRI* and *AvaI*, two enzymes with no detectable chromosome action, produced 13% and 12% DNA extraction respectively. When chromosomes digested with restriction endonucleases were subsequently incubated with proteinase K, there was a further loss of DNA ranging from 22–37% for *HaeIII* and *AluI* and from 74–78% for *EcoRI* and *AvaI*. These results with SED led to some assumptions about the action of restriction enzymes on fixed chromosomes.

Although restriction enzymes induce multiple nicks in the DNA molecule, the amount of DNA extracted from chromosomes during incubation depends on the number of chromosome breaks produced and the prevention of loss by chromosome proteins. Restriction enzymes with a high frequency of recognition sequences in DNA can give rise to a large number of short DNA fragments that are extracted from chromosomes during incubation; the decrease in chromosome staining is the result of

Table 1. Endonucleases and species tested for Re-banding[a]

Enzymes	Human	Gorilla	Chimpanzee	Orangutan	African green monkey	Chinese hamster	*Mus musculus*	*Akodon* (Cricetidae) several species	Mosquito *Aedes albopictus*
AluI	Re(C)	Re(C)	Re(C)	Re(C)	Digest	Re(C)	Re(C)	Re(C)	Digest
HaeIII	Re(C)(G)	Re(C)(gap) (G)	Re(C)(gap) (G)	Re(C)(gap) (G)	Re(C)(G)	Re(C)(G)	Re(C)	Re(C)(G)	Digest
MboI	Re(C)	Re(C)	Re(C)	Re(C)	--	--	Re(C)	Re(C)	Re(C)
DdeI	Re(C)	--	--	--	--	--	--	--	--
HinfI	Re(C)(gap)	Re(C)	Neg	Re(C)	--	--	Re(C)	--	--
RsaI	Re(C)	Re(C)	Neg	Neg	--	--	Re(C)	Re(C)	--
MspI	Re(C)	--	--	--	--	--	Re(C)	--	Digest
HpaII	Neg	--	--	--	--	--	Neg	--	Re(C)
HinPI	Neg	--	--	--	--	--	Neg		Digest
HhaI	Neg	--	--	--	--	--	Neg	--	Digest
TaqI	Re(G)	--	--	--	--	--	Re(C)(G)	--	Digest
EcoRI	Neg	--	--	--	--	--	Neg	--	--
EcoRII	Re(C)	--	--	--	--	--	Re(C)(G)	--	--
AvaI	Neg	--	--	--	--	--	--	--	--
AvaII	Neg	--	--	--	--	--	Re(C)(G)	--	--
BstNI	Neg	--	--	--	--	--	Re(C)(G)	--	--
References[b]	1, 2, 3	4	4	4	5	5	5, 6	5	7

[a] Abbreviations: Dashes = enzyme not tested; Neg = lack of chromosome digestion; Digest -- chromosome digestion without banding.

[b] References: 1. Miller et al. (1983); 2. M.S. Bianchi et al. (1985); 3. Mezzanotte et al. (1983); 4. N.O. Bianchi et al. (1985); 5. Bianchi (unpubl.); 6. Kaelbling et al. (1984); 7. N.O. Bianchi et al. (1986).

Table 2. Percentage of DNA extractions from fixed chromosome preparations by SED

Fraction	*AluI*	*HaeIII*	*EcoRI*	*AvaI*
Restriction endonuclease Fraction I	52.5	60	13.2	5.5
Proteinase K Fraction II	37.3	21	74.4	77.9
DNase I Fraction III	9.3	10	9.5	15.4
Residue Fraction IV	0.8	8	2.8	1.4

this loss of DNA. On the other hand, chromosome proteins probably prevent long DNA sequences from being extracted from chromosomes. As soon as these proteins are digested with proteinase K, the long DNA fragments are also released, and DNA remains only in the chromosome regions lacking the recognition sequences of the endonuclease. These regions can then be digested by incubation with DNase I. Finally, the small amount of radioactivity released by scraping the slides probably corresponds to cell debris retained after the successive enzyme digestions (Table 2).

The size of the fragments extracted from human chromosomes during digestion with restriction enzymes can be estimated with a mathematical model developed by Bishop et al. (1983). *TaqI* is the enzyme that produces the smallest average DNA fragment size (1,179 bp) within the group of endonucleases not inducing chromosome banding. On the other hand, among the enzymes inducing banding, *RsaI* is the endonuclease that produces the largest average fragment size (493 bp). Thus, it seems reasonable to assume that the fragments that can be extracted from fixed chromatin during digestion with restriction enzymes (fraction 1) are smaller than 1,000 bp. The data of Bishop et al. (1983) also show that even those enzymes unable to induce chromosome banding are capable of producing some proportion of short DNA fragments. These fragments are probably the cause of the loss of DNA observed after treatments with *AvaI* and *EcoRI* (Table 2).

The above thesis can be directly tested by measuring the size of the DNA fragments extracted during incubation with restriction endonucleases. An experiment of this type was performed with *MboI* in our laboratory. After SED, using *MboI* as test endonuclease, the digests of fractions I (*MboI*) and II (proteinase K) were concentrated and subsequently subjected to electrophoresis in 2% agarose plus formamide using lambda DNA digested with *HindIII* as marker. Radioactive DNA fragments were recovered from the electrophoresis gel and the relative amount of DNA segments of different size was determined by scintillation counting. Figure 1 shows the results obtained. As expected, the bulk of DNA fragments extracted after *MboI* digestion was shorter than 1000 base pairs. On the other hand, proteinase K which releases DNA retained by chromatin proteins mainly produced fragments longer than 1000 base pairs.

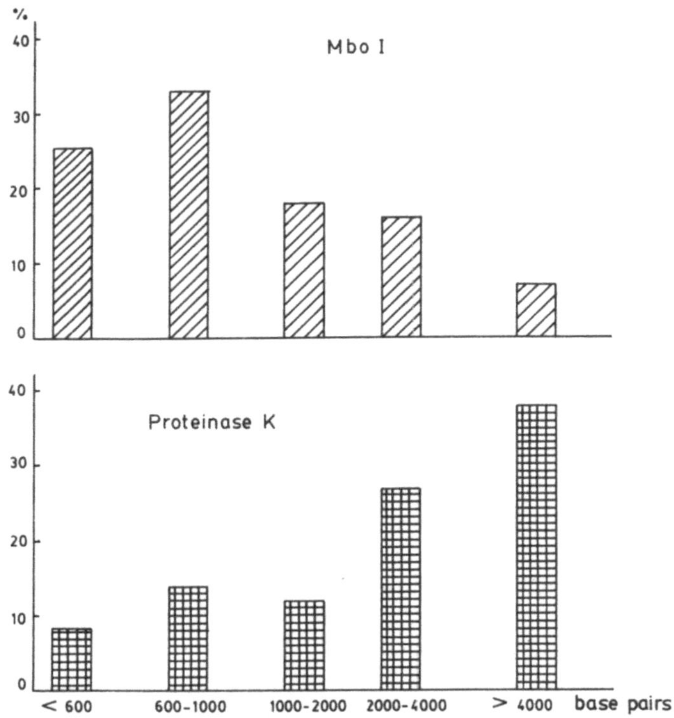

Fig. 1. DNA extraction from metaphase preparations induced by *MboI* and subsequent proteinase K digestions. The bulk of DNA fragments extracted by *MboI* is shorter that 1000 bp. Proteinase K preferentially extracts fragments longer than 1000 bp

Thus far, it can be concluded that certain restriction enzymes produce variable DNA extractions from different regions of fixed chromosomes inducing changes in the intensity of stain and banding. The amount of DNA extracted by a given enzyme is proportional to the frequency of restriction sites for the endonuclease. Short DNA fragments (well below 1000 base pairs long) are extracted from chromosomes during restriction enzyme treatments, whereas long DNA fragments (above 1000 base pairs long) are retained by chromosome proteins and released from chromosomes following protein digestion. Thus, chromosome regions showing increased stainability indicate a low frequency of restriction sites, while areas of low or very low stainability (gaps) reflect a high concentration of cleaving sites for an endonuclease.

The influence of the chromatin fibril organization upon the induction of Re-banding shall be dealt with in the next section. We think, however, it worth commenting here that regional differences in chromosome organization probably have a very limited role in the response to restriction enzymes.

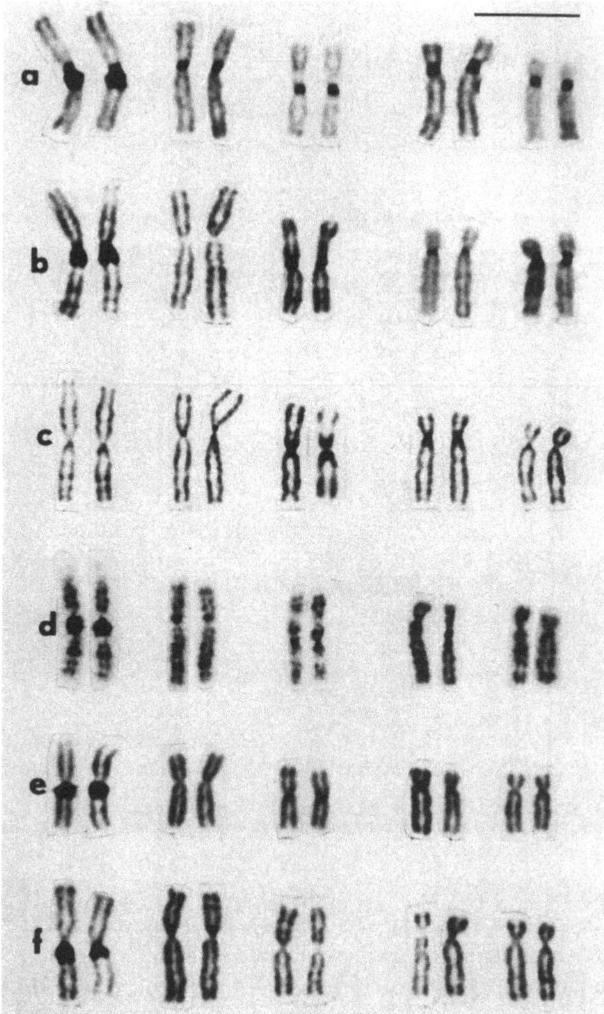

Fig. 2a–f. Human chromosome pairs 1–5: **a** conventional C-banding induced with alkali treatment; **b** Re(C)(*AluI*)-banding; **c** Re(C)(*MboI*)-banding; **d** Re(C)(G)(*HaeIII*)-banding; **e** Re(C)(*DdeI*)-banding; **f** Re(C)(*RsaI*)-banding. Bar = 10 μm

3 Re-Banding and Re-Banding Polymorphisms in Human Chromosomes

Re-banding is induced by treating metaphase spreads with a reaction mixture containing 30 U of a given restriction endonuclease diluted in 20 μl of its specific assay buffer. This reaction mixture is dropped onto each slide and carefully covered with a coverslip while avoiding the formation of air bubbles. Slides are then placed in a moist chamber and incubated for 2 h at 37°C. After incubation the coverslips are removed

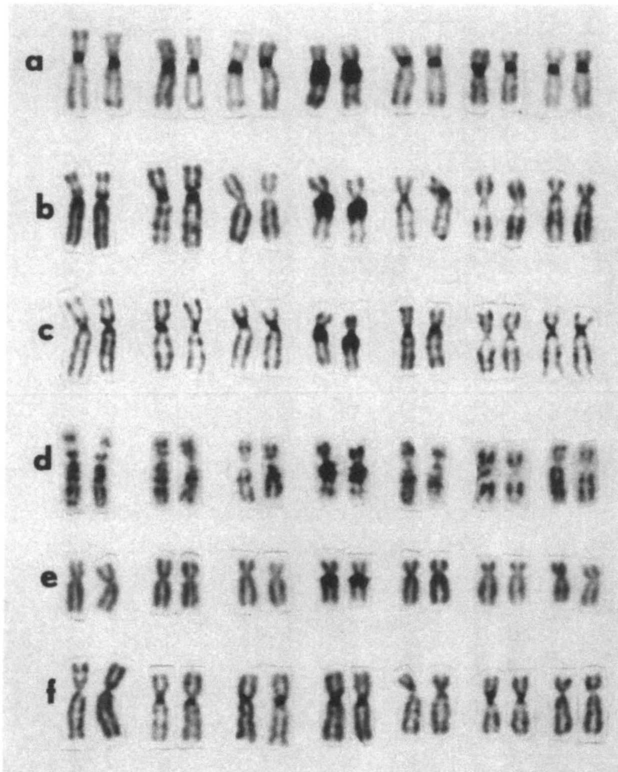

Fig. 3a–f. Human chromosome pairs 6–12: bandings and magnification as indicated in Fig. 2

and the slides are stained with Giemsa (Bianchi et al. 1984). By using this method, it has been possible to induce Re-banding in the species included in Table 1.

The detailed pattern of Re-banding in human chromosomes has been reported by Miller et al. (1983) and M.S. Bianchi et al. (1985). In this presentation we shall refer to Fig. 2–6 to illustrate the distribution of conventional alkali-induced C-banding (Figs.2–5a), the Re(C)(*AluI*)(*MboI*)(*HaeIII*)(*DdeI*)(*RsaI*)-bandings (Figs. 2–5b–f), the Re(G)(*HaeIII*)-banding (Figs. 2–5d) and the Re(C)(gap)(*HinfI*)-banding (Fig. 6). It has been recently reported that *MspI* induces selective Re(C)-banding of human chromosomes 1 (Bianchi et al. 1986); Fig. 7 illustrates this pattern.

In humans, it has been possible to identify the presence of Re(C)-banding polymorphisms between homologous chromosomes from different individuals and also between homologous chromosomes from the same individual (M.S. Bianchi et al. 1985). In some cases these polymorphisms are associated with polymorphisms in the amount of C-band material. More oftenly, however, Re(C)-banding polymorphisms are independent of the amount of C-bands. In this last case a polymorphism for a given endonuclease does not necessarily imply a similar modification for other restriction

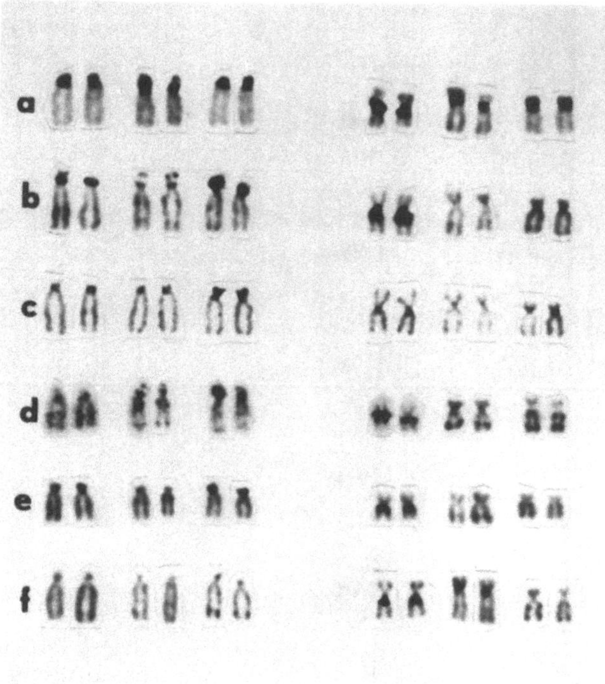

Fig. 4a–f. Human chromosome pairs 13–18: bandings and magnification as indicated in Fig. 2

enzymes. Re(C)-banding polymorphisms are characterized by a different size of the Re(C)-bands of homologous chromosomes.

Over a sample of five normal human beings, three showed Re(C) polymorphisms (M.S. Bianchi et al. 1985). Pairs 1, 3, 4, 13, 14, 9 and 18 were found to show heteromorphisms with *AluI, MboI, HaeIII, DdeI, HinfI* or *MspI*. Figures 7 and 8 illustrate some Re(C)-banding polymorphisms.

It has been reported that lymphocyte chromosomes from patients with certain neoplastic diseases show rearrangements and variation in the amount of heterochromatin with a frequency higher than that observed in normal human populations. Hence, it has been assumed that these variations may be associated with neoplastic cell transformation (Bianchi and Ayres 1971; Atkin and Brito-Babapulle 1985). Re(C)-banding allows the quantification of restriction sites for various endonucleases in the heterochromatic regions of human chromosomes. Therefore, the investigation of the Re(C)-banding polymorphisms in normal and neoplastic patients would perhaps be useful in the identification of individuals with a high risk for cancerous cell transformation.

Re(C)-banding can also be induced in G_1 prematurely condensed chromosomes (PCC) (M.S. Bianchi et al. 1985) (Fig. 9). Therefore, it may be inferred that Re-banding is independent of the cell cycle stage. The analysis of chromocenters in interphase nucleic provides additional support for this assumption. Atlhough the number of chromocenters in lymphocytes stimulated with phytohemagglutinin and digested with

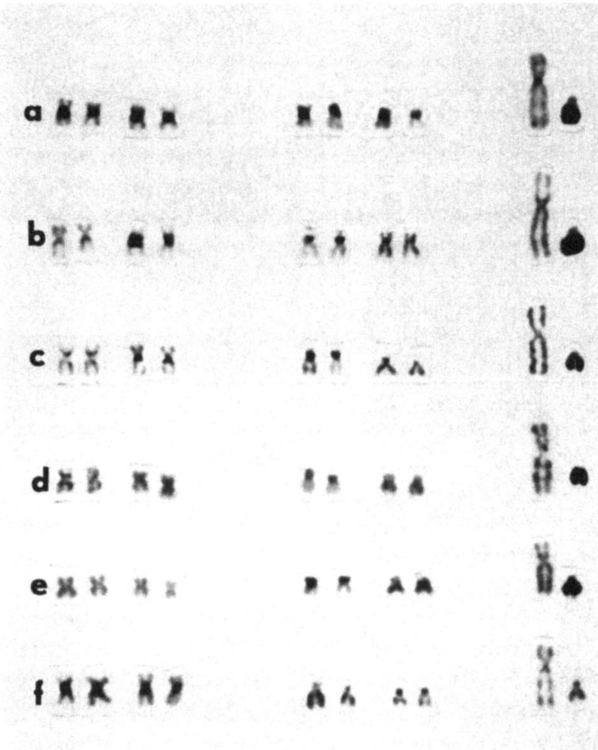

Fig. 5a–f. Human chromosome pairs 19–22 and XY: banding and magnification as indicated in Fig. 2

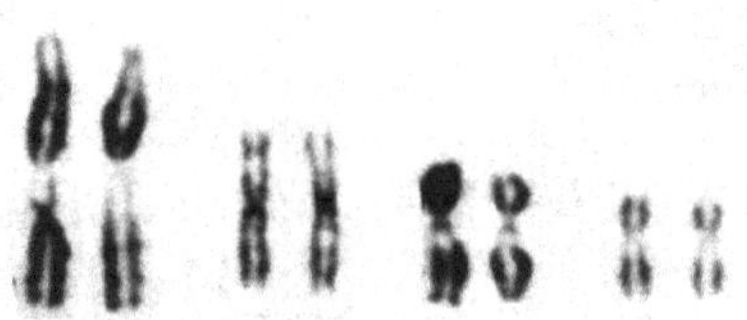

Fig. 6. Human chromosome pairs 1, 9 and 16 showing Re(gap)(*HinfI*)-banding. Pair 3 showing Re(C)(*HinfI*)-banding

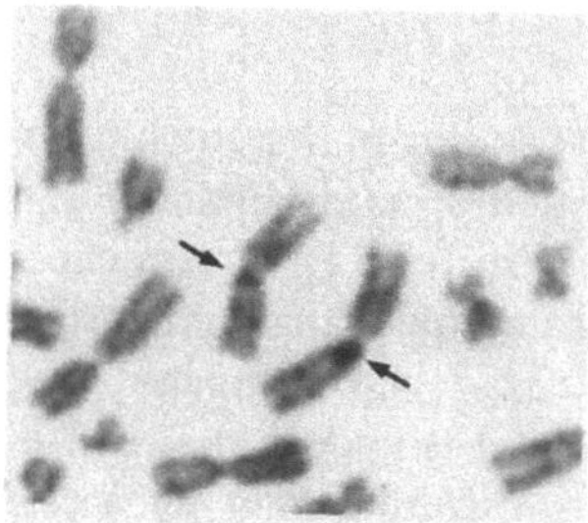

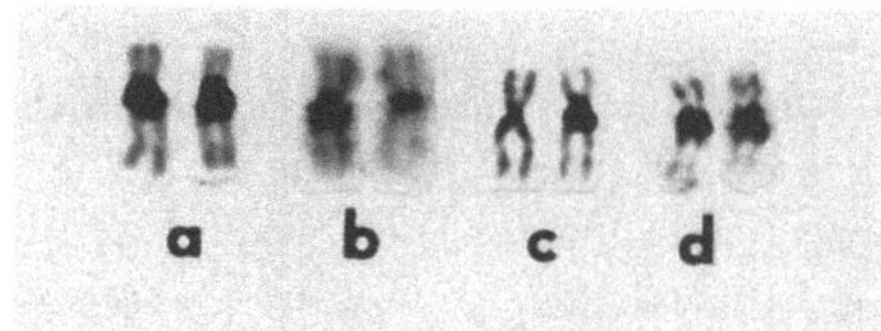

Fig. 7

Fig. 8

Fig. 7. Part of a human metaphase digested with *MspI*. *Arrows* show Re(C)-banding in pair 1; notice the banding polymorphism

Fig. 8a–d. Re(C)-banding polymorphisms in human pair 9: **a** conventional C-banding; **b** Re(C)-(*AluI*)-banding; **c** Re(C)(*MboI*)-banding; **d** Re(C)(*DdeI*)-banding. Notice the *AluI* and *MboI* polymorphisms and the lack of polymorphism with conventional C-banding and with *DdeI*

restriction enzymes vary from cell to cell, there is in many cases a good correlation between the number of Re(C)-bands in metaphase chromosomes and the number of Re(C)-chromocenters in interphase nuclei. This coincidence is particularly clear for enzymes such as *DdeI* and *HinfI*, that induce only a small number of chromosome bands and an equivalent small number of chromocenters in interphase cells. Thus, given the similar response of interphase and metaphase chromosomes to restriction enzymes it can be concluded that the level of folding of the chromatin fibril probably plays little or no role in the induction of Re-banding. This concept is further supported by the appearance of Re(C)-banding polymorphisms. The most cogent explanation for the different response of homologous chromosome regions to a given restriction enzyme is to assume a different concentration of restriction sites and not a distinct level of chromatin organization.

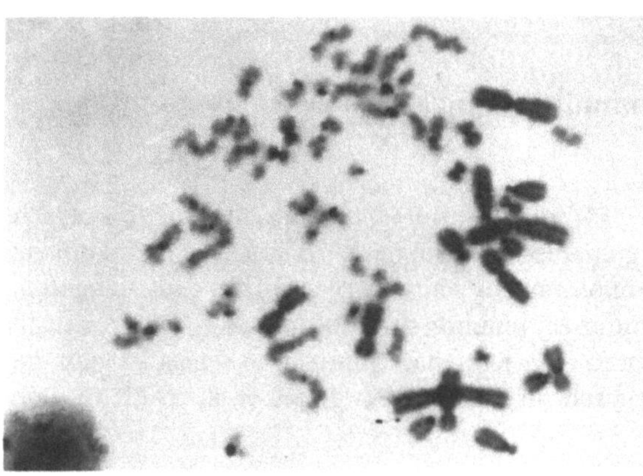

Fig. 9. Polyethyleneglycol was used to fuse metaphase CHO cells with G_1 human lymphocytes. Re-banding was induced with *AluI* endonuclease. Notice Re(C)-bands in human PCC and in some Chinese hamster metaphase chromosomes

4 Re-Banding in Big Apes Chromosomes

It has been proposed that the analysis of restriction enzyme banding in shared chromosomes (homologous chromosomes in phylogenetically related species) (Bianchi and Bianchi 1969) may provide information on the evolutionary drift of DNA sequences at restriction sites (M.S. Bianchi et al. 1985). The chromosomes of orangutan, gorilla, chimpanzee and humans have been extensively characterized and compared (see review in Seuanez 1979; Yunis and Prakash 1982). Thus far, it is known that karyotypes from human and great apes are highly homologous, differing only by a few chromosome rearrangements and by some variability in the amount and distribution of satellite DNAs. The paleontological records, the analysis of amino acid substitutions in certain proteins and the study of unique-sequence DNA divergency between species have allowed the phylogenetically history of the taxon to be reconstructed and dated (see reviews in Pilbeam 1984; Sibley and Ahlquist 1984). Accordingly, hominids seem to be the best candidate for the analysis of the behaviour of restriction DNA sites in shared chromosomes.

AluI, HaeIII, MboI, HinfI and *RsaI* endonucleases were employed to induce Re-banding in chimpanzee, gorilla and orangutan chromosomes (N.O. Bianchi et al. 1985). The results obtained can be summarized as follows: (1) The *HaeIII* G-banding pattern was the only feature common to the whole chromosome complement of humans and great apes. (2) Although these hominids share the whole chromosome complement (except for the few rearrangements differentiating the karyotypes), there is no case of similar restriction C-banding patterns between all shared chromosomes from any two given species. Thus, the three great apes exhibit a total of 12 different restriction C-banding patterns (one per enzyme and per species); if we include the data on human chromosomes, the total number of restriction C-banding patterns thus far known for hominids is 19. (3) *HinfI* did not induce banding in chimpanzee chromosomes, and *RsaI* did not induce banding in chimpanzee or orangutan chromosomes. (4) *HaeIII* induced gaps in several chromosome pairs of apes but not of humans; on the other hand, *HinfI* induced gaps in three pairs of human chromosomes but not in ape chromosomes. Figure 10 depicts the Re-banding pattern in three groups of shared chromosomes of big apes that will be used as examples in the next two sections to illustrate some points of discussion.

5 C- and Re(C)-Banding in Hominid Chromosomes and its Relationship with Human Satellite DNA

It is generally accepted that human DNA contains four (I–IV) satellite fractions which cross-react with the centromeric regions of most human chromosomes and with the centromeric regions of several chromosomes of chimpanzee, gorilla and orangutan. Distribution of human satellite fractions in hominid chromosomes has been reviewed by Seuanez (1979); it is worth mentioning here that chimpanzee seems to lack the human satellite fraction II and orangutan the fraction IV (Jones et al. 1972; Gosden et al. 1977).

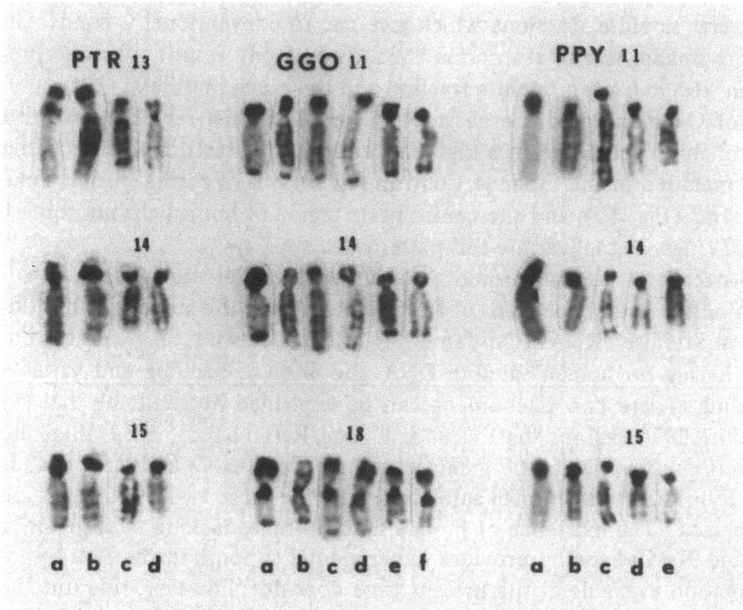

Fig. 10a–f. C-banding and Re-banding in three groups of shared chromosomes from chimpanzee (*PTR*), gorilla (*GGO*) and orangutan (*PPY*) chromosomes: **a** C-banding; **b** Re(C)(*AluI*)-banding; **c** Re(C)(G)(*HaeIII*)-banding; **d** Re(C)(*MboI*)-banding; **e** Re(C)(*HinfI*)-banding; **f** Re(C)(*RsaI*)-banding. The *numbers* above each chromosome group indicate the position of the chromosome in the karyotype of the species. Chimpanzee has been taken as the reference species; accordingly, chromosomes from this species are arranged in columns maintaining the correlative numbering, while shared chromosomes from gorilla and orangutan are placed in each row to correspond to chimpanzee chromosomes. The long arm of human chromosome 2 is shared with PTR 13, GGO 11 and PPY 11; human chromosome 12 is shared with PTR, GGO and PPY 14; human chromosome 14 is shared with PTR 15, GGO 18 and PPY 15. GGO 18 is submetacentric due to a pericentric inversion characteristic of the species

When the C- and Re(C)-banding patterns for each chromosome pair in each hominid species were compared, three types of correlations could be observed. In some cases, the C- and Re(C)-banding were similar; for instance, such is the case for C- and Re(C)-(*AluI*)(*HaeIII*)(*MboI*)-bands in chimpanzee chromosomes 14 (Fig. 10), or C- and Re(C)(*AluI*)(*HaeIII*)(*DdeI*)(*RsaI*)-bands in human chromosome 1 (Fig. 2a,b, d–f). Because the C-banded region of these chromosomes comprise three satellite fractions for chimpanzee chromosome 14 and four fractions for human chromosome 1 (Seuanez 1979), it can be assumed that these satellite families lack or have a very low concentration of restriction sites for the enzymes inducing Re(C)-banding. Several other examples of this type can be found in hominid chromosomes.

A second type of correlation is seen in those chromosomes having smaller Re(C)-bands than conventional C-bands. C- and Re(C)(*HaeIII*)-banding of orangutan chromosome 14 (Fig. 10), and C- and Re(C)(*RsaI*)-banding of human chromosomes 9 (Fig. 3a,f) provide examples of dimorphism between C- and restriction C-bands. These chromosomes contain several satellite fractions which give rise to conventional C-bands. On the other hand, the smaller size of restriction C-bands probably results from the presence of restriction sites in a given satellite fraction and their lack in others.

The presence of C-banding and the absence of Re(C)-banding represents a third pattern, which probably occurs due to a high concentration of restriction sites in the satellite fractions included in the C-bands. Centromeric regions of gorilla chromosome 14 digested with *AluI* (Fig. 10), and the centromeric region of human chromosome 1 digested with *MboI* (Figs. 2a,c) illustrate this pattern.

Several chimpanzee and gorilla chromosomes show characteristic C-banded telomeric regions exhibiting variable patterns of Re(C)-banding and the absence of hybridization with human satellite cRNAs (Gosden et al. 1977). Likewise, all great ape centromeric regions having no human satellite DNA also show C-banding and variable restriction C-banding. These two phenomena can be explained by assuming that (1) there are nonsatellite DNA regions that occur as C- and Re(C)-bands, or (2) there are satellite DNA fractions specific to the great apes that occur as C- and Re(C)-bands and that do not hybridize with human satellite fractions. These two hypotheses are not mutually exclusive. The existence of human chromosomes lacking satellite DNA but showing C- and Re(C)-banding provides experimental support to the first hypothesis. The information available at the present time does not, however, rule out the second. Thus far, there is no data on the fractionation of gorilla and orangutan DNAs. In the chimpanzee genome, on the other hand, only two satellite fractions homologous to human satellites I and III have been identified (Jones et al. 1972; Prosser et al. 1973), despite the fact that chromosomes of this species cross-react with human satellites I, III and IV (Gosden et al. 1977; Seuanez 1979). It is then evident that the fractionation method used was insufficient to isolate all the satellite fractions present in the chimpanzee DNA.

6 Variability of Restriction Sites in Shared Chromosome Regions

The response of shared chromosomes to restriction enzymes emphasizes the extreme variability in the frequency of restriction sites in homologous chromosome regions from different hominids; three examples will illustrate the point.

First, the induction of gaps by restriction enzymes results from a very high frequency of restriction sites (approximately 1 per 15–70 base pairs) in the gapped region (Bianchi et al. 1984). *HaeIII* induce gaps in several pairs of great ape chromosomes, but not in human chromosomes, whereas *HinfI* induce gaps in human (Fig. 6) but not in great ape chromosomes. Second, there are not two shared chromosome regions showing the same response to a given restriction endonuclease. Third, nonhomologous chromosome regions containing similar satellite fractions may exhibit different responses to restriction enzyme digestions; for instance, centromeric regions of gorilla chromosomes 15 and 18 comprise the same satellite fractions but show different banding after *AluI* or *HaeIII* treatments. The same phenomenon is also observed in human chromosomes 1 and 9; both pairs have the same four satellite DNA fractions; the former pair, however, does not band with *MboI* while the latter does (Figs. 2b and 3b).

The variable concentration of restriction sites in shared chromosomes suggests that the frequency of restriction enzyme recognition sites is a characteristic that in some cases evolved independently for each species after phylogenetic branching. Furthermore, the variable Re(C)-banding shown by similar satellite fractions located in nonhomologous chromosomes indicates that frequencies of restriction sites have also evolved independently for each chromosome pair. These two conclusions pose the problem of explaining how the variability in the frequency of restriction sites in shared chromosome regions arose. At least three different hypotheses can be suggested: (1) gain or deletion, (2) inactivation and (3) de novo formation of restriction sites.

Gain or deletion implies the insertion or loss of multiple restriction sequences in a DNA region that undergoes little other modification. Thus far, there is no indication of a molecular mechanism leading to multiple insertions or deletions of 4- or 5-bp-long repeated sequences at 15- to 500-pb intervals. This mechanism seems, therefore, improbable.

The inactivation or activation of restriction sites can result from DNA methylation (Stasiak and Klopotowski 1979; McClelland 1981). In fact, it has been demonstrated that DNA methylation in mammals preferentially takes place in satellite DNA (Harbers et al. 1975; Gautier et al. 1977), and that shared chromosome regions of great apes containing similar satellite DNA fractions exhibit variable degrees of methylation (Schnedl et al. 1975). At first sight, the data on DNA methylation seem to suggest that the inactivation of restriction sequences may explain the variable response of shared DNA fractions to restriction enzyme treatments. However, a more detailed analysis of the problem indicates otherwise. In mammals, methylation is restricted to cytosines of the doublet CpG (Cooper 1983). Restriction sites of the enzymes used to study hominid chromosomes do not comprise the CpG bases. Therefore, activation or inhibition of restriction through methylation is probably not involved in modulating the response to *AluI, HaeIII, HinfI, RsaI* and *MboI*. (*MboI* is inhibited by adenine methylation, but his is unique to prokaryotes.)

The finding of humanlike satellite DNA fractions in all hominid species tested indicates the evolutionary conservatism of these repeated DNA families (Seuanez 1979). Thus, it seems reasonable to assume that ancestral hominids had a library of repeated sequences that were variably amplified during speciation for each chromosome pair. Moreover, because satellite DNA is genetically inert, mutations can accumulate freely in this DNA (Brutlag 1980). Through this process, random and independent changes in the frequency of restriction sites could occur for each chromosome pair. The most cogent explanation for the variable number of restriction sites occurring during speciation in similar satellite DNA fractions located in homologous or in nonhomologous chromosomes seems to be the amplification of a basic repeated sequence, which, prior to amplification, changed the frequency of restriction sites through mutation. A mechanism of this type implies a de novo and variable formation of restriction sequences for each chromosome pair. In fact, there are clear examples of amplification of satellite DNA fractions restricted to only one chromosome pair of a given hominid species. Human chromosome 1 and gorilla chromosome 17 are two examples of independent amplification that may have occurred after speciation branching.

There is now evidence suggesting that amplification and de novo formation of restriction sequences may also explain the variable frequency of restriction sites in shared chromosome regions containing nonsatellite DNA. Measurement of the DNA content in individual chromosome pairs of five species of akodont (Cricetidae) rodents has shown that shared chromosomes may contain variable amounts of nonsatellite DNA (Bianchi et al. 1983). Therefore, amplification of nonsatellite DNA fractions may take place during speciation and perhaps may be preceded or accompanied by random mutations changing the frequency of restriction sites.

7 Visualization of the Sites of DNA Methylation by Re-Banding

Methylation of cytosine is believed to play an important role in gene regulation, but much remains to be determined about this role and about the distribution of 4-methylcytosine (5mC) at the DNA and chromosome levels. Animal cells methylate the cytosine of the dinucleotide CpG. In vertebrates, 5mC constitutes 0.7–2.8 mol% (mol 5mC per 100 bases), whereas in arthropods, cytosine methylation is only about 0.03%. Coelenterates, molluscs and echinoderms show intermediate levels of methylation (Ehrlich and Wang 1981; Cooper 1983).

Immunological identification of 5mC has been used to detect different levels of methylation between euchromatin and heterochromatin and between the heterochromatic regions of different chromosomes from the same complement (Miller et al. 1974; Schnedl et al. 1975). Immunofluorescence, however, cannot differentiate between undermethylation and low levels of CpG and the quality of antibody preparations may vary with respect to their specificity and activity. Therefore, the low levels of 5mC reported for the polytene chromosomes of *Drosophila* (Eastman et al. 1980) and *Sciara* (Wei et al. 1981) are still considered controversial (Cooper 1983).

The restriction enzymes *HhaI* (GCGC), *HinPI* (GCGC) and *HpaII* (CCGG) fail to cleave DNA when 5mC is in the restriction site. On the other hand, *MspI*, an isoschizomer of *HpaII*, has CpG in its restriction sites and cuts the DNA whether or not the C is methylated (McClelland 1981). One potential complication in the use of *MspI* is that this enzyme is influenced by flanking sequences and methylation of the outer cytosine in the restriction site.

Recently (Bianchi et al. 1986), the above endonucleases were used to digest chromosome preparations from human and mosquito (*Aedes albopictus*) cells. Similar Re-banding treatments were also performed on cells of the same species grown in the presence of 5-azacytidine (5-azaC) to inhibit methylation. The results obtained indicate that these methods are useful for detecting interspecies and intrachromosomal variations in the level of 5mC and CpG sequences. We shall summarize in this section the most important findings obtained with *HpaII* and *MspI* enzymes.

The cytological analysis of human chromosome preparations treated with *HpaII* showed chromosome staining similar to that of control slides. DNA extractions with this enzyme were 4.5% for human cells. On the other hand, *MspI* produced chromosome digestion in human chromosome preparations, inducing the appearance of Re(C)-bands, moderate Re(G)-bands and a decrease in the intensity of stain relative to that of control slides. Re(C)(*MspI*)-banding was located in the paracentromeric region of

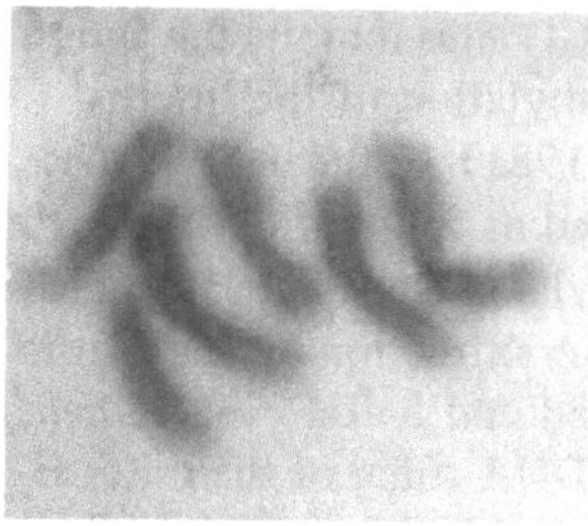

Fig. 11

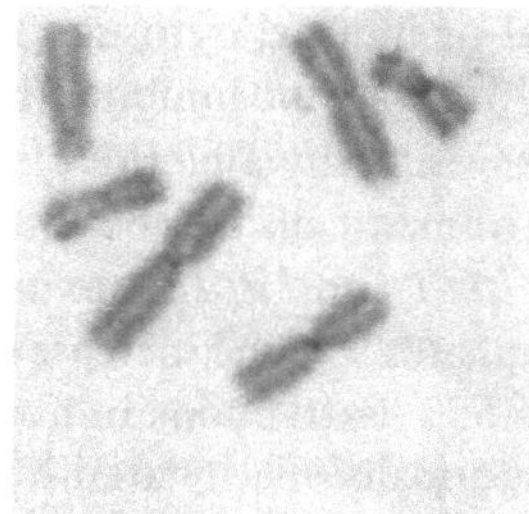

Fig. 12

Fig. 11. Mosquito metaphase treated with *MspI.* Notice the chromosome digestion and the lack of banding.

Fig. 12. Mosquito metaphase digested with *HpaII.* Notice the Re(C)-banding in the centromere region of several chromosomes

the long arm of chromosomes 1 (Fig. 7). Since *MspI* cleaves unmethylated and methylated DNA, the Re(C)-bands induced by this enzyme were probably due to a moderately low concentration of restriction sites in the Re(C)-banded regions of chromosome 1. *MspI* produced DNA extractions of about 40% for human cells. This percentage, which contrasts with the low level of extraction obtained after digestions with the *HpaII* isoschizomer, indicates that a large proportion of the internal C in the CCGG restriction sequences are methylated. *MspI* also induced Re(gaps) in the short arms of human chromosomes 13–15 and 21–22 (nucleolar organizing regions), but *HpaII* did not. These findings support the assumption of high levels of methylated CpG doublets in human ribosomal cistrons (Miller et al. 1983; Bianchi et al. 1986).

MspI induced 25–30% DNA extractions and a clear decrease of stain without banding in *A. albopictus* chromosome spreads (Fig. 11). *HpaII* induced also 25–30% DNA removal and a decrease of stain from mosquito chromosome spreads; this enzyme, however, gave rise to the appearance of small Re(C)-bands in the centromere region of four to five mosquito chromosomes (Fig. 12). The Re(C)(*HpaII*)-banding, combined with the lack of *MspI* bands, seems to suggest that the resistance to *HpaII* digestion of centromeric mosquito regions is due to high levels of CpG methylation and not to low levels of CCGG restriction sites. Demethylation with 5-azaC will give further support to this assumption.

Replacement of cytidine by 5-azaC in the DNA inhibits C-methylation due to the impossibility of the N in the 5-position to accept methyl groups, and also through a blockage of methyltransferases (Jones and Taylor 1981). In human and mosquito cells the incorporation of 5-azaC produced drastic changes in the response of cell to *HpaII* and *MspI* (incorporation of 5-azaC was obtained by growing the cells in the presence of 8 μM of the base analogue for two to three cycles).

The extraction of DNA from 5-azaC-labelled human cells was similar for *HpaII* and *MspI*; in both cases it was approximately 50–55%. Moreover, both enzymes elicited Re(C)-banding in chromosomes 1.

Methylation of the external or internal cytosines of CCGG sites inhibits the action of *HpaII*. On the other hand, *MspI* is not inhibited by methylation of the internal C but is blocked when the outer C is methylated (McClelland 1981). In the light of these facts, the above results with human cells can be interpreted as follows: (1) the high levels of DNA extraction by *HpaII* and *MspI* from demethylated DNA indicate a high frequency of CCGG restriction sites, (2) the increase in DNA extraction by *MspI* from demethylated DNA (40% and 55% extractions in nontreated and 5-azaC-treated cells respectively) suggests that in normally methylated human DNA some of the outer sequences are methylated and consequently not cleaved by *MspI*, (3) the 12-fold increase in DNA extractions by *HpaII* from demethylated DNA indicates that almost all the internal Cs of the CCGG sites are methylated in normally methylated human DNA.

The amount of DNA extracted from mosquito cells by *HpaII* and *MspI* was 25–30% independent of whether the DNA was 5-azaC-substituted or not. However, the Re(C)-bands induced by *HpaII* in mosquito chromosomes disappeared after 5-azaC incorporation. These findings confirm that Re(C)(*HpaII*)-bands in these cells are related with high levels of cytosine methylation, and that the bulk of CCGG sequences in the mosquito genome is unmethylated.

The echinoderm pattern of DNA methylation is characterized by the presence of two well-defined compartments or domains, one heavily methylated and the other unmethylated or slightly methylated. It has recently been proposed that this is the ancestral pattern of DNA methylation from which the arthropod and vertebrate patterns evolved through the loss of the methylated domain in arthropods and through methylation of the unmethylated domain in vertebrates (Cooper 1983). The results obtained with *HpaII* and *MboI* endonucleases in mosquito cells suggest a certain degree of compartmentation in the distribution of 5mC, which can be interpreted as a remnant of the ancestral echinoderm pattern.

8 Final Remarks

Even since they were discovered, restriction endonucleases have been a powerful tool to characterize the DNA. In the usual procedure, total DNA or a given DNA fraction is digested with a restriction endonuclease. Thereafter, the severed DNA is separated by electrophoresis in families of fragments of similar length; thus, it is possible to estimate how many base pairs separate two consecutive restriction sites for a given restriction endonuclease. The use of a large set of different enzymes allows an insight on the DNA base sequence, and also on the frequency of the different restriction sites in the DNA analyzed.

As commented above, Re-banding depends on the frequency and distribution of restriction sites in the chromosome complement. Hence, it is clear that both methods, electrophoretic and cytogenetic, depend upon the same phenomenon, i.e. the existence of specific base sequences which are recognized and nicked by a restriction enzyme. Yet, besides the distinct methodological approaches, the electrophoretic and the cytogenetic methods differ in some other fundamental aspects. Thus, for instance, the former technique serves to characterize of bulk DNA fractions obtained from the whole genome. Conversely, although more qualitative, Re-banding is unsurpassed in providing information about the frequency of restriction sites in individual chromosomes, or in discrete regions of individual chromosomes.

It is clear that electrophoretic and cytogenetic methods have different levels of resolution. Hence, it is not striking to find cases of coincidence and cases of disagreement between the data obtained with each technique. One example of agreement and one example of disagreement will illustrate the point.

In human cells, according to Frommer et al. (1982), satellite II DNA is cleaved by *HinfI* into a large number of very short fragments (10–80 bp); the digestion products of satellite III and IV DNA yield a range of fragments varying from 15–250 bp. On the other hand, *HinfI* digestion of satellite I DNA produces three major fragments of 770, 850 and 950 bp. The frequency of cleavage sites in satellite II, III and IV DNA is probably much higher than in the rest of the DNA. Consequently, chromosome regions rich in these satellite DNAs are more extensively digested by *HinfI* and appear as unstained gaps (chromosomes 1, 9, 16) (Fig. 6). On the other hand, the average length of fragments induced by *HinfI* in satellite I DNA is not much different from that in nonsatellite DNA (Bishop et al. 1983). Hence, the loss of DNA in chromatin regions containing satellite I DNA and in the rest of the chromatin should be about the same; accordingly, positively stained banding is not observed in human chromosomes, or appears only occasionally on chromosome 3 (Fig. 6).

Human satellite I DNA has been found to be resistant to *HaeIII* digestion (Mitchell et al. 1979). Therefore, all human chromosomes having this satellite DNA would be expected to show some sort of banding after treatment with *HaeIII*. It has been reported that 17 of the 23 chromosomes of the human haploid set contain satellite I sequences in the centromeric regions (Jones et al. 1974; Gosden et al. 1975), yet only six chromosomes pairs show banding after *HaeIII* digestion (Figs. 2–5c). The location of satellite DNA in human chromosomes has been determined by in situ hybridization with RNA probes complementary to satellite DNA (Jones et al. 1974; Gosden et al. 1975, 1981; Manuelidis 1978; Beauchamp et al. 1979). It has been shown that 40% of the molecules present in satellite III DNA can cross-reanneal with 10% of the molecules of satellite I and II DNA (Mitchell et al. 1979). Therefore, it is not surprising that several of the satellite I locations detected by in situ hybridization may well be the result of cross-reactions with satellite II, III or IV. These technical difficulties may be the cause of the contradiction between the results with restriction enzyme-induced banding of chromosomes and restriction enzyme digestion of satellite DNA fractions.

In this review we have tried to emphasize the value of the Re-banding techniques for the analysis of chromosome organization. It is evident that the cytogenetic use of restriction enzymes has provided new information on the complexity of the eukaryotic chromosome structure. However, still more challenging are the problems that can be answered by the use of Re-banding in adequate experimental models.

Which is the exact mechanism and the meaning of the restriction site variation between homologous and shared chromosome regions? How far back in the genealogy can the Re-banding polymorphisms be traced? Is there any relationship between cancer cell transformation and the chromosome variation in restriction sites? These are some of the questions which deserve further attention and which undoubtedly are of immediate applicability for a better understanding of chromosome organization.

References

Atkin NB, Brito-Babapulle V (1985) Chromosome 1 heterochromatin variants and cancer: a reassessment. Cancer Genet Cytogenet 18:325–331

Beauchamp RS, Mitchell AR, Buckland RA, Bostock CJ (1979) Specific arrangements of human satellite III DNA sequences in human chromosomes. Chromosoma 71:153–166

Bianchi MS, Bianchi NO, Pantelias GE, Wolff S (1985) The mechanism and pattern of banding induced by restriction endonucleases in human chromosomes. Chromosoma 91:131–136

Bianchi NO, Ayres JP (1971) Heterochromatin location on chromosomes of normal and transformed cells from African green monkey (*Cercopithecus aethiops*). Exp Cell Res 69:236–239

Bianchi NO, Bianchi MS (1969) Origin of the pattern and chronology of chromosome replication in vertebrates. Genetics (Suppl) 61:275–287

Bianchi NO, Redi C, Garagna C, Capanna E, Manfredi-Romanini MG (1983) Evolution of the genome size in Akodon (Rodentia, Cricetidae). J Mol Evol 19:362–370

Bianchi NO, Bianchi MS, Cleaver JE (1984) The action of ultraviolet light on the patterns of banding induced by restriction endonucleases in human chromosomes. Chromosoma 90:133–138

Bianchi NO, Bianchi MS, Cleaver JE, Wolff S (1985) The pattern of restriction-enzyme-induced banding in the chromosomes of chimpanzee, gorilla and orangutan and its evolutionary significance.J Mol Evol 22:323–333

Bianchi NO, Morgan WF, Cleaver JE (1985) Relationship of ultraviolet light-induced DNA-protein cross-linkage to chromatin structure. Exp Cell Res 156:405–418

Bianchi NO, Vidal-Rioja L, Cleaver JE (1986) Direct visualization of the sites of DNA methylation in human, and mosquito chromosomes. Chromosoma 94:362–366

Bishop DT, Williamson JA, Skolnick MH (1983) A model for restriction fragment length distributions. Am J Human Genet 35:795–815

Brutlag DL (1980) Molecular arrangement and evolution of heterochromatic DNA. Ann Rev Genet 4:121–144

Comings DE, Avelino E, Okada TA, Wyandt HE (1973) The mcheanism of C- and G-banding of chromosomes. Exp Cell Res 77:469–493

Cooper DN (1983) Eukaryotic DNA methylation. Human Genet 64:315–333

Eastman EM, Goodman RM, Erlanger BF, Miller OJ (1980) 5-Methylcytosine in the DNA of the polytene chromosomes of the diptera *Sciara coprophila, Drosophila melanogaster* and *D. persimilis.* Chromosoma 79:225–239

Ehrlich M, Wang Y-H (1981) 5-Methylcytosine in eukaryotic DNA. Science 212:1350–1357

Frommer M, Prosser J, Tkachuk D, Reisner AH, Vincent PC (1982) Simple repeated sequences in human satellite DNA. Nucleic Acids Res 10:547–563

Gautier F, Bunemann H, Grotjahn L (1977) Analysis of calf-thymus satellite DNA: evidence for specific methylation of cytosine in C-G sequences. Eur J Biochem 80:175–183

Gosden JR, Mitchell AR, Buckland RA, Clayton RP, Evans JH (1975) The location of four human satellite DNAs on human chromosomes. Exp Cell Res 92:148–158

Gosden JR, Mitchell AR, Seuanez HN, Gosden CM (1977) The distribution of sequences complementary to human satellite DNAs I, II and IV in the chromosomes of chimpanzee (*Pan troglodytes*), gorilla (*Gorilla gorilla*) and orangutan (*Pongo pygmaeus*). Chromosoma 63:253–271

Gosden JR, Lawrie SS, Gosden CM (1981) Satellite DNA sequences in the human acrocentric chromosomes: information from translocations and heteromorphisms. Am J Hum Genet 33: 243–251

Harbers K, Harbers B, Spencer JH (1975) Nucleotide clusters in deoxyribonucleic acids. XII. The distribution of 5-methylcytosine in pyrimidine oligonucleotides of mouse L-cell satellite DNA and main band DNA. Biochem Biophys Res Commun 66:738–746

Horz W, Hess I, Zachau HG (1974) Highly regular arrangement of a restriction-nuclease-sensitive site in rodent satellite DNAs. Eur J Biochem 45:501–512

Jones KW, Purdom IF, Prosser J (1974) The chromosomal location of human satellite DNA I. Chromosoma 49:161–171

Jones KW, Prosser J, Corneo G, Ginelli E, Bobrow M (1972) Satellite DNA constitutive heterochromatin, and human evolution. In: Pfeiffer RA (ed) Modern aspects of cytogenetics: constitutive heterochromatin in man. Schattauer, Stuttgart, p 45

Jones PA, Taylor SM (1981) Hemimethylated DNA prepared from 5-azacytidine-treated cells. Nucleic Acids Res 9:2933–2947

Kaelbling M, Miller D, Miller O (1984) Restriction enzyme banding of mouse metaphase chromosomes. Chromosoma 90:128–132

Lica L, Hamkalo B (1983) Preparation of centromeric heterochromatin by restriction endonuclease digestion of mouse L 929 cells. Chromosoma 88:42–49

Lima-de-Faria A, Isaksson M, Olsson E (1980) Action of restriction endonucleases on the DNA and chromosomes of *Muntiacus muntjak*. Hereditas 92:267–273

Manuelidis L (1978) Complex and simple sequences in human repeated DNAs. Chromosoma 66: 1–21

McClelland M (1981) The effect of sequence specific DNA methylation on restriction endonuclease cleavage. Nucleic Acids Res 9:5859–5866

Mezzanotte R, Bianchi U, Vanni R, Ferrucci L (1983) Chromatin organization and restriction nuclease activity on human metaphase chromosomes. Cytogenet Cell Genet 36:562–566

Miller DA, Firschein IL, Dev VG, Tantravahi R, Miller OJ (1974) The gorilla karyotype: chromosome lengths and polymorphisms. Cytogenet Cell Genet 13:536–550

Miller DA, Choi YC, Miller OJ (1983) Chromosome localization of highly repetitive human DNAs and amplified ribosomal DNA with restriction enzymes. Science 219:395–397

Mitchell AR, Beauchamp RS, Bostock CJ (1979) A study of sequence homologies in four satellite DNAs of man. J Mol Biol 135:127–149

Pilbeam D (1984) The descent of hominoids and hominids. Sci Am 250:84–96

Prosser J, Moar M, Bobrow M, Jones KW (1973) Satellite sequences in chimpanzee (*Pan troglodytes*). Biochim Biophys Acta 319:122–134

Sahasrabuddhe C, Pathak S, Hsu T (1978) Response of mammalian metaphase chromosomes to endonuclease digestion. Chromosoma 69:331–338

Schnedl W, Dev VG, Tantravahi R, Miller DA, Erlanger BF, Miller OJ (1975) 5-Methylocytosine in heterochromatic regions of chromosomes: chimpanzee and gorilla campared to the human. Chromosoma 52:59–66

Seuanez HN (1979) The phylogeny of human chromosomes. Springer, Berlin Heidelberg New York

Sibley CG, Ahlquist JE (1984) The phylogeny of the hominoid primates as indicated by DNA-DNA hybridization. J Mol Evol 20:2–15

Stasiak A, Klopotowski T (1979) Four-standard DNA structure and DNA base methylation in the mechanism of action of restriction endonucleases. J Theor Biol 80:65–82

Wei LH, Erlanger BE, Eastman EM, Miller OJ, Goodman R (1981) Inverse relationship between transcriptional activity and 5-methylcytosine content of DNA in polytene chromosomes of *Sciara coprophila*. Exp Cell Res 135:411–415

Yunis JJ, Prakash O (1982) The origin of man: a chromosomal pictorial legacy. Science 215: 1525–1530

13 Chromosome Aberrations Induced by Restriction Endonucleases

G. OBE[1], V. VASUDEV[1,2], and C. JOHANNES[1]

1 Introduction

Bacterial restriction endonucleases (RE) of type II bind at specific sequences in DNA and generally cleave both strands of the DNA inside their binding site, i.e. they produce DNA double-strand breaks (DSB). The DSBs are blunt or cohesive and have 3′-OH and 5′-phosphate ends. Generally, sequences of 4 to 6 base pairs are recognized by REs. Some REs have recognition sites with internal ambiguities of up to 6 bases, and some cut the DNA outside their recognition site, up to 13 nucleotides away. Kessler et al. (1985) described more than 500 REs with known and nearly 80 REs with unknown recognition sites (see Chap. 11, this Vol.).

Since REs recognize specific sequences in DNA and produce DSBs they represent indispensible tools for the molecular biologist. In case it would be possible to treat eukaryotic chromosomes inside the living cells with REs and in case the REs would be able to cut the DNA in the chromatin of living cells and thereby produce chromosomal aberrations, they would be of immense interest for mutation research in different ways. The following questions could be addressed:

1. What types of chromosomal aberrations are produced? Have REs S-phase-dependent or -independent activities?
2. What is the ultimate lesion for the production of chromosomal aberrations? Are REs which produce DSB with blunt ends more effective in the production of chromosomal aberrations than REs producing cohesive ends?
3. What are the dose-effect characteristics of RE-induced chromosomal aberrations?
4. What is the correlation between recognition sites of different REs and their capacity to produce chromosomal aberrations? Is it possible to influence the frequencies of RE-induced chromosomal aberrations by influencing the accessibility of the chromatin for the REs, or by influencing the recognition sites?

---

1 Institut für Genetik, Freie Universität Berlin, Arnimallee 5–7, D–1000 Berlin 33
2 Department of Zoology, University of Mysore, Manasa Gangotri, Mysore, India

Cytogenetics. Ed. by G. Obe and A. Basler

2 Chromosomal Aberrations Produced by REs

2.1 Uptake of REs by Eukaryotic Cells

REs are very effective in producing chromosomal aberrations in living cells (Figs. 1–12). Table 1 shows the REs which have been shown to induce chromosomal aberrations. As can be seen from Table 1, REs with different recognition sites producing blunt or cohesive ends are positive. Comparative analyses with different REs did not confirm the assumption that REs producing DSBs with blunt ends are more effective than REs producing DSBs with cohesive ends in the production of chromosomal aberrations (Obe et al. 1985). Important questions concerning the interpretation of comparative studies on the chromosome breaking activity of different REs are (1) how do the enzymes enter the cells, and (2) enter different REs the cells to the same extent? Both of these questions are unanswered up to now. Experiments in which the cells were treated with Sendai-virus preparations to mediate a better penetration of the enzymes (Bryant 1984; Natarajan and Obe 1984) do not lead to higher aberration frequencies than experiments in which cell pellets are directly treated with the REs, i.e. without an additional treatment to permeabilize the cells (Obe et al. 1985). Since the cells are trypsinized before they are pelleted, it was assumed that treatment with trypsin mediates the uptake of the REs. This was ruled out by experiments in which cell pelletes were generated by scraping the cells from the bottom of the petri dishes with a rubber policeman. The aberration frequencies were similar to those in trypsinized cells (Obe and Natarajan 1985). One possibility is that components of the buffers in which the REs are shipped can mediate the penetration of the REs into the cells. If this is true it could lead to differences in the extent as to which the cells take up the REs, because these buffers can be different for different enzymes. Experiments in which cells were treated with REs and recovered in medium containing bromodeoxyuridine (BrdUrd) for 18 to 22 h showed that the damaged cells were exclusively uniformly stained, i.e. they were in their first posttreatment metaphase (M1), differentially stained metaphases (M2) had aberration frequencies in the control range. Data of this type may indicate that only a fraction of cells take up the enzymes and these may show aberrations, the other cells which do not take up the REs grow as if not treated at all (Gustavino et al. 1986).

2.2 REs Induce Chromosomal Aberrations Independent of the S-Phase

Experiments with Chinese hamster ovary (CHO) cells in which different recovery times were used after treatment show that Alu I induced aberrations independent of the S-phase of the cell cycle (Obe and Winkel 1985). Treatment in the G1 phase of the cell cycle led to chromosome-type aberrations, treatment in the S-phase led to both, chromosome- and chromatid-type aberrations (sometimes in one metaphase), and treatment in the G2 phase of the cell cycle led to chromatid-type aberrations. Typical aberrations produced by Alu I in different phases of the cell cycle are shown in Figs. 1–12. These data show that REs behave like ionizing radiations, they belong to the direct acting (S-phase-independent) chromosome breaking agents. Systematic

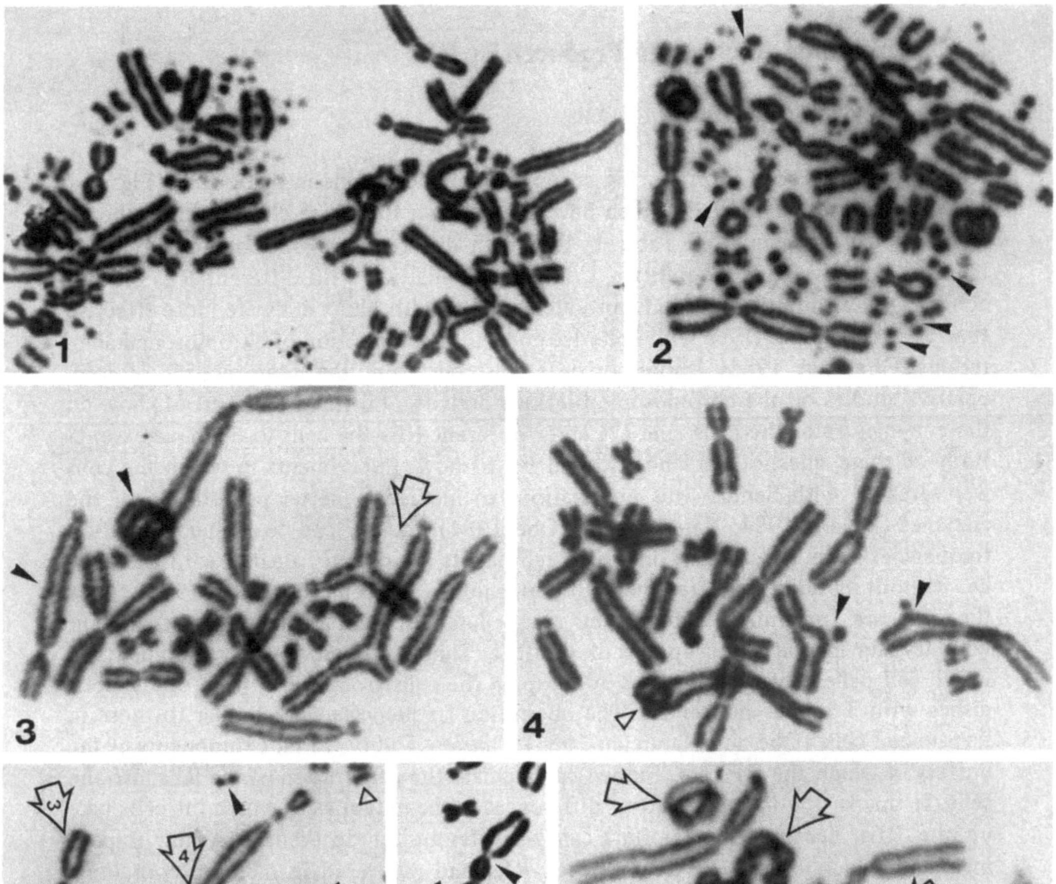
1
2
3
4
5
3
4
3
6
7
8
9
10
11
12

studies concerning this aspect have been done with Alu I (Obe and Winkel 1985) and to a lesser extent with Pvu II (Natarajan and Obe 1984). In most analyses, recovery times of 16–22 h were used, i.e. the cells were treated in the G1 phase. In all these cases the predominant aberrations were of the chromosome type, which shows the S-independent nature of the production of aberrations by REs in these cases, too.

In human peripheral lymphocytes, Alu I was shown to produce chromosomal aberrations and these were of chromosome-type when the cells were treated in G1 and of chromatid-type when the cells were treated in G2 (Obe et al. 1986a).

Figs. 1–12. Different types of aberrations induced by Alu I in CHO cells

Fig. 1. The *left* metaphase shows dicentric chromosomes, rings, minutes and fragments, the *right* metaphase shows different types of chromatid interchanges. At the time of treatment the left cell was in the G1 phase and the right cell was in the S-phase of the cell cycle

Fig. 2. Polycentric chromosomes, ring chromosomes, minutes and fragments. The minutes, which represent very small acentric rings, are indicated by *arrowheads.* At the time of treatment this cell was in the G1 phase of the cell cycle

Fig. 3. Dicentric chromosome and ring chromosome (*arrowheads*) and two triradials which represent chromatid aberrations (*arrow*). At the time of treatment this cell was in G1 or in the very early S-phase of the cell cycle

Fig. 4. Interstitial deletions (*filled arrowheads*) and a chromatid intrachange (*open arrowhead*). At the time of treatment this cell was in the S-phase of the cell cycle

Fig. 5. Polycentric chromosomes (*arrows,* the *numbers* indicate the number of centromeres in the polycentrics), minutes (*filled arrowheads*) and fragments (*open arrowheads*). At the time of treatment this cell was in the G1 phase of the cell cycle

Fig. 6. Triradial between a dicentric and a tricentric chromosome. The centromeres are indicated by *filled arrowheads.* The *open arrowhead* indicates the exchange which led to the formation of the triradial. At the time of treatment this cell was in G1 or in the very early S-phase of the cell cycle

Fig. 7. Ring chromosomes (*arrows*), dicentric chromosomes and fragments. The big ring chromosome in the *middle* has multiple, interlocked areas which result from sister chromatid exchanges. At the time of treatment this cell was in the G1 phase of the cell cycle

Fig. 8. Chromatid interchanges (*arrows*). At the time of treatment this cell was in the S-phase of the cell cycle

Fig. 9. Interstitial deletion (*filled arrowhead*) and duplication-deletion (*open arrowhead*). At the time of treatment this cell was in the S-phase of the cell cycle

Fig. 10. Chromatid interchange (*open arrowhead*) between a monocentric and a dicentric chromosome (*filled arrowheads*). At the time of treatment this cell was in the S-phase of the cell cycle

Fig. 11. Hexacentric chromosome (*filled arrowheads*), a dicentric chromosome, minutes and fragments. At the time of treatment this cell was in the G1 phase of the cell cycle

Fig. 12. Hexacentric chromosome (*filled arrowheads*) and fragments. At the time of treatment this cell was in the G1 phase of the cell cycle

Table 1. List of restriction endonucleases (RE) found to induce chromosomal aberrations in mammalian cells[a]

RE tested	Recognition sites (5′ to 3′) and cleavage sites (slashes)	Cells treated with the REs	References[b]
Alu I	AG/CT	Chinese hamster ovary (CHO)	5–8, 10, 13, 14
		Chinese hamster V79 cells	11
		Human diploid fibroblasts	4
		Human peripheral lymphocytes	9
Asu III	A T G /CG C G C	CHO	3
		Mouse fibroblasts (PG19)	3
Bam HI	G/GATCC	CHO	2, 3, 5
		PG19	3
Bsp	G C GAGCA/C T T	CHO	10
Bst NI	A CC/ GG T	CHO	10
Cfo I	GCG/C	CHO	13
Dra I	TTT/AAA	CHO	12, 15
Eco RI	G/AATTC	CHO	13
Eco RV	GAT/ATC	CHO	3, 12
Hpa I	GTT/AAC	CHO	10
Hpa II	C/CGG	CHO	10
Msp I	C/CGG	CHO	10
Nci I	G CC/ GG C	CHO	4
Nun II	GG/CGCC	CHO	3
Pvu II	CAG/CTG	CHO	3
		V79	1
		PG19	3
Rsa I	GT/AC	CHO	10
Sca I	AGT/ACT	CHO	12
Sma I	CCC/GGG	CHO	3
Taq I	T/CGA	CHO	13

[a] The 5′ to 3′ strands of the recognition sites are given. *Slashes* indicate the cleavage sites.

[b] References: 1. Bryant (1984); 2. Gustavino et al. (1986); 3. Natarajan and Obe (1984); 4. Natarajan and Obe (unpubl.); 5. Obe and Johannes (1987); 6. Obe and Kamra (1986); 7. Obe and Natarajan (1985); 8. Obe and Winkel (1985); 9. Obe et al. (1986a); 10. Obe et al. (1985); 11. Obe et al. (1986b); 12. Obe et al. (unpubl.); 13. Stoilov et al. (1986); 14. Vasudev and Obe (1987); 15. Vasudev and Obe (unpubl.).

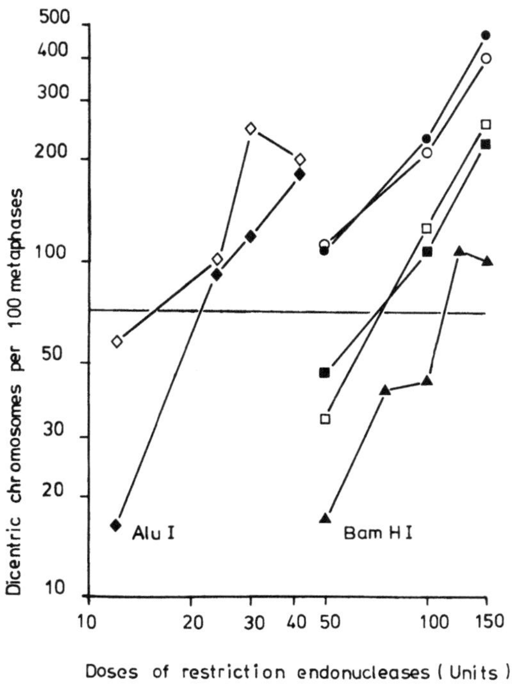

Fig. 13. Dose-effect relationships for the frequencies of dicentric chromosomes as calculated from all polycentric chromosomes (for calculation see Table 3) obtained by treating CHO cells with Alu I or with Bam HI. The recovery time after the treatment was 18 or 22 h, only M1 metaphases were analyzed. Pooled data from two to three independent experiments are given. The continuous line indicates the number of dicentric chromosomes per 100 metaphases found after irradiation of CHO cells with 4 Gy of 50 kV X-rays. (Data from Obe and Johannes 1987).
◇ Treatment of 1×10^6 cells with Alu I, recovery time 22 h;
◆ treatment of 4×10^6 cells with Alu I, recovery time 22 h;
○ treatment of 1×10^6 cells with Bam HI and 3.2 M $(NH_4)_2SO_4$, recovery time 18 h;
● treatment of 1×10^6 cells with Bam HI and 3.2 M $(NH_4)_2SO_4$, recovery time 22 h;
□ treatment of 1×10^6 cells with Bam HI, recovery time 18 h;
■ treatment of 1×10^6 cells with Bam HI, recovery time 22 h;
▲ treatment of 4×10^6 cells with Bam HI, recovery time 18 h

2.3 Quantitative Analyses

Results of experiments aimed at the production of chromosomal aberrations with REs are characterized by considerable quantitative variabilities. The causes for these variabilities are manifold, such as the physiological state of the cell population treated, or variations in the number of cells treated, to mention only two factors which may be of special importance. Nevertheless, respective studies with CHO cells have shown that the frequencies of chromosomal aberrations produced by REs follow a dose-effect relationship. Recently, we analyzed dose-effect relationships of the frequencies of polycentric chromosomes under experimental conditions in which only M1 cells were analyzed (Obe and Johannes 1987). Even in these analyses the outcome of independent experiments showed considerable quantitative variabilities. The results obtained with Alu I or Bam HI are shown in Fig. 13. The dose-effect relationships are linear or nearly so. The linear regressions calculated from these data have correlation coefficients which are all greater than 0.8 (Obe and Johannes 1987). What can be seen in Fig. 13 is that treatment of high cell numbers (4×10^6) leads to less aberrations than treatment of low cell numbers (1×10^6). Using the regression lines calculated from the aberration frequencies obtained by treatment of 1×10^6 cells, we were able to compare the chromosome breaking activity of Alu I and Bam HI with 50 kV X-rays, whose chromosome breaking activities were analyzed under similar experimental conditions with the same type of cells. One Gy of X-rays produced the same amount of polycentric chromosomes as 2 U of Alu I or 7.9 U Bam HI (Obe and Johannes 1987). In Fig. 13 the value of polycentric chromosomes obtained by irradiating CHO cells with 4 Gy X-rays is given for comparison. These data clearly show that quantitative analyses concerning the chromosome breaking activity of REs are possible and generate meaningful results.

The dose-effect characteristics obtained for polycentric chromosomes show a strong linear component after treatment with REs and a quadratic component after treatment with X-rays (Obe and Johannes 1987), indicating that with respect to the dose-effect characteristics of polycentric chromosomes, REs act rather like neutrons than like X-rays.

3 Importance of the Type of DSB Produced for the Chromosome Breaking Activity of REs

Comparative analyses with different types of REs producing DSB with blunt or cohesive ends do not support the idea that REs producing DSB with blunt ends are more effective in inducing chromosomal aberrations than REs producing cohesive ends (Obe et al. 1985). Dra I which produces blunt DSB is clearly less active than Alu I which also produces blunt ends. Dra I is even less active than Bam HI which produces cohesive ends of 4 bases. Data from comparable experiments (treatment of 4×10^6 cells, recovery after RE treatment 18 to 22 h), show that about 70 polycentric chromosomes in 100 metaphases are produced by 20 U Alu I, 100 U Bam HI and 250 U Dra I.

4 Importance of Recognition Sites for the Chromosome Breaking Activity of REs

4.1 Comparative Analyses

The data discussed in Section 3 point to the recognition sites as being responsible for the quantitative differences obtained with respect to the aberration frequencies produced by different REs. The high effectivity of Alu I to break mammalian chromosomes may be explained by the enormous repetition of Alu sequences in the mammalian genome (Jelinek and Schmid 1982; Evans 1984). As discussed in Section 2, it

Table 2. Dicentric chromosomes induced in CHO cells following heat treatment and posttreatment with Alu I[a]

Treatments: Heat	Treatments: Alu I	Recovery time between heat treatment and treatment with Alu I	Dicentric chromosomes per 100 metaphases
–	–	0	2
–	+	0	77
+	–	0	0
+	+	0	23
+	+	1 h	47
+	+	5 h	24
+	+	22 h	64

[a] Heat treatment (46°C for 6 min) was followed by recovery times of different duration before the cells were treated with 30 U Alu I. Following the treatments, the cells were recovered in the presence of BrdUrd for up to 22 h and only uniformly stained M1 metaphases were analyzed. Pooled data from two to six independent experiments are given (Vasudev and Obe 1987).

cannot be ruled out that different activities of REs with respect to their chromosome breaking activity could partly result from differences in the extent to which the cells take up different enzymes.

4.2 Influence of the Configuration of the Chromatin: Heat Shock

We tried to influence the recognition sites of REs in the chromatin and to see whether this can influence the chromosome breaking activity of the REs. In one series of experiments we treated CHO cells with heat (46°C for 6 min) and posttreated such cells with Alu I, following different recovery times after the heat shock (Vasudev and Obe 1987). As can be seen in Table 2, when cells are treated up to 5 h after the heat shock with Alu I, less aberrations are produced when compared to cells which were not pretreated with heat. Recovery after the heat shock for 22 h no longer reduces the ability of Alu I to induce chromosomal aberrations. The interpretation of these data is based on the finding of others that heat shock in mammalian cells leads to the accumulation of nonhistone proteins in the chromatin, and to a protection of the chromatin against attack by nucleases (Tomasovic et al. 1978; Warters et al. 1980). With respect to chromosome breaking activity of Alu I these data may indicate that the protection of DNA, i.e. of recognition sites in the chromatin by the heat-induced accumulation of proteins, leads to less aberrations. This draws the attention to the fact that the suprastructure of the chromatin itself may be of great importance for the capacity of different REs to produce chromosomal aberrations in that it can protect the DNA from being cut by the REs. A direct correlation between recognition sites and RE-induced chromosomal aberrations, therefore, cannot be expected.

4.3 Influence of the Configuration of the Chromatin: Salt

A general change in the suprastructure of the chromatin can be achieved by treatment with different salts. Treatment of CHO cells with Alu I in the presence of a high concentration of $(NH_4)_2SO_4$ (3.2 M; Obe and Kamra 1986), NaCl (3.2 M, Kamra and Obe, in prep.) and $CaCl_2$ (1.6 M; Tuschy, unpublished results) leads to significantly higher aberration frequencies when compared to the frequencies obtained in the absence of salt. Since Alu I-treated cells have higher aberration frequencies even when the cells are posttreated with $(NH_4)_2SO_4$ (Obe and Kamra 1986) the effect of the salt cannot be explained by assuming a better penetration of the enzyme in the presence of the salt. Another possible explanation would be that Alu I changes the specificity of its recognition site under the influence of high salt concentrations. Such a "relaxed activity" which would lead to more recognition sites being cut by the enzyme is not known for Alu I and was not found in the presence of $(NH_4)_2SO_4$ (see Obe and Kamra 1986). The salt activity cannot be explained by an influence on the cell cycle. In our previous work we did not label the cells with BrdUrd, thus we cannot be sure that M1 metaphases were analyzed exclusively. In the meantime we have data generated from M1 cells and they show the same phenomenon, namely a significant elevation of the aberration frequencies under the influence of salt (Tuschy, unpublished results). With Bam HI we also found a significant effect of salt (Fig. 13). Bam HI is known to have "relaxed activities" under different experimental conditions (George et al. 1980; Malyguine et al. 1980; Kolesnikov et al. 1981); whether this is the case in the presence of $(NH_4)_2SO_4$ is not known.

The finding that posttreatment of Alu I-treated cells with $(NH_4)_2SO_4$ leads to an elevation of the Alu I-induced chromosomal aberrations may indicate that Alu I is active inside the cell nucleus for a longer time. In one experiment we found that even after a recovery time of 40 min in the absence of Alu I, salt treatment still led to an elevation of the aberration frequencies (Obe and Kamra 1986).

In biochemical experiments, high salt concentrations (1.6 M NaCl as compared to 0.6 M NaCl) have been shown to strip off protein from chromatin (Clark and Felsenfeld 1971). Wilkins and Hart (1974) have shown that an access of chromatin from UV-irradiated cells to micrococcus endonuclease is considerably increased (about 60%) when the respective assay is performed in 2 M NaCl as compared to less than 0.15 M salt. These results are discussed by the authors to mean that sites which are normally inaccessible to the micrococcus endonuclease are masked by nucleoprotein. This protein can be removed from the chromatin by the high salt concentration and this in turn makes more sites available for the action of the endonuclease. These data could serve as an in vitro model for our analyses with intact cells.

4.4 Modification of Recognition Sites: Substitution

Another possibility to probe into the relationship between recognition sites and the capacity of an RE to induce chromosomal aberrations is to influence the sites directly by substituting thymine (T) with bromouracil (B). Substitution of T with B can be achieved by allowing the cells to incorporate BrdUrd instead of thymine. We pre-

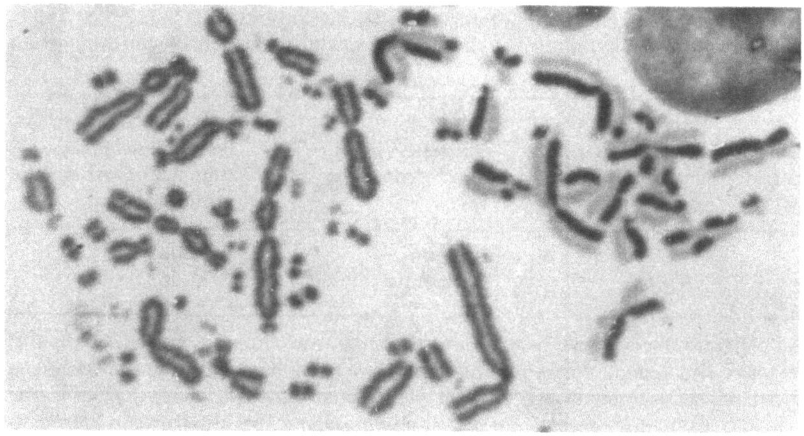

Fig. 14. Chromosomal aberrations induced by Dra I in CHO cells with DNA of different constitution at the time of treatment. The uniformly stained metaphase was TT-TT and the differentially stained metaphase was TB-BT at the time of treatment. The uniformly stained metaphase has multiple aberrations (polycentric chromosomes, minutes and fragments), the differentially stained metaphase is free of aberrations (see Table 3 and text)

treated CHO cells with BrdUrd for 15 h and treated these cells with Dra I (recognition site: TTT/AAA). After a recovery time of 16 h in the absence of BrdUrd, we analyzed differentially and uniformly stained metaphases in the same preparations with respect to chromosomal aberrations. Metaphases with differentially stained chromosomes should originate from cells which were uniformly substituted with B during the 15-h exposure to this compound (TB-BT). Since these cells grew for one cell cycle after the RE treatment in the absence of BrdUrd, their chromosomes have the constitution TB-TT and can be stained differentially. The uniformly stained metaphases remained unsubstituted during the 15 h in the presence of BrdUrd and have the constitution TT-TT. Figure 14 and Table 3 show the result of such an analysis. There are clearly less aberrations in differentially than in uniformly stained metaphases and this shows that substitution of T by B, i.e. a change in the recognition site, leads to less aberrations. Preliminary results from this laboratory show that Alu I treatment of BrdUrd-substituted cells leads to aberration frequencies similar to those obtained after treatment of unsubstituted cells. Alu I has the recognition site AG/CT and Dra I TTT/AAA. The important difference with respect to T is that it is one base away from the site to be cut in Alu I but adjacent to this site in Dra I. Petruska and Horn (1980) analyzed the digestion of DNA from Syrian hamster melanoma cells nearly 100% substituted or unsubstituted with BrdUrd, using Hpa I (GTT/AAC) and Mbo I (/GATC). They found that BrdUrd substitution leads to an eightfold inhibition of cleavage by Hpa I and a fivefold stimulation of cleavage by Mbo I. With respect to the position of T, a similar situation pertains to Dra I, in that it is at the cleavage site. Petruska and Horn (1980) discussed their results to mean that the bromine atom in the BrdUrd holds the RE away from the site of cleavage when it is situated directly at the cleavage site. The inhibitory effect of substitution of T by B with respect to the chromosome breaking

Table 3. Dicentric chromosomes as calculated from all polycentric chromosomes induced by 200 U Dra I in CHO cells in which the thymine (T) in the chromosomal DNA was either substituted with BrdUrd (B) or not substituted[a]

Chromosome constitution at the time of treatment with Dra I	Chromosome constitution at the time of analysis	Staining characteristics of the chromosomes	Dicentric chromosomes per 100 metaphases (number of metaphases analyzed)
TT-TT	TT-TT	Uniform	89 (100)
TB-BT	TB-TT	Differential	7 (100)

[a] The number of dicentric chromosomes associated with polycentric chromosomes with more than two centromeres was calculated by substrating one centromere from the polycentric and taking the number of the remaining centromeres as the number of dicentric chromosomes (e.g. tricentric chromosome = two dicentric chromosomes; hexacentric chromosome = five dicentric chromosomes) (unpublished results from one experiment).

activity of Dra I can be discussed in the same sense. The stimulatory effect of B-substitution on the digestion by Mbo I is discussed to mean that the bromine leads to a tighter binding of the RE. Our cytological results with Alu I are not comparable with this because in Mbo I the T is farther away from the site to be cut than in Alu I.

Berkner and Folk (1979) showed that DNAs containing uracil, 5-hydroxymethyluracil, 5-bromouracil or glycosylated 5-hydroxymethylcytosine were generally less susceptible to degradation by different REs than unsubstituted DNAs.

5 DSBs are the Ultimate Lesions for the Production of Chromosomal Aberrations

DNA is the primary target for the induction of chromosomal aberrations, but since mutagens produce a large variety of different lesions in the chromosomal DNA the question is: what is the ultimate lesion which gives rise to chromosomal aberrations? The dose-effect characteristics of dicentric chromosomes produced by ionizing radiations with different relative biological effectiveness and energy transfer characteristics indicate that DSBs are the ultimate lesions for the production of chromosomal aberrations. Experiments with an endonuclease from *Neurospora crassa* indicate that transformation of DNA single-strand breaks to DSBs in the chromatin of living cells, leads to chromosomal aberrations (Obe et al. 1982).

Bryant (1984) was able to show that treatment of Chinese hamster V79 cells with Pvu II and Bam HI leads to the production of DNA strand breaks, which can be expected to be mainly DSB. It has been shown that REs, including Alu I, are able to cleave DNA-RNA hybrids (Molloy and Symons 1980), and this could lead to DNA single-strand breaks (SSB) in areas of the chromatin which are active in the transcription of RNA. Posttreatment of RE-treated cells with *Neurospora* endonuclease should lead to an elevation of the aberration frequencies in case SSBs are present. Indeed, respective experiments revealed a significant elevation of the aberration frequencies

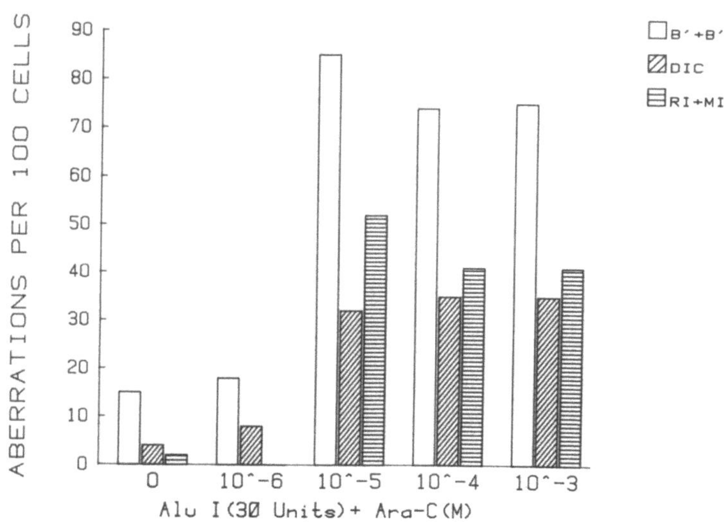

Fig. 15. Chromosomal aberrations per 100 metaphases induced in CHO cells by Alu I in the absence and presence of different concentrations of cytosine arabinoside (ara C). Recovery time after the treatments was 22 h. B′ + B″ = chromatid and chromosome breaks; DIC = dicentric chromosomes; RI + MI = ring chromosomes and minutes. (Data from Obe and Natarajan 1985)

when cells were treated with Alu I and posttreated with the *Neurospora* endonuclease. Respective analyses show that this did not result from the *Neurospora* enzyme but from the high concentration of $(NH_4)_2SO_4$ in which the enzyme is suspended (Obe and Kamra 1986). These experiments cannot rule out that some SSBs can be produced. If some SSBs are induced, this could explain the few cases in which chromosome-type aberrations and chromatid-type aberrations, namely triradials, are found in the same cells after treatment in G1. Figure 3 is an example of a very rare case in which two triradials occur together with a dicentric chromosome and a ring chromosome in one metaphase. These cases could also be explained by assuming that such cells were exposed to the enzyme in the very early S-phase of the cell cycle, when most of the DNA is still unreplicated and that the cell cycle of these cells is extremely prolongated.

Apart from the possibility of the production of a few SSBs, it can be expected that the main type of lesion produced by REs is the DSB. Most of these DSBs will be sealed back by ligases and will not give rise to chromosomal aberrations. This can be evidenced by the inhibition of strand-break repair with cytosine arabinoside (ara C). Treatment of cells with Alu I in the presence of ara C leads to significantly higher aberration frequencies (Obe and Natarajan 1985; Fig. 15), this has also been shown with Pvu II in CHO and mouse cells (Natarajan and Obe 1984).

Repair of RE-induced breaks can be deduced from another, more indirect evidence. A comparison of the intercellular distribution of polycentric chromosomes with the theoretical Poisson distributions show that there are generally too many cells with no aberration and cells with one and two aberrations are too rare. Cells with higher

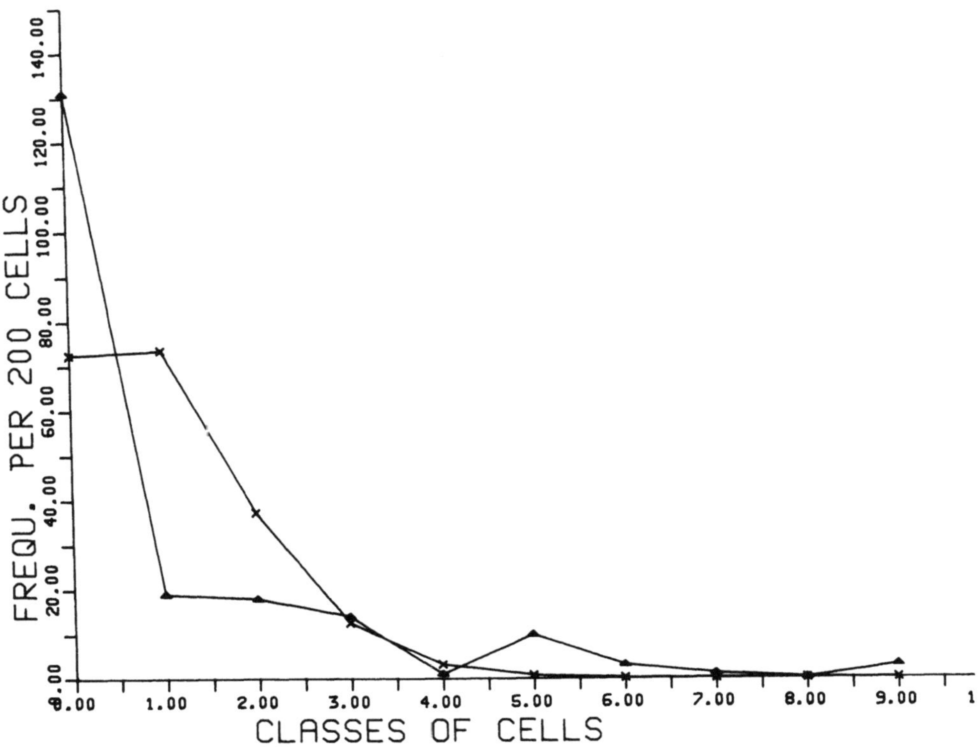

Fig. 16. Intercellular distribution of dicentric chromosomes as calculated from the polycentric chromosomes (for calculation see Table 3) in comparison with the theoretical Poisson distribution. 4×10^6 cells were treated with 150 U Bam HI. The cells were recovered for 18 h in the presence of BrdUrd and M1 metaphases were analyzed. Pooled data from two independent experiments. ▲ Empirical distribution; × theoretical Poisson distribution. (Data from Obe and Johannes 1987)

numbers of aberrations are too frequent when compared with the theoretical Poisson distribution. This is especially pronounced when cells were treated with high doses, i.e. when the aberration frequencies are high (see Fig. 16). Similar results were obtained for chromatid translocations and for chromatid breaks in human lymphocyte chromosomes treated with an alkylating agent (Obe 1969). These typical deviations from the theoretical Poisson distributions can be interpreted by assuming repair activities. Cells with few lesions will appear in the 0-class when these lesions are repaired. Such a mechanism will lead to a deficit of cells with one or few aberrations. Cells with more aberrations can be assumed to be disturbed in their repair capacities (Obe 1969; Köhler et al. 1976).

In Chinese hamster V79 cells, Alu I has been shown to induce mutations at the hypoxanthine phosphoribosyltransferase locus, but not at the Na^+/K^+ ATP ase locus (Obe et al. 1986b); this shows again that REs induce exclusively DNA strand breaks.

6 Conclusions

Restriction endonucleases are an ideal tool to break mammalian chromosomes in different stages of the cell cycle. They represent the only chromosome breaking agent which produces mainly one type of DNA damage, namely DNA double-strand breaks. These enzymes act like scissors which can be used to cut the chromosomes and thereby allow us to gain more insight into the mechanism by which chromosomal aberrations are formed.

Acknowledgements. Our work reported in this chapter was supported by the Deutsche Forschungsgemeinschaft (Ob 49/2-1). One of us (VV) is grateful for a DAAD fellowship.

References

Berkner KL, Folk WR (1979) The effects of substituted pyrimidines in DNAs on cleavage by sequence-specific endonucleases. J Biol Chem 254:2551–2560

Bryant PE (1984) Enzymatic restriction of mammalian cell DNA using Pvu II and Bam HI: evidence for the double-strand break origin of chromosomal aberrations. Int J Radiat Biol 46: 57–65

Clark RJ, Felsenfeld G (1971) Structure of chromatin. Nature New Biol 229:101–106

Evans HJ (1984) Structure and organisation of the human genome. In: Obe G (ed) Mutations in man. Springer, Berlin Heidelberg New York Tokyo, pp 58–100

George J, Blakesley RW, Chirikjian JG (1980) Sequence-specific endonuclease Bam HI. J Biol Chem 255:6521–6524

Gustavino B, Johannes C, Obe G (1986) Restriction endonuclease Bam HI induces chromosomal aberrations in Chinese hamster ovary (CHO) cells. Mutation Res 175:91–95

Jelinek WR, Schmid CW (1982) Repetitive sequences in eukaryotic DNA and their expression. Annu Rev Biochem 51:813–844

Kessler C, Neumaier PS, Wolf W (1985) Recognition sequences of restriction endonucleases and methylases – a review. Gene 33:1–102

Köhler W, Loeschcke V, Obe G (1976) Analysis of intercellular distributions of chromatid aberrations. Mutat Res 34:427–436

Kolesnikov VA, Zinoviev VV, Yashina LN, Karginov VA, Baclanov MM, Malygin EG (1981) Relaxed specificity of endonuclease Bam HI as determined by identification of recognition sites in SV 40 and pBR 322 DNAs. FEBS Lett 132:101–104

Malyguine E, Vannier P, Yot P (1980) Alteration of the specificity of restriction endonucleases in the presence of organic solvents. Gene 8:163–177

Molloy PL, Symons RH (1980) Cleavage of DNA.RNA hybrids by type II restriction enzymes. Nucleic Acids Res 8:2939–2946

Natarajan AT, Obe G (1984) Molecular mechanisms involved in the production of chromosomal aberrations III. Restriction endonucleases. Chromosoma 90:120–127

Obe G (1969) Die interzelluläre Verteilung chemisch induzierter Chromatidbrüche und -translokationen bei menschlichen Leukozyten in vitro. Z Naturforsch 24b:1207–1208

Obe G, Johannes C (1987) Chromosomal aberrations induced by the restriction endonucleases Alu I and Bam HI: comparison with X-rays. Biol Zentralbl 106:175–190

Obe G, Kamra OP (1986) Elevation of Alu I-induced frequencies of chromosomal aberrations in Chinese hamster ovary cells by *Neurospora crassa* endonuclease and by ammonium sulfate. Mutat Res 174:35–46

Obe G, Natarajan AT (1985) Chromosomal aberrations induced by the restriction endonuclease Alu I in Chinese hamster ovary cells: influence of duration of treatment and potentiation by cytosine arabinoside. Mutat Res 152:205–210

Obe G, Winkel E-U (1985) The chromosome-breaking activity of the restriction endonuclease Alu I in CHO cells is independent of the S-phase of the cell cycle. Mutat Res 152:25–29

Obe G, Natarajan AT, Palitti F (1982) Role of DNA double-strand breaks in the formation of radiation-induced chromosomal aberrations. In: Natarajan AT, Obe G, Altmann H (eds) DNA repair, chromosome alterations and chromatin structure. Elsevier/North-Holland Biomedical Press, Amsterdam New York, pp 1–9

Obe G, Palitti F, Tanzarella C, Degrassi F, De Salvia R (1985) Chromosomal aberrations induced by restriction endonucleases. Mutat Res 150:359–368

Obe G, Jonas R, Schmidt S (1986a) The restriction endonuclease Alu I induces chromosomal aberrations in human peripheral lymphocytes in vitro. Mutat Res 163:271–275

Obe G, Von der Hude W, Scheutwinkel-Reich M, Basler A (1986b) The restriction endonuclease Alu I induces chromosomal aberrations and mutations in the hypoxanthine phosphoribosyltransferase locus, but not in the Na^+/K^+ ATPase locus in V79 hamster cells. Mutat Res 174: 71–74

Petruska J, Horn D (1980) Bromodeoxyuridine substitution in mammalian DNA can both stimulate and inhibit restriction cleavage. Biochem Biophys Res Commun 96:1317–1324

Stoilov L, Mullenders LHF, Natarajan AT (1986) Influence of bromodeoxyuridine substitution of thymidine on sister-chromatid exchanges and chromosomal aberrations induced by restriction endonucleases. Mutat Res 174:295–301

Tomasovic SP, Turner GN, Dewey WC (1978) Effect of hyperthermia on nonhistone proteins isolated with DNA. Radiat Res 73:535–552

Vasudev V, Obe G (1987) Effect of heat treatment on chromosomal aberrations induced by the alkylating agent trenimon or the restriction endonuclease Alu I in Chinese hamster ovary (CHO) cells. Mutat Res 178:81–90

Warters RL, Roti Roti JL, Winward RT (1980) Nucleosome structure in chromatin from heated cells. Radiat Res 84:504–513

Wilkins RJ, Hart RW (1974) Preferential DNA repair in human cells. Nature (London) 247:35–36

14 DNase I, Chromosome Aberrations, and Cancer

M. Zajac-Kaye[1]

1 Introduction

The concept that malignancy is caused by a change in cellular genetic material was postulated as early as 1902 by Theodore Boveri. He wrote in his book "...the essence of my theory is not the abnormal mitosis, but a certain abnormal chromatin-complex, no matter how it arises. Every process which brings about this chromatin would lead to a malignant tumor" (Boveri 1914). Chromosome abnormalities associated with a variety of cancer cells, however, have been studied only since the 1960s. In the last decade the use of improved banding techniques has shown that the chromosome defects are present in most neoplasias (Yunis 1983) and that specific chromosome lesions consistently accompany certain types of tumors (Rowley 1983). The relevance of chromosome aberrations to carcinogenesis has also been under extensive investigation in various cellular systems where agents such as chemicals, radiation, and viruses have induced chromatid breaks and exchanges (Benedict 1972; Sugiyama 1975; Evans 1977; German 1983). Although the molecular mechanism underlying such chromosome alterations remains unknown, it is hypothesized that these changes play a significant role in the development of the neoplastic process.

Recent data has identified a number of normal cellular genes which are "activated" in various types of human cancer where they are presumed to function as oncogenes (Cooper 1982; Bishop 1983). Some of these oncogenes have been found either directly or in the vicinity of chromosome break points (Dalla-Favera et al. 1982; Shtivelman et al. 1985). The observation that genes with oncogenic potential are found in a region where frequent, well-characterized chromosome aberrations take place allows us to address an important but poorly understood question: What is the molecular mechanism by which such precise chromosome rearrangements are generated and, in particular, do cellular oncogenes or the regions which surround them provide a specific site for a chromosome break point to occur.

The cellular *myc* gene, located on chromosome 8, is consistently involved in chromosome translocations in Burkitt's lymphoma (Dalla-Favera et al. 1982; Taub et al. 1982). This locus is surrounded by chromatin sites which are hypersensitive to digestion with DNase I raising the possibility that scission of chromatids could be

1 Medicine Branch, National Cancer Institute, National Institute of Health, Bethesda, MD 20892, USA

Cytogenetics. Ed. By G. Obe and A. Basler

produced by enzyme action. The availability of specific sites recognized by endonucleolytic enzymes could explain why the same chromosome aberrations are consistently found in certain types of neoplasia. Loci specificity for DNase I action could result from a regulated exposure of only certain parts of the genetic material to enzymatic attack.

Chromosome aberrations induced by radiation, viral infection, or a variety of chemical compounds have been thought to be due to a direct physical or chemical attack on genetic material. Since so many diverse agents can produce chromosome aberrations and neoplasia, it seems likely that some common mechanism may be involved. Diverse carcinogenic agents such as chemicals, radiation, and viruses could induce their effect either by the production of new hydrolytic enzymes (as in the case of viruses) or by destruction of lysosomal organelles, releasing bound hydrolytic enzymes into the cellular cytoplasm. Although direct interactions with DNA or interference with repair mechanisms cannot be excluded, enzyme activation could represent an important contributory mechanism in the chromosome damage induced by chemicals, radiation, or viruses. The significance of endonucleolytic enzyme activation and its possible role in chromosome aberrations and neoplasia has, so far, been overlooked.

In this chapter, I will discuss the effect of an endonucleolytic enzyme, DNase I, exogenously introduced into living mammalian cells. I will first briefly review what is known about the mechanism of DNase I action and then present evidence indicating that DNase I when delivered into the cell cytoplasm by means of liposomes is (1) able to reach the cell nucleus giving rise to mutations and chromosome aberrations and (2) can neoplastically transform mammalian cells in culture.

2 Mechanism of DNase I Action on DNA in Chromatin

Pancreatic DNase I introduces single-stranded breaks in duplex DNA generating 5′-P and 3′-OH termini. When intact chromatin is digested with DNase I and the resulting DNA denatured and subjected to gel electrophoresis, a pattern of 10 base multiple ladder fragments are obtained (Noll 1974). This indicates that there are sites on DNA in the nucleosome which are exposed, making them accessible to DNase I action, and that there is a periodicity of these exposed sites. Similarly, when chromatin is digested with DNase I and then the DNA is extracted and subjected to electrophoresis under nondenaturing conditions, the resulting duplex fragments give a series of bands, again at a periodicity of 10. These findings indicate that both strands of the duplex molecules are cut and that even though the recognition region on both strands are separate they are located in close proximity.

The DNA fragments generated by DNase I treatment have protruding single-stranded ends as demonstrated by S1 nuclease treatment (Lutter 1977; Sollner-Webb and Felsenfeld 1977). A simplified diagram of the DNase I pattern of digestion of DNA in chromatin is shown in Fig. 1.

An intriguing feature of DNase I is that it preferentially affects active transcriptional regions of chromosome. It was demonstrated by Weintraub and Groudine (1976) that when only 10% of a chick erythroblast nucleus is digested with DNase I,

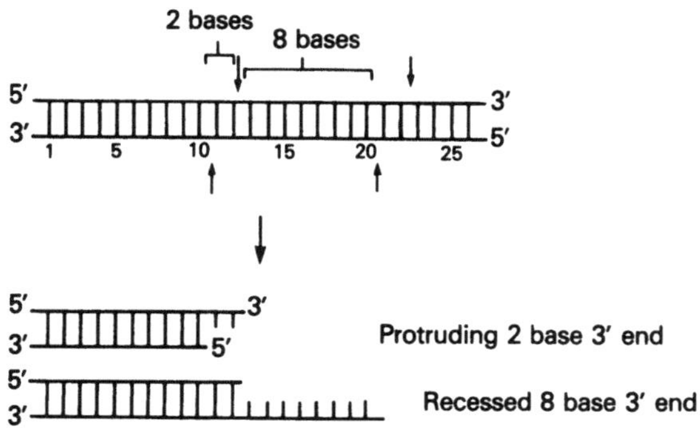

Fig. 1. DNase I digestion pattern of DNA in chromatin. (After Lewin 1985)

50% of the DNA which originally hybridized to a cDNA probe for a globin message was lost. Subsequent experiments revealed that all active genes, whether coding for abundant or for rare mRNA, contain regions of DNA which are susceptible to DNase I (Garel et al. 1977). Every active gene has a DNase I hypersensitive site in the region immediately upstream from the promoter (Wu et al. 1979; Elgin 1981). In addition to DNase I, hypersensitive sites are known to exist for other enzymes such as DNase II and micrococcal nuclease (Nedospasov and Georgiev 1980). Thus, the aberrant release of otherwise dormant endonucleolytic enzymes may not just create random breaks in the chromatin structure, but may specifically interact with hypersensitive sites which are believed to have a major regulatory role in normal gene expression.

3 Chromosome Aberrations and Somatic Mutations Induced with DNase I-in-Liposomes

DNase I can be successfully delivered into Syrian hamster embryo (SHE) cells by means of liposomes (Zajac-Kaye and Ts'o 1984). The biological response upon entry into the living cells was demonstrated by the dose-dependent cytotoxicity exerted by the enzyme. When SHE cells are treated with increasing doses of DNase I encapsulated in liposomes, the survival curve initially shows no reduction in cloning efficiency, but then a sharp decline in a narrow range of concentrations of DNase I-in-liposomes is observed (Fig. 2). Since the amount of lipid varied for each of the DNase I concentrations, treatment with an increasing amount of empty liposomes was shown to have no effect on cell survival (Fig. 2). In addition, 6 mg of exogenously added DNase I, buffer alone, or as much as 12 mg lipids had no effect on cell survival. The results of the dose-dependent cell killing indicated that DNase I delivered by liposomes could enter cells in a biologically active form.

What is then the fate of the DNase I present in the cellular cytoplasm? Can this enzyme find its way to the nucleus and act on the cellular genetic material? To answer these questions, mutations at the hypoxanthine phosphoribosyl transferase (HPRT) and Na^{+}/K^{+} ATPase loci which could be tested by resistance to 6-thioguanine (6TG) and ouabain, respectively, were examined. Gene alteration in Na^{+}/K^{+} ATPase can change this plasma membrane enzyme in such a way that it can lose its sensitivity to

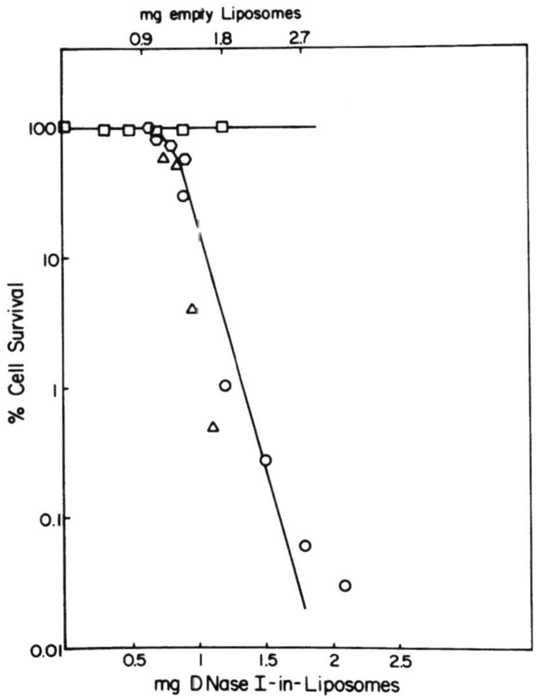

Fig. 2. Effect of DNase I-in-liposomes dose on cell survival of Syrian hamster embryo cells. O , ○, △ Treatment with DNase I-in-liposomes. □ Treatment with empty liposomes. (After Zajac-Kaye and Ts'o 1984)

ouabain but retain its ability to control ionic balance (Lea and Winter 1977); however, deletion of this gene would be fatal. On the other hand, mutation to 6TG resistance is known to be associated with either deletion or an alteration in the X-chromosome-linked HPRT enzyme required for the purine salvage pathway (Caskey and Kruh 1979). An agent such as DNase I, an endonuclease that cuts the DNA molecule, was expected to induce deletion mutation more efficiently than base substitution mutation or other types of gene alterations. Thus, the induction of mutation at the HPRT locus would be more likely than at the Na^+/K^+ ATPase locus. This prediction was confirmed since DNase I-in-liposomes induced mutation at the HPRT locus in a dose-dependent manner (Zajac-Kaye and Ts'o 1984) but was not able to induce mutations at the Na^+/K^+ ATPase locus, regardless of the DNase I-in-liposome dose or the time of expression used for mutant selection (Zajac-Kaye 1982). This observation suggests that DNase I may lead to gross genetic damage rather than simple gene alteration. If this notion is correct DNase I should be able to induce chromosome aberrations upon entry into the cell nucleus.

The incidence of chromosome aberrations was studied in SHE cells 15 h after treatment with DNase I-in-liposomes, DNase I alone, empty liposomes, and buffer only. Chromosome aberrations were induced with DNase I-in-liposomes but not with the appropriate controls (Table 1). In 250 control metaphases only one chromosome aberration was found. By contrast in 162 metaphases examined from the cultures treated with DNase I-in-liposomes, there were 25 aberrant metaphases with 44 chromosome aberrations such as gaps and breaks, isochromatid breaks, dicentric

Table 1. Induction of chromosome aberrations in Syrian hamster embryo cells by DNase I-in-liposomes. (After Zajac-Kaye and Ts'o 1984)

Treatment	Relative cell survival (%)	No. of metaphases analyzed	No. of cells with aberrations	Total No. of aberrations[b]	Aberrant metaphases (%)	Aberrations (%)
Untreated[a]	100	250	1	1	0.4	0.4
DNase I-in-liposomes	82.0	61	4	5	6.6	8.2
	56.4	58	5	15	8.6	27.6
	54.5	43	16	22	37.2	53.5

[a] Includes treatment of cells with buffer, empty liposomes, and DNase I alone.
[b] Aberrations include: gaps, chromatid breaks, chromatid exchanges, isochromatid breaks, dicentrics, and complexed reunions of many chromosomes or small fragments.

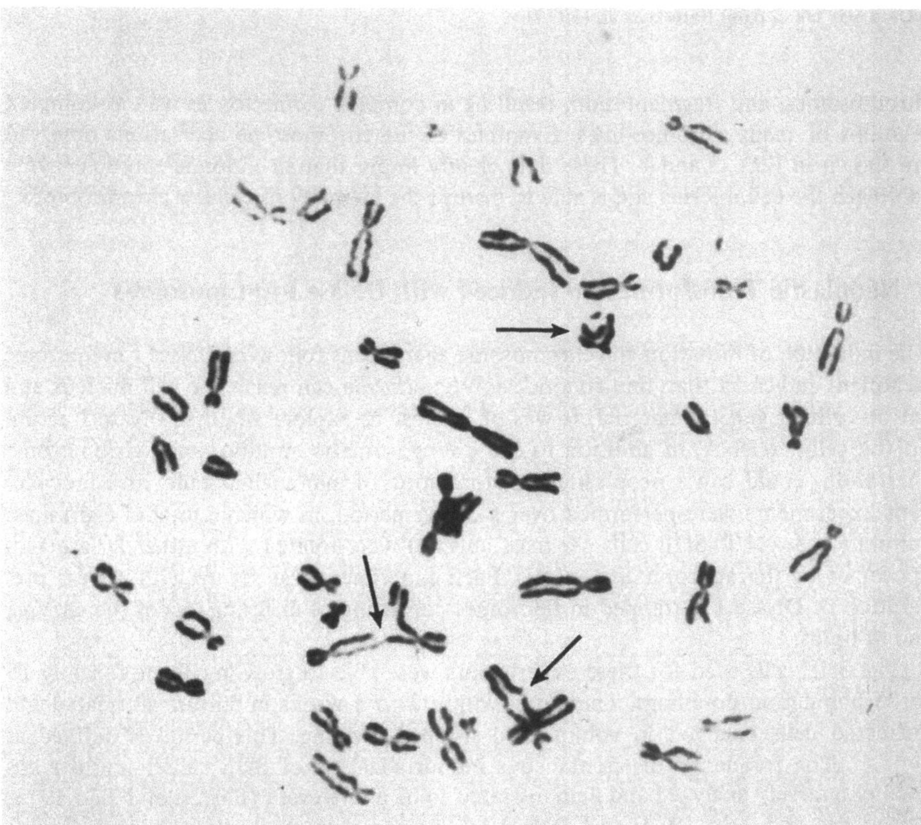

Fig. 3. Chromosome aberrations observed 15 h after treatment of SHE cells (passage 4) with DNase I encapsulated in liposomes. The *arrows* point to chromosome and chromatid breaks and fragmentation of chromosome material

Table 2. Summary of phenotypic characteristics of SHE cells transformed by DNase I-in-liposomes. (After Zajac-Kaye and Ts'o 1984)

Cell lines	Growth in culture	Cloning efficiency in 0.3% soft agar[a]	Tumorigenicity (animals with tumors/ animals injected)[b]
SHE (passage 4)	Monolayers	0	0/27[c]
DL-1	Multilayers	0.25	9/9
DL-2	Multilayers	0.174	6/6
DL-3	Multilayers	0.25	7/7
DL-4	Multilayers	0.27	6/6
DL-5	Multilayers	0.29	6/6

[a] Values are average percentages of cloning efficiency from ten dishes; 1×10^5 cells were tested per 60-mm dish.
[b] Three-day-old hamsters were injected with 2×10^6 cells; latent period was 3–6 weeks.
[c] Data was taken from Barrett et al. (1979).

chromosomes, and fragmentation, resulting in complex formation as well as complex reunions of many chromosomes. Examples of the chromosome aberrations observed are shown in Figs. 3 and 4. These data clearly imply that an endonucleolytic enzyme can reach the cell nucleus and is able to disrupt the integrity of cellular chromosomes.

4 Neoplastic Transformation Induced with DNase I-in-Liposomes

The induction of mutation and chromosome aberrations following DNase I-in-liposome treatment indicates that this endonucleolytic enzyme can reach the cell nucleus and act on cellular genetic material. It was of interest to explore whether DNase I action on the cellular DNA, in addition to cell killing, somatic mutation, and chromosome aberration, could cause neoplastic transformation of mammalian cells. Five independent experiments were performed over a 2-year period, in which a total of eight mass culture ($2–3 \times 10^6$ SHE cells per mass culture) were treated with either DNase I-in-liposomes or the appropriate controls. Each experiment was set up with a fresh preparation of DNase I entrapped in liposomes consisting of differing lots of the enzyme and of lipids.

The SHE cells used for these experiments were able to grow in culture for only 15 to 25 population doublings. Therefore, within 2 to 3 weeks in culture all treated and untreated cells enlarged in volume and stopped dividing. This period is defined as "crisis". The frequency of spontaneous immortalization of SHE cells in culture has been extensively analyzed and demonstrated to be a rare event (Barrett and Ts'o 1978; Zajac-Kaye and Ts'o 1984). All DNase I-in-liposomes-treated cultures escaped crisis within 1 to 3 weeks and became permanently established cell lines. Upon recovery from crisis, five independent DNase I-in-liposomes-treated cell lines were examined every two to three passages for the acquisition of the following neoplastic phenotypes:

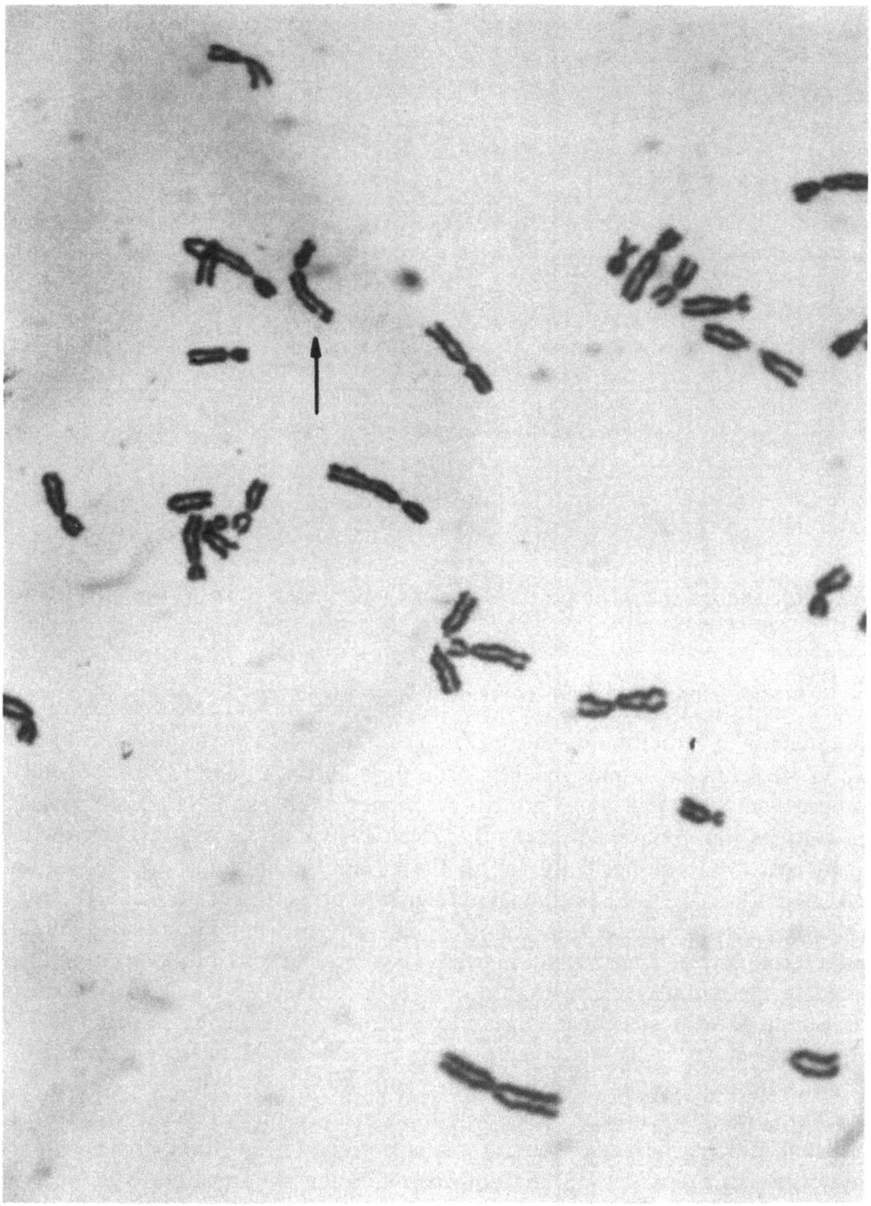

Fig. 4. Chromosome aberrations observed 15 h after treatment of SHE cells (passage 4) with DNase I encapsulated in liposomes. The *arrow* points to a chromatid gap

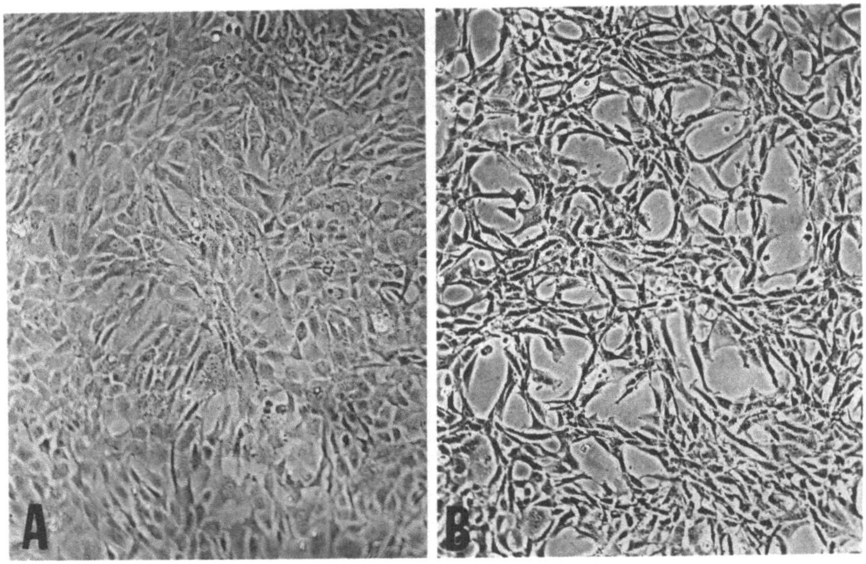

Fig. 5A,B. Morphological characteristics of (A) Syrian hamster embryo cells (passage 4); B DNase I-in-liposomes transformed live (DL-2). (After Zajac-Kaye and Ts'o 1984)

(1) loss of contact inhibition (which resulted in piling up and criss-crossing of cells with an increase in saturation density); (2) colony formation when plated at low cell density in 1% serum-containing medium or in 0.3% soft agar; and (3) tumorigenicity in newborn hamsters. The above growth properties of all five DNase I-in-liposomes-treated cultures were published previously (Zajac-Kaye and Ts'o 1984). The growth in 0.3% soft agar and tumorigenicity for all the DNase I-in-liposomes-treated cell lines (named DL-1 through DL-5) as compared to growth properties of normal SHE cells at early passage are summarized in Table 2. The DNase I-in-liposomes-treated lines showed morphological transformation with criss-crossing and piling up of cells, as compared to the monolayer growth of normal SHE cells (Fig. 5; Table 2). In addition, all DL cell lines expressed anchorage-independent growth, giving rise to 174–290 transformants per 10^5 cells in all mass cultures tested. Also, all lines were tumorigenic giving rise to fibrosarcoma in 100% of injected animals.

Together, these data indicate that cells from mass cultures, treated with DNase I encapsulated in liposomes, can establish permanent cell lines with in vitro and in vivo neoplastic growth properties. In addition, chromosome abnormalities were observed in the DNase I-in-liposomes-transformed cells and in the tumor line derived from them, but not in normal SHE cells. The most common abnormalities found in the DL-1 cell line were dicentric chromosomes and double-minute chromosomes. In the extreme examples many chromosome aberrations, such as dicentrics, triads, rearrangements, and double-minute chromosomes were present in single metaphases (Zajac-Kaye and Ts'o 1984).

An intriguing observation, however, was that in spite of demonstrating a direct perturbation to the genetic material, a prolonged latency period was required before

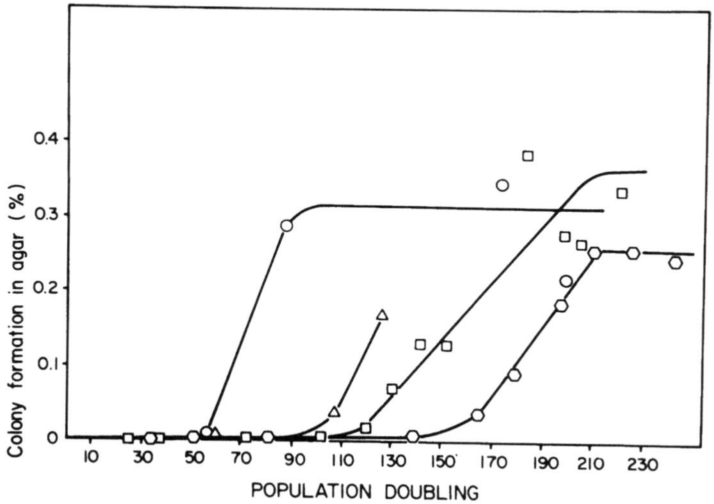

Fig. 6. Difference in the progression time for the expression of anchorage-independent growth in cell lines initially treated with various doses of DNase I-in-liposomes. ○ DL-3 (4.4% cell survival); △ DL-2 (50% cell survival); □ DL-4 (73% cell survival); O DL-5 (81% cell survival)

the treated cells expressed the full neoplastic phenotype. In fact, a wide range of population doublings were required for the expression of anchorage-independent growth in the treated cell lines. To determine if there is a relation between the initial insult given to SHE cells by DNase I-in-liposomes and the latency time in culture for the expression of the anchorage-independent phenotype, the percent of cloning efficiency in soft agar of the cell lines treated with differing amounts of DNase I-in-liposomes was plotted as a function of population doubling. As seen in Fig. 6, the greater the amount of enzyme used for the initial cell treatment generating the highest amount of cytotoxicity, the fewer population doublings were required for the acquisition of an anchorage-independent phenotype. As described above, an increase in the DNase I delivered generated an increased number of chromosome aberrations shortly after enzyme treatment, suggesting that the number of genes affected may directly influence subsequent event(s) leading to the expression of the anchorage-independent phenotype.

5 Discussion

A major goal of cancer research is to find a common mechanism to explain how a variety of unrelated oncogenic agents can induce cells to escape normal growth control and become neoplastically transformed. These diverse oncogenic agents, such as chemicals, radiation, or viruses, are also known to induce chromosome aberrations, suggesting a direct action on nuclear chromatin. Another possible mechanism for producing deletions or rearrangements of genetic material by a variety of oncogenic agents is the release or production of an enzyme capable of directly attacking DNA.

Oncogenic retroviruses such as the Friend leukemia virus (Nissen-Mayer and Nes 1980), myeloblastosis virus (Samuel et al. 1979), murine sarcoma virus (Tsou et al. 1978) as well as DNA viruses such as the Epstein-Barr virus (Clough 1979; Ooka et al. 1984) and herpes simplex virus (Hoffmann and Cheng 1978) have DNase encoded by the viral genome. Moreover, when human embryo fibroblasts are infected with herpes simplex virus, the activities of lysosomal DNase and virus-coded DNase were shown to be elevated (Sablina 1978). In cells that did not support a productive infection, an increase in the activity of lysosomal DNase alone was observed. In both cell types, those supporting a productive infection and in those that did not, chromosome aberrations were observed (Sablina 1978), suggesting that chromosome aberrations could be generated by the release of either a presynthesized lysosomal enzyme or by newly synthesized DNase following virus infection.

Chemical carcinogens such as dimethylbenz(a)anthracene, 2-acetylaminofluorene, and other polycyclic hydrocarbons concentrate in lysosomes as detected by fluorescence microscopy (Allison and Mallucci 1964; Tanigaki et al. 1967). No nuclear fluorescence was observed in these studies although the presence of carcinogenic chemicals in the nucleus could be below the level of detection; thus, a direct action of the chemicals on cellular DNA could not be excluded. Other studies have also revealed a release of endonucleolytic enzymes from isolated liver lysosomes by either ultraviolet or ionizing radiation (Wills and Wilkinson 1966). Again, an effect of radiation on cellular DNA could not be excluded. Since damage to lysosomes either by radiation or chemical carcinogens cannot be accomplished without excluding their effect on cellular DNA, one way to examine the hypothesis that released or newly synthesized endonucleases may play a role in the induction of chromosome aberrations and neoplastic transformation is to deliver biologically active exogenous DNase into living mammalian cells and follow its course of action. Indeed, active DNase I can be successfully delivered into cells by means of liposomes as demonstrated by a dose-dependent cytotoxicity exerted by the enzyme (Fig. 2). In addition, DNase I can reach the cell nucleus as demonstrated by the rapid induction of chromosome aberrations and somatic mutation at the HPRT locus. The observation that DNase I failed to induce point mutation at the Na^{+}/K^{+} ATPase locus, but was able to induce resistance to 6TG, suggests that the action of DNase I may lead to genetic damage through chromosome rearrangement rather than through simple gene mutation. This notion is supported by the detection of chromosome aberrations shortly after DNase I-in-liposomes treatment (Table 1; Figs. 3 and 4). The chromosome damage included chromatid and chromosome gaps and breaks, leading to nonhomologous chromosome pairing that may result in exchanges of chromosome material. Chromosome rearrangements may also lead to the activation of cellular protooncogenes present in mammalian cells. Alternatively, the endonucleolytic cuts generated by DNase I may result in small deletions of genes necessary for the regulation of normal cell function.

The data obtained with DNase I indicate that damage to DNA is not sufficient to cause the immediate expression of a neoplastic phenotype. The time of expression of anchorage-independent growth, however, was related to the initial dose of DNase I used for treatment (Fig. 6). This result, in conjunction with the observation that an increase in the DNase I dose generated increased numbers of chromosome aberrations (Figs. 3 and 4; Table 1), suggests that the number of genes affected is proportional to the DNase I dose delivered, which in turn may have a direct impact on subsequent event(s) necessary for tumor progression.

The successful delivery of active DNase I into SHE cells via liposomes indicates that this technique can also be a useful tool for the introduction of a variety of restriction enzymes into mammalian cells. It had been shown that restriction enzymes are able to induce chromosome aberrations when introduced into permeabilized cells by inactivated Sendai virus preparations (Natarajan and Obe 1984). The delivery of restriction enzymes by means of liposomes could allow a detailed examination of chromosome changes and their relationship to the development of the neoplastic transformation process, as well as allowing the manipulation of cellular DNA in living mammalian cells.

In summary, to induce DNA changes that lead to somatic mutation, chromosome aberrations and neoplastic transformation in SHE cells, a biological agent was used, which is a DNA hydrolytic enzyme, normally stored in the cellular lysosomal apparatus. An enticing hypothesis proposes that damage to the lysosomal membrane by a variety of oncogenic agents and subsequent release of the destructive enzymes may not only cause cell death but also alter normal behavior of surviving cells, leading to neoplastic disease.

Acknowledgments. I wish to thank Drs. Frederic Kaye, Edward Gelmann, André Veillette, and Marc Lippman for critical reading of the manuscript. This work was done as part of a Ph.D. thesis in the laboratory of P.O.P. Ts'o at The Johns Hopkins University.

References

Allison AC, Mallucci L (1964) Uptake of hydrocarbon carcinogens by lysosomes. Nature (London) 203:1024–1027

Barrett JC, Ts'o POP (1978) Evidence for the progressive nature of neoplastic transformation in vitro. Proc Natl Acad Sci USA 75:3761–3765

Benedict WF (1972) Early changes in chromosomal number and structure after treatment of fetal hamster cultures with transforming doses of polycyclic hydrocarbons. J Natl Cancer Inst 49: 585–590

Bishop JM (1983) Cellular oncogenes and retroviruses. Annu Rev Biochem 52:301–354

Boveri TH (1914) Zur Frage der Entstehung maligner Tumoren. Fischer, Jena (Engl Transl: The origin of malignant tumors. William & Wilkens, Baltimore, 1929)

Caskey CT, Kruh GD (1979) The HPRT locus. Cell 16:1–9

Clough W (1979) Deoxyribonuclease activity found in Epstein-Barr virus producing lymphoblastoid cells. Biochemistry 18:4517–4521

Cooper GM (1982) Cellular transforming genes. Science 217:801–806

Dalla-Favera R, Bregni M, Erikson J, Patterson D, Gallo R, Croce C (1982) Human c-myc gene is located on the region of chromosome 8 that is translocated in Burkitt lymphoma cells. Proc Natl Acad Sci USA 79:7924–7827

Elgin SCR (1981) DNAase I-hypersensitive sites of chromatin. C2ll 27:413–415

Evans HJ (1977) Molecular mechanisms in the induction of chromosome aberration. In: Scott D, Bridges BA, Sobels FH (eds) Progress in genetic toxicology. Elsevier/North Holland Biomedical Press, Amsterdam New York, pp 57 74

Garel A, Zolan M, Axel R (1977) Genes transcribed at diverse rates have a similar conformation in chromatin. Proc Natl Acad Sci USA 74:4867–4871

German J (ed) (1983) Chromosome mutation and neoplasia. Liss, New York

Hoffman PJ, Cheng YC (1978) The deoxyribonuclease induced after infection of KB cells by herpes simplex virus type 1 or type 2. J Biol Chem 253:3557–3562

Lea JR, Winter CG (1977) Ligand-dependent tryptic inactivation of the ouabain sensitivity of ADP-ATP exchange catalysed by canine renal NaK-ATPase. Biochem Biophys Res Commun 76:772–777

Lewin B (1985) Genes. John Wiley & Sons, New York

Lutter LC (1977) Deoxyribonuclease I produces staggered cuts in DNA of chromatin. J Mol Biol 117:53–69

Natarajan AT, Obe G (1984) Molecular mechanism, involved in the production of chromosomal aberrations. III Restriction endonucleases. Chromosoma 90:120–127

Nedespasov SA, Georgiev GP (1980) Non-random cleavage of SV40 DNA in the compact minichromosome and free in solution by micrococcal nuclease. Biochem Biophys Res Commun 92:532–539

Nissen-Meyer J, Nes IF (1980) Characterization of an endonuclease activity associated with Friend-murine leukemia virus. Biochim Biophys Acta 609:148–157

Noll M (1974) Internal structure of the chromatin subunit. Nucleic Acid Res 1:1573–1578

Ooka TM, De Turenne G, De The, Daillie J (1984) Epstein-Barr virus-specific DNAase activity in nonproducer Raji cells after treatment with 12-O-Tetradecanoyl-phorbol-13-Acetate and sodium butyrate. J Virol 49:626–628

Rowley JD (1983) Consistent chromosome abnormalities in human leukemia and lymphoma. Cancer Invest 1:267–280

Sablina OV (1978) Role of nucleases in the production of chromosome abnormalities in virus-infected cells. Genetica 14:1919–1927

Samuel KP, Papas TS, Chirikjian JG (1979) DNA endonucleases associated with the avian myeloblastosis virus DNA polymerase. Proc Natl Acad Sci USA 76:2659–2663

Shtivelman E, Blifshitz RP, Gale, Canaani E (1985) Fused transcript of *abl* and *bcr* genes in chronic myelogenous leukaemia. Nature (London) 315:550–554

Sollner-Webb B, Felsenfeld G (1977) Pancreatic DNAase cleavage sites in nuclei. Cell 10:537–543

Sugiyama T (1975) Chromatid rearrangement and carcinogenesis. Gann Monogr Cancer Res 17: 393–403

Tanigaki N, Kitagawa M, Yagi Y, Pressman D (1967) The reaction of specific antibody with 2-acetylaminofluorene fixed in liver cells. Cancer Res 27:747–751

Taub R, Kirsch I, Morton C, Lenoir G, Swan D, Tronick S, Aaronson S, Leder P (1982) Translocation of the c-myc gene into the immunoglobulin heavy chain locus in human *Burkitt lymphoma* and murine plasmacytoma. Proc Natl Acad Sci USA 79:7837–7841

Tsou KC, Lo KW, Herberman RB (1978) Nucleases and adenosine 3',5'-cyclic monophosphate phosphodiesterase activities in murine sarcoma virus (Moloney)-inected mice. J Natl Cancer Inst 61:1077–1083

Weintraub H, Groudine M (1976) Chromosomal subunits in active genes have an altered conformation. Science 193:848–856

Wills ED, Wilkinson AE (1966) Release of enzymes from lysosomes by irradiation and the relation of lipid peroxide formation to enzyme release. Biochem J 99:657–666

Wu C, Bingham PH, Livak KJ, Holmgren R, Elgin SCR (1979) The chromatin structure of specific genes. Evidence for higher order domains of defined DNA sequence. Cell 16:797–806

Yunis JJ (1983) The chromosomal basis of human neoplasia. Science 221:227–236

Zajac-Kaye M (1982) The study of the mechanism of neoplastic transformation induced by DNAase I encapsulated in liposomes. PhD Thesis. The Johns Hopkins Univ, Baltimore, MD

Zajac-Kaye M, Ts'o POP (1984) DNAase I encapsulated in liposomes can induce neoplastic transformation of Syrian hamster cells in culture. Cell 39:427–437

15 DNA Damage and Cytogenetic End Points

R. A. BAAN[1]

1 Introduction

For any living organism, the preservation of the integrity of its DNA as a genetic blueprint is of prime importance. Like other cellular components, the DNA is subject to a variety of challenges of widely different origin and nature. Evidence for the priority given to the protection and preservation of the DNA structure is found in the multitude of enzymes that can be mobilized by the cell to remove or counteract undesired alterations in its DNA.

Damage inflicted upon the genetic material may lead to harmful effects in the cell or in the organism as a whole (see Fig. 1). When the damage is not repaired or is repaired in an erroneous way, mutations may arise. In higher multicellular organisms, mutations in somatic cells may play a role in certain stages of carcinogenesis, whereas mutations in germ cells may give rise to heritable diseases. The notion that damage that persists in the genetic material may play a role in the etiology of cancer, has gained support by recent reports on cellular transformation through activation of oncogenes.

This chapter is meant to serve as an introduction to the nature of various types of DNA damage, the agents that cause them and the different repair responses they evoke in the cell. Where possible, the relations between primary DNA damage and the various cytogenetic end points that are dealt with in other parts of this volume, will be discussed.

2 DNA Damage

In apparent contrast with its character as a reliable blueprint molecule, DNA can undergo numerous alterations and rearrangements that are part of the normal regulatory processes connected with gene expression (Calos and Miller 1980). DNA damage can be defined as an alteration in the chemical structure or in the sequence of individual nucleotides in a DNA molecule. These two phenomena are often closely linked, because a chemical alteration can lead to a change in base sequence, which can become manifest as a mutation.

---

1 Department of Genetic Toxicology, TNO Medical Biological Laboratory, P.O. Box 45, 2280 AA Rijswijk, The Netherlands

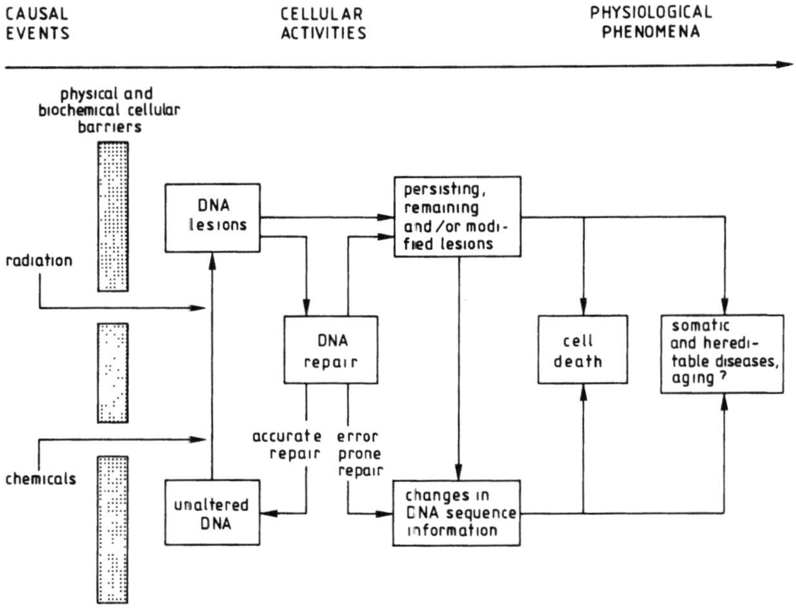

Fig. 1. Hypothetical pathways through which induced DNA damage may lead to adverse health effects. (From Vijg et al. 1986)

2.1 Damage as a Consequence of the Chemical Nature of DNA

The mechanism of base pairing through formation of hydrogen bonds between opposite bases is an important element of the self-replicative nature of DNA. Many sequence changes occur as a consequence of mispairing during DNA replication, although proof-reading and postreplication repair events restore most of these errors, the resulting spontaneous mutation frequency in bacterial systems being 10^{-9}–10^{-12} per base pair replication (Drake 1969). During the repair process itself, mispairing can also occur and lead to sequence changes in the genome.

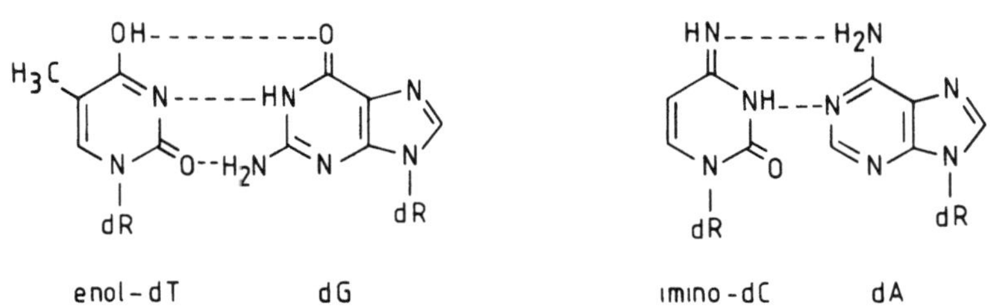

Fig. 2. Base-pairing properties of nucleobases in enol- and imino-form. The enol-form of deoxythymidine (enol-dT) can base pair with deoxyguanosine (dG), the imino-form of deoxycytidine (imino-dC) can form a base pair with deoxyadenosine (dA). Similarly, enol-dG and imino-dA pair with dT and dC respectively

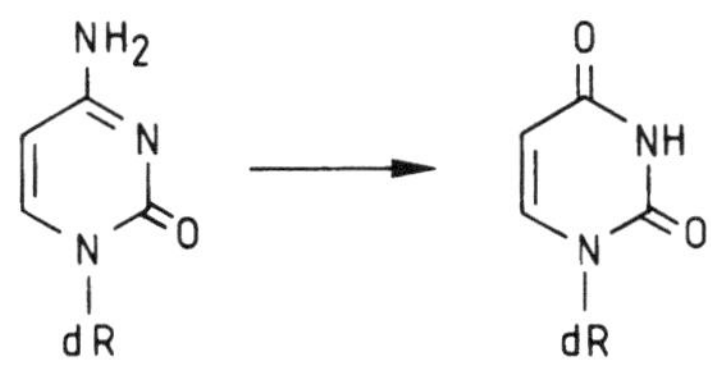

d-cytidine d-uridine

Fig. 3. Base change as a result of deamination. Hydrolytic deamination of deoxycytidine results in the formation of deoxyuridine, which can form a base pair with deoxyadenosine

Under physiological conditions, the purine and pyrimidine bases can undergo tautomeric shifts from the regular keto- and amino-forms into the enol- and imino-forms, which changes their base-pairing properties (see Fig. 2). Furthermore, loss of exocyclic amino groups (deamination) can result in alternative base-pairing preferences, hence also lead to changes in nucleotide sequence (see Fig. 3). Complete loss of a base, due to spontaneous hydrolysis, can occur under physiological conditions, although more rapidly at acidic pH or at elevated temperatures (Shapiro 1981). The resulting apurinic or apyrimidinic sites are a source of spontaneous mutations. They can also lead to a weakening or even a disruption of the sugar-phosphate backbone, resulting in a DNA strand break (Lindahl and Andersson 1972).

2.2 Damage as a Consequence of External Factors

Apart from the intrinsic chemical characteristics of the DNA structure, numerous external factors can cause DNA damage. Various types of radiation and many chemical compounds are reactive towards DNA.

Among the best-studied model systems to investigate the biological effects of genomic damage is the irradiation of cells with ultraviolet light (UV). When cells are exposed to 260-nm UV, a wavelength near the absorbance maximum of DNA, the most abundant lesion is the pyrimidine dimer, in which a cyclobutane ring has formed between neighbouring pyrimidines, through saturation of the C5-C6 bonds (see Fig. 4a). Apart from the pyrimidine dimers, various other photoproducts are also formed (Demple and Linn 1982; Haseltine 1983).

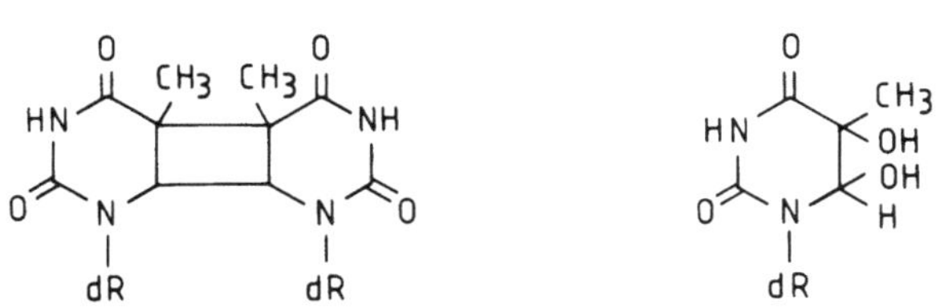

a b

Fig. 4. a Cyclobutyl pyrimidine dimer, a major photoproduct in DNA irradiated with ultraviolet light. **b** 5,6-dihydroxy-5,6-dihydro-thymine (thymine glycol), a damaged base induced in DNA by ionizing radiation

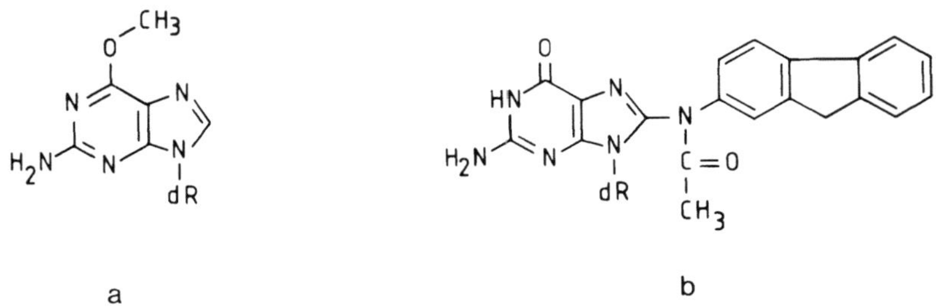

Fig. 5a,b. Examples of adducts formed in DNA as a result of exposure to direct-acting or metabolically activated chemicals. **a** O^6-methyldeoxyguanosine, induced by direct-acting methylating agents, e.g. methylnitrosourea. **b** N-(deoxyguanosin-8-yl)-2-acetylaminofluorene, induced by 2-acetylaminofluorene (2AAF), a compound that requires bioactivation

The effects of ionizing radiation on DNA can be broadly divided into those resulting from direct interaction of the radiation energy with DNA components and those where excited chemical species or radicals serve as intermediates in energy transfer (Ward 1975). The total spectrum of radiation damage is, therefore, quite complex and not yet completely established. As an example of the many types of radiation-induced base damage one can mention thymine-glycol (Hariharan and Cerutti 1972) (see Fig. 4b).

Both direct and indirect interaction of ionizing radiation with DNA can lead to breaks in either DNA strand. These single-strand breaks do not correlate very well with the lethal effects of the ionizing radiation. Double-strand breaks, which can be regarded as a combined occurrence of independent single-strand breaks at opposite sites or as originating from a direct single effect of radiation, are more harmful for the cell and correlate more closely with lethality (Bonura and Smith 1976). The relation between strand breaks and cytogenetic effects will be discussed later on.

For many years, the interaction of chemicals with the genetic material has been studied intensively (Singer and Kusmierek 1982). Many alkylating agents are reactive electrophiles that attack nucleophilic centres in the DNA molecule. Adducts are thus formed, e.g. at the N7 or O^6 position of guanine (see Fig. 5a). Bifunctional alkylating agents can form crosslinks within the genome or with neighbouring proteins, thereby hampering proper transcription and replication. Of great importance was the discovery of the metabolic conversion of nonreactive compounds into reactive electrophilic species (Miller 1978). Model systems in which enzyme fractions from homogenized rat liver were included to activate various types of chemicals, have proven useful in the study of the mechanism of adduct formation at different sites in the DNA (see Fig. 5b). The presence of adducts in DNA can give rise to inhibition of replication, interference with regular base pairing or to destabilization of chemical bonds which may lead to depurination and strand breakage.

It is known that the action of metabolizing enzymes, e.g. those linked to cytochrome P450, involves the formation of reactive radicals as a side effect of a detoxification or activation pathway. It has been suggested, therefore, that not only the adducts formed by direct reaction with electrophilic derivatives of the metabolized chemical, but also the damages induced by these radicals, so-called oxygen damage,

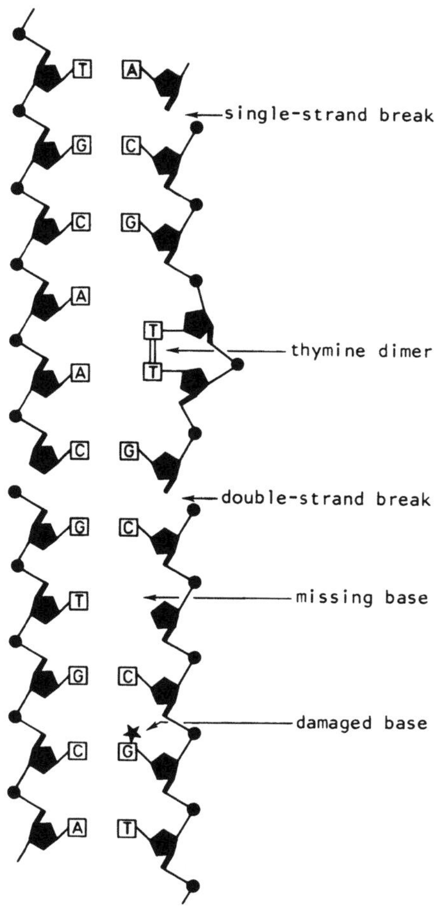

Fig. 6. Schematic representation of various types of DNA damage

represent an important part of the overall DNA damage (Cerutti 1985). This would imply that chemical conversions catalyzed by this type of enzymes could result in DNA lesions similar to those induced by ionizing radiation. Figure 6 shows a rather severely damaged piece of DNA in which several of the lesions mentioned above have been indicated.

3 Cellular Responses to Induction of DNA Damage

A multitude of enzymes can be activated by the cell to prevent or counteract undesired effects of modifications in DNA (Friedberg 1985). Two general types of damage repair can be discerned: direct removal of the damage without interruption of the sugar-phosphate backbone and excision repair, in which a segment containing the damage is excised from the DNA strand. In the latter case, repair synthesis and ligation follow the excision step in order to restore the original DNA structure.

Various types of enzymes involved in direct reversal of DNA damage have been characterized. The linkage between UV-induced pyrimidine dimers can be removed by the enzyme DNA photolyase under the influence of photoreactivating light (wavelength > 300 nm) (Rupert 1975). This enzyme has been isolated from various types of plants and animals, but its presence in human cells is still controversial.

An interesting class of enzymes that restore alkyl-damage by direct reversal are the methyltransferases. These enzymes have been detected in many different prokaryotic and eukaryotic cell types. Studies on their mechanism of action have shown that the enzymes directly accept a methyl group from the O^6 position of methylguanine, but lose activity in doing so (Lindahl et al. 1982).

Base insertion is a third example of damage repair without interruption of the sugar-phosphate backbone. Apurinic sites can arise spontaneously, as mentioned above, or as a result of alkylations leading to an instable adduct, e.g. at the N7 of guanine. Furthermore, they can be introduced by enzymes; various glycosylases have been discovered, which catalyze the removal of damaged bases through disruption of the glycosylic bond (Lindahl 1979). An enzyme called purine insertase has been identified and found to catalyze the re-insertion of guanine and adenine at depurinated sites in polynucleotides (Livneh et al. 1979).

The excision repair pathways are the most widely studied cellular responses to DNA damage induction. These processes involve the removal of the DNA lesion through excision of a segment from the damaged DNA strand. Different types of enzymes play a role in excision repair: in some instances, damage-specific glycosylases first remove a damaged base, thereby creating an apurinic or apyrimidinic site (Duncan 1981). Endonucleases then excise a fragment containing the damaged site. In other instances, when the damage is not recognized by a specific glycosylase, the repair occurs through endonucleolytic cleavage of the sugar-phosphate backbone at two sites flanking the damaged base, or by one such nick followed by exonucleolytic degradation of the damaged fragment (Lindahl 1982). In each of these cases, the repair process is completed with the action of DNA polymerase and DNA ligase, which restore the double-strand structure of the DNA.

These repair pathways have been schematically outlined in Fig. 7.

4 Detection of DNA Damage

Many methods allow detection and quantitation of DNA damage at the molecular level. Such methods are of particular importance for biomonitoring experiments in which (human) exposure to genotoxicants can be established on the basis of the presence of lesions in the DNA (Baan et al. 1986).

Base modifications due to deamination or adduct formation can often be recognized in the cell by specific glycosylases, which leads to the formation of apurinic or apyrimidinic sites (Lindahl 1979). These sites are alkali-labile and can be converted into strand breaks, which in turn can be accurately measured by means of the alkaline elution technique (Kohn and Ewig 1973). Of course, the major effect of ionizing radiation, i.e. the direct generation of DNA strand breaks, can also be monitored with this highly sensitive elution method. DNA damage can further be detected indirectly,

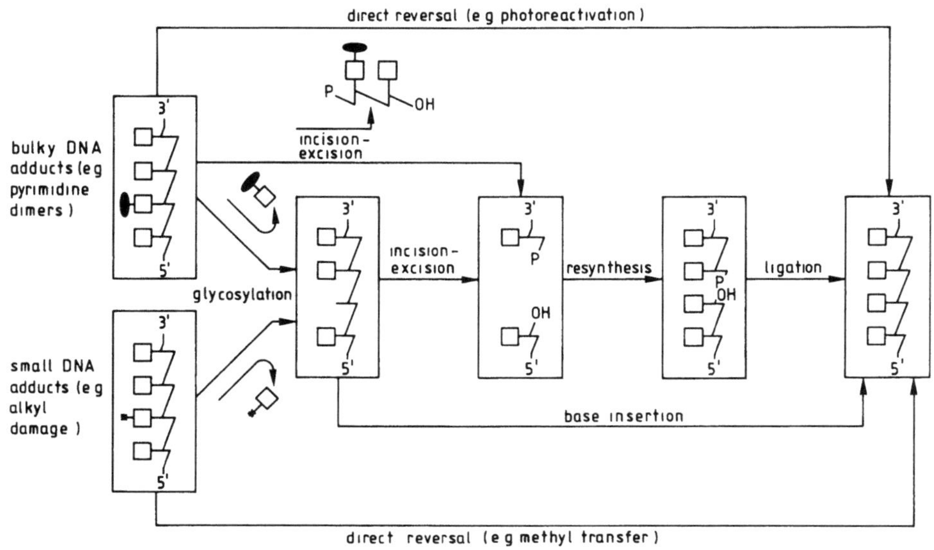

Fig. 7. Schematic outline of some of the major pathways of DNA damage repair. (From Vijg et al. 1986)

through analysis of other aspects of the repair response it induces in the cell. Lesion-dependent enzymatic incisions leading to strand breaks, or repair synthesis which involves incorporation of nucleotides in DNA can be monitored with various methods.

Chemical modifications in DNA, e.g. adducts induced by electrophilic chemicals and radiation-induced damage such as thymine dimers or thymine glycols, can be detected with highly sensitive methods. A widely used technique is the analysis of radiolabelled adducts in DNA isolated from cultured cells or animals treated with radioactive genotoxicants (Baird 1979). For practical purposes in biomonitoring, the use of radioactive material is obviously unsuitable. Two other methods are briefly described here. The immunochemical approach to DNA adduct detection involves the use of adduct-specific antisera or monoclonal antibodies that are suitable for the detection of adducts in isolated DNA or in single cells (Baan et al. 1986). Due to the characteristics of the antibodies, this method is adduct-specific. Only in certain instances does the chemical similarity of adducts permit the detection of different lesions with the same antibody preparation. Another method, which is highly sensitive and applicable for different types of adducts, is the analysis of DNA damage by means of postlabelling (Reddy et al. 1984). This technique involves labelling of degraded DNA with a ^{32}P-phosphate group. Non-modified nucleotides and adducts can be separated by means of thin-layer chromatography and the adduct pattern can be visualized by means of autoradiography. Separation of adducts can also be performed on HPLC (high pressure liquid chromatography).

Changes in DNA sequence information, e.g. induced as a consequence of deamination of certain nucleobases or of adduct formation leading to mispairing during replication or erroneous repair, can be detected with various methods. The presence of DNA fragments containing such mutations can be visualized with high accuracy on two-dimensional denaturing gradient gels. Novel developments have demonstrated that a single base-pair change in fragments containing up to 2000 base pairs is detectable (Lohman et al. 1986). Sequencing of cloned fragments can provide the molecular details of the DNA alterations.

5 DNA Damage and Cytogenetic Effects

Direct or indirect effects of DNA damage can also be observed as cytogenetic end points. These can be divided into four general categories: changes in chromosome structure, changes in chromosome number, micronuclei and sister chromatid exchanges. In this section the relation between primary DNA damage and each of these end points will be discussed.

5.1 Chromosome Aberrations

Ionizing radiation is a highly efficient agent in inducing chromosome breakage which gives rise to structural aberrations. In several experiments, evidence has been presented that double-strand breaks contribute to a large extent to the induction of such aberrations (Obe et al. 1982), although single-strand breaks may also play a quantitatively minor role. Treatment of cells with restriction endonucleases specifically generating double-strand, blunt-end breaks, increased the frequency of chromosome aberrations in these cells (Natarajan and Obe 1984) (see Chap. 13, this Vol.). Chromosome breakage involves the appearance of instable ends which can be correctly rejoined by repair enzymes but sometimes form new combinations. Such exchanges in which parts from different chromosomes become linked, are thought to originate from an association complex between two DNA regions in chromosomes in which breaks are present. The aberration arises from erroneous recombination during repair. The occurrence of aberrations of this type depends on the rate of repair: short-lived breaks will generate aberrations only when there is a nearby association complex. When break repair is inhibited, e.g. with the nucleoside-analogue araC, the frequency of aberrations increases (Natarajan et al. 1986). When recombinations occur asymmetrically relative to the centromeres, the resulting aberrations are seen as dicentric chromosomes and acentric fragments. These acentrics may give rise to the formation of micronuclei (see below).

Aberrations in chromosome number arise from disturbances of the mitotic cell division process. The resulting aneuploidy can be caused by genetic factors but also by chemicals that otherwise do not interact with DNA. Various *Drosophila* mutants have been described in which the centromeres separate prematurely (Davis 1971); in yeast, a gene product required for chromosome attachment to the spindle has been identified (Thomas and Botstein 1986). Chemical induction of aneuploidy can occur along different routes. Interference with spindle formation and function or with other processes during cell division has been described for chemicals like colchicine, benomyl, diazepam and other compounds. In these cases, an epigenetic mechanism, not directly involving damage in the genome itself, is the basis for the numerical chromosome aberrations (Liang and Brinkley 1985).

5.2 Sister Chromatid Exchanges

Sister chromatid exchanges (SCEs), the reciprocal interchanges of DNA between chromatids, obviously must originate from some process involving strand breakage. SCEs are induced by many mutagenic agents, but ionizing radiation, which causes direct DNA breaks, is a poor inducer of SCEs. However, when DNA double-strand breaks are induced by restriction endonucleases in S-phase cells, SCEs can be formed rather efficiently (Natarajan et al. 1985). Chemicals that interact with DNA in different ways have been shown to induce SCEs with widely different efficiencies when compared with the number of induced mutations (Carrano et al. 1978). Apparently, the conversion of a chemically induced DNA lesion to an SCE is dependent on the nature of that lesion. The mechanism of SCE induction is therefore not readily explained and a clear relation with a particular type of DNA damage cannot be indicated.

5.3 Micronuclei

Micronuclei arise from loss of whole chromosomes or chromatid fragments that are released from the nucleus but retained in the cytoplasm. The induction of micronucleated cells can be the result of disturbances in the cell division process or of chromosome aberrations involving the formation of acentric fragments. Micronucleus formation is closely correlated with chromosome breakage (Heddle et al. 1983).

6 Concluding Remarks

The occurrence of cytologically visible damage in chromosomes can be related in most cases to two general phenomena: the presence of DNA strand breaks and their erroneous repair, and interference with cell division. Various physical and chemical agents are direct inducers of DNA strand breaks (ionizing radiation, bleomycin), whereas other agents interact with DNA in such a way that a strand break may be generated as a secondary effect, e.g. after depurination of instable adducts. Furthermore, metabolic activation processes may involve the formation of reactive chemical species such as hydroxyl radicals or peroxides, which can induce DNA lesions similar to those generated by ionizing radiation. Finally, many of the cellular repair responses triggered by DNA damage induction involve enzymatic strand breakage as an intermediate step in the repair pathway. Therefore, if strand breaks are not the primary result of interaction between a damaging agent and DNA, somewhere down the line such breaks may occur during cellular activities aimed at removal of the lesion, or they may be induced as a side effect of metabolic conversion of the agent. Structural chromosome aberrations and micronuclei arise from erroneous recombination events during the repair of these breaks. The accurate transmission of the total genetic complement of a cell to its progeny depends not only on the structural integrity of the individual chromosomes but also on the reliability of the cell division machinery. It is not surprising, therefore, that also factors interfering with various steps during this process (centromere separation, spindle fiber formation) can give rise to abnormalities that are cytologically visible as numerical chromosome aberrations or as micronucleated cells.

Acknowledgements. I thank Dr. F. Berends for critical reading of the manuscript. The comments from Prof. A.T. Natarajan and from Drs. J.D. Jansen and J. Vijg are gratefully acknowledged.

I thank Dr. J. Vijg for permission to use Figs. 1 and 7, which appeared earlier in Vijg et al. (1986).

References

Baan RA, Lohman PHM, Fichtinger-Schepman AMJ, Muysken-Schoen MA, Ploem JS (1986) Immunochemical approach to detection and quantitation of DNA adducts resulting from exposure to genotoxic agents. In: Sorsa M, Norppa H (eds) Monitoring of occupational genotoxicants Liss, New York, pp 135–146

Baird WM (1979) The use of radioactive carcinogens to detect DNA-modification. In: Grover PL (ed) Chemical carcinogens and DNA. CRC, Boca Raton, pp 59–83

Bonura T, Smith KC (1976) The involvement of indirect effects in cell-killing and DNA double-strand breakage in γ-irradiated *Escherichia coli* K-12 cells. Int J Radiat Biol 29.293 296

Calos MP, Miller JH (1980) Transposable elements. Cell 20:579–595

Carrano AV, Thompson LH, Lindl PA, Minkler JL (1978) Sister chromatic exchange as an indicator of mutagenesis. Nature (London) 271:551–553

Cerutti PA (1985) Active oxygen and promotion. In: Fischer SM, Slaga TJ (eds) Arachidonic acid metabolism and tumor promotion. Nijhoff, Boston, pp 131–168

Davis BK (1971) Genetic analysis of a meiotic mutant resulting in precocious sister-centromere separation in *Drosophila melanogaster.* Mol Gen Genet 113:251–272

Demple B, Linn S (1982) 5,6-Saturated thymine lesions in DNA: production by ultraviolet light or hydrogen peroxide. Nucleic Acids Res 10:3781–3789

Drake JW (1969) Comparative rates of spontaneous mutation. Nature (London) 221:1132

Duncan BK (1981) DNA glycosylases. In: Boyer PD (ed) The enzymes, 3rd edn, vol 14: Nucleic acids. Pt A. Academic Press, London New York, pp 565–586

Friedberg EC (1985) DNA repair. Freeman, New York

Hariharan PV, Cerutti PA (1972) Formation and repair of γ-ray-induced thymine damage in *Micrococcus radiodurans.* J Mol Biol 66:65–81

Haseltine WA (1983) Site specificity of ultraviolet light induced mutagenesis. In: Friedberg EC, Bridges BA (eds) Cellular responses to DNA damage. Liss, New York, pp 3–12

Heddle JA, Hite M, Kirkhart B, Mavournin K, MacGregor JT, Newell GW, Salamone MF (1983) The induction of micronuclei as a measure of genotoxicity. Mutat Res 123:61–118

Kohn KW, Ewig RAG (1973) Alkaline elution analysis, a new approach to the study of DNA single-strand interruptions in cells. Cancer Res 33:1849–1853

Liang JC, Brinkley BR (1985) Chemical probes and possible targets for the induction of aneuploidy. In: Dellarco VL, Boytek PE, Hollaender A (eds) Aneuploidy: Etiology and mechanisms Plenum, New York, pp 491 -505

Lindahl T (1979) DNA glycosylases, endonucleases for apurinic/apyrimidinic sites and base excision repair. Progr Nucleic Acids Res Mol Biol 22:135–192

Lindahl T (1982) DNA repair enzymes. Annu Rev Biochem 51:61–87

Lindahl T, Andersson A (1972) Rate of chain breakage of apurinic sites in double-stranded DNA. Biochemistry 11:3618–3623
Lindahl T, Demple B, Robins P (1982) Suicide inactivation of the *E. coli* O^6-methylguanine-DNA methyltransferase. EMBO J 1:1359–1363
Livneh Z, Elad D, Sperling J (1979) Enzymatic insertion of purine bases into depurinated DNA in vitro. Proc Natl Acad Sci USA 76:1089–1093
Lohman PHM, Vijg J, Uitterlinden AG, Slagboom P, Berends F (1986) DNA methods for detecting and analyzing mutations in vivo. In: Léonard A, Kirsch-Volders M (eds) Proc 16th Annu Meet European Environmental Mutagen Society, 25–30 August, Brussels, Belgium, pp 76–85
Miller EC (1978) Some current perspectives on chemical carcinogenesis in humans and experimental animals: presidential address. Cancer Res 38:1479–1496
Natarajan AT, Obe G (1984) Molecular mechanisms involved in the production of chromosomal aberrations: III. Restriction endonucleases. Chromosoma 90:120–127
Natarajan AT, Mullenders LHF, Meijers M, Mukherjee U (1985) Induction of sister-chromatid exchanges by restriction endonucleases. Mutat Res 144:33–39
Natarajan AT, Darroudi F, Mullenders LHF, Meijers M (1986) The nature and repair of DNA lesions that lead to chromosomal aberrations induced by ionizing radiation. Mutat Res 160: 231–236
Obe G, Natarajan AT, Palitti F (1982) Role of DNA double strand breaks in the formation of radiation-induced chromosomal aberrations. In: Natarajan AT, Obe G, Altmann H (eds) Progress in mutation research, vol 4: DNA repair, chromosome alterations and chromatin structure. Elsevier, Amsterdam, pp 1–9
Reddy MV, Gupta RC, Randerath E, Randerath K (1984) ^{32}P-Postlabelling test for covalent DNA binding of chemicals in vivo: application to a variety of aromatic carcinogens and methylating agents. Carcinogenesis 5:231–243
Rupert CS (1975) Enzymatic photoreactivation: overview. In: Hanawalt PC, Setlow RB (eds) Molecular mechanisms for repair of DNA, Pt A. Plenum, New York, pp 73–87
Shapiro R (1981) Damage to DNA caused by hydrolysis. In: Seeberg E, Kleppe K (eds) Chromosome damage and repair. Plenum, New York, pp 3–18
Singer B, Kusmierek JT (1982) Chemical mutagenesis. Annu Rev Biochem 52:655–693
Thomas JH, Botstein D (1986) A gene required for the separation of chromosomes on the spindle apparatus in Yeast. Cell 44:65–75
Vijg J, Roza L, Mullaart E, Lohman PHM (1986) Species specificity in the induction and repair of DNA damage. In: Ramel C, Lambert B, Magnusson J (eds) Genetic toxicology of environmental chemicals, Pt A: Basic principles and mechanisms of action. Liss, New York, pp 179–187
Ward JF (1975) Molecular mechanisms of radiation-induced damage to nucleic acids. Adv Rad Biol 5:181–293

16 Sister Chromatid Exchanges

A. T. Natarajan and L. H. F. Mullenders[1]

1 Introduction

Sister chromatid exchanges (SCEs) are cytological manifestations of DNA double-strand breakage and rejoining at homologous sites between the two chromatids of a chromosome. The occurrence of SCEs was deduced from the transformation of small ring chromosomes to large ring chromosomes following cell division (McClintock 1938). Using tritiated thymidine as marker and microautoradiography for detection, Taylor et al (1958) demonstrated the occurrence of SCEs from the silver grain pattern on the sister chromatids. This method was eventually replaced by cytochemical methods. Latt (1973) showed that if cells were grown in medium containing 5-bromo-deoxyuridine (BrdUrd) for two cycles, the sister chromatids can be distinguished by the differential quenching of the fluorescence of the fluorochrome Hoechst 33258. Reduced staining with Giemsa stain of the BrdUrd-incorporated chromatids was also found to be useful in differentiating sister chromatids (Perry and Wolff 1974).

SCEs are induced very efficiently by several mutagenic and carcinogenic agents, especially those which form covalent adducts to the DNA or interfere directly or indirectly with DNA replication. Agents which induce direct DNA strand breaks (with the exception of restriction endonucleases) are poor inducers of SCEs, whereas agents which act in an "S"-dependent way to induce chromosome aberrations, induce SCEs efficiently. The induction of SCEs has been considered to be directly correlated with the induction of point mutations (Carrano et al. 1978) or cytotoxicity (Natarajan et al. 1984). While most of the mutagens induce SCEs very efficiently, it is not necessary that all agents which induce SCEs also induce point mutations. Inhibitors of DNA synthesis, such as cytosine arabinoside, hydroxy urea as well as inhibitors of poly (ADP-ribose) synthetase induce SCEs efficiently without increasing the frequencies of point mutations (Natarajan et al. 1981). Earlier, the SCE induction test was considered as a sensitive method for detecting the mutagenic ability of chemicals. At the present time many regulatory authorities require a mandatory in vitro chromosome aberration test in preference to the SCE test. Since several reviews on SCEs have appeared in the literature (Sandberg 1982; Wolff 1982; Tice and Hollaender 1984),

1 Department of Radiation Genetics and Chemical Mutagenesis, State University of Leiden, Sylvius Laboratories, Wassenaarseweg 72, 2333, AL, Leiden, The Netherlands and
J.A. Cohen Institute of Radiopathology and Radiation Protection, Leiden, The Netherlands

Cytogenetics. Ed. by G. Obe and A. Basler

we would like to review in this chapter only those studies which have implications on the origin of SCEs.

2 Contribution of Incorporated BrdUrd on the Frequencies of SCEs

To visualize sister chromatids, the cells have to be grown in medium containing either tritiated thymidine or halogenated analogues, such as BrdUrd or CldUrd. A question which has been asked from the time of the first demonstration of SCEs using H3TdR, is whether the so-called spontaneous or baseline frequency of SCEs is due to normal cellular processes or induced by the beta radiation from the tritium and incorporated halogenated base analogue Under certain conditions the baseline frequency of SCEs increases dramatically. In such instances when analysis of the frequency of SCEs induced during the first and second cell division was performed, it turns out that the increase is mainly due to the SCEs formed during the second cell division. In the second division, the DNA-containing BrdUrd is used as template for replication. For example, cells derived from Bloom's syndrome patients have elevated frequencies of SCEs and most of these SCEs are formed in the second cell division (Shirashi et al 1982). In a Chinese hamster ovary cells mutant (EM9) there is a 12-fold increase in the spontaneous frequency of SCEs in comparison to the wild-type cells and the majority of this increase occurs during the second cell cycle (Dillehay et al. 1983). Inhibitors of poly(ADP-ribose) synthetase, such as 3-aminobenzamide induce SCEs effectively and this induction depends upon the inhibitor being present during the second cell division (Natarajan et al. 1981). The extent of this increase is directly correlated with the amount of BrdUrd incorporated in the DNA (Natarajan et al. 1983, Zwanenburg and Natarajan 1984). There are several possibilities for the occurrence of SCEs during second division. Incorporation of BrdUrd can change the structural integrity of the chromatin in such a way that it induces lesions which generate SCEs upon replication. Inhibitors of poly-(ADP-ribose) synthetase slow down the repair of these lesions, thus enhancing the probability of generating SCEs. The only types of DNA lesions which could be detected in BrdUrd-substituted DNA were alkali-labile sites (Zwanenburg et al. 1984).

One way to study this problem of involvement of BrdUrd in SCE formation is to employ very low levels of BrdUrd incorporation. The routine FPG (fluorescence plus Giemsa) technique requires around 30% substitution with BrdUrd to get a reasonable sister chromatid differentiation. Recently, monoclonal antibodies against BrdUrd have become available. Using these antibodies attempts have been made to differentiate sister chromatids following very low levels of BrdUrd substitution. Pinkel et al. (1985) demonstrated that SCEs can be detected in CHO cells with as low as 0.6% BrdUrd substitution. They found that the baseline frequency was around six SCEs per cell and this frequency remained constant until 20% substitution and increased further with higher substitution levels. We have investigated this problem using antibodies from a different source (Natarajan et al 1986; Fig. 1). In order to obtain a desired level of substitution, the cells were incubated in medium containing an appropriate mixture of BrdUrd (10 μM), thymidine (10 μM), 5-fluorodeoxyuridine (1 μM) and deoxycytidine (100 μM). The frequency of SCEs obtained at the lowest level of

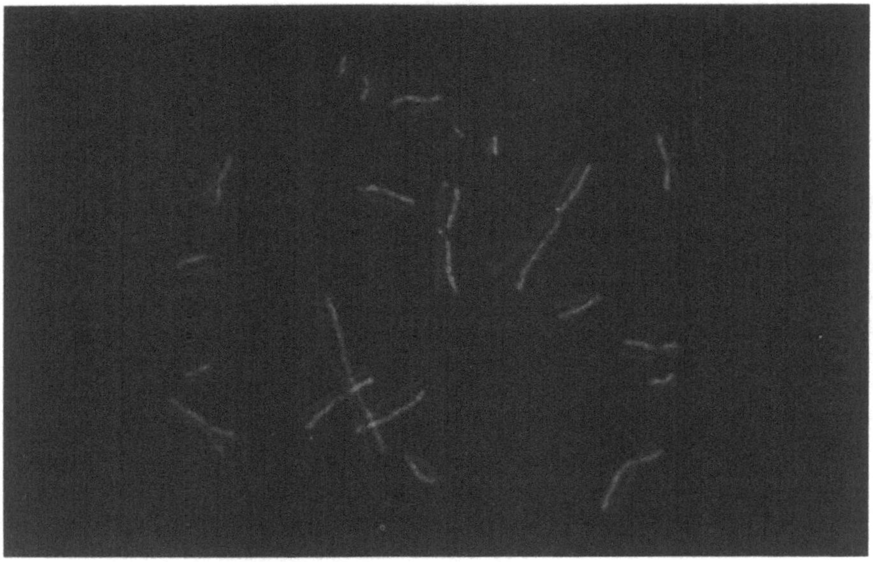

Fig. 1. Sister chromatid differentiation in BrdUrd substituted chromosomes as detected by immunochemical staining using monoclonal antibodies against BrdUrd (FITC staining)

BrdUrd substitution (1%) was around four. There was an increase in the frequency of SCEs with the increase in substitution with BrdUrd. Unlike Pinkel et al. (1985) we did not observe a plateau up to 20% substitution. The frequency of four SCEs per cell represents two SCEs per cell cycle. This value is very similar to the one reported in the literature using ring chromosomes in Chinese hamster cells without incorporation of BrdUrd (Sutou 1981). The dose response curve (Fig. 2) shows that the influence of the incorporated BrdUrd appears at two levels. The steep initial increase at low levels of incorporation may be related to conformational changes in the organization of the chromatin and further increase seen at higher levels may reflect the influence of this analogue on the repair of DNA lesions.

As indicated earlier, inhibitors of poly(ADP-ribose) synthetase increase the frequencies of SCEs and this increase is dependent on the extent of BrdUrd substitution. Therefore, it was of interest to study the influence of 3-aminobenzamide (3AB) on the frequencies of SCEs in cells with very low BrdUrd substitution levels. CHO cells were incubated in appropriate medium to obtain BrdUrd incorporation levels of 1, 3 and 10% in the first cell cycle and treated with 10 mM 3AB in the second cell cycle. The results are presented in Table 1. It can be seen that 3AB increases the frequencies of SCEs even at very low levels of BrdUrd substitution. If perturbation in the architecture of the chromatin leads to formation of SCEs e.g. by introducing delay in DNA replication, the inhibition of poly-(ADP-ribose) synthetase should potentiate this delay.

Two CHO mutants with increased spontaneous frequencies of SCEs have been studied with regard to their response at very low BrdUrd substitution levels using an

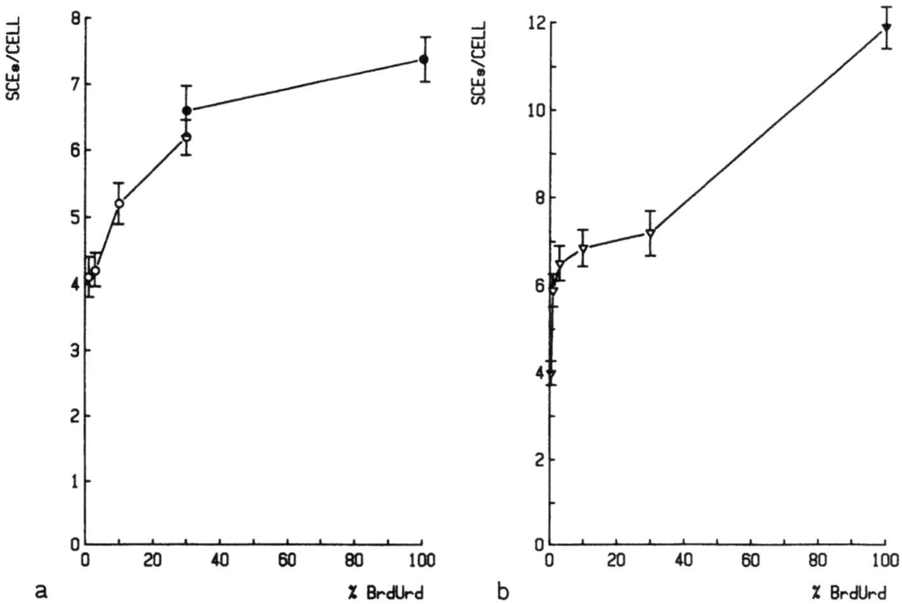

Fig. 2. a Frequencies of SCEs observed in wild-type CHO cells at different levels of BrdUrd substitution. ○ Staining with FITC; ● staining with FPG techniques. **b** Frequencies of SCEs observed in the UV-sensitive CHO mutant cell line (43-3B) at different levels of BrdUrd substitution. ▽ Staining with FITC; ▼ staining with FPG techniques

Table 1. Frequencies of SCEs in CHO cells with different levels of BrdUrd substitution and the influence of 3AB on these frequencies (Natarajan et al., unpubl. results)[a]

Degree of substitution	Frequencies of SCEs (SE)[b]		
(%)	−3AB	+ 3AB	Increase
1	4.1 (1.4)	14.7 (2.2)	10.6
3	4.2 (1.3)	16.5 (2.0)	12.3
10	5.2 (1.5)	20.4 (2.1)	15.2

[a] Fifty cells were scored for each point.
[b] SE = standard error of the mean.

antibody technique to detect SCEs. EM9, which has a very high level of spontaneous SCEs responded with lower frequency of SCEs at a 0.6% substitution level However the frequency did not reach the spontaneous level observed of wild-type cells (Pinkel et al 1985). CHO mutant 43 3B, which has about double the frequency of SCEs at a 100% substitution level in comparison to the wild type, responded with a very low frequency of SCEs at a 0.5% substitution level, namely about four per cell. This frequency is similar to that observed in wild-type cells (Natarajan et al. 1986).

However, the increase in the frequency of SCEs is very steep up to the 3% level. This indicates that the increased spontaneous levels of SCEs in these cells are due to

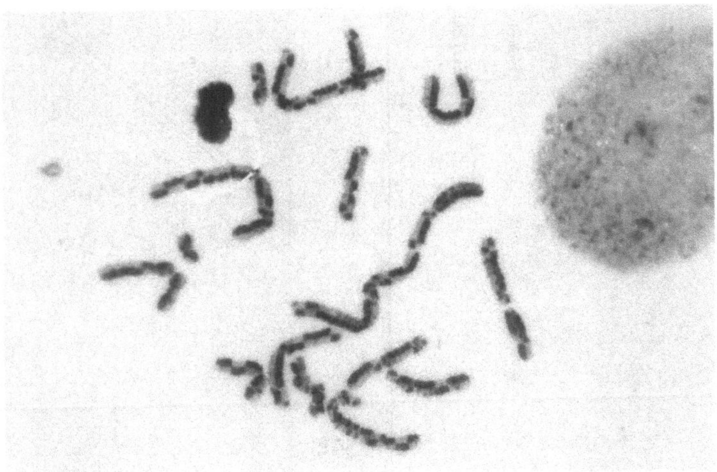

Fig. 3. SCEs induced by the restriction endonuclease Cfo I in the S-phase of the cell cycle of CHO cells

incorporated BrdUrd. This cell line is known to be defective in excision repair and most probably is unable to repair lesions arising due to BrdUrd incorporation.

3 The Lesions that Lead to SCEs

It has long been recognized that SCEs originate during DNA replication. Mutagenic agents which act in an S-dependent manner and inhibitors of DNA synthesis induce SCEs very efficiently. Inhibitors of DNA synthesis do not induce any lesions in the DNA per se, however, they may give rise to strand breaks in replicating DNA. For the formation of SCEs, one would require an exchange between replicated or replicating sister chromatids at homologous points. For such an exchange DNA double-strand breaks (DSBs) are necessary. It has been demonstrated recently that when restriction endonucleases are introduced into cells, they are capable of inducing chromosome aberrations as a consequence of the DSBs formed (Bryant 1984; Natarajan and Obe 1984). If DSBs are required for the formation of SCEs, then one should be able to induce SCEs by introducing restriction endonucleases into the cells during the S-phase. When CHO cells prelabelled with BrdUrd for 12 h and treated with restriction endonucleases, such as CFO 1, during the S-phase the frequencies of SCEs increased (Natarajan et al. 1985; Table 2; Fig. 3). When cells were synchronized and treated with restriction endonucleases, it was found that the SCEs were preferentially induced in cells that were in their early or mid-S-phase of the cell cycle. These results indicate that DNA DSBs introduced during S-phase can lead to SCEs. However ionizing radiations which are potent inducers of chromosome aberations and DNA strand breaks are poor inducers of SCEs. There are basic differences between the strand breaks induced by ionizing radiations and restriction endonucleases. Radiation does not

Table 2. Frequencies of SCEs induced by different restriction endonucleases in Chinese hamster ovary cells. (Data from Natarajan et al. 1985; Stoilov et al 1986)

Enzyme	Concentration (units)	SCEs/cell	SCE (range)
Control	–	10.7	5–16
Pvu II	100	13.3	5–45
Sma I	100	14.0	10–50
Hpa I	100	31.9	12–50
Cfo I	100	24.7	14–50
Hae III	100	17.1	7–22
Alu I	30	15.8	5–30
EcoR I	60	15.8	5–30
Taq I	50	16.2	5–25

exclusively induce DSBs and those induced are very low in number per genome at biologically relevant doses and are distributed randomly. In contrast, DSBs induced by restriction endonucleases are specific, much higher in number and non-randomly distributed depending on the distribution of recognition sites and on the accessibility of different regions of DNA for the action of the enzyme. A probable location of the DSB induced by REs is at the replication fork, where the replicating DNA is less masked, partially devoid of histones and thus more accessible for the endonuclease than other parts of the chromatin.

Acknowledgement. This work was financially supported in part by the EEC Environmental Research Programme, contract No. 139-77-ENV-N.

References

Bryant PE (1984) Enzymatic restriction of in situ mammalian cells DNA using Pvu II and Bam H1. Evidence for double strand break origin of chromosomal aberrations. Int J Radiat Biol 46: 57–65

Carrano AV, Thompson LH, Lindl PA, Minkler JL (1978) Sister chromatid exchanges as indicator of mutagenesis. Nature (London) 271:551 553

Dillehay LE, Thompson LH, Minkler JL, Carrano AV (1983) The relationship between sister chromatid exchange and perterbations in DNA replication in mutant EM9 and normal CHO cells. Mutat Res 109:283–296

Latt SA (1973) Microfluorometric detection of deoxyribonucleic acid replication in human metaphase chromosomes. Proc Natl Acad Sci USA 70:3395–3399

McClintock B (1938) The production homozygous deficient tissues with mutant characteristics by means of aberrant mitotic behaviour of ring shaped chromosomes. Genetics 23:315–376

Natarajan AT, Obe G (1984) Molecular mechanisms involved in the production of chromosomal aberrations Chromosoma 90:120 127

Natarajan AT, Csukas I, Zeeland AA van (1981) Contribution of incorporated 5-bromodeoxyuridine in DNa to the frequencies of sister chromatid exchanges induced by inhibitors of poly (ADP-ribose)-polymerase. Mutat Res 84:125–132

Natarajan AT, Zeeland AA van, Zwanenburg TSB (1983) Influence of inhibitors of poly(ADP-ribose) polymerase on DNA repair, chromosomal alterations and mutations. In: Miwa M,

Hayaishi O, Shall S, Smulson M, Sugimura T (eds) ADP-ribosylation, DNA-repair and cancer. Japan Scientific Societies Press, Tokyo, pp 227–242
Natarajan AT, Simmons JWIM, Vogel EW, Zeeland AA van (1984) Relationship between cell killing, chromosomal aberrations, sister chromatid exchanges and point mutations induced by monofunctional alkylating agents in Chinese hamster cells. A correlation with different ethylation products in DNA. Mutat Res 128:31–40
Natarajan AT, Mullenders LHF, Meijers M, Mukherjee M (1985) Induction of sister chromatid exchanges by restriction endonucleases. Mutat Res 144:33–39
Natarajan AT, Rotteveel AHM, Pieterson J van, Schliermann MG (1986) Influence of incorporated 5-bromodeoxyuridine on the frequencies of spontaneous sister-chromatid exchanges, detected by immunological methods. Mutat Res 163:51–55
Perry P, Wolff S (1974) New Giemsa method for differential staining of sister chromatids. Nature (London) 261:156–158
Pinkel D, Thompson LH, Gray JW, Vanderlaan M (1985) Measurement of sister chromatid exchanges at very low bromodeoxyuridine substitution levels using monoclonal antibody in Chinese hamster ovary cells. Cancer Res 45:5795–5798
Sandberg AA (ed) (1982) Sister chromatid exchanges. Liss, New York, p 706
Shirashi Y, Yoshida TH, Sandberg AA (1982) Analysis of single and twin sister chromatid exchanges in endoreduplicated in normal and Bloom syndrome B-lymphoid cells. Chromosoma (Berl) 87:1–8
Stoilov L, Mullenders LHF, Natarajan AT (1986) Influence of bromodeoxyuridine substitution of thymidine on sister chromatid exchanges and chromosomal aberrations induced by restriction endonucleases. Mutat Res 174:295–301
Sutou S (1981) Spontaneous sister chromatid exchanges in Chinese hamster cells in vivo and in vitro. Mutat Res 82:331–341
Taylor JH, Woods PS, Hughes WL (1958) The organization and duplication of chromosomes as revealed by autoradiographic studies using tritium labelled thymidine. Proc Natl Acad Sci USA 43:122–128
Tice RR, Hollaender A (1984) Sister chromatid exchanges Plenum, New York, p 1033
Wolff S (1982) Sister chromatid exchanges. John Wiley & Sons, New York, p 306
Zwanenburg TSB, Natarajan AT (1984) 3-aminobenzamide-induced sister chromatid exchanges are dependent on incorporated bromodeoxyuridine in DNA. Cytogenet Cell Genet 38:278–281
Zwanenburg TSB, Mullenders LHF, Natarajan AT, Zeeland AA van (1984) DNA lesions, chromosomal aberrations and G2 delay in CHO cells cultured in medium containing bromo- or chlordeoxyuridine. Mutat Res 127:155–168

17 Cytogenetic Surveillance of Occupational Exposures

M. SORSA[1] and J. W. YAGER[1,2]

1 Introduction

It has become feasible by use of cytogenetic methods to detect human exposures at biologically significant levels and in so doing to consider the complexity of human exposure patterns with various interacting factors and modifying responses.

During the last few years, interest in occupational exposure and occupational diseases has greatly increased. One of the reasons is probably the growing interest in environmental quality. This increased interest readily translates into concern that industrial effluents and products may cause illness in the general population and that workers are the persons likely to be the first to show ill-health effects caused by hazardous exposures. A second reason relates to expanding social consciousness and the contention that if the information is available workers are entitled to know the potential health hazards of their occupational exposures and such risks could be minimized.

In the following the application of various cytogenetic research methods in occupational settings will be discussed.

1.1 Objectives of Cytogenetic Surveillance

The main conceptual basis for the application of cytogenetic assays, i.e the measurement of certain types of chromosome damage in human somatic cells, is that damage to the genetic material of cells represents initial events in a process that may eventually lead to manifestations of ill health. Thus cytogenetic surveillance can serve as an early indicator, enabling prevention of adverse effects.

Most experience with the use of cytogenetic surveillance derives from "high-exposure" occupational situations. In order to make cytogenetic surveys of low-level exposures and to determine the meaning of such results for health effects in the population as a whole or for individuals poses additional extrapolation problems. The presence of chromosome aberrations can signal a potential problem in the popula-

1 Institute of Occupational Health, SF-00250 Helsinki, Finland
2 Present address: Department of Biomedical and Environmental Health Science, University of California, Berkeley, CA 94720, USA

Cytogenetics. Ed. by G. Obe and A. Basler

Table 1. Characteristic advantages and disadvantages of cytogenetic surveillance methods with peripheral lymphocytes (PL)

Advantage	Disadvantage
The method allows detection of induced genetic damage in human cells	Prospective follow-up studies in exposed groups are seldom carried out
PHA stimulates T-lymphocytes, which comprise about 70% of total lymphocytes	Interindividual variation in responses may be caused by differences in lymphocyte populations
Damage induced in lymphocytes in different organs may be picked up in PLs	PLs may not represent target cells for ill-health manifestations
Accumulated damage can be picked up: 90% of T-lymphocytes have a half life of about 3 years	Accumulation of damage may disturb interpretation of results in acute exposure studies
Radiation-induced chromosome-type aberrations remain constant for years	Aberrations induced by most chemicals represent only the misrepaired lesions (or other distortions of the DNA helix)
Establishing costs of cytogenetics laboratory are not very high and the number of helping personnel needed is low	Tedious technique requiring standardized laboratory techniques and skilled workers
Dose and quality of radiation can be extrapolated from aberration data	Dose-effect relationships with chemical mutagens still unknown

tion, but extrapolation of the risk in quantitative terms is extremely difficult. Thus, chromosome changes are an indicator of cellular genetic damage in a population, but they presently cannot be used to predict future health effects for a given individual. The main relevancy of the finding of induced chromosome damage in peripheral lymphocytes lies in the indicative meaning that exposures being able to damage human genetic material have taken place.

2 Applicable Methodologies

2.1 Methods Detecting Chromosome Damage in Somatic Cells

Rigorous study design is necessary in the application of all cytogenetic biomonitoring methods since many interindividual factors that are not related to the specific chemical exposure(s) of interest may affect the cytogenetic parameters studied (see Sect. 2.2).

Recommended methods for the three types of cytogenetic tests i.e., structural chromosome aberrations, sister chromatid exchange, and micronucleus assay, follow in order with regard to their establishment. That is, the method for chromosome aberration analysis is relatively well established, while the method for sister chromatid exchange analysis is at an intermediate stage of applicability, and that for the micronucleus assay is under current development.

Table 2. Some typical characteristics of ionizing radiation and mutagenic chemicals in the induction of chromosome damage in peripheral lymphocytes (PL)

Parameter	Ionizing radiation	Mutagenic chemicals
Typical lesion	DNA double-strand breaks	Base alkylations, DNA cross-links, intercalation
Dependancy on DNA synthesis (S)	S-independent	S-dependent (except few radiomimetic chemicals)
Types of chromosome aberrations induced	G_0,G_1 : chromosome-type S: chromosome- and chromatid-type G_2 : chromatid type	Mostly chromatid type
Induction of SCEs	Ineffective as SCE inducer	Effective SCE inducers
Relationship to repair in PL	Lesions repaired or misrepaired immediately. Aberrations produced remain permanent	Misrepaired lesions can be expressed in vitro

These three endpoints are all analyzed in peripheral blood lymphocytes of occupationally exposed populations since these cells are most easily accessible in healthy workers. Further, these cells may serve as indices of exposure to other somatic as well as germinal cells in the body since the lymphocytes circulate throughout the body in G_0 and, therefore, may be exposed to a parent compound and/or active metabolites at several locations. Thus, the most feasible sample for cytogenetic analyses of worker populations is whole blood obtained by routine venipuncture (see Table 1 for advantages and disadvantages). References to current methods in the following section give detailed descriptions of methods, sources of variability, and criteria for scoring and reporting of data.

2.1.1 Structural Chromosome Aberrations

The methods of determining induced chromosome aberrations in human lymphocytes were well-developed by the mid-1970s (Evans and O'Riordan 1975); currently recommended methodology, classification, and scoring of structural aberrations and statistical methods have been recently described (Evans 1984; Forni 1984). The reader is referred to these previous methodological descriptions.

Chromosome aberrations consist of overt breakage and rearrangements of chromosomes visualized in the metaphase plate. Chromosome aberrations are most sensitive to agents that can directly break the DNA duplex such as radiation and a few radiomimetic chemicals. However, since most chemical mutagens which cause chromatid-type damage, are S phase-dependent, the highest yield of aberrations is likely to be seen in the metaphases following first division (Table 2).

2.1.2 Sister Chromatid Exchanges

Sister chromatid exchanges involve breakage of double-stranded DNA in both chromatids followed by an exchange of whole DNA duplexes. These events occur in S-phase. SCEs are most efficiently induced by substances that form covalent adducts to the DNA, distort the DNA helix, or interfere with DNA precursor metabolism or repair; therefore, they are thought to be most sensitive to chemical mutagens (Perry and Evans 1975; Latt et al. 1981). However, the exact molecular mechanism of SCE formation is still unknown.

The method for preparation of metaphases for sister chromatid exchange analysis is analogous to that for chromosome aberrations except that 5-bromodeoxyuridine (BudR) is added to the cell culture medium and incorporated for two rounds of replication. After chromosomes are fixed and spread on slides, differential staining of the sister chromatids is accomplished enabling visualization of the exchanges (Perry and Wolff 1974).

2.1.3 Micronuclei

Micronuclei arise from chromosome fragments or lagging whole chromosomes that are not incorporated into daughter nuclei at the time of cell division; the assay detects both clastogens and agents that affect the spindle apparatus. Cells in first interphase are scored for the presence of micronuclei.

The method for determination of micronuclei in human peripheral lymphocytes utilizes similar conditions as described previously, except that treatment with colchicine is deleted. Treatment with hypotonic salt solution (Countryman and Heddle 1976; Heddle et al. 1978) was modified to retain the intact cytoplasm (Högstedt 1984) thereby reducing artifacts and increasing scoring precision. A recent modification involves the use of cytochalasin B to block cells in cytokinesis (Fenech and Morley 1985). The resulting binucleated cells are then scored. This modification, which allows assessment of cellular kinetics, is likely to increase sensitivity of micronucleus analysis although there is little information on this point to date.

2.2 Confounders and Contraindications of Cytogenetic Population Monitoring

In cytogenetic studies two major types of variations have been documented (Archer et al. 1981; Sorsa 1984). The first includes technical factors associated with slide-reading discrepancies and culture conditions, specifically with the type of medium, temperature, and BudR concentration which are known to affect the results (Latt et al. 1981). Also, sampling times can alter chromosome aberration yields, and possibly also SCE incidence through changes in populations of T- and B-lymphocytes.

The lesions induced in the DNA of lymphocytes by chemical exposure that lead to formation of structural chromosome aberrations, sister chromatid exchange, and micronuclei must persist in vivo until the blood is withdrawn and then in vitro until the cultured lymphocyte begins DNA synthesis. It is, therefore, important to score

Table 3. Main confounders of occupational cytogenetic studies and ways to control them

Confounders	Control efforts
Exposure conditions	
Identification of correct chemical exposure	Factory record checking
Estimate of dose of exposure	Industrial hygiene survey
Individual variations	
Genetic factors	Unknown before analysis
Life-style factors (smoking, special diets)	Match for smoking, check for drug use, interview in person
Health factors (viral infections, X-ray diagnostics)	Check medical records, interview in person, exclude from study
Culture conditions	
Culture time	CAs: 1st division metaphases only SCEs: 2nd division metaphases only
Culture medium and chemicals	Keep constant, note batches, score M.I. or I/II M
Time between sampling and culture	Keep constant, refrigerate, do not exceed 24 h
Persistence of the mutagen in the blood sample	In vitro experimentation
Analysis and scoring	
Scorer variation	Coded slides, one scorer or quality control slide sets
Interpretation of damage scored	Strict scoring criteria, consultation
Reference groups	
Both methodological and concurrent group controls needed	Individual prematching

cells after the first division (in the case of chromosome aberrations or micronuclei) or after the second division (sister chromatid exchanges) in order to obtain the best estimate of induced damage.

Scoring constitutes an extremely important element in cytogenetic toxicology. Slides must be randomized and coded to avoid, as far as possible, scorer bias. Consistent scoring criteria, quality control, and standardized statistical analyses and reporting should be maintained (Brøgger et al. 1984; Scott et al. 1983).

The second category of variability is due to conditions associated with the subjects such as age, sex, medication, infections, etc. Individual variations can also be caused by genetic susceptibility to environmental agents (Table 3).

It is critical to obtain a concurrent control group that is matched as closely as possible on innate factors such as sex and age as well as factors such as smoking status, viral infections and vaccinations, alcohol and drug intake, and exposure to X-rays. Additionally, it is necessary to obtain qualitative (job category, years exposed) and quantitative (e.g., breathing zone air samples for chemical analysis and specific metabolites, if possible) estimates of exposure to the putative genotoxic agent(s) in the workplace. Special consideration should be paid to proper statistical treatment of the results (Carrano and Moore 1982).

Since cytogenetic biomonitoring studies are tedious and difficult in many respects, careful preplanning is very important. Experimental confirmation of the chromosome damaging potential of the exposing agent(s) is a prerequisite to performing human cytogenetic studies.

3 Occupational Exposures and Cytogenetic Damage

3.1 Some Well-Documented Cases

Over 100 cytogenetic studies have been reported from various occupational exposure groups (Ashby and Richardson 1985). However, very few of the exposing agents have been confirmed by several independent findings, and using appropriate methods. As examples, perhaps among the best-documented occupational chemicals with positive evidence on their chromosome damaging effects to date are ethylene oxide, styrene, benzene, and alkylating anticancer agents. Since all these are quite different both in their mode and setting of exposure and in their genotoxic characteristics, these cases will be discussed briefly in the following.

3.1.1 Ethylene Oxide

Ethylene oxide is a direct-acting alkylating agent and a well-known mutagen in several test systems. On the basis of positive animal carcinogenicity data, it is considered to be carcinogenic with "sufficient evidence" to animals, but the status of human data was evaluated as "inadequate" (IARC 1985). At least 14 studies utilizing cytogenetic endpoints in humans have been completed (IARC 1985). Four of five studies found an increase in chromosome aberrations. The fifth study by Clare et al. (1985) found no increases in chromosome aberrations at exposure concentrations that were generally below the detection limit of 0.05 ppm. Of ten occupational studies completed, utilizing sister chromatid exchange as an endpoint, seven found significant increases and at least one study showed a tendency to a dose-response relationship. Of the three studies negative for sister chromatid exchanges, one gave no information on exposure concentrations and the remaining two reported air concentrations of < 1 ppm and < 5 ppm 8-h TWAs. Two studies utilizing the micronucleus test showed an elevated frequency of micronucleated cells in bone marrow and in peripheral lymphoctes in persons exposed to ethylene oxide at levels of exposure of 1–5 ppm TWA.

To date, occupational cytogenetic studies of exposure to ethylene oxide undoubtedly represent the best example of a situation in which these methods have provided a link between experimental evidence obtained in short-term and animal tests with epidemiological studies that are ongoing (Hogstedt 1986). In the case of ethylene oxide, the first data on potential long-term hazards came from very early cytogenetic studies in workers (Ehrenberg and Hällström 1967). The most valuable studies have been those that assessed exposure and accounted for epidemiological factors. Ethylene oxide is a model direct-acting alkylating compound, therefore its action is presumably less influenced by the inherent variability in metabolic activation capabilities between individuals and within the same individual under different physiological conditions or exposure concentrations that may obscure response in the case of many other indirectly-acting chemicals.

3.1.2 Styrene

Styrene (vinyl benzene) is an organic solvent used in the production of plastics and resins (IARC 1979). For styrene, the animal carcinogenicity data was evaluated to be of "limited" evidence, while the studies for human epidemiology were considered inadequate by 1982 (IARC 1982a). All 12 cytogenetic studies of worker exposure have incorporated qualitative and quantitative exposure estimates that include air sampling and often biological monitoring data with urinary metabolites. Of the eight studies completed through 1981, four showed increases in chromosome aberration frequencies that were two to ten times higher than in controls (Norppa et al. 1981). One of the four studies that utilized SCEs found a positive response; one study measuring micronuclei in lymphocytes found them to be significantly increased. Mean 8-h TWA exposures ranged from < 1 ppm to about 40 ppm with a range of 1–300 ppm. In four later studies, two have shown increases in chromosome aberrations at exposures of around 50–80 ppm TWA (Camurri et al. 1983; Watanabe et al. 1983). A marginal increase in SCEs was found by Camurri et al. (1983), while the other study that utilized SCEs was negative (Watanabe et al. 1983). Both studies that measured micronuclei (Högstedt et al. 1983; Nordenson and Beckman 1984) found significant increases.

Taken as a whole, these studies indicate that working in the reinforced plastics industry is associated with increases in chromosome aberrations that appear to occur beginning at exposures of > 40 ppm 8-h TWA, whereas SCEs appear less sensitive and are found only at higher exposure levels. Although the data on micronuclei are few, they suggest that detection of micronuclei may occur at exposure levels < 40 ppm 8-h TWA.

3.1.3 Benzene

Benzene is an established leukemogen in humans (IARC 1982a) and a carcinogen at several sites in animals (Maltoni et al. 1983). Benzene is a ubiquitous chemical finding its major uses as an industrial solvent and in manufacturing many other chemical products. The most significant of the cytogenetic studies performed on workers exposed to benzene have been reviewed recently (Sarto et al. 1984).

Of eleven studies, three showed significant increases in chromosome aberrations at relatively high exposure concentrations ranging from 25–532 ppm, while four studies showed increases at exposures ranging from 2.2–25 ppm. Of the remaining studies, two showed increases with no measure of exposure concentration and two showed no significant increases after about 1 year after exposure to 2–50 ppm had ended, while the other study that showed no significant increase in chromosome aberrations had no exposure data. As pointed out previously (Dean 1978), lack of data on the actual levels of environmental exposure hampers the evaluation of these studies.

Benzene is a weak alkylating agent (Lutz and Schlatter 1977), therefore, it is perhaps not surprising that both studies utilizing sister chromatid exchange analysis have shown no significant increases. Exposures in these studies were also low, 2.2–12.8 ppm (Sarto et al. 1984) and 2–50 ppm and up to 9 ppm (Watanabe et al. 1980).

Detection of biological effects by cytogenetic monitoring of workers exposed to benzene preceded the epidemiological detection of benzene as a leukemogen by several years. The exact mechanism of action of benzene remains a subject of investigation. As previously mentioned (Vainio 1985), negative cytogenetic studies of persons exposed to benzene should be interpreted with caution; it is also likely that increases in chromosome changes in exposed persons should be regarded as an indicator of a genotoxic hazard.

3.1.4 Alkylating Anticancer Agents

Chemotherapy has gained increasing importance in the treatment of malignancies. Many cancer chemotherapeutic agents have been shown to be carcinogenic, either in rodent bioassays or in inducing second cancers in patients (see IARC 1982b). In most instances, the patients receive a combination chemotherapy and, therefore, it is seldom possible to associate the carcinogenic effect to a specific drug. Also, the occupational risks of handling anticancer agents is a typical mixed-exposure situation.

On the basis of experimental studies, both on mutagenicity as well as carcinogenicity, the alkylating-type cytostatics are of special concern (Sorsa et al. 1985). Patient data on induction of chromosome aberrations and of SCEs show that exposure to alkylating agents can be expressed in peripheral lymphocytes, and the damage is preserved for several months or even years after treatment (Gebhart et al. 1980).

Hospital personnel handling anticancer drugs may be exposed to them by inhaling the aerosols generated and measured from workroom air (de Werk et al. 1983), or through skin contact with the drug or urine of patients (Hirst et al. 1984; Venitt et al. 1984). Most studies measuring exposure of the personnel have used the urine mutagenicity method, both with negative and positive results (see Sorsa et al. 1985). After improvements of the handling procedures the urine mutagenicity results have been mainly negative.

A few cytogenetic studies have been performed among oncology unit personnel showing that even these low level exposures to anticancer agents can be detected as increased chromosome damage in lymphocytes. As compared with office workers, the group of oncology nurses showed a significantly increased frequency of SCEs in their lymphocytes (Norppa et al. 1980). Even at an individual level, there was a tendency for a correlation between the SCE frequency and the amount of alkylating cytostatics handled (Sorsa et al. 1982). Waksvik et al. (1981) also reported an increase in SCEs and in chromosome-type gaps among nurses handling cytostatic drugs. However, no difference between the group working with dilution and dispension of anticancer agents and the control group was seen in another study from a different cancer clinic (Barale et al. 1985).

The frequency of chromosomally aberrant lymphocytes was found to be significantly higher in the group of oncology nurses than in the control groups (laboratory workers, hospital clerks) (Nikula et al. 1984). In this study the aberrations were typically chromosome-type breaks, which were significantly increased among the nurses as compared to the reference groups. The authors concluded that the observed increase in chromosome aberrations may have been due to occupational exposure to cytostatic agents.

The available hygienic measurements as well as studies on mutagenic activity in urine and cytogenetic studies all point to the possibility that hospital personnel handling cancer chemotherapeutics may be exposed to low levels of these drugs. The occupational exposure to anticancer drugs is a typical situation where the level of exposure is usually low but may simultaneously occur for a multitude of chemicals and it may continue for decades. The health risks have been clearly demonstrated among patients receiving these drugs but the scientific data concerning the cancer risks to health care personnel are totally lacking. However, two epidemiological studies have suggested an increased risk of spontaneous abortions of congenital malformations in offspring of nurses who had worked with anticancer agents without protection (Hemminki et al. 1985; Selevan et al. 1985). Still, the evidence from biological monitoring and cytogenetic studies has by now led to considerable improvements in working conditions, and this has obviously also decreased the possibility of undesirable ill-health manifestations.

Table 4. Occupational chemicals and work-related exposures established with sufficient evidence to be carcinogenic in humans, and available data on their chromosome damaging effects in workers

Compound/Exposure	Evaluation ref. by IARC	Animal carcinogenicity	Cytogenetic findings in humans[a,b]		
			CA	SCE	MN
4-Aminobiphenyl	Suppl 4 (1982) 37	Sufficient	..	..	..
Arsenic and certain arsenic compounds	Suppl 4 (1982) 50	Inadequate	+ (1)	..	..
Asbestos	Suppl 4 (1982) 52	Sufficient	+ (2)	+ (3)	..
Benzene	Suppl 4 (1982) 56	Limited	+ (4)	– (5)	..
Benzidine	Suppl 4 (1982) 57	Sufficient	+ (6)	..	..
Bis(chloromethyl)ether	Suppl 4 (1982) 64	Sufficient	+ (7)	..	..
Chromium and certain chromium compounds	Suppl 4 (1982) 91	Sufficient	+ (8) – (9)	+ (8, 10) – (9, 11)	..
Coal tars	35 (1985) 83	Sufficient	+ (12)	..	..
Coal-tar pitches	35 (1985) 83	Sufficient	– (13)	..	..
Mineral oils (certain)	33 (1984) 87	Sufficient	+ (14)	..	..
2-Naphthylamine	Suppl 4 (1982) 166	Sufficient	..	..	..
Shale oils	35 (1985) 161	Sufficient	..	+ (15)	..
Shoots and soot extracts	35 (1985) 219	Sufficient	..	..	..
Vinyl chloride	Suppl 4 (1982) 260	Sufficient	+ (16)	+ (17)	..
Auramine manufacture	Suppl 4 (1982) 53	–	..	..	..
Boot and shoe manufacture	Suppl 4 91982) 138	–	..	..	..
Coal gasification	34 (1984) 65	–	+ (18)	..	..
Coke production	34 (1984) 101	–	..	+ (19)	..
Furniture manufacture	Suppl 4 (1982) 140	–	..	..	..
Isopropyl alcohol manufacture	Suppl 4 (1982) 151	–	..	..	..
Nickel refining	Suppl 4 (1982) 167	–	+ (20)	– (20)	..
Rubber industry	Suppl 4 (1982) 144)	–	+ (21) – (22)	+ (21) – (22)	..
Hematite mining	Suppl 4 (1982) 254	–	..	..	..
Mixed handling of alkylating anticancer agents (evaluated separately)					
Chlorambucil	Suppl 4 (1982) 77	Sufficient	+ (23,24)	+ (24,25) – (26)	
Cyclophosphamide	Suppl 4 (1982) 99	Sufficient			
Melphalan	Suppl 4 (1982) 154	Sufficient			
Myleran	Suppl 4 (1982) 68	Limited			

[a] Numbers in brackets refer to references given in the following list.
[b] + Positive result; – negative result; .. no data.

References to Table 4

1. Nordenson I, Beckman L (1982) Occupational and environmental risks in and around a smelter in northern Sweden. VII. Reanalysis and follow-up of chromosomal aberrations in workers exposed to arsenic. Hereditas 96:175–181
2. Srb V, Kubzova Z, Musil M, Havel V, Kubiková K, Hartmann M (1985) Testing a genotoxic activity in the shops with an asbestos risk. II. Cytogenetic examination of lymphocytes of human peripheral blood 1982–1983. Pracov Lek 37:76–80
3. Rom WN, Livingston GK, Casey KR, Wood SD, Egger MJ, Chin GL, Jerominski L (1983) Sister chromatid exchange frequency in asbestos workers. J Natl Cancer Inst 70:45–48

4. Forni A, Pacifico E, Limonta A (1971) Chromosome studies in workers exposed to benzene or toluene or both. Arch Environ Health 22:373–378
5. Watanabe T, Endo A, Kato Y, Shima S, Watanabe T, Ikeda M (1980) Cytogenetics and cytokinetics of cultured lymphocytes from benzene-exposed workers. Int Arch Occup Environ Health 46:31–41
6. Qi R, Zhang S, Yan H, Qi J, Li L (1981) Studies on peripheral lymphocyte chromosome changes in subjects exposed to carcinogenic aromatic amines. Tianjiu Yiyao 9:684–685
7. Sram RJ, Samkova I, Hola N (1983) High-dose ascorbic acid prophylaxis in workers occupationally exposed to halogenated ethers. J Hyg Epid Microbiol Immunol 27:305–318
8. Sarto F, Comianto I, Bianchi V, Levis AG (1982) Increased incidence of chromosomal aberrations and sister chromatid exchanges in workers exposed to chromic acid (CrO) in electroplating factories. Carcinogen 3:1011–1016
9. Husgafvel-Pursiainen K, Kalliomäki P-L, Sorsa M (1982) A chromosome study among stainless steel welders. J Occupat Med 24:762–766
10. Stella M, Montaldi A, Rossi R, Rossi G, Levis AG (1982) Clastogenic effects of chromium on human lymphocytes in vitro and in vivo. Mutat Res 101:151–164
11. Nagaya T (1986) No increase in sister-chromatid exchange frequency in lymphocytes of chromium platers. Mutat Res 170:129–132
12. Sram RJ, Dobias L, Pastorkova A, Rossner P, Janca L (1983) Effect of ascorbic acid prophylaxis on the frequency of chromosome aberrations in the peripheral lymphocytes of coal-tar workers. Mutat Res 120:181–186
13. Heussner JC, Ward JB Jr, Legator MS (1985) Genetic monitoring of aluminium workers exposed to coal tar pitch volatiles. Mutat Res 155:143–156
14. Sram RJ, Jolá N, Ketosovec F, Novákova A (1985) Cytogenetic analysis of peripheral blood lymphocytes in glass workers occupationally exposed to mineral oils. Mutat Res 144:277–280
15. Kahn H, Rüütel P, Lember S (1986) Genotoxic effect of shale oils on the human organism. In: Hernberg S, Kohn K (eds) Occupational toxicology. Inst Occupat Health, Helsinki, Finl (in press)
16. Anderson D, Richardson CR, Weight TM, Purchase IFH, Adams WGF (1980) Chromosomal analyses in vinyl chloride exposed workers. Results from analysis 18 and 42 months after an initial sampling. Mutat Res 79:151–162
17. Anderson D, Richardson CR, Purchase IFH, Evans HJ, O'Riordan ML (1981) Chromosomal analyses in vinyl chloride exposed workers: comparison of the standard technique with the sister-chromatid exchange technique. Mutat Res 83:137–144
18. Hola H, Landa K, Rognicková I, Sram RJ (1985) Cytogenetic analysis of peripheral lymphocytes in workers occupationally exposed to soft-coal pressure gasification. Mutat Res 147: 298 (Abstr)
19. Miner JK, Livingston GK, Lyon KL (1983) Lymphocyte sister chromatid exchange frequencies in coke oven workers. J Occupat Med 25:30–33
20. Waksvik H, Boysen M, Hogetveit AC (1984) Increased incidence of chromosomal aberrations in peripheral lymphocytes of retired nickel workers. Carcinogen 5:1525–1527
21. Mäki-Paakkanen J, Sorsa M, Vainio H (1984) Sister chromatid exchanges and chromosome aberrations in rubber workers. Teratog Carcinog Mutag 4:189–200
22. Degrassi F, Fabri G, Palitti F, Paoletti A, Ricordy R, Tanzarella C (1984) Biological monitoring of workers in the rubber industry. I. Chromosomal aberrations and sister-chromatid exchanges in lymphocytes of vulcanizers. Mutat Res 138:99–103
23. Nikula E, Kiviniitty K, Leisti J, Taskinen PJ (1984) Chromosome aberrations in lymphocytes of nurses handling cytostatic agents. Scand J Work Environ Health 10:71–74
24. Pohlova H, Cerná M, Rössner P (1986) Chromosomal aberrations, SCE and urine mutagenicity in workers occupationally exposed to cytostatic drugs. Mutat Res 174:213–217
25. Sorsa M, Norppa H, Vainio H (1982) Induction of sister chromatid exchanges among nurses handling cytostatic drugs. In: Indicators of genotoxic exposure. Banbury Rep 13. Cold Spring Harbor Lab, pp 341–350
26. Barale R, Sozzi G, Toniolo P, Borghi O, Reali D, Loprieno N, Della Porta G (1985) Sister chromatid exchanges in lymphocytes and mutagenicity in urine of nurses handling cytostatic drugs. Mutat Res 157:235–240

4 Predictive Value of Cytogenetic Population Biomonitoring to Cancer Hazard

4.1 Human Carcinogens as Chromosome Damaging Agents in Worker Populations

The International Agency for Research on Cancer (IARC) has by 1986 evaluated 288 chemicals or exposures about which the animal carcinogenicity data is convincingly documented ("sufficient evidence") or some data is available on human carcinogenicity (Vainio et al. 1985). Among these, 30 compounds, consisting of industrial chemicals, drugs, complex mixtures or cultural habits, and nine industrial processes have been epidemiologically shown carcinogenic in humans with sufficient evidence. Of these 39 proven human carcinogens, 23 clearly concern occupational exposures; additionally, there is data on work-related exposures concerning anticancer agents, even though studies on human carcinogenicity have been conducted on patients receiving these drugs.

In Table 4, the available cytogenetic data from worker studies have been compiled concerning exposures to these known occupational carcinogens. As a whole, the data are rather limited and for most compounds or exposures there is only one cytogenetic study available. The problems concerning study designs and appropriateness of cytogenetic analyses are also obvious. Even with these limitations, an overall agreement can be seen. Carcinogens or exposures associated with carcinogenicity have usually been reported to also induce chromosome damage in peripheral lymphocytes, whenever studies have been performed. For explaining negative findings, it should be noted that the cytogenetic parameters may not be sensitive to exposures below usual occupational standards, and the peripheral lymphocytes may not always be the proper target for detecting the damage.

A few discrepant results appear in Table 4. Such contradictory results are always possible, especially considering the enormous complexity of human exposure situations. For chromium exposure the negative findings are derived from exposure to welding fumes (stainless steel welders), while positive results have been obtained with chromate platers; even though hexavalent chromium is involved with both exposures, the compounds and particulate fumes are different and may, therefore, explain the discrepancy. The contradictory findings concerning workers in the rubber industry are most probably explainable by differences in the exposure situations and job categories. The positive findings reported relate to mixing and weighing operations in the rubber industry, while the negative results concern vulcanizers of rubber (see references of Table 4). Comparisons between discrepant results, especially "borderline" results, obtained in otherwise well-conducted studies are thus largely related to the fact that no two human exposure situations are alike. Very often the source of discrepancy is in misclassification of the exposure, both qualitatively and quantitatively (Johnson 1986).

4.2 Cytogenetic Surveillance in Prevention of Occupational Cancer

For the chemicals and exposures listed in Table 4, cytogenetics has mostly been used to show existing hazards, already proven epidemiologically. The only exception in the list is exposure to benzene, which was shown to be clastogenic in humans long before its leukemogenicity was proven.

For cancer prevention actions to be taken, the other occupational exposure situations discussed in the previous section are more important; ethylene oxide, classified to show "limited" evidence as a human carcinogen, styrene, about which only "inadequate" evidence has been available, and alkylating anticancer agents about which there are no reports so far about cancer hazards possibly associated with occupational exposures. As discussed earlier, all these exposures are well documented for their chromosome damaging effects, and the positive findings in any worker population should be considered as an adverse sign of exposure.

Even though no direct connection between the chromosome damaging effects and the possible malignant potential of an agent can be shown at the individual level, chromosome damage in somatic cells of an exposed group should be considered a warning sign for possible genotoxic end effects (see various conclusions in a recent monograph, Berlin et al. 1984). Thus, positive findings of induced somatic chromosome damage should lead to actions tending to decrease the exposures.

4.3 Individual Health Hazard

On an individual basis, data from cytogenetic studies cannot be used to predict the increased likelihood of cancer or adverse reproductive outcomes for that person. Many factors remain unaccounted for at present, e.g., variability arising from innate inter- and intraindividual variation in such factors as metabolic activation of chemicals and repair capacity, as well as variation in concurrent exposures to chemicals and other substances related to life-style such as diet, alcohol intake, smoking, and various hobbies. Little is known about the effect of exposure to multiple chemicals and the production of multiple active metabolites that may occur in different proportions in different individuals. There is evidence, most recently from data concerning the localization of mammalian and human oncogenes and the role of chromosome rearrangements in their activation, that chromosome damage may be related to an increased probability of some malignant diseases (Gilbert 1983; Rowley 1983). However, too many intervening variables remain unknown to be able to make reliable predictions in the individual case. Additionally, since cancer represents a spectrum of diseases with multiple causations and with multiple steps between its initiation and final expression, it may be somewhat naive to expect that one or even a few endpoints will be able to predict all possible cancer outcomes. Further, with cytogenetic surveillance we are attempting to measure the earliest *reversible* indicators of exposure to mutagenic carcinogens in order to prevent exposures that may lead to adverse health outcomes.

5 Future Perspectives in Cytogenetic Surveillance

It is clear that cytogenetic monitoring provides a good method to directly measure changes in the genetic material of humans. In order to further elucidate the strength of the relation between these changes and the potential for adverse outcomes that may result from exposure to mutagenic and carcinogenic chemicals, it is necessary to conduct well-designed, prospective, longitudinal studies; in order to efficiently apply cytogenetics in such studies, development of reliable, automated scoring systems are needed. Such well-controlled studies are also likely to generate much more detailed information on the effect of interrelationships and variability between exposure(s), formation of active metabolites, and repair capacity on chromosome changes and to associate these with health parameters even at the individual level.

Acknowledgments. The authors wish to thank Ms. Leila Kuusela for patience and skillful typing of the manuscript. Dr. Yager was a Visiting Research Fellow at the Institute of Occupational Health through an award from the Academy of Finland, Medical Research Council.

References

Archer PB, Bender M, Bloom AD, Brewen JG, Carrano AV, Preston RJ (1981) Report of panel 1: Guidelines for cytogenetic studies in mutagen-exposed human populations. In: Bloom AD (ed) Guidelines for studies of human populations exposed to mutagenic and reproductive hazards. March of Dimes Birth Defects Foundation, New York, pp 1–35

Ashby J, Richardson CR (1985) Tabulation and assessment of 113 human surveillance cytogenetic studies conducted between 1965 and 1984. Mutat Res 154:111–133

Barale R, Sozzi G, Toniolo P, Borghi O, Reali D, Loprieno N, Della Porta G (1985) Sister-chromatid exchanges in lymphocytes and mutagenicity in urine of nurses handling cytostatic drugs. Mutat Res 157:235–240

Berlin A, Draper M, Hemminki K, Vainio H (eds) (1984) Monitoring human exposure to carcinogenic and mutagenic agents. IARC Sci Publ No 59. Int Ag Res Cancer, Lyon, Fr

Brøgger A, Norum R, Hansteen I, Clausen KO, Skardal K, Mitelman F, Kolnig A Strombeck B, Nordenson I, Andersson G, Jakobsson K, Mäki-Paakhanen J, Norppa H, Järventaus H, Sorsa M (1984) Comparison between five Nordic laboratories on scoring of human lymphocyte chromosome aberrations. Hereditas 100:209–218

Camurri L, Codeluppi S, Pedroni C, Scarduelli L (1983) Chromosomal aberrations and sister-chromatid exchanges in workers exposed to styrene. Mutat Res 119:361–369

Carrano AV, Moore DH (1982) The rationale and methodology for quantifying sister chromatid exchanges in humans. In: Heddle JA (ed) Mutagenicity: new horizons in genetic toxicology. Academic Press, London New York, pp 267–304

Clare MS, Dean BJ, De Jong G, Van Sittert NJ (1985) Chromosome analysis of lymphocytes from workers at an ethylene oxide plant. Mutat Res 156:109–116

Countryman PJ, Heddle JA (1976) The production of micronuclei from chromosome aberrations in irradiated cultures of human lymphocytes. Mutat Res 67:321–322

Dean BJ (1978) Genetic toxicology of benzene, toluene, xylenes and phenols. Mutat Res 47: 75–97

Ehrenberg L, Hällström T (1967) Haematologic studies on persons occupationally exposed to ethylene oxide. In: IAEA: Int Atom Energ Ag Rep SM 92/96, pp 327–334

Evans HJ (1984) Human peripheral blood lymphocytes for the analysis of chromosome aberrations in mutagen tests. In: Kiley BJ, Legator M, Nichols W, Ramel C (eds) Handbook of mutagenicity test procedures. Elsevier, Amsterdam, pp 405–427

Evans HJ, O'Riordan ML (1975) Human peripheral blood lymphocytes for the analysis of chromosome aberrations in mutagen tests. Mutat Res 31:135–148

Fenech M, Morley AA (1985) Measurement of micronuclei in human lymphocytes. Mutat Res 141:29–36

Forni A (1984) Chromosomal aberrations in monitoring exposures to mutagen-carcinogens. In: Berlin A, Draper M, Hemminki K, Vainio H (eds) Monitoring human exposure to carcinogenic and mutagenic agents. IARC Sci Publ No 59. Int Ag Res Cancer, Lyon, Fr, pp 325–337

Gebhart E, Lösing J, Wopfner F (1980) Chromosome studies on lymphocytes of patients under cytostatic therapy. I. Conventional chromosome studies in cytostatic interval therapy. Human Genet 55:53–63

Gilbert F (1983) Chromosomal aberrations and oncogenes. Nature (London) 303:415

Heddle JA, Lue CB, Saunder EF, Benz R (1978) Sensitivity to five mutagens in Fanconi's anemia as measured by the micronucleus method. Cancer Res 38:2983–2988

Hemminki K, Kyyrönen P, Lindbohm M-L (1985) Spontaneous abortions and malformations in the offspring of nurses exposed to anaesthetic gases, cytostatic drugs, and other potential hazards in hospitals, based on registered information of outcome. J Epid Comm Health 39: 141–147

Hirst M, Mills DG, Tse S, Levin L, White DF (1984) Occupational exposure to cyclophosphamide. Lancet 1:186–188

Högstedt B (1984) Micronuclei in lymphocytes with preserved cytoplasm. Mutat Res 130:63–72

Högstedt B, Akesson B, Axell K, Gullberg B, Mitelman F, Pero RW, Skerfving S, Welinder H (1983) Increased frequency of lymphocyte micronuclei in workers producing reinforced polyester resin with low exposure to styrene. Scand J Work Environ Health 9:241–246

Hogstedt C (1986) Future perspectives, needs and expectations of biological monitoring of exposure to genotoxicants in prevention of occupational disease. In: Sorsa M, Norppa H (eds) Monitoring of occupational genotoxicants. Progress in clinical and biological research, vol 207. Liss, New York, pp 231–243

IARC – International Agency for Research on Cancer (1979) Monographs on the evaluation of the carcinogenic risk of chemicals to humans, vol 19. Styrene, polystyrene and styrene butadiene copolymers. Some monomers, plastics and synthetic elastomers and acrolein. Int Ag Res Cancer, Lyon, Fr, pp 231–259

IARC (1982a) Monographs on the evaluation of the carcinogenic risk of chemicals to humans, vol 29. Some industrial chemicals and dyestuffs. Int Ag Res Cancer, Lyon, Fr, pp 93–148

IARC (1982b) Monographs on the evaluation of the carcinogenic risk of chemicals to humans, Suppl 4. Chemicals, industrial processes and industries associated with cancer in humans. Int Ag Res Cancer, Lyon, Fr, pp 1–292

IARC (1985) Monographs on the evaluation of the carcinogenic risk of chemicals to humans, vol 36. Some allyl and allylic compounds, aldehydes, epoxides, and peroxides. Int Ag Res Cancer, Lyon, Fr, pp 231–259

Johnson ES (1986) Duration of exposure as a surrogate for dose in the examination of dose response relations. Br J Ind Med 43:427–429

Latt SA, Allen J, Bloom SE, Carrano AV, Falke E, Kram D, Schneider E, Shreck R, Tice R, Whitfield B, Wolff S (1981) Sister chromatid exchanges: a report of the Gene-Tox program. Mutat Res 87:17–62

Lutz WK, Schlatter CH (1977) Mechanism of the carcinogenic action of benzene: irreversible binding to rat liver DNA. Chem-Biol Interact 18:241–245

Maltoni C, Conti B, Cotti G (1983) Benzene: A multipotential carcinogen. Results of long-term bioassays performed at the Bologna Institute of Oncology. Am J Ind Med 4:589–630

Nikula E, Kiviniitty K, Leisti J, Taskinen PJ (1984) Chromosome aberrations in lymphocytes of nurses handling cytostatic agents. Scand J Work Environ Health 10:71–74

Nordenson I, Beckman L (1984) Chromosomal aberrations in lymphocytes of workers exposed to low levels of styrene. Human Hered 34:178–182

Norppa H, Sorsa M, Vainio H, Gröhn P, Heinonen E, Holsti L, Nordman E (1980) Increased sister chromatid exchange frequencies in lymphocytes of nurses handling cytostatic drugs. Scand J Work Environ Health 67:229–301

Norppa H, Vainio H, Sorsa M (1981) Chromosome aberrations in lymphocytes of workers exposed to styrene. Am J Ind Med 2:299–304

Perry P, Evans HJ (1975) Cytological detection of mutagencarcinogens exposure by sister chromatid exchange. Nature (London) 258:121–125

Perry P, Wolff S (1974) New Giemsa method for differential staining of sister chromatids. Nature (London) 251:156–158

Rowley JD (1983) Human oncogene locations and chromosomal aberrations. Nature (London) 301:290–291

Sarto F, Cominato I, Pinton AM, Vrovedani PG, Merier E, Peruzzi M, Bianchi V, Levis AG (1984) A cytogenetic study on workers exposed to low concentrations of benzene. Carcinogen 5: 837–832

Scott D, Danford N, Dean B, Kirkland D, Richardson C (1983) In vitro chromosome aberration assays. In: Dean B (ed) Report of the UKEMS sub-committee on guidelines for mutagenicity testing. UKEMS, Swansea, pp 41–64

Selevan SG, Lindbohm M-L, Hornung RW, Hemminki K (1985) A study of occupational exposure to antineoplastic drugs and fetal loss in nurses. N Engl J Med 313:1173–1178

Sorsa M (1984) Monitoring of sister chromatid exchanges and micronuclei as biological endpoints. In: Berlin A, Draper M, Hemminki K, Vainio H (eds) Monitoring human exposure to carcinogenic and mutagenic agents. IARC Sci Publ No 59. Int Ag Res Cancer, Lyon, Fr, pp 339–349

Sorsa M, Norppa H, Vainio H (1982) Induction of sister chromatid exchanges among nurses handling cytostatic drugs. In: Indicators of genotoxic exposure. Banbury Rep 13. Cold Spring Harbor Lab, pp 341–350

Sorsa M, Hemminki K, Vainio H (1985) Occupational exposure to anticancer drugs – Potential and real hazards. Mutat Res 154:135–149

Vainio H (1985) Current trends in the biological monitoring of exposure to carcinogens. Scand J Work Environ Health 11:1–6

Vainio H, Hemminki K, Wilbourn J (1985) Data on the carcinogenicity of chemicals in the IARC monographs programme. Carcinogen 6:1653–1665

Venitt S, Crofton-Sleigh C, Hunt J, Speechley V, Briggs K (1984) Monitoring of exposure of nursing and pharmacy personnel to cytotoxic drugs: urinary mutation assays and urinary platinum as markers of absorption. Lancet 1:74–77

Waksvik H, Klepp O, Brøgger O (1981) Chromosome analyses of nurses handling cytostatic agents. Cancer Treat Rep 65:607–610

Watanabe T, Endo A, Kato Y, Shima S, Watanabe T, Ikeda M (1980) Cytogenetics and cytokinetics of cultured lymphocytes from benzene-exposed workers. Int Arch Occup Environ Health 46:31–41

Watanabe T, Endo A, Kumai M, Ikeda M (1983) Chromosome aberrations and sister chromatid exchanges in styrene-exposed workers with reference to their smoking habits. Environ Mutagen 5:299–309

Werk NA de, Wadden RA, Chion NL (1983) Exposure of hospital workers to airborne antineoplastic agents. Am J Hosp Pharm 40:597–601

18 Clastogenicity of Methylated p-Benzoquinones: Chemical Warfare in Nature?

G. A. FOLLE and S. LÓPEZ DE GRIEGO[1]

1 Introduction

Quinones occur widely in nature. These pigments are mainly found in higher plants, lichens, fungi and microorganisms as well (Thomson 1957). In the animal kingdom they make a striking appearance in the exocrine secretions of certain arthropods. This phylum possesses the most diverse and probably best evolved chemical defense mechanisms known. Since the early work of Fisher (1670) the study of these secretions has expanded rapidly in the last decades and up to now more than 100 compounds have been isolated and identified (Weatherston and Percy 1970).

Benzoquinones have been extensively found in arthropods' defensive substances including numerous species belonging to the classes Arachnida, Diplopoda and Insecta. As early as 1889 Gilson reported that the secretion from the odoriferous glands of the tenebrionid *Blaps mortisaga* was an oily fluid containing crystalline yellow needles. Some years later Chittenden (1896) observed that flour infested with the beetles *Tribolium castaneum* or *Tribolium confusum* (Coleoptera, Tenebrionidae) acquired a pink coloration and became unpalatable. Further research work demonstrated that this coloration was caused by a volatile secretion given off by the insects containing a mixture of methyl- and ethyl-benzoquinones. One of the few arachnids known to secrete methylated p-benzoquinones as a defense mechanism is the opilionid *Acanthopachylus aculeatus* Kirby.

Opilionids (Arachnida: Opiliones) share an unique characteristic among the arachnids in that they possess a pair of dorsal exocrine glands near the lateral edge of the prosoma. These secretory organs (variously called scent, odoriferous or stink glands) have been generally considered as a defense mechanism against predators and parasites. However, they may also play a role in conferring protection from microorganisms, in sexual, alarm or aggregation behaviors and allowing intraspecific recognition as well (Roth and Eisner 1962; Holmberg 1983). Although all three suborders of Opilionids (Cythophthalami, Palpatores and Laniatores) produce defense secretions, chemical studies on their composition have so far been restricted to the last two suborders (Eisner et al. 1977b). The Palpatores secrete naphtoquinones and short-chain acyclic compounds while the effluent of Laniatores generally contains alkylated benzoqui-

1 Division of Human Cytogenetics and Quantitative Microscopy, Instituto de Investigaciones Biológicas Clemente Estable, Av. Italia 3318, Montevideo, Uruguay

Cytogenetics. Ed. by G. Obe and A. Basler

nones and phenols (Blum and Edgar 1971; Meinwald et al. 1971; Jones et al. 1976; Eisner et al. 1977b; Wiemer et al. 1978).

Acanthopachylus aculeatus (Fig. 1) is a peculiar opilionid with a wide geographical distribution in South America and particularly easy to detect in the territory of Uruguay (Capocasale 1968) and neighboring areas of Brazil and Argentina. In 1955 Estable et al. discovered that this arachnid secreted a highly volatile yellowish aqueous fluid with pungent odor through two cephalothoracic cutaneous glands (Fig. 1, arrow). They named this secretion "gonyleptidine" after the taxonomic classification of the arachnid formerly regarded as a member of the Gonileptidae family. These researchers detected that gonyleptidine had remarkable antibiotic properties being bacteriostatic in vitro against Gram-positive and Gram-negative bacteria and protozoa as well. The initial evidence leading to this discovery was the casual observation of the striking effects exerted by minute amounts of this substance reaching microorganisms by air transport from arachnids located nearby in the laboratory. Gonyleptidine solutions ranging in concentrations from 1 μg/ml to 100 μg/ml showed intense antibiotic activity against various strains of *Staphylococcus aureus, B. cereus, B. subtilis, B. anthracis, E. coli* and also *Trypanosoma cruzi*. In this early work, gonyleptidine batches were obtained by collecting freshly secreted droplets from the arachnids with capillary tubes. Discharge of the fluid could be enhanced by pressing the animal's cephalothorax near the first pair of legs. This procedure rendered for each specimen only about 0.01 ml of a yellow suspension containing 2–3 mg of pigment per ml. Taking into account that after a second or third extraction the animal failed to yield a significant amount of effluent, the final volume of gonyleptidine obtained in this way from a large number of opilionids was always limited. Therefore, the identification of the active principle or principles present in gonyleptidine became imperative for research purposes.

The chemical nature and properties of the arachnid secretion were studied by Fieser and Ardao (1956) through chromatographic and infrared absorption analysis. They detected that gonyleptidine was composed of three different methyl p-benzoquinones, namely: 2,3-dimethyl-2,5-cyclohexadiene-1,4-dione (B3); 2,5-dimethyl-2,5-cyclohexadiene-1,4-dione (B5) and 2,3,5-trimethyl-2,5-cyclohexadiene-1,4-dione (B7) as shown in Fig. 2. These compounds showed high sensitivity to light, thermolability and crystallized below 5°C in long, yellow needles. This study also indicated that B3 was the major component of the natural secretion (73.20%) and the minor ones B5 (11.34%) and B7 (15.46%). Depending on the type of bacteria assayed, methyl p-benzoquinones exhibited different antibiotic action when compared with gonyleptidine. Average activities for B3, B5 and B7 in terms of multiples of gonyleptidine activity were respectively −4, +2, +2 against Gram-positive bacteria and +4, +4, +2 against Gram-negative ones (Estable et al. 1955). It must also be pointed out that direct contact with *A. aculeatus* secretion produces intense skin irritation in humans (Capocasale 1968) and induces necrosis when injected repeteadly into laboratory animals (Freyre et al. 1958).

Saez and Drets (1958) studied the cytogenetic effects of gonyleptidine on *Allium cepa* meristem root tips and on germ cells of *Laplatacris dispar* (Orthoptera). *A. cepa* root tips were immersed in flasks containing aqueous solutions of the arachnid secretion at concentrations ranging from 1–1000 μg/ml. The insects were placed in sealed

Fig. 1. *Acanthopachylus aculeatus* secreting gonyleptidine (*arrow*)

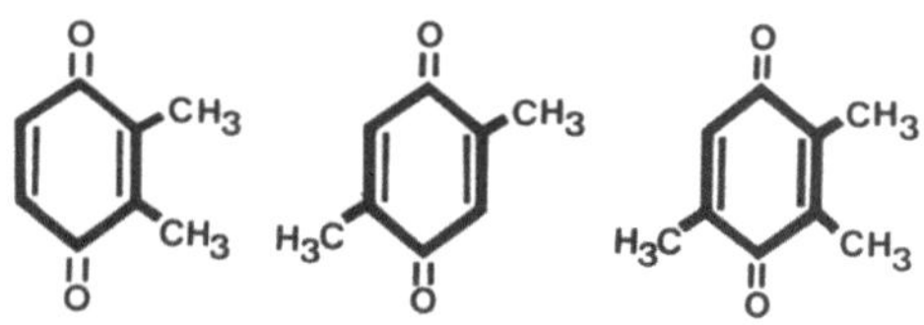

B3 B5 B7 2

Fig. 2. Methyl p-benzoquinones produced by *A. aculeatus*

glass chambers in order to make them inhale the vapors of pure gonyleptidine. In both experimental procedures they used different exposure and recovery periods of time. Meristem cells showed intense cytogenetic lesions such as chromosome stickiness, poliploidy, c-metaphases, diplochromosomes, interference with cell division and also various degrees of nuclear alterations. Similar findings were observed in *L. dispar* germ cells besides chromosome fragmentations, translocations, anaphase bridges and increased number of quiasmata. Based on these experimental data the authors concluded that "gonyleptidine is a substance endowed with mutagenic properties". In the following sections we present and discuss additional cytogenetic evidence obtained by assaying B3, B5 and B7 in human lymphocyte cultures and mouse bone marrow cells. Some aspects of the animal biology and the possible implications of its secretion in nature are also considered.

2 Cytogenetic Findings

2.1 Chromosomes

Human leukocyte cultures were treated with methyl p-benzoquinones following pulse and continuous exposures. In the first case, cells were subjected to a 10 μg/ml dose

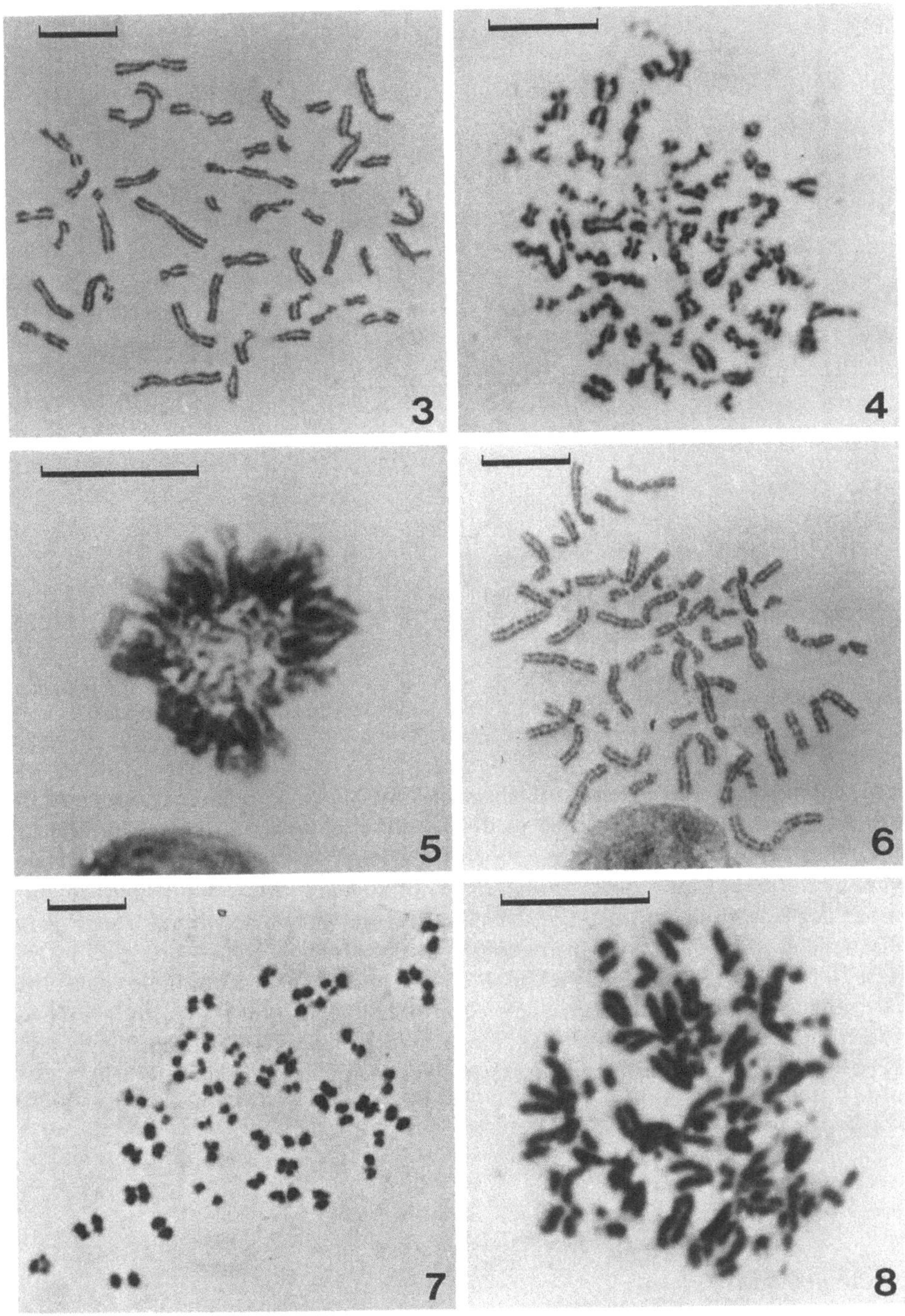

Figs. 3–8. Unless otherwise indicated cells were exposed to a 90-min pulse of B3 (10 μg/ml) and a recovery period of 24 h. Bars = 10 μm. (Figures 3, 6 and 8 from Drets et al. 1982)

either for 60 or 90 min with recovery periods of 8 and 24 h respectively. Continuous exposures were carried out by adding methyl p-benzoquinones 6 or 24 h before cell fixation at concentrations ranging from 0.01 to 500 μg/ml.

Analysis of B3 (10 μg/ml) treated metaphases revealed a wide range of cytogenetic effects (Folle et al. 1980; Drets et al. 1982). Chromatid breaks and gaps were frequently observed in all experimental series (Fig. 3), whereas chromosome-type lesions were detected only occasionally. Metaphases with a high number of breaks showed chromatin despiralization and complex rearrangements (Fig. 4). In some cases induced damage was intense enough to produce chromosome pulverization. Another conspicuous finding was the presence of chromosome exchanges such as triradials and quadriradials and to a lesser extent dicentrics and ring configurations. Table 1 shows the percentage of damaged metaphases (including all types of chromosome lesions recorded), chromosome exchanges and number of breaks per cell. Breaks and exchanges were more frequent in the longer pulse (90 min) and continuous (24 h) exposure assays. Other alterations included poliploid cells, metaphase coalescence (Fig. 5) and chromosome stickiness. Higher doses (⩾ 50 μg/ml) intensified these multiple effects and also produced cell division arrest as indicated by a low count of mitotic figures.

Leukocyte cultures exposed continuously to B3 (10 μg/ml) for the last 6 h prior to harvesting showed induction of in vitro cross-banding in metaphase chromosomes (Fig. 6). Selected benzoquinone-banded chromosomes were analyzed using Bandscan, an interactive program for on-line linear scanning of banded chromosomes (Drets 1978). This quantitative study supplied good evidence that induced banding could correspond to G-band patterns according to previous data obtained in our laboratory (Drets and Seuanez 1974).

The presence of various degrees of chromosome shortening was a striking finding in our experiments (Fig. 7). This colchicinelike effect of methyl p-benzoquinones increased sharply at higher doses. When human lymphocytes were subjected to B3 (10 μg/ml) for 12 h before harvesting the cells in the absence of colchicine, up to 20.6% of total cells recorded (n = 500) showed extreme chromosome shortening resembling fragmentation (anatelophase cells). Measurements carried out on comparative partial karyotypes including pair No. 1 and chromosome groups D and G from control metaphases and anatelophase cells revealed an average reduction of approximately 70% in chromosome length (Drets 1983).

An additional set of experiments conducted to assay B5 and B7 with comparable doses and exposure times demonstrated similar cytogenetic effects to those described

Fig. 3. Leukocyte metaphase presenting multiple chromatid breaks

Fig. 4. Heavily damaged metaphase showing numerous breaks, complex configurations and a moderate degree of chromosome shortening

Fig. 5. Metaphase coalescence induced by methyl p-benzoquinone

Fig. 6. Chromosome cross-banding obtained by pretreating leukocyte cultures with B3 (10 μg/ml; 6-h continuous exposure)

Fig. 7. Extreme chromosome shortening with chromatid segregation; cf. Fig. 3 reproduced at the same magnification

Fig. 8. Mouse chromosome fragmentation induced by B3 in vivo (10 μg/g body wt.; 7-h exposure)

Table 1. Percentage of metaphases and chromosome-induced alterations versus treatment. (From Drets et al. 1982)[a,b]

Duration and type of treatment	Damaged metaphases (%)	Metaphases with exchanges (%)	Number of breaks per cell
60-min pulse 8 h recovery (n = 100)	23	3	0.39
90-min pulse 24 h recovery (n = 104)	27	4	0.71
6-h continuous exposure (n = 121)	56	3	0.55
24-h continuous exposure (n = 94)	64	4	0.74
Control (n = 250)	7	0	0.08

[a] B3 (10 μg/ml).
[b] Leukocytes obtained from the same donor.

for B3. However, these compounds seemed to differ in their ability to produce chromosome lesions. In our laboratory, Rosensaft (personal communication 1984) scoring 175 leukocyte metaphases treated with B5 (90-min pulse; 10 μg/ml) found only 0.35 breaks per cell (control: 0.09) which is nearly half the value obtained for B3 under similar experimental conditions (see Table 1).

The major component of gonyleptidine was also assayed in vivo in mouse bone marrow cells. In these experiments the chemical was injected intraperitoneally at concentrations ranging from 2.5 to 50 μg per gram body weight in mice previously stimulated with a yeast-glucose solution according to the technique described by Lee and Elder (1980). Scoring of bone marrow metaphases confirmed cytogenetic data from human lymphocytes. A high number of chromatid and chromosome breaks was induced by B3 (number of breaks per cell: 0.55) at a dose of 10 μg/g body wt. and exposure times from 7 to 48 h (Fig. 8). Besides, chromosome exchanges were more frequently found in experiments combining higher doses (50 μg/g body wt.) and short treatments (1 h). Some interesting effects observed included the induction of C- and G-banding, intense chromosome shortening and peculiar ringlike configurations formed by sticky chromosomes.

2.2 Sister Chromatid Exchanges

For SCE analysis human lymphocytes were grown in culture medium containing 10 μg/ml 5-bromodeoxyuridine (BrdU, Sigma) under safe light conditions. B3 was pulse added for 90 min at a concentration of 10 μg/ml 24 h prior to cell fixation (B3 cells). Differential staining was obtained by the FPG technique (Perry and Wolff 1974) slightly modified.

The scoring of 152 B3 metaphases yielded 2516 exchanges ($\bar{X}$ = 16.53; SD = 7.16) while control cells (n = 152) revealed only 814 SCEs ($\bar{X}$ = 5.33; SD = 3.0). Statistical analysis (t-test) showed a highly significant difference (t = 18.49; $p < 0.001$) between these two groups. Histograms showing the distribution of control and treated metaphases according to their SCE counts can be seen in Fig. 9a,b. Although a few control

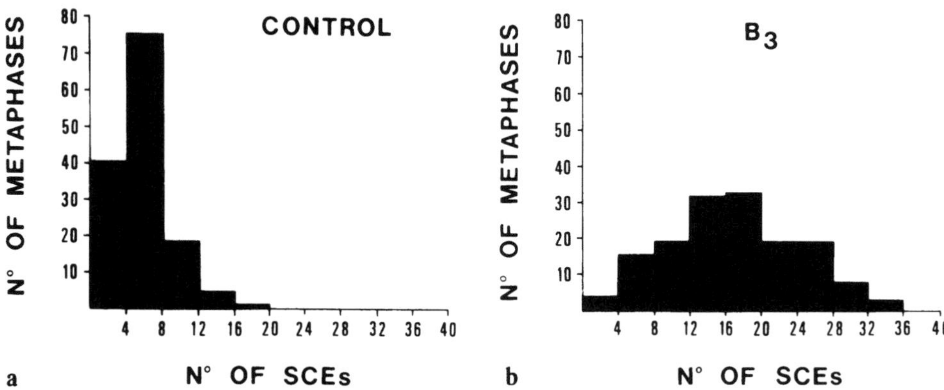

Fig. 9a,b. Distribution of control (a) and B3-treated metaphases (b) according to their SCE yields (see text)

metaphases had high SCE levels, the majority of them (96.05%) fell within the range of 0 to 12 exchanges. Benzoquinone-treated metaphases, on the other hand, showed a rather symmetrical distribution between 0 and 36 SCEs. One B3 metaphase carrying 28 sister chromatid exchanges is illustrated as an example in Fig. 10.

In order to classify the SCEs, human chromosomes were divided into ten different groups (1, 2, 3, B, C, D, E, F, G and Y) and the localization of exchange events in chromosome long arms or short arms determined. Table 2 shows the distribution of SCEs in the chromosome arms (p or q) of control and B3 metaphases for each chromosome group (percentages are also included). In both types of cells approximately three-quarters of the exchanges (control: 74.94%; B3: 74.84%) were located at the long arms, whereas only one-fourth (control: 25.06%; B3: 25.16%) at the short arms. The quotients p+q B3/p+q control revealed an increase of SCE counts by a factor of three in almost all chromosome groups. The lower quotients were observed in the F- and G-groups which combined represented merely 1.67% of total SCEs detected in B3 cells.

A comparison between observed and expected exchange frequencies according to chromosome relative length is presented in Table 3. The application of a chi square analysis, assuming that the distribution of exchange points is proportional to chromosome (or chromosome group) relative length, indicated a highly significant difference ($p < 0.001$) between the observed and expected number of exchanges for both control and B3 cells. Treated cells showed more SCEs than expected in pairs No. 1, 2 and 4–5 while the E-, F- and G-groups less than expected. Similar results were obtained in control cells with the exception of pairs No. 1 and 2 in which observed and expected frequencies did not seem to differ. It is interesting to note that in some B3 cells we detected the simultaneous occurrence of chromatid breaks and sister chromatid exchanges. A study carried out to determine the localization of these lesions in relation to exchange points revealed that just 20.45% of the chromatid breaks were associated with SCEs. As an example, Fig. 11 shows a B3 metaphase with several SCEs and chromatid breaks where only one of them (arrow) occurred at an exchange site. Chromosome lesions induced by B3 gave rise to rearrangements like triradials and

Table 2. Localization of sister chromatid exchanges in the chromosome arms (including percentages) of control and B3-treated metaphases for each chromosome group

Pair No.	Control				p+q	B3				p+q	p+q B3/
	p	%	q	%	%	p	%	q	%	%	p+q control[a]
1	30	3.685	41	5.037	8.722	124	4.929	144	5.723	10.652	3.77
2	29	3.563	53	6.512	10.075	94	3.736	158	6.280	10.016	3.07
3	34	4.177	32	3.931	8.108	75	2.981	122	4.849	7.830	2.98
4–5	31	3.808	103	12.654	16.462	86	3.418	301	11.963	15.381	2 89
6–12 + X	73	8.968	237	29.115	38.083	228	9.062	704	27.981	37.043	3.01
13–15	–	–	90	11.057	11.057	2	0.080	285	11.327	11.407	3.19
16–18	3	0.368	38	4.668	5.036	18	0.715	130	5.167	5.882	3.61
19–20	4	0.491	7	0.860	1.351	6	0.239	13	0.517	0.756	1.73
21–22	–	–	9	1.106	1.106	–	–	23	0.914	0.914	2.56
Y	–	–	–	–	–	–	–	3	0.119	0.119	–
Total	204	25.060	610	74.940	100.000	633	25.160	1883	74.840	100.000	

[a] The quotients p+q B3/p+q control point out the increased ratios of SCEs in exposed cells.

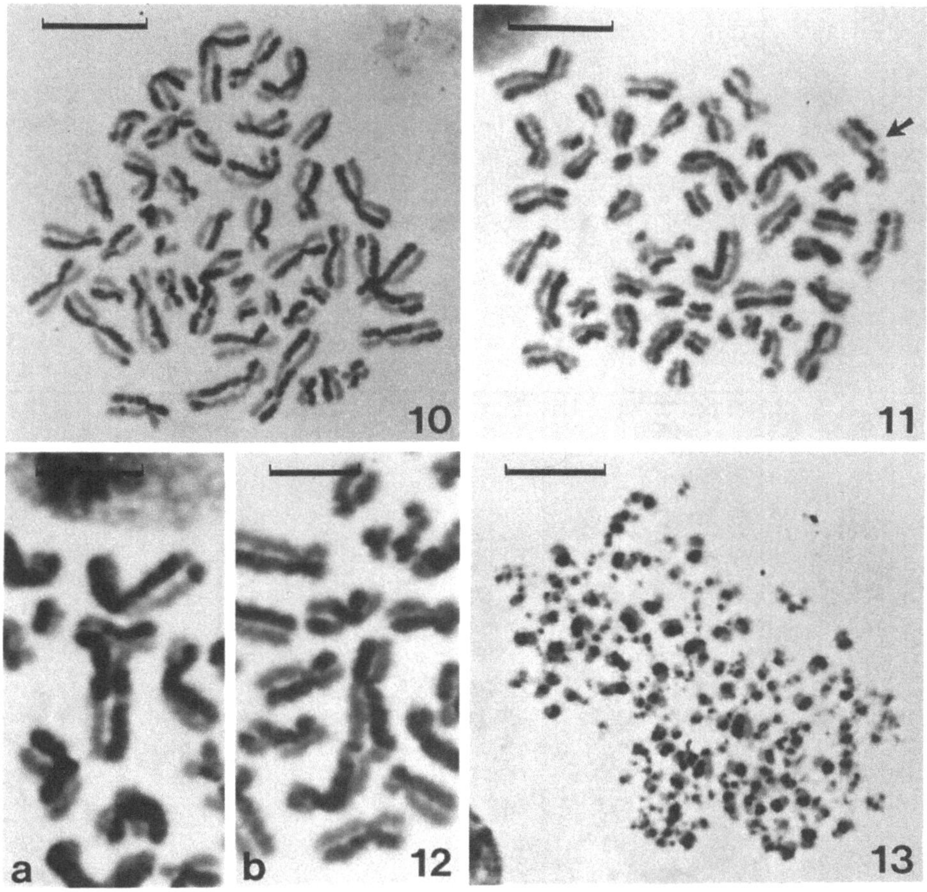

Figs. 10–13. In all cases cells were grown in 10 μg/ml 5-bromodeoxyuridine and submitted to B3 (10 μg/ml; 90-min pulse) 24 h before cell fixation. Bars = 10 μm, excluding Fig. 12a,b in which they equal 5 μm

Fig. 10. Metaphase with 28 sister chromatid exchanges induced by B3

Fig. 11. Breaks and SCEs in a lymphocyte metaphase; one chromatid break is located at an exchange point (*arrow*)

Fig. 12a,b. Triradial (**a**) and quadriradial (**b**) configurations showing sister chromatid exchanges

Fig. 13. Pulverized metaphase with manifest differential chromatid staining

Table 3. SCE observed and expected frequencies in accordance with chromosome relative length for both control and B3-exposed metaphases

Pair No.	Relative length	Percent length	Control			B3		
			Observed	Expected	χ^2 [a]	Observed	Expected	χ^2 [a]
1	16.88	8.1448	71	66.299	0.333	268	204.923	19.416
2	16.04	7.7394	82	62.999	5.731	252	194.723	16.848
3	13.66	6.5911	66	53.652	2.842	197	165.832	5.858
4–5	24.76	11.9469	134	97.247	13.890	387	300.584	24.844
6–12 + X	74.82	36.1013	310	293.865	0.886	932	908.309	0.618
13–15	21.52	10.3836	90	84.522	0.355	287	261.251	2.538
16–18	19.08	9.2063	41	74.939	15.371	148	231.631	30.195
19–20	10.46	5.0470	11	41.083	22.026	19	126.983	91.826
21–22	7.88	3.8022	9	30.950	15.567	23	95.663	55.193
Y	2.15	1.0374	–	8.444	–	3	26.101	20.446
Total	207.25	100.000	814	814.000	77.001	2516	2516.000	267.782

[a] Chi square analyses indicate the degree of fitness between observed and expected SCE values (see Text).

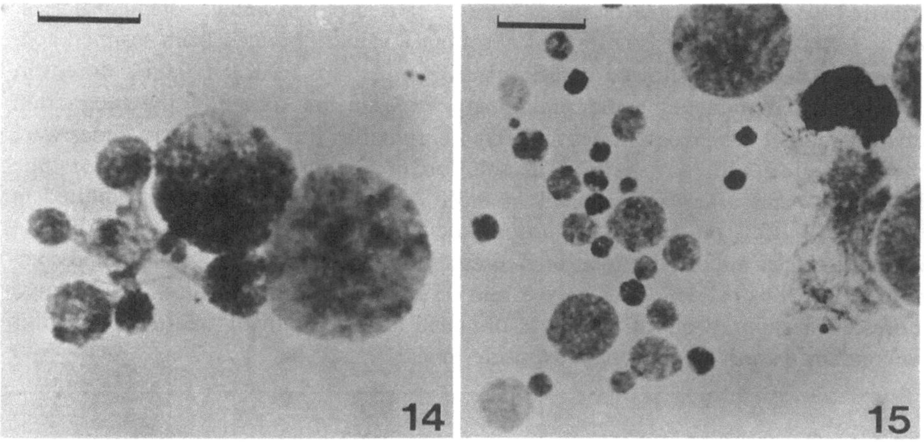

Figs. 14–15. Treatment was similar to that indicated for Figs. 3–5 and 7. Bars = 10 μm

Fig. 14. Complex arrangement of micronuclei connected by filaments to the main nucleus

Fig. 15. Group of micronuclei induced by methyl p-benzoquinone; no connections with main nuclear structures are evident

quadriradials presenting sister chromatid exchanges (Fig. 12a,b) and also to pulverized BrdU-substituted chromosomes with conspicuous differential staining (Fig. 13).

2.3 Nuclei

Production of micronuclei either in vitro or in vivo was a constant finding in all experimental protocols. These nuclear aberrations exhibited a wide variation in size as well as in morphology ranging from simple chromatin masses connected or unconnected by filaments or bridges to the main nucleus to complex formations of tangled micronuclei (Fig. 14). In some occasions we were able to detect groups of micronuclei lacking connections with adjacent nuclear structures as can be seen in Fig. 15. Recently, we assayed B3, B5 and B7 at doses of 10 and 20 μg/ml in pulse (90 min) and continuous (24 h) exposures to investigate the morphological types and frequencies of micronuclei induced by these chemicals. Statistical analysis of recorded data showed a significant variation in the distribution of the frequencies of micronuclear configurations found under each experimental condition suggesting that the different compounds and doses assayed probably gave rise to micronuclei through several cytogenetic mechanisms. It must also be pointed out that in these experiments some anatelophase cells showed extremely shortened chromosomes surrounded by nuclear membranelike structures. This could indicate that the encircling of these peculiar chromosomes could be an additional mechanism leading to micronuclei formation. A complete report on these observations will be published elsewhere.

In order to verify the chromatin loss due to micronuclei induction, DNA content in leukocyte nuclei treated with B3 (10 μg/ml; pulse and continuous exposures) was estimated by microphotometric measurements carried out in Feulgen-stained slides at 565 nm. Statistical analysis (*t*-test) of absorbance values obtained from scanned control and treated nuclei showed significant differences ($t > 1.90$; $p < 0.05$). Moreover, we have also conducted some preliminary experiments to study the interaction between methyl p-benzoquinones and fixed chromatin. For this purpose, slides were initially treated with a 7×10^{-3} M solution of B3 for different periods of time (1–30 min) and afterwards stained in 0.5% quinacrine dihydrochloride diluted in McIlvaine's buffer, ph 4.5. Microfluorometric scanning of quinacrine-stained human leukocyte nuclei indicated a relative fluorescence reduction of about 30% (when compared with control slides) after only 5 min of exposure to the chemical. A maximum fluorescence suppression of 45% was obtained after 10 min of treatment and this level remained constant for longer exposure times.

3 Discussion

Results obtained scoring metaphase chromosomes, sister chromatid exchanges and nuclei in human leukocytes as well as cytogenetic evidence recorded in mouse bone marrow cells suggested that all three components of *A. aculeatus* secretion are potent clastogenic compounds. To our knowledge, little research work has been devoted to study the possible mutagenic properties of benzoquinones and related compounds. Consequently, the molecular mechanisms leading to aberration production by these chemicals are at present not well understood. Quinones easily combine with proteins most probably by addition reactions involving free amino and sulphydryl groups and it has been demonstrated that enzyme systems could be inhibited in this way (Fieser and Fieser 1956; Thomson 1957). Whether these compounds produce chromosome aberrations by reacting primarily with proteins or DNA is an interesting problem that remains to be elucidated. Methylated p-benzoquinones, like numerous well-defined chemical mutagens, induced a dramatic increase of chromatid breaks and derived rearrangements with low frequencies of chromosome-type aberrations. The high incidence of chromatid breaks and gaps suggested that nuclear damage could have occurred during the S- and G2-periods of the cell cycle (Drets et al. 1982). Chromatin stickiness, metaphase coalescence and mitotic arrest observed either in pulse or continuous exposures were interpreted as cytotoxic effects.

Lüers and Obe (1972) studied the action of p-benzoquinone in human leukocytes and *Drosophila* cells. These authors concluded that this compound was not mutagenic since they could not detect chromosome exchanges in their experiments. As mentioned, methyl p-benzoquinones induced both in vitro and in vivo chromosome aberrations and exchanges under different experimental conditions. The benzoquinoid antibiotic Dnacin B1 inhibits DNA synthesis and also raises the melting temperatures of various double-stranded DNAs (Tanida et al. 1982). Moreover, cleavage of PM2 DNA strands probably through generation of hydroxyl radicals was demonstrated with the reduced form of this agent. Dnacin B1 also induced lambda prophage in *E. coli* TE-219 strains and showed antitumor properties in mice with leukemia P338. Future detection of similar DNA-modifying properties in methylated p-benzoquinones could account for at least some of the cytogenetic effects described and also confirm DNA as the primary target of these chromosome breaking agents. In fact, microfluorometric studies indicated that some kind of DNA-benzoquinone interaction could occur. Fluorescence reduction on slides pretreated with B3 suggested that this compound could interfere with quinacrine binding to DNA. It is worth mentioning that some complex quinonoid pigments from pathogenic fungi such as xanthomegnin and luteosporin have been shown to be genotoxic in the HPC/DNA repair test though no definite mutagenic activity of these chemicals was detected in the *Salmonella* microsome test (Mori et al. 1983).

The production of in vitro cross-banding by methyl p-benzoquinones was an interesting phenomenon that cannot be easily explained. Hsu et al. (1973) reported the induction of G- and R-bands in Chinese hamster cells treated with DNA cross-linking or intercalating agents and Shafer (1973) obtained G-banding by pretreating human leukocyte cultures with actinomycin-D. G-banding is currently produced with proteolytic enzymes (e.g. trypsin) and according to Sumner (1973) this longitudinal differentiation could be a consequence of a varying concentration of protein disulphide and sulphydryl groups along the chromosome arms. As described previously benzoquinones readily combine with proteins through reactions implicating sulphydryl and amino groups (Fieser and Fieser 1956).

Induction of sister chromatid exchanges deserve some additional comments. B3 produced a triplication of SCEs in nearly all chromosome groups and the overall distribution of exchange points in the chromosome arms was almost identical in control and treated metaphases. In B3 cells we detected more SCEs than expected in the 1, 2 and 4–5 groups while less than expected in the smaller chromosomes. The excess of exchanges in the A- and B-groups has already been reported in human leukocytes (Crossen et al. 1977) as well as the deficit of SCEs in the E-, F- and G-groups (Chaganti et al. 1974; Latt 1974; Galloway and Evans 1975). In the present study no definite attempt was made to correlate exchange points to specific chromosome regions revealed by banding techniques since no efficient quantitative method for SCE localization was available. There are several indications that SCE might occur preferentially in the band-interband regions of G- or Q-banded chromosomes or in the junctions between eu- and heterochromatin (Latt 1974; Carrano and Wolff 1975; Bostock and Christie 1976; Kato 1979). Zack et al. (1976) studied the localization of SCEs through image computer analysis in BrdU-substituted human chromosomes sequentially stained with quinacrine dihydrochloride and Hoechst 33258 but could not obtain accurate results due to the poor delineation of fluorescent bands and exchanges afforded by the computer system. In this respect, our laboratory has recently developed a new computer program that can detect precisely the position of exchanges along the chromosome arms in FPG-differentiated chromosomes (see Chap. 3, this vol.). Further application of this program could indicate whether there are different SCE localization patterns in normal and benzoquinone-treated human chromosomes.

According to Schmid (1976), "micronuclei originate from chromatin which for different reasons has been lagging in anaphase". Two main causes can lead to chromatin lagging: chromosome breakage and malfunction of the spindle apparatus. The evidence supplied by our observations indicated that several mechanisms may be involved in the production of micronuclei by B3, B5 and B7. On the one hand, anaphase lagging of acentric fragments and dicentrics induced by these clastogens could give rise to micronucleated cells. The detection of poliploid and anatelophase cells, on the other hand, might be an indication of induced mitotic spindle impairment. Finally, the encircling of extremely shortened chromosomes by nuclear membranelike structures, as previously described, may be postulated as an alternative mechanism in micronuclei formation.

A. aculeatus usually lives in colonies that gather approximately 50 individuals although in some cases colonies of up to 300 individuals have been reported (Capocasale and Bruno-Trezza 1964). The degree of association of these arachnids apparently could be modified by the temperature and humidity of the environment and varies also according to the time of the year. It is interesting to note that these colonies are always free from other animals and only another opilionid (*Pachyloides thorelli thorelli*) was occasionally detected dwelling in the same ecological niche. Capocasale and Bruno-Trezza (1964) studied the behavior patterns of these arachnids in the laboratory and concluded that they are not aggressive, preferring to escape rather than attack when threatened by predators. It must be stressed, however, that discharge of the quinonoid secretion can repell or even kill aggressors as was experimentally confirmed in one case when an *A. aculeatus* specimen was assaulted by another arachnid (*Argiope argentata* Fabr.). This is good evidence that this species utilizes its secretion as a chemical defense mechanism against predators, as detected in other opilionids, beetles and termites which secrete similar compounds (Maschwitz et al. 1972; Tschinkel 1975; Eisner et al. 1977a,b; Jones et al. 1977; Kanehisa and Murase 1977; Roach et al. 1980).

Chemical defense mechanisms, in spite of the incomplete data available, seem to be unevenly distributed within the arthropods. The occurrence of defense secretory organs exhibits a considerable variation among the different taxons. Defense mediated by chemicals should be positively correlated with the probability of attack by predators and negatively with the existence of other defense mechanisms (Pasteels et al. 1983). This is in good agreement with the above mentioned observations concerning the behavior of *A. aculeatus* specimens. The defense strategy of this species is mostly based in the expelling of gonyleptidine since it lacks alternative ways of protection. Two additional aspects must also be considered in relation to the arthropod's defensive exocrine products: the diversity of secreted compounds and the highly specialized forms of defense attained by some species. Up to now, numerous chemicals have been detected in these secretions including benzoquinones, naphtoquinones, terpenoids, phenols, short acyclic compounds, aldehydes, steroids, aliphatic esters and acids, aromatic compounds and other minor constituents. As an example, the distribution of exocrine defensive quinones in 54 arthropod species is shown in Table 4. Although rather simple combinations of two or more compounds have generally been reported in defense secretions, Pasteels et al. (1983) have remarked that complex mixtures can often be disclosed when minor constituents are looked for systematically. Brand et al. (1973) detected 14 compounds in the staphylinid *Drusilla canaliculata* (five hydroquinones, one quinone, four aldehydes and four hydrocarbons) while 18 different chemicals were identified in the secretion of the stink bug *Nezara viridula* (Gilby and Waterhouse 1965). This complexity is at present not well understood although synergism between some compounds has been demonstrated suggesting a biological significance of these mixtures. The following examples illustrate well the

Table 4. Distribution of exocrine defensive quinones in 54 species of insects, arachnids and termites[a]

p-Benzoquinone	16
p-Toluquinone	46
2,3-dimethyl p-benzoquinone	2
2,5-dimethyl p-benzoquinone	1
2,3,5-trimethyl p-benzoquinone	2
2-Ethyl p-benzoquinone	24
2-Methoxy-3-methyl p-benzoquinone	18
2,3-Dimethoxyquinone	1
Hydroquinone	15
Methyl hydroquinone	1
2-Ethyl hydroquinone	1
2-Alkyl-1,4-naphtoquinone	2

[a] Extracted in part from Roth and Eisner (1962) and Weatherston and Percy (1970).

highly specialized mechanisms of chemical-mediated defense developed by certain species. *Vonones sayi* (Arachnida: Opiliones) has an exocrine secretion similar to that of *A. aculeatus* but instead of three methylated p-benzoquinones it contains only two (B3 and B7). This arachnid, when attacked, discharges initially an oral effluent that appears as two clear droplets in the anterolateral margins of the body. In a second step a quinonoid brownish fluid is produced and immediately both secretions are mixed. Finally, the animal dips its forelegs into the mixture and dabs them against the offender. Experimental work conducted with aggressive ants clearly demonstrated the deterrent action of the mixture applied in such an unusual way (Eisner et al. 1971). The bombardier beetle *Brachinus* (Carabidae) produces a secretion containing quinones (p-benzoquinone and p-toluquinone) which are formed in the outer portion of a two-chambered gland in an explosive chemical reaction involving phenolic precursors and hydrogen peroxide and catalyzed by peroxidases and catalases. The fluid is propelled by the pressure of the liberated oxygen and sprayed, hot and with an audible sound, directly to the aggressor (Schildknecht and Holoubek 1961). Numerous other highly-evolved chemical devices have been observed within the arthropods in addition to those described here. The development of these valuable survival tools by menaced species may well represent an evolutionary response to predator activity.

In conclusion, methylated p-benzoquinones have proved to be strong inducers of chromosome and nuclear damage in plant, insect, human lymphocytes and mouse bone marrow cells through different experimental procedures including immersion, inhalation, addition to leukocyte cultures and injection. However, some interesting problems have not yet been solved. In a previous contribution (Drets et al. 1982), an original hypothesis concerning the presumptive environmental action of *A. aculeatus* clastogenic compounds was postulated. In this hypothesis, we suggested the possible mutagenic effects of volatile methyl p-benzoquinones on the genome of other species sharing the same habitat exposed to the continuous secretion of the arachnid. Unfortunately, no complete answer to this problem is available. In this respect, a detailed examination of the potential genotoxic effects of numerous substances secreted by arthropods as a defense mechanism is needed. Further research in this area could indicate that chemical warfare might not only play a role as a defense strategy but could also influence the spontaneous mutation rate of other living beings and consequently their evolutionary processes.

References

Blum MS, Edgar AL (1971) 4-Methyl-3-heptanone: Identification and role in opilionid exocrine secretions. Insect Biochem 1:181–188

Bostock CJ, Christie S (1976) Analysis of the frequency of sister chromatid exchanges in different regions of chromosomes of the Kangaroo rat (*Dipodomys ordii*). Chromosoma 56:275–287

Brand JM, Blum MS, Fales HM, Pasteels JM (1973) The chemistry of the defensive secretion of the beetle *Drusilla canaliculata* (*Col. staphylinidae*). J Insect Physiol 19:369–382

Capocasale R (1968) Nuevos aportes para el conocimiento de la distribución geográfica de los opiliones de Uruguay. Neotropica 14:65–71

Capocasale R, Bruno-Trezza L (1964) Biologia de *Acanthopachylus aculeatus* (Kirby, 1819) (Opiliones: Pachylinae). Rev Soc Urug Entomol 6:19–32

Carrano AV, Wolff S (1975) Distribution of sister chromatid exchanges in the euchromatin and heterochromatin of the Indian muntjac. Chromosoma 53:361–369

Chaganti RSK, Schonberg S, German J (1974) A manyfold increase in sister chromatid exchanges in Bloom's syndrome lymphocytes. Proc Nat Acad Sci USA 71:4508–4512

Chittenden FH (1896) Insects affecting cereals and other dry vegetable foods. US Dep Agric Entomol Bull 4:112

Crossen PE, Drets ME, Arrighi FE, Johnston DA (1977) Analysis of the frequency and distribution of sister chromatid exchanges in cultured human lymphocytes. Human Genet 35:345–352

Drets ME (1978) Bandscan – A computer program for on-line linear scanning of human banded chromosomes. Comput Progr Biomed 8:283–294

Drets ME (1983) Extreme chromosome shortening resembling fragmentation induced by 2,3-dimethyl p-benzoquinone in human lymphocytes. Environ Mutagen 5:923–927

Drets ME, Seuanez H (1974) Quantitation of heterogeneous human heterochromatin: Microdensitometric analysis of C- and G-bands. In: Coutinho EM, Fuchs F (eds) Basic life sciences, vol 4. Physiology and genetics of reproduction. Plenum, New York, pp 29–52

Drets ME, Folle GA, Aznarez A (1982) Clastogenic action of a dimethyl p-benzoquinone of animal origin. Mutat Res 102:159–172

Eisner T, Kluge AF, Carrel JE, Meinwald J (1971) Defense of phalangid: Liquid repellent administered by led dabbing. Science 173:650–652

Eisner T, Jones TH, Aneshansley DJ, Tschinkel WR, Silberglied RE, Meinwald J (1977a) Defense mechanisms of arthropods, 57. Chemistry of defensive secretions of bombardier beetles (Brachinini, Metriini, Ozaenini, Paussini). J Insect Physiol 23:1383–1386

Eisner T, Jones TH, Hicks K, Silberglied RE, Meinwald J (1977b) Quinones and phenols in the defensive secretions of neotropical opilionids. J Chem Ecol 3/3:321–329

Estable C, Ardao MI, Pradines-Brasil N, Fieser LF (1955) Gonyleptidine. J Am Chem Soc 77: 4942

Fieser LF, Ardao MI (1956) Investigation of the chemical nature of Gonyleptidine. J Am Chem Soc 78:774–781

Fieser LF, Fieser M (1956) Organic chemistry. Reinhold, New York, pp 709–731

Fisher S (1670) Some uncommon observations and experiments made with an acid juice to be found in ants. Philos Trans R Soc, London, p 2063

Folle GA, Drets ME, Aznarez A (1980) Clastogen action of benzoquinones of animal origin on human and mouse chromosomes. In: Proc 1st Latinam Symp Environmental Mutagens, Teratogens and Carcinogens, 9–11 June 1980, Puebla Mexico, p 38

Freyre HA, Rovira M, Ardao MI (1958) Toxicidad de metil-1,4 benzoquinonas y metil-1,4 benzohidroquinonas. Arch Soc Biol (Montevideo) 24:82–88

Galloway SM, Evans HJ (1975) Sister chromatid exchanges in human chromosomes from normal individuals and patients with ataxia telangiectasia. Cytogenet Cell Genet 15:17–29

Gilby AR, Waterhouse DF (1965) The composition of the scent of the green vegetable bug *Nezara viridula*. Proc R Soc London Ser B 162:105–120

Gilson G (1889) Les glands odoriféres du *Blaps mortisaga* et de quelques autres espèces. Cellule 5: 3–20

Holmberg RG (1983) The scent glands of Opiliones: a review of their function. Proc 9th Int Arachnology Congr, Panama, VIII (Abstr)

Hsu TC, Pathak S, Shafer DA (1973) Induction of chromosome cross-banding by treating cells with chemical agents before fixation. Exp Cell Res 79:484–487

Jones TH, Conner WE, Kluge AF, Eisner T, Meinwald J (1976) Defensive substances of Opilionids. Experientia (Basel) 32:1234–1235

Jones TH, Meinwald J, Hicks K, Eisner T (1977) Characterization and synthesis of volatile compounds from the defensive secretions of some "daddy longlegs" (*Arachnida: Opiliones: Leiobunum spp.*). Proc Natl Acad Sci USA 74:419–422

Kanehisa K, Murase M (1977) Comparative study of the pygidial defensive systems of carabid beetles. Appl Entomol Zool 12:225–235

Kato H (1979) Preferential occurrence of sister chromatid exchanges at heterochromatin-euchromatin junctions in the Wallaby and Hamster chromosomes. Chromosoma 74:307–316

Latt SA (1974) Localization of sister chromatid exchanges in human chromosomes. Science 185: 74–76

Lee MR, Elder FFB (1980) Yeast stimulation of bone marrow mitosis for cytogenetic investigation. Cytogenet Cell Genet 26:36–40

Lüers H, Obe G (1972) Zur Frage einer möglichen mutagenen Wirksamkeit von p-Benzochinon. Mutat Res 15:77–80

Maschwitz U, Jander R, Burkhardt D (1972) Wehrsubstanzen und Wehrverhalten der Termite *Macrotermes carbonarius*. J Insect Physiol 18:1715–1720

Meinwald J, Kluge AF, Carrel JE, Eisner T (1971) Acyclic ketones in the defensive secretion of a "daddy longlegs" (*Leiobunum vittatum*). Proc Natl Acad Sci USA 68:1467–1468

Mori H, Kawai K, Ohbayashi F, Kitamura J, Nozawa Y (1983) Genotoxicity of quinone pigments from pathogenic fungi. Mutat Res 122:29–34

Pasteels JM, Grégoire JC, Rowell-Rahier M (1983) The chemical ecology of defense in arthropods. Annu Rev Entomol 28:263–289

Perry P, Wolff S (1974) New Giemsa method for the differential staining of sister chromatids. Nature (London) 251:156–158

Roach B, Eisner T, Meinwald J (1980) Defensive substances of Opilionids. J Chem Ecol 6/2: 511–516

Roth LM, Eisner T (1962) Chemical defenses of Arthropods. Annu Rev Entomol 7:107–136

Saez FA, Drets ME (1958) The action of Gonyleptidine on the mitotic and meiotic chromosomes and on the interphase nucleus. Port Acta Biol 5/3–4:287–296

Schildknecht H, Holoubek K (1961) Die Bombardierkäfer und ihre Explosionschemie. V. Mitteilung über Insekten-Abwehrstoffe. Angew Chem 73:1–7

Schmid W (1976) The micronucleus test for cytogenetic analysis. In: Hollaender A (ed) Chemical mutagens, vol 4. Plenum, New York, Chap 36, pp 31–53

Shafer DA (1973) Banding human chromosomes in culture with actinomycin D. Lancet 1:828–829

Sumner AT (1973) Involvement of protein disulphides and sulphydryls in chromosome banding. Exp Cell Res 83:438–442

Tanida S, Hasegawa T, Yoneda M (1982) Mechanism of action of Dnacin B1, a new benzoquinoid antibiotic with antitumor properties. Antimicrob Ag Chemother 22/5:735–742

Thomson RH (1957) Naturally occurring quinones. Butterworth's, London, pp 6–54

Tschinkel WR (1975) Comparative study of the chemical defensive system of tenebrionid beetles. Chemistry of the secretions. J Insect Physiol 21:753–783

Weatherston J, Percy JE (1970) Arthropod defensive secretions. In: Beroza M (ed) Chemicals controlling insect behavior. Academic Press, London New York, pp 95–144

Wiemer DF, Hicks K, Meinwald J, Eisner T (1978) Naphtoquinones in defensive secretion of an opilionid. Experientia (Basel) 34:969–970

Zack GW, Spriet JA, Latt SA, Grandlund GH, Young IT (1976) Automatic detection and localization of sister chromatid exchanges. J Histochem Cytochem 24/1:168–177

19 Scientific Justification of Testing Chromosome Mutations and Regulatory Requirements for the Assessment of Mutagenicity

A. BASLER[1]

1 Heritable Cytogenetic Damage

A new research era in human cytogenetics began in the mid-1950s with progress in the development of human tissue culture techniques (Hsu 1952; Tjio and Levan 1956; Moorhead et al. 1960). The improvement of these methods to analyze human chromosomes and the observation that numerical chromosome mutations (Ford et al. 1959a,b; Jacobs and Strong 1959; Jacobs et al. 1959; Lejeune et al. 1959) and structural aberrations (Lejeune et al. 1963) were linked with disease in man, initiated a new branch in human genetics, namely clinical cytogenetics. The work done during the last 3 decades on chromosome analysis in adults, newborns and abortuses permits an estimate of the load of chromosome mutations in man. The results of different studies of population cytogenetics are summarized as follows (Sankaranarayanan 1979): 150,000 spontaneous abortions occur per 1,000,000 conceptions. About 50% of them are chromosomally abnormal. Of 850,000 live-births 17,000 die perinatally and about 6% are carriers of a chromosome mutation. Furthermore, of the 833,000 surviving children 5,165 are chromosomally abnormal. From these estimates it can be calculated that 6 of 1,000 newborns carry chromosome anomalies and at least 80 of 1,000 zygotes are chromosomally abnormal. These data are minimal values, because the clinically unrecognizable preimplantation and early postimplantation loss is not included in this calculation (for details see Chap. 7, this Vol.).

Jacobs (1972) compared the karyotype of newborns with chromosome anomalies and their parents and found that 100% of the aneuploidies of the sex chromosomes and autosomes and 25% of the structural aberrations were new mutations. Considering this and the above mentioned data, it has been estimated that at least 4 of 1,000 newborns carry a new genome or chromosome mutation. These mutations are often called "spontaneous" mutations, and this means that the inducing mechanisms are unknown. It was postulated that a part of these new mutations in human germ cells may be induced by exogenous agents, including drugs and chemicals (Lüers 1955a,b; Barthelmeß 1956; Röhrborn 1965). However, for a long time, even toxicologists underestimated or neglected the role of chemically induced germ cell mutations. One reason for this was the lack of epidemiological evidence that genetic diseases were the

1 Bundesgesundheitsamt, Postfach 33 00 13, D–1000 Berlin 33

Cytogenetics. Ed. by G. Obe and A. Basler

Table 1. Chromosome anomalies in spontaneous abortions, perinatal deaths and live births

	Spontaneous abortions < 12 weeks	< 28 weeks	Perinatal death	Live births
No. of cytonetic studies	1,498	986	713	56,952
Chromosome anomalies (%)	61.5	30.9	5.2	0.6
References	Boué et al. (1975)	Ruzicska and Czeizel (1970); Creasy et al. (1976)	Machin (1974); Sutherland et al. (1974)	After Hook and Hamerton (1977)

result of human exposure to chemical mutagens. Human geneticists, on the other hand, assumed that mutations are induced in germ cells but mainly selected during embryogenesis. This assumption was based on cytogenetic analyses of abortus material and newborns (Table 1): in the first trimenon, 61.5% of karyotyped spontaneous abortions were found to be chromosomally abnormal (Boué et al. 1975); 30.9% occurred in the second trimenon (Ruzicska and Czeizel 1970; Creasy et al. 1976). The incidence of chromosome anomalies among perinatal deaths was 5.2% (Machin 1974; Sutherland et al. 1974), whereas only 0.6% of live births carried chromosome mutations (Hook and Hamerton 1977).

The direct evidence of induced germ cell mutations, the selection at various stages of prenatal development and the outcome of chemically induced mutations was experimentally examined using 2,3,5-triethyleneimonobenzoquinone-1,4 (Trenimon) (Basler et al. 1976) and mitomycin C (Basler 1980). In the experiments with mitomycin C, for example, female mice were treated with 5 mg/kg in the preovulatory phase just before ovulation and mated with untreated males. About 43% of all zygotes showed induced chromosome aberrations. In further experiments, females were dissected at different stages past conception (p.c.). The preimplantation loss (14.6%), the number of dead implants (44.6%) and the percentage of living embryos (44.8%) without (40.4%) and with (0.4%) aberrations at day 16 p.c. was examined. As shown in Fig. 1, a strong selection against genetically abnormal embryos takes place at the pre- and postimplantation stages of embryogenesis. The largest part of chemically induced chromosome aberrations is eliminated, but a small number of embryos with chromosome mutations is able to survive. In the late embryonic stages, 1% of living embryos were carriers of chromosome anomalies.

It may be concluded that the genetic load of the human population by agents inducing chromosome mutation is low, because induced germ cell mutations are eliminated mainly during oogenesis or spermatogenesis and at various stages of embryogenesis. An unacceptable high risk remains, on the other hand, for the afflicted pregnant woman and her family. Frequently, these mutations manifest themselves as abortions, sometimes as genetic diseases. Therefore, there is a necessity to identify chemicals with mutagenic properties and to avoid human exposure. However, neither toxicologists nor the society and the legislators paid much attention to mutation

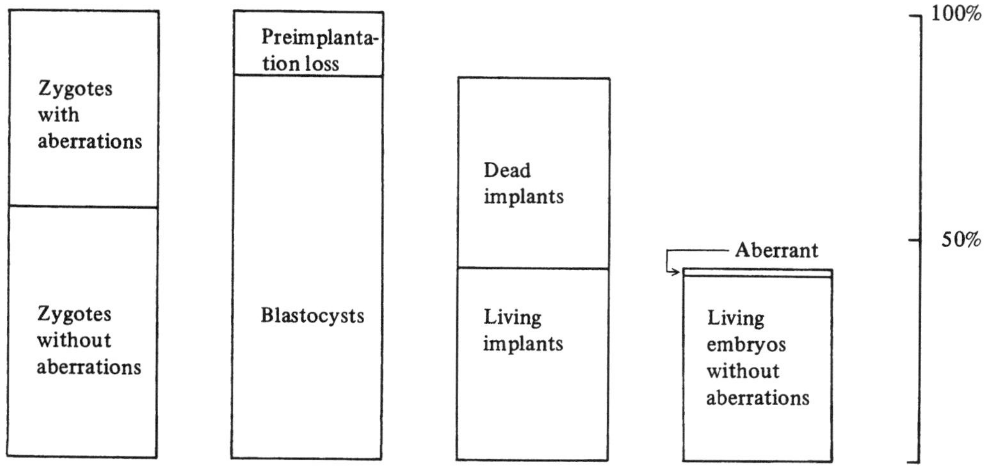

Fig. 1. Elimination of chemically induced chromosome aberrations in mice. Females were treated with mitomycin C (5 mg/kg, i.p.) 3 h prior to ovulation and mated with untreated males

research in the 1950s and 1960s. The situation changed in the 1970s with findings that gene and chromosome mutations are not only responsible for genetic disease, but may also be responsible for the induction of cancer.

2 Cytogenetic Short-Term Tests as Screening Assays for Carcinogenicity

For a long time, the detection of possible human carcinogens was based only on epidemiological studies or long-term animal experiments. Long-term animal studies, however, suffer from some disadvantages. There is only limited laboratory capacity to investigate all suspected carcinogens or even all new chemicals. The studies are extremely expensive and take 3 to 4 years to complete. Alternative strategies for the identification of carcinogens were in great demand. Hope arose when the multistage concept of carcinogenesis was substantiated by short-term tests for the identification of initiators and promotors. Between 1970 and 1975 Ames and colleagues developed a simple bacterial test to detect chemical mutagens. Using this test, about 300 carcinogens and non-carcinogens of a wide variety of chemical types had been tested in this *Salmonella* microsome test at the time of its first publication. Because of the high correlation of 90% between mutagenicity in this assay and carcinogenicity (McCann et al. 1975) it was supposed "that with few exceptions carcinogens are mutagens" (Ames et al. 1975). There was considerable optimism that the bacterial short-term tests would be able to replace the long-term carcinogenicity tests. Making this conclusion it was overlooked that the high correlation was obtained only in retro-validation studies evaluating well-known carcinogens and non-carcinogens. In analyses conducted later, a greater percentage of false-positive as well as false-negative results were obtained. Genotoxicity of the false-negatives could be assessed, however, in most cases in other in vitro mutagenicity tests using eukaryotic cells.

Table 2. List of chemicals causally associated with cancer in humans and results obtained with the *Salmonella* microsome test and/or cytogenetic tests

Chemical carcinogen	Gene mutations in prokaryotes	Structural aberrations	Genome mutations
4-Aminobiphenyl	+		
Arsenic and arsenic compounds	–	+	
Asbestos	–	+	
Azathioprine	+	+	
Benzene	–	+	
Benzidine	+	+	
N,N-bis(2-chloroethyl)-2-naphthylamine (Chlornaphazine)	+		
Bis(chloromethyl)ether and technical-grade chloromethyl methyl ether	+		
1-4-butanediol dimethanesulphonate (Myleran)	+	+	
Chlorambucil	+	+	
Chromium and certain chromium compounds	+	+	
Cyclophosphamide	+	+	
Diethylstilboestrol	–	–	+
Melphalan	+	+	
Methoxsalen with ultraviolet A therapy (PUVA)	+	+	
Mustard gas	+	+	
2-Naphthylamine	+	+	
Nickel (dust and aerosols of certain nickel compounds)	–	?	
Soots, tars and oils	+	+	
Treosulphan		+	
Vinyl chloride	+	+	
Zinc chromate	+		

Ishidate (1983) performed, for example, chromosome aberration tests in vitro using a Chinese hamster fibroblast cell line. They published data on a total of 587 substances, including known carcinogens. Accordingly, over 90% of the carcinogens were positive in the chromosome aberration test using cultured mammalian cells. Furthermore, the genotoxicity of specific compounds which had been negative in the Ames test were only detected in assays analyzing chromosome mutations. In Table 2, chemicals are listed that are causally associated with cancer in humans [chemicals of group 1a according to IARC (1982) and chemicals of category A1 according to DFG (1985)]. Furthermore, data are included which were obtained by applying either the *Salmonella* microsome test and/or cytogenetic tests. As can be seen, all human carcinogens were positive in either the chromosome or the Ames tests, or both. In most cases, a concordance between the results of the Ames test and cytogenetic data was obtained. However, the following carcinogens were all negative in the Ames test (Soderman 1982) but induced chromosome mutations: Arsenic and arsenic derivatives are negative in the *Salmonella* microsome test, but they induce chromosome aberrations in mammalian cells in vitro (Léonard and Lauwerys 1980), including human lymphocytes (Nordenson et al. 1981). Asbestos is also negative in *Salmonella typhimurium* and

E. coli, but induces chromosome aberrations in cell cultures (Huang et al. 1978). Benzene produces chromosome aberrations in mammalian cells in vitro (Koizuma et al. 1974) and in vivo (Philip and Jensen 1970; Siou et al. 1981), but it is not a bacterial mutagen. Diethylstilboestrol (DES) did not induce mutations in a variety of bacterial systems. Conflicting data exist with regard to the induction of structural chromosome aberrations (de Serres and Ashby 1981; Ivett and Tice 1981). There is no doubt that DES reacts with the mitotic apparatus and induces both polyploidies (Ishidate et al. 1981) and aneuploidies (Banduhn and Obe 1985). Nickel gave negative results in mutagenicity tests with prokaryotes. Data obtained in cell cultures have been equivocal. Chromosome aberrations were found to be slightly increased at high doses and after long treatment times (Léonard et al. 1981).

These data suggest that several carcinogens can be detected only in chromosome aberration tests. This clearly supports the concept that a bacterial gene mutation test is insufficient for screening for carcinogens. It should, therefore, be combined with other screening tests using eukaryotic cells or organisms.

A further indication that bacterial and cytogenetic tests complement each other came from a recently published review. Natarajan and Obe (1986) compared the results of 148 chemicals tested with both the micronucleus test and the Ames test. Of these 50 were positive and 40 were negative in both of these systems. The two test systems responded differently to 58 chemicals. The following chemicals which induce chromosome aberrations or spindle defects were detected by the micronucleus test and not by the bacterial assay: aminopterin, colchicine, cycloheximide, cytosine arabinoside, ethanol, procarbacine (natulan), methotrexate, vinblastine and vincristine.

In a collaborative study different in vitro assays were performed to evaluate short-term tests for carcinogens (Ashby et al. 1985). Several carcinogens and non-carcinogens were selected in order to ascertain which among the available eukaryotic in vitro assays would be the best complement of the Ames test for screening to detect potential carcinogens. It was demonstrated that only those tests analyzing chromosome aberrations, gene mutations in mammalian cells, aneuploidy in yeast and cell transformation gave encouraging results with regard to the compounds investigated. From these results it was concluded that the accuracy of eukaryotic in vitro assays for screening carcinogens is as good as the Ames test. It was furthermore acknowledged that out of these tests the performance of a chromosome aberration test would have some advantages, if only one test is demanded as an assay complementary to the *Salmonella* microsome test. The main argument to select this test is that the test does not merely confirm the data obtained with a bacterial gene mutation test, but supplements the spectrum because data are obtained using an independent genetic end point.

All these results gave rise to the concept of testing new chemicals in at least two mutagenicity assays, a bacterial and a non-bacterial one and to investigate two different genetic end points, i.e. point mutations and chromosome mutations.

3 Legislative Aspects: Cytogenetic Assays as Part of a Test Battery

It is accepted worldwide that gene mutations and/or chromosome mutations in somatic cells may be responsible for the induction of cancer in man and that germ cell mutations can cause genetic disease in newborns. To protect humans, all industrial countries require the submission of toxicological data, including data on mutagenicity before a new product is marketed. The requirements depend on the nature of the product. However, for the same product the test guidelines sometimes differ between countries. Bearing in mind that numerous short-term tests have been developed to detect genetic damage, it is not surprising that several guidelines, test strategies and decision approaches on the subject of detecting mutagenic and carcinogenic chemicals have been developed during the last years (EEC 1979; Williams 1980; OECD 1981; ICPEMC 1982, 1983a,b; IPCS 1985).

The philosophy of these test strategies will be explained in detail by the example of how chemical compounds used in industry are tested. This will be supplemented by notes for guidance on the testing of medicinal products.

3.1 Mutagenicity Testing of Industrial Chemicals

Toxicological test requirements and test procedures for industrial chemicals were set up by the Organization for Economic Cooperation and Development (OECD 1981) and the European Economic Community (EEC 1979) in order to harmonize the above mentioned different rules and legislations. The EEC Council Directive 79/831/EEC was approved on Sept. 18, 1979; it was amended for the sixth time by Directive 67/548/EEC. In the Federal Republic of Germany the Act on Protection against Dangerous Substances [ChemG (Chemicals Act) 1980] came into force on January 1, 1982. "The purpose of this Act is to protect man and the environment from the harmful effects of dangerous substances by means of compulsory testing and notification of substances and compulsory classification, labelling and packaging of dangerous substances and preparations, and by means of prohibitions and restrictions as well as specific legal provisions concerning toxicity and occupational safety (§ 1 Chemicals Act)."

A number of substances, however, are exempt from compulsory notification. Medicines, cosmetics, foods and additives, feed and pesticides have already become subject to more stringent rules of approval and must therefore not be notified under the Chemicals Act. Furthermore, there is no obligation of notification for substances that have been placed on the market in an EEC member state before September 18, 1981. Substances, having been produced in another EEC member state and notified there in accordance with an analogous procedure after that date, do not require a repeat notification. For other chemicals, with a few more exceptions (§ 5 Chemicals Act), notification is compulsory.

The basic concept of the Act consists in the obligation of the manufacturer or importer to inform national and supranational authorities. It is the manufacturer's obligation to inform the responsible authorities of those data on a new substance which are of importance for the national government and the European Communities. For this purpose, the manufacturer or importer must submit to the notification authority, 45 days before marketing a new substance for the first time, the following data:

1. the identifying features;
2. details of its use;
3. harmful effects during such use;
4. the quantities of the substance which he proposes to put into circulation or introduce each year;
5. the methods of proper disposal, possible re-use and neutralization; and
6. the documents on the tests performed.

To enable those responsible for marketing and the government authorities to estimate the risks of new substances, manufacturers and importers must perform a basic set of tests. The rules for performing such testing are graduated in accordance with the annual amount of the chemical produced. This graduated system of testing is based on the argument of the legislators that when major quantities are placed on the market, more extensive testing is necessary because of the quantitatively increased risk. Data from the toxicological base set assays may give sufficient information on the potential toxicity of these compounds to ensure occupational safety. Chemicals which are produced in major volumes, transported and used in different companies, necessarily need a better assessment of their possible hazards. At least, a full evaluation is necessary if larger amounts of the chemical are produced and the environment or the consumer become exposed to it.

The performance of toxicological tests is prescribed by the Chemicals Act. However, the Act does not stipulate how the characteristics to be described in the different types of documentation on testing are to be determined. Details have been regulated by Annexes VII and VIII to Council Directive 79/831/EEC (Table 3). At the basic level, if 1 t per year is produced, two tests are required for mutagenicity, one of which should be a bacteriological one, with and without metabolic activation, and the other, a non-bacteriological one. In the case of positive results, the government authorities may require additional mutagenicity tests, if 10 t per year, or 50 t cumulatively are produced.

Where the annual amount of the substance placed on the market within the EEC has exceeded 100 t or has reached a total of 500 t since the beginning of its importation or production (Level 1), the performance of two additional tests is obligatory, also in case of negative results at the basic level. If the results of these tests are also negative, further tests are not necessary at this level. If the results of one or both tests are positive, at least two verification tests are necessary. A confirmation of positive results should lead to a long-term carcinogenicity study at this level.

Where the amount of a substance introduced into the market reaches 1,000 t per year or a cumulative amount of 5,000 t, the applicant should inform the competent authority of such fact. He should also establish a program of tests to be conducted by him to enable the competent authority to estimate the risks of the substance for man and the environment. While the base set and Level 1 only ask for mutagenicity studies, which at the same time are considered as a prescreening for carcinogenicity, Level 2 requires a carcinogenicity study unless there are valid reasons not to do so.

The annexes fix the extent of testing to be done for chemicals produced in different quantities, but they do not prescribe the tests to be used. As a completion, methods of toxicological testing have been described by scientists of the EEC member

Table 3. Council Directive 79/831/EEC; requirements for mutagenicity and carcinogenicity

Base set (1 t/year)

Annex VII

§ 4.3.1 Mutagenicity (including carcinogenic prescreening test)

§ 4.3.2 The substance should be examined during a series of two tests, one of which should be bacteriological, with and without metabolic activation, and one non-bacteriological.

Level 1 (100 t/year or a total amount of 500 t)

Annex VIII

Additional mutagenesis studies (including screening for carcinogenesis)

A. If results of the mutagenesis tests (of the base set) are negative, a test to verify mutagenesis and a test to verify carcinogenesis screening are obligatory.

If the results of the mutagenesis verification test are also negative, further mutagenesis tests are not necessary at this level; if the results are positive, further mutagenesis tests are to be carried out (see B).

If the results of the carcinogenesis screening verification test are also negative, further carcinogenesis screening verification tests are not necessary at this level; if the results are positive, further carcinogenesis screening verification tests are to be carried out (see B).

B. If the results of the mutagenesis tests are positive (a single positive test means positive), at least two verification tests are necessary at this level. Both mutagenesis tests and carcinogenesis screening tests should be considered here. A positive result of a carcinogenesis screening test should lead to a carcinogenesis study at this level.

Level 2 (1,000 t/year or a total amount of 5,000 t)

Annex VIII

Carcinogenicity study

---

states. The methods of testing chemicals for mutagenicity (Table 4) which have already been approved are to be included in Annexes V and VIII of Council Directive 79/831/EEC. These methods are more or less in accordance with the OECD test guidelines (OECD 1981); in part, even the same experts participated in the preparation of both documents.

The question arises which combinations of these tests should be applied at which stage. The tests which may be used depend on numerous factors, such as physicochemical properties of the test substance, its metabolic profile and pharmacodynamic characteristics, its pharmacokinetics and the results of other toxicity studies. There-

Table 4. Methods for the assessment of mutagenicity (including prescreening of carcinogenicity) listed in Annexes V and VIII of Council Directive 79/831/EEC

Tests for gene mutations:
Salmonella typhimurium reverse-mutation assay
Escherichia coli reverse-mutation assay
Gene mutation, *Saccharomyces cerevisiae*
In vitro mammalian cell gene mutation
Sex-linked recessive lethal test in *Drosophila melanogaster*
Mouse spot test
Tests for chromosome mutations:
Mitotic aneuploidy, *Saccharomyces cerevisiae*
In vitro mammalian cytogenetic test
In vivo mammalian bone marrow cytogenetic test
Micronucleus test
Mammalian germ-cell cytogenetics
Mouse heritable translocation
Rodent dominant lethal test
Indicator tests for DNA effects:
Mitotic recombination, *Saccharomyces cerevisiae*
DNA damage and repair; unscheduled DNA synthesis
Sister chromatid exchange in vitro
Other indicator tests for carcinogenic potential:
In vitro mammalian cell transformation test

fore, a rigid schedule for the selection of tests is inappropriate. Nevertheless, some general principles can be given (Fig. 2).

The initial assessment of a possible mutagenic activity of a substance consists of assays for gene mutations in bacteria (*Salmonella typhimurium, Escherichia coli*) and for chromosome mutations in mammalian cells, preferably in vitro. For scientific reasons (e.g. chemicals with highly bactericidal properties), other tests are used as well.

When both tests in the base set were negative, one test for gene mutations in eukaryotic cells and another for chromosome aberrations in somatic cells in vivo should be performed as supplementary studies at Level 1. If one test has been positive in the base set, the supplementary studies should include at least one test detecting the same genetic end point. Some methods are available which provide an indication of an effect on DNA (mitotic recombination, UDS, SCE) but which do not have a mutagenic event as their "end point". It may be appropriate to perform an indicator test as a second one to obtain additional data and these may be useful in the interpretation of positive findings in the base set.

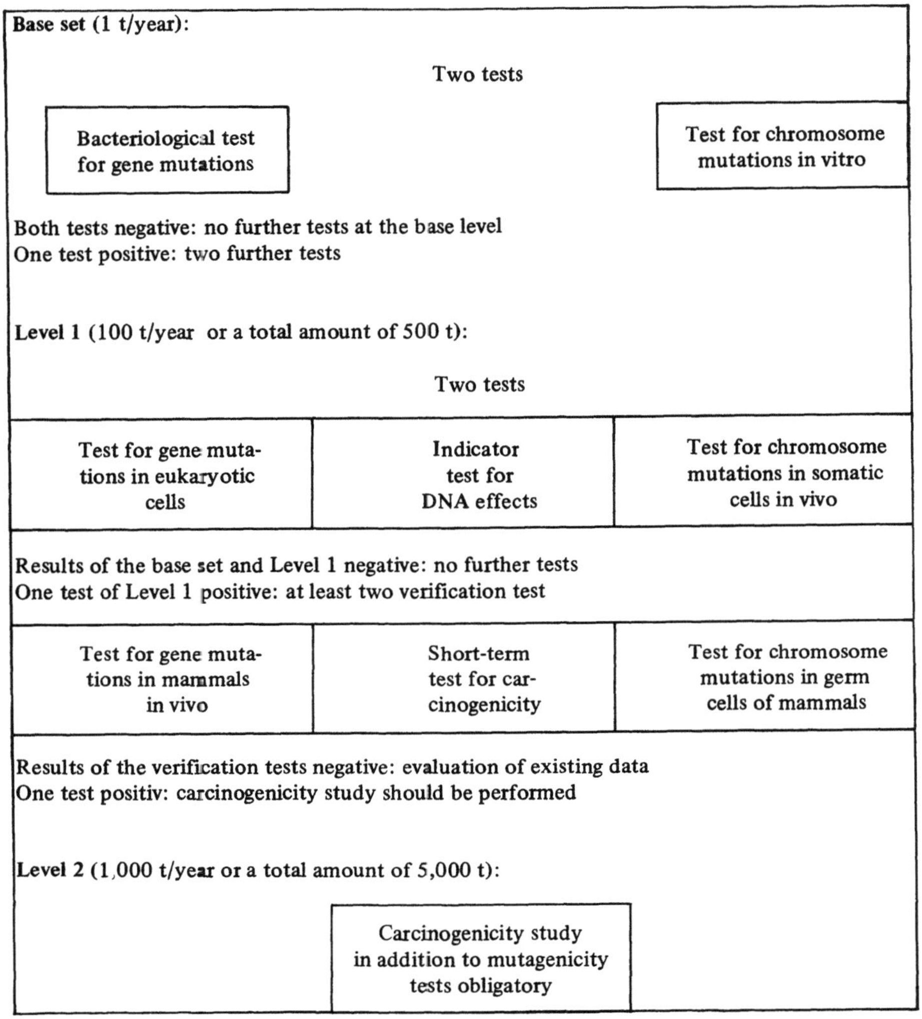

Fig. 2. Mutagenicity and carcinogencity testing of industrial chemicals according to Council Directive 79/831/EEC

An estimation of the risk should to be possible with the increasing extent of testing. While at the basic level, the potential mutagenic effects are studied, the findings have to be verified by onward testing. Thus, clastogenic effects are studied in vitro at the basic level, but tests for chromosome mutations in somatic cells in vivo are to be performed at Level 1. In the event of positive findings, heriditary effects should be studied in mammalian germ cells.

Point mutations are studied first in bacteria, then in eukaryotic cells and finally, in somatic cells in vivo. The mouse-specific locus test measures heritable effects in whole mammals produced mainly by gene mutations. This method may be used when the possible genetic risk of a substance, e.g. a drug, to man is to be estimated. However, for industrial chemicals there is generally no need of a risk estimation of this type. And, in view of the large number of animals necessary, this test is not recommended. Nevertheless, also gene mutation-inducing industrial chemicals can be classified as germ-cell mutagens. This is the case if there is a positive finding in somatic cells plus sufficient evidence that the substance in its active form interacts with germ-cell DNA following systemic administration by an appropriate route.

Positive findings of mutagenicity tests, regarded also as an indication of carcinogenicity, warrant additional short-term tests for carcinogenicity, such as cell transformation tests. If this test is positive, a carcinogenicity study should be performed already at Level 1.

It should be stressed that this procedure is not obligatory. It only demonstrates the procedure according to the Council Directive and serves as a guidance for applicants. Furthermore, when the findings obtained at the basic level have become available, the onward procedure can be discussed in a dialogue between the competent governmental authority, on the one hand, and the notifier, on the other.

The Chemicals Act means an obligation not only with respect to the evaluation of dangerous substances and preparations, but also with respect to their classification, labelling and packaging. It is the purpose of these regulations to provide the general public and persons handling such substances with essential information on their dangerous properties and on the possibilities for avoiding such risks. This labelling is to describe all possible risks, including heritable genetic risks, which may arise from the common handling and use of the dangerous substances or preparations. If, however, classification, labelling and packaging do not suffice to afford protection, the Act has provided for corresponding action by authorities to avert possible risks which include bans on the manufacture, marketing or use of substances.

3.2 Mutagenicity Testing of Medicinal Products

As outlined in the foregoing, it is an accepted fact that many chemicals used in industry are dangerous to man. Minimal investigations performed within the base set are often thought to give sufficient information on the mutagenic properties to ensure occupational safety. This procedure is of course inappropriate for medicinal products. A complete evaluation of possible genetic hazards for man is required before an exposure of patients for therapeutical reasons can be envisaged. Thus, before medicinal products undergo clinical trials, any new substances have to be investigated for mutagenic properties in a battery of tests. The methods used should be well validated. The tests listed in Annexes V and VII of Directive 79/831/EEC (Table 4) are regarded as appropriate.

Council Directive 83/570/EEC recommends tests relating to the placing of medicinal products on the market. This Directive is supplemented by annexes which give some guidance on how to perform such investigations of medicinal products. It has been recognized that stringent requirements are not appropriate and that the combination of tests applied depends on the specific characteristics of the substance to be investigated. The applicant has to explain the chosen test strategy and to justify the reasons for the selection of tests used. Unless valid reasons speak against this, he should select one test from each of the following four categories:

1. test for gene mutations in bacteria;
2. test for chromosome aberrations in mammalian cells in vitro;
3. test for gene mutations in an eukaryotic system;
4. in vivo test for genetic damage.

Further information may be gained by carrying out supplementary investigations, including population monitoring during clinical trials. If it is intended to perform such investigations, consultation with the competent regulatory authorities is appropriate.

It is the aim of this test procedure to evaluate possible mutagenic properties. In the case of positive results, a second quite separate question will arise: What is the significance of these results in terms of its genetic hazard for man? If there is any hazard, most medicinal products, with few exceptions, are discarded. Mutagenic drugs will rarely undergo clinical trials; cytostatics and drugs used for the treatment of life-threatening diseases being an exception. Furthermore, risk-benefit considerations are to be performed. The intended use of the drug, the degree of exposure, the age and reproductive status of the patient and the risks of alternative substances have to be taken into consideration.

4 Concluding Remarks

Chemically induced chromosome mutations were first discovered in the 1940s in studies of Oenothera and other plants (Oehlkers 1943). During the following decades, fundamental knowledge of chromosome structure as well as of cytogenetics were gained through experiments on higher plants and cell cultures. For a variety of reasons, however, these systems did not appear to be all equally well suited for mutagenicity testing and, in particular, the estimation of the risk for man. The main reason for this is the lack of a mammalian metabolism of chemicals. This problem was solved in part by combining the sensitive and practicable in vitro mutagenicity tests with an exogenous metabolizing system by adding mammalian microsomes or hepatocytes to cell cultures. Since then, cytogenetic in vitro methods have been applied not only to test chemicals for potentially mutagenic properties, but also to screen them for carcinogenicity. Today, these methods, including the analysis of chromosome mutations, have become an integral part of toxicological evaluation to define the genotoxic potency. To estimate the ability, in vivo tests are required when in vitro assays have yielded positive results. However, only very few mammalian germ-cell tests can be performed in order to assess heritable genetic risks. Furthermore, because of the limited knowledge of interorgan relationships (i.e. between lymphocytes and germ cells), even human monitoring can only be used with limitations for a quantitative genetic risk assessment. Further scientific research is required to link experimental results, human biomonitoring data and the occurrence of genetic disease in humans for a better human risk assessment and for scientifically justified decision-making by the responsible government authorities.

References

Ames BN, McCann J, Yamasaki E (1975) Methods for detecting carcinogens and mutagens with the Salmonella/mammalian-microsome mutagenicity test. Mutat Res 31:347–364

Ashby J, de Serres FJ, Draper M, Ishidate M, Margolin BH, Matter BE, Shelby MD (1985) Evaluation of short-term tests for carcinogens. Report of the international programme on chemical safety's. Collaborative study on in vitro assays. Elsevier, Amsterdam

Banduhn N, Obe G (1985) Mutagenicity of methyl 2-benzimidazolecarbamate, diethylstilbestrol and estradiol: Structural chromosomal aberrations, sister-chromatid exchanges, C-mitoses, polyploidies and micronuclei. Mutat Res 156:199–218

Barthelmeß A (1956) Mutagene Arzneimittel. Arzneim Forsch 6:157–168

Basler A (1980) Die Wirkung chemischer Mutagene auf die Oogenese von Säugetieren. Habil, Univ Düsseldorf

Basler A, Buselmaier B, Röhrborn G (1976) Elimination of spontaneous and chemically induced chromosome aberrations in mice during early embryogenesis. Human Genet 33:121–130

Boué J, Boué A, Lazar P (1975) Retrospective and prospective epidemiological studies of 1500 karyotyped spontaneous human abortions. Teratology 12:11–26

ChemG – Chemikaliengesetz (1980) Gesetz zum Schutz vor gefährlichen Stoffen. Bundesgesetzbl, Jahrg 1980, Teil I, pp 1718–1728

Creasy MR, Crolla JA, Alberman ED (1976) A cytogenetic study of human spontaneous abortions using banding techniques. Human Genet 31:177–196

DFG – Deutsche Forschungsgemeinschaft (1985) Maximale Arbeitsplatzkonzentrationen und biologische Arbeitsstofftoleranzwerte 1985. VCH, Weinheim

EEC – European Economic Community (1979) Council directive 79/831/EEC amending for the sixth time directive 67/548/EEC on the approximation of the laws, regulations and administrative provisions relating to the classification, packaging and labelling of dangerous substances. Off J Eur Commun L 259, 15.10.1979, pp 10–19

EEC – Annex V, Pt B: Methods for the determination of toxicity. Off J Eur Commun L 251, 19.09.1884, pp 94–145

EEC – Annex VI, Pt II D: Guide to the classification and labelling of dangerous substances and preparations. Off J Eur Commun L 257, 16.09.1983, pp 13–33

EEC – Annex VII. Information required for the technical dossier ("Base Set") referred to in article 6(1). Off J Eur Commun L 259, 15.10.1979, pp 21–25

EEC – Annex VIII. Additonal information and tests required under article 6(5). Off J Eur Commun L 159, 15.10.1979, pp 26–28

EEC – Toxicological methods of Annex VIII (in preparation)

EEC (1983) Council directive 83/570/EEC amending directives 65/65/EEC, 75/318/EEC and 75/319/EEC on the approximation of provisions laid down by law, regulations or administrative action relating to proprietary medicinal products. Off J Eur Commun L 332, 28.11.1983, pp 1–10

EEC (1984) Council recommendation concerning tests relating to the placing on the market of proprietary medicinal products. Off J Eur Commun C 293, 05.11.1984, pp 8–9

EEC – Annex II. Testing of medicinal products for their mutagenic potential. Off J Eur Commun C 293, 05.11.1984, pp 11–14

Ford CE, Hamerton JL (1956) The chromosomes of man. Nature (London) 178:1020–1023

Ford CE, Jones KW, Miller OJ, Mittwoch U, Penrose LS, Ridler M, Shapiro A (1959a) The chromosomes in a patient showing both mongolism and the Klinefelter syndrome. Lancet 1:709–710

Ford CE, Jones KW, Polani PE, De Almeida JC, Briggs JH (1959b) A sex-chromosome anomaly in a case of gonadal dysgenesis (Turner's syndrome). Lancet 1:711–713

Hook EB, Hamerton JL (1977) The frequency of chromosome abnormalities detected in consecutive newborn studies. Differences between studies. Results by sex and by severity of phenotypic involvement. In: Hook EB, Porter IH (eds) Population cytogenetics. Studies in humans. Academic Press, London New York, pp 63–79

Huang SL, Saggioro D, Michelmann H, Malling HV (1978) Genetic effects of crocidolite asbestos in Chinese hamster lung cells. Mutat Res 57:225–232

Hsu TC (1952) Mammalian chromosomes in vitro. The karyotype of man. Hereditas 43:167 172
IARC – International Agency for Research on Cancer (1982) Monographs on the evaluation of the carcinogenic risk of chemicals to humans. IARC Monogr Suppl 4. Int Ag Res Cancer, Lyon, Fr
ICPEMC – International Commission for Protection against Environmental Mutagens and Carcinogens (1982) Mutagenesis testing as an approach to carcinogenesis. Mutat Res 99:73–91
ICPEMC (1983a) Screening strategy for chemicals that are potential germ-cell mutagens in mammals. Mutat Res 114:117–177
ICPEMC (1983b) Regulatory approaches to the control of environmental mutagens and carcinogens. Mutat Res 114:179–216
IPCS – International Programme on Chemical Safety (1985) Environmental health criteria 51. Guide to short-term tests for detecting mutagenic and carcinogenic chemicals. WHO, Geneva
Ishidate M (ed) (1983) Chromosomal aberration test in vitro. Realize, Tokyo
Ishidate M, Sofuni T, Yoshikawa K (1981) Chromosomal aberration tests in vitro as a primary screening tool for environmental mutagens and/or carcinogens. Gann Monogr Cancer Res 27: 95–108
Ivett JL, Tice RR (1981) Diethylstilbestrol-diphosphate induces chromosomal aberrations but not sister chromatid exchanges in murine bone marrow cells in vivo. Environ Mutagen 3:445–452
Jacobs PA (1972) Chromosome mutations: Frequency at birth in humans. Human Genet 16: 137–140
Jacobs PA, Strong JA (1959) A case of human intersexuality having a possible XXY sex-determining mechanism. Nature (Lodnon) 183:302–303
Jacobs PA, Baikie AG, Court Brown WM, Strong JA (1959) The somatic chromosomes in mongolism. Lancet 1:710
Koizuma A, Dobachi Y, Tachibana Y, Tsuda K, Katsunuma H (1974) Cytokinetic and cytogenetic changes in cultured human leukocytes and HELA cells induced by benzene. Ind Health Jpn 12: 23–29
Lejeune J, Gautier M, Turpin R (1959) Étude des chromosomes somatiques de neuf enfants mongoliens. CR Acad Sci Paris 248:1721–1722
Lejeune J, Lafourcade J, Berger R, Vialatte J, Boeswillwald M, Seringe P, Turpin R (1963) Trois cas de délétion partielle du bras court d'un chromosome 5. CR Acad Sci Paris 257:3098–3102
Léonard A, Lauwerys RR (1980) Carcinogenicity, teratogenicity and mutagenicity of arsenic. Mutat Res 75:49–62
Léonard A, Gerber GB, Jacquet P (1981) Carcinogenicity, mutagenicity and teratogenicity of nickel. Mutat Res 87:1–15
Lüers H (1955a) Genetische Spätschäden nach Behandlung mit zytostatischen Stoffen. 9. Österr Ärztetagg 1955. Springer, Wien, pp 120–125
Lüers H (1955b) Zur Frage der Erbgutschädigung durch tumortherapeutische Cytostatica. Z Krebsforsch 60:528
Machin GA (1974) Chromosome abnormality and perinatal death. Lancet 1:549–551
McCann J, Choi E, Yamasaki E, Ames BN (1975) Detection of carcinogens as mutagens in the Salmonella/microsome test: Assay of 300 chemicals. Proc Natl Acad Sci USA 72:5135–5139
Moorhead PS, Nowell PC, Mellman WJ, Battips DM, Hungerford DA (1960) Chromosome preparations of leukocytes cultured from human peripheral blood. Exp Cell Res 20:613–616
Natarajan AT, Obe G (1986) How do in vivo mammalian assays compare to in vitro assays in their ability to detect mutagens? Mutat Res 167:189–201
Nordenson I, Sweins A, Beckman L (1981) Chromosome aberrations in cultured human lymphocytes exposed to trivalent and pentavalent arsenic. Scand J Work Environ Health 7:277–281
OECD – Organisation for Economic Co-Operation and Development (1981) Guidelines for testing of chemicals. OECD, Paris
OECD. Introduction to the OECD guidelines on genetic toxicology testing and guidance on the selection and application of assays. OECD, Paris (in preparation)
Oehlkers F (1943) Die Auslösung von Chromosomenmutationen in der Meiose durch Einwirkung von Chemikalien. Z Ind Abst Vererbungsl 81:313–341
Philip P, Jensen M (1970) Chromosome aberrations induced in rat bone marrow cells. Acta Pathol Microbiol Scand Sect A 78:489–490

Röhrborn G (1965) Über mögliche mutagene Nebenwirkungen von Arzneimitteln beim Menschen. Human Genet 1:205–231

Ruzicska P, Czeizel A (1970) Cytogenetic studies on mid-trimester abortuses. Human Genet 10: 273–297

Sankaranarayanan K (1979) The role of non-disjunction in aneuploidy in man. An overview. Mutat Res 61:1–28

Serres FJ de, Ashby J (eds) (1981) Evaluation of short-term tests for carcinogens. Report of the international collaborative program. Elsevier/North Holland Biomedical Press, New York

Siou G, Conan L, el Haitem M (1981) Evaluation of the clastogenic action of benzene by oral administration with 2 cytogenetic techniques in mouse and Chinese hamster. Mutat Res 90: 273–278

Soderman JV (ed) (1982) CRC handbook of identified carcinogens and noncarcinogens: Carcinogenicity-mutagenicity database. CRC, Florida

Sutherland GR, Bauld R, Bain AD (1974) Chromosome abnormality and perinatal death. Lancet 1:752

Tjio JH, Levan A (1956) The chromosome number of man. Hereditas 42:1–6

Williams GM (1980) Batteries of short-term tests for carcinogen screening. In: Williams GM, Kroes R, Waaijers HW, Poll KW van de (eds) The predictive value of short-term screening tests for carcinogenicity evaluation. Elsevier, Amsterdam, pp 327–346

Subject Index

Mosaics 170, 171, 174, 175, 214

MIX
Papier aus verantwortungsvollen Quellen
Paper from responsible sources
FSC® C105338

If you have any concerns about our products,
you can contact us on
ProductSafety@springernature.com

In case Publisher is established outside the EU,
the EU authorized representative is:
Springer Nature Customer Service Center GmbH
Europaplatz 3, 69115 Heidelberg, Germany

Printed by Libri Plureos GmbH
in Hamburg, Germany